AF817245

Design Research Foundations

Series Editors

Ilpo Koskinen, School of Design, The Hong Kong Polytechnic University, Hong Kong

Pieter E Vermaas, Department of Philosophy, Delft University of Technology, Delft, The Netherlands

Assistant Editor

Clementine Thurgood, Faculty of Health, Arts and Design, Swinburne University of Technology, Melbourne, Victoria, Australia

The goal of the series is to provide a platform for publishing state of the art research on foundational issues in design and its applications in industry and society. Suitable topics range from methodological issues in design research to philosophical reflections on the specificities of design rather than actual design work or empirical cases only. The definition of design behind the series is inclusive. In terms of disciplines, it ranges from engineering to architecture. In terms of design work, it ranges from conceptual issues in design through design experiments and prototypes to evaluative studies of design and its foundations.

Proposals should include:

A proposal form, as can be found on this page
A short synopsis of the work or the introduction chapter
The proposed Table of Contents
The CV of the lead author(s)
If available: one sample chapter

We aim to make a first decision within 1 month of submission. In case of a positive first decision the work will be provisionally contracted: the final decision about publication will depend upon the result of the anonymous peer review of the complete manuscript. The series editors aim to have the complete work peer-reviewed within 3 months of submission.

The series discourages the submission of manuscripts that contain reprints of previous published material and/or manuscripts that are below 150 pages/75,000 words.

For inquiries and submission of proposals authors can contact the series editors, Pieter Vermaas via: p.e.vermaas@tudelft.nl; or Ilpo Koskinen via: ilpo.koskinen@polyu.edu.hk

More information about this series at http://www.springer.com/series/13775

Pieter E. Vermaas • Stéphane Vial
Editors

Advancements in the Philosophy of Design

Springer

Editors
Pieter E. Vermaas
Department of Philosophy
Delft University of Technology
Delft, The Netherlands

Stéphane Vial
PROJEKT Design Lab
University of Nîmes
Nîmes, France

ISSN 2366-4622 ISSN 2366-4630 (electronic)
Design Research Foundations
ISBN 978-3-030-10351-4 ISBN 978-3-319-73302-9 (eBook)
https://doi.org/10.1007/978-3-319-73302-9

© Springer International Publishing AG 2018
Softcover re-print of the Hardcover 1st edition 2018
This work is subject to copyright. All rights are reserved by the Publisher, whether the whole or part of the material is concerned, specifically the rights of translation, reprinting, reuse of illustrations, recitation, broadcasting, reproduction on microfilms or in any other physical way, and transmission or information storage and retrieval, electronic adaptation, computer software, or by similar or dissimilar methodology now known or hereafter developed.
The use of general descriptive names, registered names, trademarks, service marks, etc. in this publication does not imply, even in the absence of a specific statement, that such names are exempt from the relevant protective laws and regulations and therefore free for general use.
The publisher, the authors and the editors are safe to assume that the advice and information in this book are believed to be true and accurate at the date of publication. Neither the publisher nor the authors or the editors give a warranty, express or implied, with respect to the material contained herein or for any errors or omissions that may have been made. The publisher remains neutral with regard to jurisdictional claims in published maps and institutional affiliations.

Printed on acid-free paper

This Springer imprint is published by the registered company Springer International Publishing AG part of Springer Nature.
The registered company address is: Gewerbestrasse 11, 6330 Cham, Switzerland

Contents

Towards a Philosophy of Design

Pieter E. Vermaas and Stéphane Vial

Keywords Philosophy of design · Design · Design research · Interdisciplinarity · Design for values · Social design

With this volume we present 24 contributions to the philosophy of design. Design is an emerging topic in philosophy and not yet one on which work is shaped by a common set of questions or by an academically entrenched discipline of philosophy of design. We therefore consider it an effort in itself that we can present 24 contributions. Throughout the years we have approached in our careers design from our separate disciplinary perspectives and probed whether design was becoming a more general topic of philosophical reflection. One of us (Pieter) is working in a philosophy department and analyzed design as part of a larger project within the philosophy of technology. This has led to a predecessor volume on the philosophy of design (Vermaas et al. 2008), to analyses of design (Houkes and Vermaas 2010), to joint work with design researchers on the structure of design (e.g., Vermaas and Dorst 2007), and to the creation of the Design Research Foundations book series, in which this volume has appeared. The second of us (Stéphane) has worked first as a 'philosophy applied to design' teacher (Vial 2015c) and now is working in a design department and in a design research center. He analyzed design from a phenomenological perspective and contributed to developing the knowledge of design in France. These efforts led to a monograph about how to design affects, structures, and frames experience (Vial 2010) and to the founding of the French-speaking journal *Sciences du Design* edited by Stéphane (Vial 2017).[1]

Our separate work may be taken as proof that design has found its way to philosophy, yet when teaming up we discovered a more substantial interest. A call for

[1] http://www.sciences-du-design.org

P. E. Vermaas (✉)
Department of Philosophy, Delft University of Technology, Delft, The Netherlands
e-mail: p.e.vermaas@tudelft.nl

S. Vial
PROJEKT Design Lab, University of Nîmes, Nîmes, France
e-mail: stephane.vial@unimes.fr

© Springer International Publishing AG 2018

P. E. Vermaas, S. Vial (eds.), *Advancements in the Philosophy of Design*, Design Research Foundations, https://doi.org/10.1007/978-3-319-73302-9_1

contributions on the topic following a successful conference track on philosophy of design in the 2015 *11th European Academy of Design Conference* in Paris, France, produced close to a hundred reactions. The 24 contributions in this volume grew out of this large response. Still philosophy of design is at this moment not an entrenched discipline. The authors of the contributions to this volume are not working within a common discipline and are not drawing from shared earlier thinking or shared approaches and methods. Rather authors are originating from a spectrum of different disciplines, ranging from philosophy to design research, and from product design to architecture. This makes the current volume a diverse one presenting work on design from different perspectives, raising different issues about design, and having different expectations of what is to be achieved by a philosophy of design. Avoiding the trap to use our editorial roles for defining what true issues, methods and goals are for the discipline, we present the 24 contributions primarily in an open and constitutive way. The value of this volume lies in opening up the philosophy of design, and not in closing it down by announcing what its final structure will be. However, we will introduce the work included in the volume and say something about the different research traditions it originates from. And we can offer our personal perspectives and interests in the philosophy of design, for sharing our enthusiasm, and for opening the debate on what a philosophy of design can do and deliver. Our academic perspectives can be said to be grounded in philosophy (Pieter and Stéphane) but also physics (Pieter) and psychology (Stéphane).

1 From Philosophy and Physics to Design

The academic perspective that comes with physics includes a continuing urge to explore unknown phenomena and the belief that these unknown phenomena can eventually be captured, described, and understood. It also comes with a more traditional philosophy of science perspective on knowledge, on science, and on scientific progress. For instance, Pieter's PhD on the philosophy of the enigmatic physical theory of quantum mechanics (Vermaas 1999) was not aimed at emphasizing the enigma but at analyzing it and exploring ways to understand and describe the reality to which quantum mechanics may refer. And the assumption that this understanding and description is possible was beyond doubt.

When analyzing design with this physics mindset, claims in design research about its specificity become challenges rather than warnings that further analysis is spurious. Nigel Cross' (2006) claim that design defines a third culture separate to C.P. Snow's two cultures of the natural sciences and the arts, leads with this mindset to the task to characterize this specificity. The views that design expertise is close to a *sui generis* skill that people can acquire only through studio teaching and lots of practice, translates into attempts to understand this expertise. And taking up these challenges does not imply squeezing design into the mold of science or of art, but to looking in detail to how design researchers themselves describe design, through cases and through design methods. For, in fact, design researchers are already for decades capturing the specificity of design problems and the specificity of the reasoning designers employ to address these problems. And the more recent *design thinking* movement may be seen as a movement that describes in detail (or in simplicity) how design reasoning works, for this movement includes an effort to scale design from a skill mastered by a few trained experts to an approach that can be employed by all, which in turn requires explaining design reasoning to all.

Similar challenges appear when considering design research itself. When Christopher Frayling (1993) claims that design enables a type of *research through design* that differs from more traditional forms of scientific research, the physics mindset drives towards the challenge of capturing this new way of doing research. Second, from a philosophy of science perspective questions can be raised about the epistemic value of design methods. Can these methods be validated for their claims about how design problems are to be addressed, and is there design specific research methodology needed for this validation? In design research the general view seems to be that the use of design methods by designers is too erratic for ever allowing systematic research on their effectiveness, yet some efforts are already made within design research towards taking up validation (e.g., Seepersad et al. 2006). Finally, when design researchers claim that research on design has quality indicators different to such indicators for scientific research, the challenge is to find those indicators. In architectural research this challenge has been addressed (Van der Hoeven 2011).

When Pieter took up a position in philosophy of technology at the Philosophy Department of Delft University of Technology, this physics perspective led to a first attempt to a generic characterization of design as the development of *use plans* for technical artefacts for realizing goals (Houkes and Vermaas 2010). Aligning with Herbert Simon's (1996, p. 111) general definition that "everyone designs who devises courses of action aimed at changing existing situations into preferred ones", this characterization puts emphasis on the teleology of designing, and sees the description of the artefact itself as a subsidiary activity in designing. This characterization was, however, neither meant as definite nor as complete, since it is also part and parcel of physics that a model or theory of a phenomenon is presented within the research community as a first effort rather than as the ultimate truth. Hence the characterization of design is available for improvement, yet meant as sufficiently detailed for addressing the set of questions in the philosophy of technology the Philosophy Department of Delft University of Technology was working on. The key observation underlying these questions was that technical artefacts are philosophically interesting entities since they have a *dual nature* by necessarily combining in their description structural and intentional concepts (Kroes and Meijers 2006). This dual nature defines the challenge to analyze how designers forge the relation between the structural and the intentional in technical artefacts. And the reasoning in design was taken to be a process by which designers realized this combination and bring teleology in the material world through translating goals of clients in descriptions of physical entities (Houkes and Vermaas 2010).

Design is increasingly taking center stage in the philosophy of technology. The emerging efforts to arrive at responsible innovation and at technologies that are respecting or even realizing our moral and societal values, advance design not only as a process for translating goals of clients in physical entities, but also as a process for realizing moral and societal values, from privacy and transparency in ICT to the good life in architectural urban designs (Friedman et al. 2006; Van den Hoven et al. 2015). This design for values approach in philosophy of technology matches quite closely with the advancement in design research of design thinking and the emerging generalization of design from products to product-service systems. Again from a philosophy of science perspective questions do emerge. What is the structure of design methods for designing product-service systems and for designing for values? What is the specificity of these design methods as compared to methods in the sciences? How can these methods be validated, now not only from a research-methodological point of view but also from a societal one? In previous decades philosophy of technology has abound in challenging the idea that humans can shape reality to their wishes by engineering design, leading to notions as a *technological fix* taken as an (ineffective) attempt to resolve a societal issue by technological means (Volti 1992) and the *design fallacy* as the (false) idea that designers can determine the use of the products they design (Ihde 2008). Contemporary design thinking is in that sense just a new form of an old modernist ideal, which requires a critical analysis in a philosophy of design.

2 From Philosophy and Psychology to Design

The academic perspective that comes with philosophy and psychodynamic psychology includes both a strong taste for concepts, for clear and distinct ideas, and the belief that secondary psychic processes (as judgment, reasoning, thinking, et cetera) are infiltrated by primary psychic processes (as wishes, anxiety, and fantasies). This is why it also comes with a more continental phenomenology perspective on subjectivity and human experience. For instance, Stéphane's PhD on the structure of the digital revolution (Vial 2012) was a philosophical inquiry into the technical structures of perception. It focused on how technical artefacts condition the way in which the world appears to us and in which phenomena are given to us, especially in an information age where the process of appearance is reshaped by digital technology. This approach can be characterized as a techno-transcendental phenomenology (Vial 2013) and marks a departure from postphenomenology (Ihde 1990; Verbeek 2005), which stipulates that technologies mediate our relationship to things. Techno-transcendental phenomenology instead seeks to render more perceptible the overall transcendental technical nature of appearance, as historically determined by an era's technical culture. Just as Jacob von Uexküll (1934) urged us to imagine each animal as surrounded by a sort of soap bubble that represents its milieu related to its biological circumstances, we must imagine human beings from a given historico-technical period as occupying a sort of phenomenological soap bubble, or techno-perceptual vessel, that is profoundly unique and determined by that period. For instance, being born and raised in the current digital "soap bubble" is phenomenologically different from being born and raised in the mechanical "soap bubble" of the Nineteenth Century. This is the main idea of the "ontophany theory" as presented in (Vial 2013).

In such a perspective, design can be considered as a creative phenomenology or a phenomenology by practice: it produces unprecedented modes of appearance through various types of effects (Vial 2010; Vial 2015a). What is important in design is not how objects look but how they produce effects that condition experience.

The techno-transcendental phenomenological perspective on design was an attempt to define what design is from a 'reception' point of view, which is usually not so much considered by design researchers. The international design research movement traditionally restricts the scope of the design act to the "conception" part of it, which deals with the specific logics and processes that designers adopt when doing design. Kees Dorst noted that "within design research, the emphasis on the process of design is still overwhelming" (Dorst 2008). Alain Findeli (2010, p. 289) showed in a clear manner that "the 'conception' part is only one of the two main moments or constituents of a design project, the 'reception' part being the other one"; indeed, "the design act is incomplete if we do not address what happens to the project's output once it starts its life in the social world". From such a perspective, design as a topic for philosophy could be considered both as a *process* (conception) and an *experience* (reception). Design* – the star indicates the broadened meaning – is not only something we do, it is something we live. This is the angle that was adopted at the University of Nîmes, the university where Stéphane took up his current position, through the PROJEKT Design lab, a research center for design and social innovation.

This phenomenological perspective was in a natural way combined with the shared claim for the specificity of design, which in Nîmes is called the "epistemological originality of design" (Findeli 1998; Vial 2015b). The task to characterize this specificity does not only concern design as a *process* but also design as an *experience*. What is the specificity of a design experience? How to define it? In which way differentiates it from an art experience or a scientific experience? These are the kind of questions that must be addressed by a design phenomenology. A first conceptual attempt was developed with the notion of "effect of design" (Vial 2010; Vial 2015a), which was used as a tool to define three criteria to differentiate design and non-design from a reception point of view. This research would need more development in a global philosophy of design in order to elaborate more on what is design as an experience.

At the University of Nîmes, in the PROJEKT research center, we made the choice for a cross-disciplinary approach by building a team composed of philosophers, semioticians, ethnographers, sociologists, and communication experts, who also are for the most of them design practitioners or, at least, intimately connected with design practice and practitioners. Our angle is social design in a broad sense, ranging from service design to design for policy. This choice comes with a couple of questions. What is social design? Is design 'social' by nature? What does it mean for design to be 'social'? How to define 'social'? Is there in social design a philosophy of society and of social change? If yes, how is it different from how the social sciences approach society and social change? If yes, what kind of philosophy is it?

Is it about ontology, logics, ethics, or politics? Can social design be an applied philosophy or a realization of philosophy? And so on.

3 What Design and Philosophy of Design Can Do

The questions we discussed in the two previous sections are not defining an exhaustive list of issues that a philosophical reflection on design should focus on. These questions have their origin in the disciplines the two of us work in and the disciplines we come from, hence cover at most a part of this list. Design is an emerging topic for philosophy, and the time has come to let interested researchers set the agenda and let it evolve to a strong discipline of philosophy of design, as we have strong disciplines of philosophy of science, philosophy of art, and philosophy of technology. And for exploring the issues that a philosophy of design has to take up, we may draw from other philosophical disciplines as we do and as is done by the authors in this volume. Yet for avoiding possible biases, we propose to not consider philosophy of design as a branch of the philosophies of science, of art, or of technology, but approach it as a discipline of its own, with its own schools of thought, its major concepts, its controversies, and at one day, its own history.

An argument for propelling this development of a philosophy of design can be derived by comparing the questions we ended up with from our respective backgrounds. This argument fits the contemporary approach to evaluating academic research by its relevance to society, and concerns the possibility of addressing societal issues by design. Both the philosophy of technology and the phenomenological perspectives lead to efforts to actively address these issues by, respectively, design for values and social design. Yet from the philosophy of technology perspective one can also raise doubts about this possibility by referring to the earlier mentioned criticisms of the idea that humans can shape reality to their wishes by design. Hence, given that designers are currently embraced as the new innovators in technology and society, an analysis is needed of what design actually can do.

It may be argued that design for values and social design are not instances of a technological fix in the original sense (Weinberg 1966) since these new design approaches do not aim at solving societal problems with merely technological means. Design for values and social design may be taken as different by addressing societal problems with insights from ethics and the social sciences. Yet this does not yet establish that design for values and social design will always be successful. Even more traditional engineering design may fail in solving its problems (e.g., Petroski 1992). Hence, it may be expected that also design for values and social design can be unsuccessful, since societal issues are well-known as 'wicked problems' (Rittel and Webber 1973). Wicked problems are essentially unique problems for which: (i) there is no definite formulation (stakeholders cannot agree on the definition); (ii) solutions are not true-or-false, but better or worse; (iii) solutions are numerous and, when implemented, change the way to formulate the problem (Ritchey 2013). This possibility raises questions for a philosophy of design.

Research-methodological questions concern the effectiveness and efficiency of design for values and of social design. This involves developing the concepts and criteria to determine whether design projects in the societal realm are successful. And it involves developing research methods to assess how efficient methods for design for values and social design are (Vermaas 2016). Addressing such questions would require more empirical work about how often design for values and social design fail and a more transparent proposition to society about how successful such design may be taken to be.

With the expectation that design for values may fail, follow-up questions emerge ranging from epistemic to ethical ones. A first set of questions may concern determining conditions under which design for values and social design may be assumed to be (more) successful. A second set of questions is about the analysis of failures of such design, understanding whether factors that may lead to failures are related to the extension of design from the technical realm of engineering to the societal realm. A philosophy of design may also aim on this point at more constructive results, as means and suggestions for improving the effectiveness of methods for design for values and for social design.

A third set of questions concerns the epistemic and moral grounds to offering design for values and social design. Given that it may fail, what are factual and moral conditions that reasonably should be satisfied before design for values and social design can be offered for addressing societal problems? Also if it would be successful always, design for values and social design may mean that society is subjected to large scale experiments (Van de Poel 2013) in which social structures are modified. Specifically because design thinking methods allow challenging and changing the initial problem statements of clients through social and emphatic research methods and through reframing, doing design for values and doing social design may involve not delivering what clients initially aimed at. This means that society has to accept unexpected solutions, for instance when design is taking up wicked societal issues that may require to changing the problem definition while addressing it. Determining what the moral grounds for such experiments and changes are will be a challenge for a philosophy of design.

4 Overview

The 24 contributions to this volume include chapters that take up the question of what design can do, and these are brought together in the final part of the volume. Before giving an overview of the topics taken up in this volume, we draw attention to the methodological diversity of the different contributions. As we said, the authors of these contributions are originating from a spectrum of different disciplines, ranging from philosophy to design research, and from product design to architecture. And this diversity is reflected by the diversity in the methodology and argumentative styles used in the contributions. Drawing hard distinctions between the contributions by means of disciplinary background would amount to an unproductive typecasting of authors, to ignoring that some authors are already combining approaches from both philosophy and design research, and to overlooking that some of the chapters are the result of collaborations between philosophers and design researchers. Still one can encounter in this volume chapters that are more pronouncedly philosophical and chapters that are more clearly design research, which may sometimes not be comfortable for the reader since we here are facing different intellectual traditions.

Chapters within philosophical traditions typically draw more extensively from philosophical resources and are in their use of resources of design research more selective, focusing on a few key texts rather than on the latest state of the art. Such chapters are moreover putting emphasis on the structure of the arguments that are advanced and typically describe and discuss cases of design in more cursory ways. Furthermore, philosophical chapters are generally more speculative, they build arguments on concepts and develop theoretical frameworks that can have an empirical side which is not always easily visible, or that are not directly related to a design practice. Such chapters can sometimes appear as somewhat abstract for readers accustomed to the intellectual traditions in design and design research. As Alfred North Whitehead used to say, philosophy is "the endeavour to frame a coherent, logical, necessary system of general ideas in terms of which every element of our experience can be interpreted" (Whitehead 1978, p. 3). Building a system of general ideas in terms of which every element of design practice and experience could be interpreted is, to say it mildly, a difficult task and for sure a project that is still at the beginning. Nevertheless, Whitehead's characterization of philosophy implies that philosophers working on design would benefit from being educated in design or from being simply acculturated to it. Reversely, it implies that designers and design researchers in philosophy of design would benefit from developing their ability and taste to speculative philosophy. "The study of philosophy is a voyage towards the larger generalities" (Whitehead 1978, p. 10). It requires us to accept changing our relation to language. "Every science must devise its own instruments. The tool required for philosophy is language. Thus philosophy redesigns language in the same way that, in a physical science, pre-existing appliances are redesigned" (Whitehead 1978, p. 11). This is why the technical language of philosophy is the basic component of all philosophical methods, which usually comes with conceptual abstraction.[2] By doing so, some chapters of this volume try to build contributions to philosophy of design either by philosophizing about design or philosophizing about concepts that can make sense for design.

Chapters following approaches from design research are instead regularly focusing on specific cases of designs or of design processes, and put emphasis on understanding the richness of such cases for deriving conclusions from this understanding. Instead of building a system of general ideas in terms of which every element of design practice could be interpreted, such chapters try to draw a general idea from a design practice case or set of cases. By doing so, these chapters contribute to philosophy of

[2] "The technical language of philosophy represents attempts of various schools of thought to obtain explicit expression of general ideas presupposed by the facts of experience" (Whitehead 1978, p. 12).

design either by exemplifying an idea through design projects or by reusing a philosophical idea in a design practice.

As there is no generally accepted methodology and writing style in philosophy, we anticipate that there will also not be one single argumentative approach emerging in the philosophy of design; rather we expect that design research will enrich the pluriformity of such approaches in the philosophy of design. Yet, when considering this volume, the reader may benefit from realizing that this pluriformity does exist.

The volume starts in a more basic manner with four contributions on design concepts. In Chapter "A Philosophical Approach for Distinguishing "Green Design" from Environmental Art", *Sue Spaid* takes a look at how design's outcomes differ from those of artistic actions by distinguishing "Green Design" from Environmental Art. In Chapter "Scratching the Surface: "Appearance" as a Bridging Concept between Design Ontology and Design Aesthetics", *Annina Schneller* offers a conceptual essay about the special ingredient that makes an artefact a design object and asserts that any definition of design objects necessarily includes their appearance. In Chapter "The Varieties of Good Design", *Salu Ylirisku* and *Mattias Arvola* adopt six senses of goodness as discerned by the logician Georg Henrik von Wright for analyzing the concept of good design. Finally Chapter "Collisions, Design and The Swerve" by *Jamie Brassett* and *John O'Reilly* is devoted to an examination of the role, value, and applicability of the concept of collision to design through The Swerve, Lucretius's clinamen, and how it helps understanding design as a creative process.

The second part of the volume is about the thinking processes that constitute design. *Thomas Wendt* opens in Chapter "Arational Design" with a critical analysis of a construal of design thinking as rational thinking and argues for an arational understanding of design. Chapter "A Case for Graphic Design Thinking" by *Katherine Gillieson* and *Stephan Garneau* argues for a broad view of graphic design thinking as a distinct approach to problem-solving by presenting seven characteristics pertaining to graphic design thinking in particular. In Chapter "The Role of Abduction in Production of New Ideas in Design", *Lauri Koskela*, *Sami Paavola*, and *Ehud Kroll* consider abduction in design, drawing from work by C.S. Peirce, in philosophy of science and in design research. Chapter "Lively Objects: Designing Science Exhibits with John Dewey" by *Kim Kullman* discusses John Dewey's work on experience and experiment, and applies it for analyzing reasoning in the design of exhibitions at the Exploratorium science museum in San Francisco. Chapter "Sketch Representation and Design as Generative Transformation" by *James Andrew Self* and *Gabriela Goldschmidt* analyses design as a generative, transformative act and discusses the role of sketching during the conceptual ideation phase of design. And in Chapter "Models in Engineering Design: Generative and Epistemic Function of Product Models" *Claudia Eckert* and *Rafaela Hillerbrand* end this second part by considering the role of models in design and argue that this role is much broader than that of representation as is emphasized in the philosophy of science literature on models.

The third part of the volume brings together three chapters on design aesthetics and design phenomenology. *Virginia Tassinari* opens in Chapter "Notes for an

Aesthetics of Social Innovation: A Reading through the Lenses of Jacques Rancière's Philosophy" with a reflection on how such a sustainable aesthetics, rooted in Jacques Rancière's concept of partition of the sensible and in the DESIS network experience, could be considered as a driver of social and behavioral change. Chapter "Conceptualizing Aesthetics in Design: A Phenomenological Framework" by *Mads Nygaard Folkmann* analyses aesthetics as an approach to understand how design frames experience and enlarges aesthetics beyond the classical limited sensual dimension by including also experiential and cultural aspects. Chapter "Phenomenology in Spatial Design Disciplines: Could it offer a bridge to sustainability?" by *Emina Kristina Petrović*, *Bruno Marques*, *Natasha Perkins*, and *Guy Marriage* ends this third part: it proposes that the philosophy of phenomenology is both applicable and necessary for a deeper and more integrated approach to spatial design disciplines in a world that aspire to be sustainable.

The fourth part is on design research and design epistemology. It starts with Chapter "The Specificity of Design Research: How practice-based design knowledge can enter the Great Archive of Science" by *Paolo Volonté*, *Lucia Rampino*, and *Sara Colombo*, which is an inquiry on the specific nature of design research considered as research-through-design. In Chapter "Design Research as a Meta-Discipline" *Anne Caplan* analyses research enabled through design projects using Henri Lefebvre's notion of meta-philosophy. *Dingmar van Eck* relates in Chapter "On Testing Engineering Design Methods: Explanation, Reverse Engineering, and Constitutive Relevance" work in philosophy of science on mechanistic explanations with design methods for reverse engineering and design optimization, and argues that this relation provides a constraint for assessing these design methods. In Chapter "Research in Interior Architecture: Interdisciplinary Viewpoints and Research Approaches" *Ann Petermans*, *Jan Vanrie*, and *Kris Pint* consider interior architecture and argue that the diversity of bodies of knowledge used in this discipline calls for a similar diversity in approaches to studying it. The interdisciplinary nature of current design resurfaces in Chapter "The Philosophical Underpinnings of Design Theory" where *Anne-Françoise Schmid* analyses more generally the philosophical and epistemological frameworks needed for understanding design.

The fifth part is on sustainability and ecology in design. In Chapter "Effects of Design and Sustainable Design of Technical Artefacts", *Karina Vissonova* proposes to form a consistent understanding of what should fall under the 'sustainable design' kind and what should not. Chapter "Ecological Design as an Ecology of Love: Epistemological and Ethical Implications" by *Gonzalo Salazar* and *Seaton Baxter* is an attempt to synthesize a new epistemology of design called an 'ecology of design' and argues that design become ecological only when its praxis is mainly commanded by the emotion and ecology of love. Finally, Chapter "Scales of Design: Ecodesign and the Anthropocene" by *Victor Petit* and *Bertrand Guillaume* offers an essay on the encounter between design and the global environment in the Anthropocene and more particularly look at the issue of scales in the context of the ecological crisis.

The final part of the volume on design, politics, and society brings us back to our question of what design can do. Chapter "Governmentality, Technologies, & Truth

Effects in Communication Design" by *Katherine Hepworth* argues that communication design knowledge and artefacts are inherently governmental through Foucault's theories of discursive technologies, truth effects, and governmentality. In Chapter "The Black Book: Emilio Ambasz's University of Design"*Matthew Holt* describes a proposal in the 1970s by the architect and product designer Emilio Ambasz to establish a new research institution that puts design center stage to our understanding of our human made reality, and analyzes the responses these efforts elicited in academia. Finally *Paul A. Rodgers* and *Craig Bremner* end in Chapter "The Design of Nothing: A Working Philosophy" with a sharp critique of what design actually has brought and can bring to humanity.

We hope that together these 24 contributions may give the reader a rich source to the emergent discipline of philosophy of design. Still, as we said, this philosophy is not yet an entrenched discipline with a shared basis or common approaches and methods. The contributions therefore do not make up a coherent contribution to the philosophy of design, but may better be seen as a substantial effort of exploring and trailblazing the scope of the discipline and demonstrating its richness and broadness. The contributions moreover cannot be taken as exhausting the scope of the philosophy of design; topics as the ethics of design and the ontology of the designed world are not or under-represented. Hence, in addition to offering 24 contributions to the philosophy of design, we hope this volume will be a basis for further work towards establishing a flourishing, rich, and more coherent and complete discipline of philosophy of design.

References

Cross, N. (2006). *Designerly ways of knowing*. London: Springer.

Dorst, K. (2008). Design research: A revolution-waiting-to-happen. *Design Studies, 29*(1), 4–11.

Findeli, A. (1998). Will design ever become a science? In P. Strandman (Ed.), *No guru, no method: Discussion on art and design* (pp. 63–69). Helsinki: UIAH.

Findeli, A. (2010). Searching for design research questions: Some conceptual clarifications. In R. Chow, G. Joost, & W. Jonas (Eds.), *Questions, hypotheses & conjectures: Discussions on projects by early stage and senior design researchers* (pp. 286–303). Bloomington: iUniverse.

Frayling, C. (1993). Research in art and design. *Royal College of Art Research Papers, 1*(1), 1–5.

Friedman, B., Kahn, P. H., Jr., & Borning, A. (2006). Value sensitive design and information systems. In P. Zhang & D. Galletta (Eds.), *Human-computer interaction in management information systems: Foundations* (pp. 348–372). Armonk: M.E. Sharpe.

Houkes, W., & Vermaas, P. E. (2010). *Technical functions: On the use and design of artefacts*. Dordrecht: Springer.

Ihde, D. (1990). *Technology and the lifeworld: From garden to earth*. Bloomington: Indiana University Press.

Ihde, D. (2008). The designer fallacy and technological imagination. In P. E. Vermaas, P. Kroes, A. Light, & S. A. Moore (Eds.), *Philosophy and design: From engineering to architecture* (pp. 51–60). Dordrecht: Springer.

Kroes, P. A., & Meijers, A. W. M. (2006). The dual nature of technical artefacts. *Studies in History and Philosophy of Science, 37*, 1–4.

Petroski, H. (1992). *To engineer is human: The role of failure in successful design* (2nd ed.). New York: Vintage Books.

Ritchey, T. (2013). Wicked problems. *Acta Morphologica Generalis, 2*(1).
Rittel, H., & Webber, M. (1973). Dilemmas in a general theory of planning. *Policy Sciences, 4*, 155–169.
Seepersad, C. C., Pedersen, K., Emblemsvåg, J., Bailey, R., Allen, J. K., & Mistree, F. (2006). The validation square: How does one verify and validate a design method? In K. E. Lewis, W. Chen, & L. C. Schmidt (Eds.), *Decision making in engineering Design* (pp. 303–314). ASME.
Simon, H. A. (1996). *The sciences of the artificial* (3rd ed.). Cambridge, MA: MIT Press.
Uexküll Von, J. (1934). *Mondes animaux et monde humain*. Paris: Denoël.
Van de Poel, I. (2013). Why new technologies should be conceived as social experiments. *Ethics, Policy & Environment, 16*(3), 352–355.
Van den Hoven, J., Vermaas, P. E., & van de Poel, I. (Eds.). (2015). *Handbook of ethics, values and technological design*. Dordrecht: Springer.
Van der Hoeven, F. (2011). Mind the evaluation gap: Reviewing the assessment of architectural research in the Netherlands. *Architectural Research Quarterly, 15*, 177–187.
Verbeek, P.-P. (2005). *What things do: Philosophical reflections on technology, agency, and design. Penn State: Penn State*. University Press.
Vermaas, P. E. (1999). *A philosopher's understanding of quantum mechanics: Possibilities and impossibilities of a modal interpretation*. Cambridge: Cambridge University Press.
Vermaas, P. E. (2016). A logical critique of the expert position in design research: Beyond expert justification of design methods and towards empirical validation. *Design Science, 2*, e7.
Vermaas, P. E., & Dorst, K. (2007). On the conceptual framework of John Gero's FBS-model and the prescriptive aims of design methodology. *Design Studies, 28*, 133–157.
Vermaas, P. E., Kroes, P., Light, A., & Moore, S. A. (Eds.). (2008). *Philosophy and design: From engineering to architecture*. Dordrecht: Springer.
Vial, S. (2010). *Court traité du design*. Paris: Presses Universitaires de France.
Vial, S. (2012). *The structure of the digital revolution*. PhD in Philosophy thesis defended on November 21, 2012. Paris: Paris Descartes University.
Vial, S. (2013). *L'être et l'écran: Comment le numérique change la perception*. Paris: Presses Universitaires de France. Second revised edition 2017.
Vial, S. (2015a). The effect of design: A phenomenological contribution to the quiddity of design presented in geometrical order. *Artifact, III*(4), 4.1–4.6. Online: https://scholarworks.iu.edu/journals/index.php/artifact/article/view/5137/25631.
Vial, S. (2015). *Le design*. Paris: Presses Universitaires de France. Online: https://www.cairn.info/le-design--9782130620433.htm. Second revised edition 2017.
Vial, S. (2015c). Philosophy applied to design: A design research teaching method. *Design Studies, 37*, 59–66. Online: http://www.sciencedirect.com/science/article/pii/S0142694X15000022.
Vial, S. (2017). A Look at design research in France through design journals: Building a design discipline. *She Ji: The Journal of Design, Economics, and Innovation, 3*(2), 146–156: http://www.sciencedirect.com/science/article/pii/S2405872616300478
Volti, R. (1992). *Society and technological change* (2nd ed.). New York: St. Martin's Press.
Weinberg, A. M. (1966). Can technology replace social engineering? *Bulletin of the Atomic Scientists, 22*(10), 4–8.
Whitehead, A. N. (1978). *Process and reality: An essay in cosmology*. New York: The Free Press. (first original ed. 1929).

Part I
Design Concepts

A Philosophical Approach for Distinguishing "Green Design" from Environmental Art

Sue Spaid

Abstract In this paper, I begin by analyzing several environmental design projects that are difficult to distinguish from environmental art projects, so as to tease out obvious distinctions between these two fields' practical aspirations. I then employ Arthur Danto's Theory of Action, as described in his 1979 essay "Basic Actions and Basic Concepts," to show how design's outcomes differ from those of artistic actions, even though both effectively entail actions. Unlike design actions, artistic actions prompt interpretations or greater reflection, since artwork meanings are comparatively polyvalent. I next discuss what Bruno Latour describes as the semiotic question of meaning, in particular, the relationship between the designer's guiding principles and his/her design's implicit values, which articulate those principles. I then discuss the importance of design's entwining conception and making. Lastly, I return to the urgency awaiting environmental designers, whose most successful nature-based solutions, whether sustainable architecture, large-scale public works, or edible foodstuff will result from either efforts to recover "lost" practices or innovative strategies for translating nature's processes. "Green designers," especially, owe it to their public to tap what Latour terms design's normative question, so as to optimize resource management and sustainable design.

Keywords Nature-based solutions · Environmental · Normative · Good design

1 Introduction to the Problem

These days, there's a growing interest in what is primarily termed nature-based solutions, but is also known as "green infrastructure." In 2016, the European Commission (EC) selected scores of demonstration projects that it plans to fund to the tune of €40 million, which conveys the significance of this burgeoning field, especially since the EC aims to target income-generating projects that can be implemented widely, so as to multiply their economic benefits across Europe. This

S. Spaid (✉)
Independent Scholar, Belgium, UK
e-mail: suespaid@gmail.com

© Springer International Publishing AG 2018

P. E. Vermaas, S. Vial (eds.), *Advancements in the Philosophy of Design*, Design Research Foundations, https://doi.org/10.1007/978-3-319-73302-9_2

represents an especially positive opportunity, on the heels of a decade that has already witnessed the adoption of scores of environmental practices from phytoremediation to hybrid cars, renewable energies, and aquaponic/aeroponic urban farms; all designed to address consumer concerns regarding land loss, greenhouse gas emissions, and sustainable living. For the purposes of this paper, environmental design (a.k.a. "green design") engenders environmentally-friendly products (independent of scale) whose designs can be taught, improved upon, and applied broadly, whereas environmental artworks are typically prototypes, one-off solutions, initiated by artists to redress a particular site's environmental issues, which they expect others, especially designers, to eventually copy.

Recent attempts at "green design" include: architects Stefano Boeri's Milan twin towers hosting 730 trees (2014), Jean Nouvel's Sydney skyscraper draped with Patrick Blanc's hanging gardens (2013), Herzog and de Meuron's similarly-decorated Pérez Art Museum Miami (2013), Renzo Piano's California Academy of Sciences (2008), and Dominique Perrault's Bibliothèque national de France (1996); Stefan Sagmeister's edible typefaces; product designer Thomas Heatherwick's London Garden Bridge; Kona Design's Pasona-Tokyo's indoor office farm (2010); plus vertical farms developed by: Singapore's Sky Green (2012); LA Urban Farms (2015); and ZipGrow for the U.S. Pavilion at Expo 2015 Milano. Even commercial real-estate developers are recognizing the demand for roof gardens, locally-grown food, and buildings that capture and store rainwater, recycle grey water, and cleanse stormwater; leaving thousands more examples of green infrastructure as either under consideration or in the pipeline around the globe.

At first glance, it doesn't really seem as though environmental design oriented toward nature-based solutions presents any philosophical conundrums, let alone contributes anything of value to philosophy of design discussions. There's no real reason to doubt that these projects qualify as design. Furthermore, I don't imagine too many people squabbling over whether designers actually deserve credit for projects in which horticulturalists, botanists, and engineers play more significant roles than do the award-winning architects, product designers, and graphic designers who hire them for their expertise. Few would charge designers as "unethical" for seeking out collaborators, so long as they remunerate experts who both beef up the winning proposal and carry out a firm's bid by managing construction projects that design firms themselves are ill-equipped to implement.

Some might worry that greening-strategies are rather a trendy maneuver, proposed by greedy designers who just "follow the money," climbing aboard a green gravy train, primarily because this is what futuristic Asian developers and future-oriented European nations fund these days. Some see green infrastructure as a design strategy especially geared toward making people feel happy, since scientific studies show that human beings feel their happiest (best) when surrounded by nature, or at least images of nature (Zelenski and Nisbet 2014). Socio-biologist E. O. Wilson gave this an evolutionary twist in his book *Biophilia*, which proposes that human beings' "urge to affiliate with other forms of life is to some degree innate" (Wilson 1984, 85). Others may deem such projects mostly window-dressing, since their motivation is clearly monetary, luring wealthy investors and buyers, rather than contributing anything of lasting ecological value. That is, today's designers may be greening cities, but some are doing so at a cost to diminishing resources that might be better spent otherwise, if only there existed methodologies for targeting and rewarding genuinely "green" strategies that ought to be implemented instead.

So despite the apparent progress, genuine innovation, and enhanced environments on our horizon, numerous philosophy of design issues lurk behind the scenes. Some may find it surprising that fine artists first tested many of the strategies that designers are proposing these days. In fact, the current popularity of vegetative walls could be attributed to Jeff Koons' 13-meter tall *Puppy*, which premiered at documenta 9 in 1992, but has graced the grounds of Guggenheim Museum Bilbao since 1997. For *Park up a Building* (1996), Vito Acconci temporarily suspended a tree-lined staircase alongside the Centro Gallego de Arte Contemporáneo in Spain. In fact, avant-garde artists have been testing innovative ecoventions (ecology + invention) since the early 1960s, most of which were funded and experienced as art. Artists' efforts later captured the imaginations of designers, whether architects, product developers, and even graphic designers, keen to reorient their practices, so as to attract clients increasingly interested in sustainable resource management (Spaid 2002). Others may be surprised to learn that artists rarely mind it when designers copy their ideas. In fact, many hope they do! This already poses a philosophical difference of note.

Other philosophical issues include design's distinct capacity for communicating its function, which is not necessarily avant-garde art's goal. In fact, environmental design's clarity and greater visibility enable it to influence public awareness. As a result, the message environmental design conveys (typically its function) had better be tested and true, rather than mere window dressing. On a certain level, environmental design owes a greater fidelity to truth than ordinary design, when it showcases supposedly viable green infrastructure to the world. For its part, environmental art is the design world's testing ground and will likely remain a fairly underground phenomenon until more art historians undertake this movement as their field of investigation and museums feature environmental art projects (Spaid 2016a).

Perhaps environmental art's success both as a testing ground for novel strategies and as an "ideas-whisperer" befits its underground status, where it feeds "in-the-know" architects and "hip" designers schemes that might feel far less compelling were they present in clear view, where "inspired by" might be deemed at best, appropriation, or at worst, poaching. I don't imagine environmental art's more experimental methods (whether "radical" tactics or high-risk undertakings) for resolving environmental problems ever becoming part of a design practice built on expertise, rather than whim. Bruno Latour considers design to be low risk, as compared to building something, but I don't see how one can sever the design action from its outcomes. In fact, I worry that his characterizing design as low risk underestimates the damages incurred by confident designers, who not knowing their limits fail to involve or refuse to listen to the appropriate experts. Design is hardly a harmless affair. And "green design," especially, stands to do far more long-range damage should their outcomes not live up to their promised functions.

In the next section, I analyze several environmental design projects that are difficult to distinguish from environmental art projects, so as to tease out obvious distinctions between these two fields' practical aspirations. I then employ Arthur Danto's Theory of Action, as described in his 1979 essay "Basic Actions and Basic Concepts," to show how design's outcomes differ from those of artistic actions, even though both effectively entail actions. Unlike design actions, artistic actions prompt interpretations or greater reflection, since artwork meanings are comparatively polyvalent. In Section Four, I discuss what Latour describes as the semiotic question of meaning, in particular, the relationship between the designer's guiding principles and his/her design's implicit values, which articulate those principles. I next discuss the significance of design's entwining conception and making. Lastly, I return to the urgency awaiting environmental designers, whose most successful nature-based solutions, whether sustainable architecture, large-scale public works, or edible foodstuff are likely to flourish as a result of recovered "lost" practices or innovative strategies for translating nature's processes. Optimizing resource management and sustainable design requires tapping the normative constraints of good and bad design.

2 Distinguishing Practical Design from Practical Art

Some have suggested that describing design as practical action differentiates it from artworks. Problem is, some practical actions turn out to be "practical" artistic actions, such as ecoventions (ecology + invention), institutional critique, activist art, and public engagement practices. Consider Joseph Beuys' 1971 forest action or his later *7000 Oaks* (1982–1987) (Fig. 1) for which he and hundreds of volunteers planted 7000 oak trees in Kassel, Germany, as his contribution to documenta 7. Artworks like these definitely have practical consequences. Social-design scholar Victor Margolin worries that the increasingly practical nature of artworks only complicates matters. Operating under the impression that artists tell stories, while designers "do," he asks, "How do we think about art that moves from discourse to action, art whose intent is to produce a useful result" (Margolin 2006)? Ignoring a fifty-year history of artists initiating practical actions, he claims that artists' practical pursuits are in conflict with art's *discursive* role.

> ...[B]y what criteria do we evaluate this work?...In the never-ending debates on the difference between art and design, the distinction usually comes down to the primacy of discourse in artistic practices...But when artists want to achieve social results without identifying themselves as designers, how should the critical community respond?.Once artists enter a realm of action, it is difficult to characterize their projects differently from those of other actors such as landscape designers or even architects...the discursive has spilled over into the practical, and the practical has become more discursive (Margolin).

Fig. 1 Joseph Beuys, *7000 Eichen*, 1982–1987/2016, W. 22nd St., New York City, USA Photo Credit: Theresa Hackett

To my lights, Margolin's restricting the "realm of action" to design proves problematic, since it leads him to make several erroneous points, namely: (1) Given the "primacy of discourse in artistic practice," design is distinguishable from art as practical. (2) The intent to achieve "social results" is the purview of design, and (3) The critical community should be wary of artists' practical actions, if doers also claim to be artists while doing so. He must not have realized that artists began performing practical actions in 1962, when Beuys first called for an "action" to clean up the Elbe River and Alison Knowles performed *Make a Salad* at the Institute for Contemporary Art in London (Spaid 2012, 16–17).

The point that proves most troublesome for our purposes is Margolin's limiting the discursive to art, especially since many scholars function as architects, even though their designs never get built, yet one no longer functions as an artist if one's actions, whether discursive or practical, fail to be valued as art! Similarly, one can function as an artist, even if one's designs don't work out as intended, though designers' designs that fail also fail categorically as design. So the quality of "discursive" belongs no less to design than to art. Regarding Margolin's second point, the history of artists' actions, artists farming as art, community art, and participatory art proves that artists have been applying their skills to achieve "social results" since the sixties, so one wonders why he specifically attributes "social results" to the design field (Spaid 2012, 218–229). In fact, "green design" and "social architecture" are rather recent trends, partly inspired by Samuel Mockbee's "Rural Studio Program." The Modernist buildings of Le Corbusier and Mies van der Rohe no doubt launched "social experiments" that have been in play now for sixty years, though their popularity rises and falls as the public's appreciation for their experiments waxes and wanes.

To address Margolin's third point, I would argue that design and art are judged by totally different standards, so there's no good reason for the critical community to be wary, unless of course, designers start calling what they do "art." Design *can* be art, though only when it offers an innovative solution that others later copy, not when it appropriates art historical strategies![1] One point that differentiates discursive design from discursive art is that artists' designs (whether in the form of a drawing, photo/text proposal or digital file) are often meant as ends in themselves (as exhibited art), while designers' designs remain the means to some end, since they must be built to function as design. The primary exception of course is "visionary" architecture, or drawings of buildings deemed impossible to construct, such as those associated with Lebbeus Woods, Hugh Ferriss and the early designs of Coop Himmelblau, Superstudio, and Archigram. I now try to clarify the differences between practical design and practical art.

When some design (constructed, manufactured, or built by others) is reduced to a singular idea, under what conditions is this idea art and not design?[2] In a way, the answer is easy, since the answer lies in the way these two fields are practiced. Not surprisingly, similar conditions differentiate art from science. Science and design are hardly fields guided by whim, intuition, or supposition. Each follows general principles, best practices, and guidelines that are tested, vetted, and circulated by respected peers. When Chicago-based artist-collaborative Haha first hatched their plan (1992) to transform storefronts into hydroponic vegetable farms so as to minimize the need for washing and peeling vegetables, thereby keeping nutrients in tact for eaters plagued by HIV, people no doubt thought their plan was radical. Nearly twenty-five years later, hydroponic, aeroponic, and aquaponic systems that tender similar results are popping up everywhere, so as to transform concrete jungles into edible landscapes.

People tend to distinguish design, which *solves* problems, from art that *creates* problems, though this distinction fails for those artists, whose practical projects must at least attempt to solve problems, if not solve them altogether. As with the above HaHa example, artists' design solutions have yet to be tested, vetted, or circulated, so their success as design remains uncertain, but such projects are driven by a singular goal to tackle a particular problem. Even stranger, artists whose practical solutions turn out to be wildly popular or scientifically successful are often disinclined to repeat them, lest they become known for specializing in one area. Many artists would prefer their sought-after design solutions to become prototypes that serve as models for others to copy, so that they can continue experimenting. It

[1] I have in mind here Frank Lloyd Wright's *Falling Water* as a design object that doubles as art.

[2] To see how art can be reduced to ideas, see Rob Pruitt's *101 Art Ideas You Can Do Yourself* (1999).

thus seems that what characterizes practical art is the absence of verifiable standards for evaluating its outcomes, unlike the testing new designs for cars or washing machines undergo. The problem of the artist-initiated design solution proves complex, especially since a work's greatness as design (sustainable, cost-effective, easily implementable, well-built) often depends on the very qualities people invoke to differentiate design from art, while its success as art depends on hotly-contested values such as "originality," "ingenuity," "influence," and "generosity."

In light of the differences between art and design, it helps to look at how practitioners trained in design versus art have approached urban farming. In contrast to art-farms, which tend to meld the "I can," "I will," and "I know," design-farms tend to replace the "I know" with the "I show," perhaps because designers want their projects, which typically employ less experimental farming techniques, to stand out and communicate attractive solutions that passersby will either want to emulate or purchase for their own use.

Whether an art-farm or a design-farm, ownership (title/deed, owner's labor/resources/time) and authorship (creator/producer), as they pertain to edible property, prove nearly impossible either to enforce or assert. Since 2005, architect Fritz Haeg has worked with nearly a dozen home owners and apartment dwellers, stretching from Connecticut to Maryland, Kansas, and California all the way to Italy, Turkey, and the United Kingdom, transforming their front yards into *Edible Estates* (subsistence farms).[3] Project descriptions and blogs posted on Haeg's website suggest that most of the homeowners he's worked with originally imagined their neighbors both eyeing and stealing their bounty, which sounds like a warranted fear! Baltimore *Edible Estates* co-owner Clarence Ridgely notices that passersby consider his front-yard farm fair game, free snacks available those who first notice what's ripe (Ridgely 2008)! Even those who deem picking their neighbor's flowers rude might regard stealing the fruits of their neighbor's labor akin to showering them with complements. So long as food access seems a "God-given" right, food production will lack the sense of ownership commonly afforded gardeners. Moreover, food is grown to be eaten, while flowers are grown to be experienced, so when passersby notice produce in need of harvesting, why hesitate?

In Haeg's book *Edible Estates: Attack on the Front Lawn,* he maps out the criteria that he uses to determine which front yards provide the best hosts for his "regional prototypes," local versions of his model. He prefers homes "on a somewhat lengthy typical residential street lined entirely with uninterrupted groomed front lawns," whose front yard is "very visible from the street, with regular car traffic" (Haeg 2010). Furthermore, prospective *Edible Estate* owners ought to be "super enthusiastic about the project, and committed to and willing to continue the *Edible Estates*

[3] Between 2005 and 2013, Haeg installed or received funding for fifteen *Edible Estates*: (15) Twin Cities, US (2013), (14) Aarhus, DK (2013), (13) Holon, IL (2013), (12) Budapest, HU (2012), (11) Istanbul, TK (2012), (10) Rome, IT (2010/2011), (9) Aldrich Contemporary Art Museum, Ridgefield CT (2010), (8) New York City, NY (2009), (7) Los Angeles, CA (2008), (6) Baltimore, MD (2008), (5) Sierra Ridge Apartment Complex, Austin, TX (2008), (4) Brookwood House Council, London, UK (2007), (3) Maplewood, NJ (2007), (2) Lakewood, CA (2006) and (1) Salina, KS (2005).

prototype as long as they live in the house" (Haeg). Unfortunately, a divorce, move, and other issues have prevented several *Edible Estates* owners from proceeding as Haeg envisioned.

In Berin Golonu's article "Greening the Revolution," she claims that Haeg's front yard "organic" farms offer both practical and discursive potential. She cites as "practical benefits" their reduced water usage, minimizing pesticide runoff, and lower carbon foot prints (Golonu 2008, 44). While all of this sounds great, only the reduced carbon-footprint point is totally true, since Haeg's farms tend to consume more water than lawns and most likely require pesticides, since replacing one's lawn with food couldn't help but attract loads of pesky critters. So far as I know, Haeg is not particularly interested to create ecologically innovative lawns or farms. Rather, he aims to inspire a conversation about the ubiquitous U.S. lawn, which his design-farms (and related catalog) have achieved. If one wanted an ecologically-sound lawn, one would opt for unkempt, renegade lawns that are full of weeds and look like heck, but maximize habitat, ensure biodiversity, and require little upkeep. Some *Edible Estate* owners have remarked that their front-lawn farms actually require far more maintenance than their earlier, back-yard farms ever did (www.fritzhaeg.com).

In 2008, WORK Architecture Company (WORKac) created a functioning design-farm as part of P.S.1's Young Architects Program, which commissions architects to build new works in its courtyard to host its summer concert series. WORKac's *Public Farm (P.F.1)* engaged over thirty artists, designers, engineers, farmers, and green suppliers in its design and production. Although P.S.1's press release asserted that food harvested from its farm would be served in its restaurant, multiple telephone conversations with restaurant staff failed to verify this claim. Golonu describes *P.F.1* as a visionary environment, a farm, a playground, an art installation and an "educational model for sustainable building and design" (42). *Public Farm* architects Amale Andraos and Dan Wood see their structure as providing a "space for leisure and relaxation," as well as serving a "didactic purpose" by availing information regarding *Public Farm*'s green collaborators. In contrast to Haeg's one-lawn-at-a-time approach, these architects believe that "ecology works best at a citywide scale to effect change and that it's hard to make a real difference on an individual level" (42). Despite its many possibilities, *Public Farm* functioned more as a popular gathering spot, offering shelter from the elements, than as a "public farm" for providing sustenance, though plenty of partiers surely partook in spontaneous acts of foraging/gleaning.

Perhaps the most well known design-farm is *The Edible Schoolyard/Yale Sustainable Food Project* (since 2003), an ongoing project initiated by Yale University's Berkeley Dining Hall and famed Berkeley (CA) chef Alice Waters. *The Edible Schoolyard* gained global publicity when it was included in the traveling exhibition "Into the Open," curated by Aaron Levy, William Menkin, and Andrew Sturm, for the United States Pavilion at the 2008 Venice Biennial Architecture Exhibition, traveling the next year to Parsons/The New School and the National

Constitution Center. Represented by a vegetative wall, passersby were free to pick the growing vegetables.[4]

Another arena that seems especially well-suited for environmental design is the contemplative garden, whose primary attributes concern taste, meditation, and beauty. By contrast, farms typically require the discussion and evaluation of the best available technologies, the need for invention versus convention, function versus aesthetics, and most obviously, ongoing group negotiation. While gardens are often designed by someone working from a "drawing board," who likely assigns the work to hired hands, farms are hotbeds for negotiation and discussion among the multiple stakeholders, who eventually share in the work and the bounty. It doesn't take such a stretch of one's imagination to consider a well-maintained garden great design or beautiful art. By contrast, farms, whether designed by designers or artists are messy, and can seem dirty and noisy, requiring a far greater imagination to recognize their artistic contributions, in terms of qualifying as art or having aesthetic attributes.

By looking at these three design-farms, several features that distinguish practical design from practical art begin to emerge. (1) Built design must minimally achieve its stated function. (2) Good design communicates its function and doesn't prioritize innovation or carry out experiments on its public. (3) Like visual art, visionary design can be discursive. (4) Designers welcome becoming experts in specific fields, since regular income tied to their specialties affords them opportunities to submit far-flung Requests for Proposals (RFP's). (5) Practical art, which aims to solve something very specific, doubles as a beta test (the alpha test having occurred in the artist's studio). By contrast, practical design often has numerous goals, which are either competing or contradictory, when lumped together in one project, such as maximizing durability and minimizing resource exploitation.

3 Practical Actions' Differing Outcomes

These days, designers commonly distinguish their contribution as an action, since design work generates products either *with* or *for* others, presumably some client. As a result of designers' consultation and collaboration with others (stakeholders/consumers, fellow designers, and manufacturers/builders), designers earn acclaim for actions (ideas, strategy, design, and implementation plan) that produce desired outcomes. As a result of successful actions, designers gain credibility that leads to their selling future projects. *The Design Way* authors Harold G. Nelson and Erik Stolterman consider "design wisdom" to be "an integration of reason with observation, reflection, imagination, action, and production or making" (Nelson and Stolterman 2012, 18). Since the same could be said of "science wisdom," or "art wisdom" for that matter, they add a more specific condition:

[4] http://beta.constitutioncenter.org/ncc_press_Into_the_Open.aspx

> [D]esign wisdom has the ability to shift from an analog experience of life, to a digital or analytic perspective of the world and *back again*. This is done by means of a design process that begins initially with a complex, undifferentiated, situation, which then transitions through a process of discernment and distinction and ultimately terminates with the integration of innovative designs into a desired seamless reality for those being served directly or affected incidentally. Therefore, one of the most vital aspects of design is that the outcome of any practical digital and analytic intervention must be transformed back into the analog. This is to ensure that, with each new design addition, life continues to be experienced as a whole (18-19).

Although these authors don't offer a case study of this process, nor do they explore this shift in any great detail, they imply that "design wisdom" shifts from digital to analog, while "science wisdom" need not engage the analog experience of life and "art wisdom" need not adopt the digital or analytic perspective. In other words, scientists and engineers don't have to have a working knowledge that comes from hands-on experience to employ computer simulations that engineer solutions. They problematically stereotype artists as shunning technological tools that otherwise allow them to do their work faster in favor of hands-on approaches. To my lights, a designer's capacity to shift between digital and analytic practices fails to characterize design work in general, though these authors' advice is well taken. As demonstrated by "New Craft: Design After Design" (2016), an exhibition at the XXI Milan Design Triennale, the number of contemporary designers using found or recycled materials, selling handmade goods on Etsy, or opting for letterpress suggests that the shifts they attribute to "design wisdom" only apply to manufactured goods.

It's hardly surprising that design, whose outcomes are the products of human beings' engagement in collaborative discussion, decisions, and activities, is described as an action. Problem is, artistic actions are no less collaborative, this too fails to distinguish design from art. As early as 1958, Yves Klein exhibited an empty gallery to demonstrate the priority of social encounters over objects. That same year, the Situationist International (SI), led by Guy Debord, launched its journal *Internationale Situationniste*, whose aesthetics of everyday life inspired real-life interventions involving the creation of revolutionary situations. Beginning in the early 1960s, artists similarly adopted the term "actions" to denote artworks that facilitate environmental transformations, thus differentiating their practical actions from that era's happenings, situations, and environments. In 1971, Beuys performed *Eine Aktion im Moor* (Bog Action) and *Forest Action* (along with students in the second case) to publicize the rapid destruction of European wetlands and Germany's forests, respectively.

For over five decades, the Harrison Studio has practiced what they term "conversational drift," whereby they initiate and lead public discussions to address otherwise neglected topics of ecological interest. Newton Harrison explains: "We are storytellers. Our art is about engagement" (Spaid 2002, 21). And their form of storytelling typically initiates transformative actions, as was the case with their vision for the *Green Heart of Holland* (1995), a vast ring of Dutch cities enveloped by farmland, but under threat from encroaching housing developments. Because they involved several Dutch ecologists and landscape architects in their proposal,

Table 1 Danto's 1979 Theory of Action Applied to Art (Case 1) and Design (Case 2)

Case	Cause	Effect	"Theory of Action" Comments	*Transfiguration* Examples
1	R	R	It is an artwork when the representation is true and its being true is explained by its impact when the resulting representation is satisfied.	Andy Warhol's *Brillo Boxes*, Roy Lichtenstein's portrait of madam Cézanne, Picasso's painted tie, and Duchamp's shovel.
2	R	~R	"It is action, when the representation is true but its *being* true is explained through the impact of the person whose representation it is on the world."	Erle Loran's Cézanne diagram, Brillo cartons stacked in store rooms, a tie painted by a child, most advertisements, posters, TV programs, stories, and actions that fail as art.

some of the original ideas their proposal recommended, like the Bio-Diversity Ring that provides a protective eco-urban edge, were not only folded into the Minister of the Environment's formal proposal eight months later, but found eventual implementation (Spaid 2002, 34–36).

I next employ Arthur Danto's 1979 Theory of Action to show that art and design outcomes are quite different, despite their both being actions, since the outcomes of design actions don't also require an interpretation. In *Transfiguration of the Commonplace*, Danto characterizes artworks as belonging to a rather large class of "representationally characterizable" events (R), which also includes actions such as posters, advertisements; billboards; signs; packaging, maps, charts, graphs, logos, and illustrations (Danto 1981, 83). Although Danto's 1979 table presents four cases, I focus on the first two, which are most relevant here. Case I comments are my assessment, while Case 2 comments are Danto's (Danto 1979, 481). As Case 2 examples (a diagram, Brillo cartons, a painted tie, advertisements, posters, etc.) provided by Danto in *Transfiguration* indicate, Case 2 specifically addresses design, since practical actions that both facilitate and communicate their functions don't *prompt* "representationally characterizable" events. Case 2 characterizes designers' actions (their designs) that produce desired functions and thus don't require further interpretation or analysis. Case 1 actions cause "representationally characterizable" effects, while Case 2 actions reflect "the person whose representations it is on the world." Case 1 actions thus impart "our capacity to reflect upon our experience, or to ascribe contents to our own thoughts, as a result of our having experienced an artwork" (Spaid 2016b) (Table 1).

For the most part, Danto's Case 2 actions cohere with Latour's "post Promethean theory of action," so-called since he considers design's five advantages (over Modernism's strategy of revolutionary overhaul) to be its modesty, attention to details, meaningfulness, continuous state of redesign, and ethical-orientation. He continues, "This theory of action has arisen at just the moment when every single thing, every detail of our daily existence, from the way we produce food, to the way we travel, build cars or houses, clone cows, etc. is to be well, redesigned" (Latour 2008). At first glance, Latour's acknowledging design's semiotic thrust such that we conceive of artifacts as designed things or distinguish good design from bad suggests that he may consider Case 1 actions design. But I don't actually think this is so, since he definitely emphasizes the designer, not the designed object, as the actor fulfilling or carrying out some desired task. He continues, "To design something allows us to raise not only the semiotic question of meaning but also the normative question of good and bad design. This is true of DNA manipulation, as well as of climate control, gadgets, fashion, cities, or natural landscapes, a perfect case of design from beginning to end. Artificiality is our destiny, but it does not mean accepting the modernist definition of an artifact as the invasion of matters of fact over the softer flesh of human frailty forever" (Latour 2008). It is no wonder that Case 2 actions continuously undergo redesign, whereas Case 1 actions remain in constant motion like a chain reaction. Recall how many times you may have heard an artist say, "I just make the stuff. I put it out there. And then it has a life of its own."

4 Semiotic Question of Meaning

Earlier, I noted that few would view as unethical those architects whose RFP's benefitted from the additional wisdom of engineers, botanists, landscape architects, interior designers, and others who both improve their firm's proposal and then manage its implementation. I've noted that artists too, like the Harrison Studio who collaborated with Dutch ecologists and landscape architects, routinely work collectively. In fact, practical actions typically result from collaborative practices that convene various stakeholders, practitioners, and experts to tease out the winning solution. But the big question remains, who rightly deserves to put their name on the proposal? Who signs the drawings? To whom does the credit belong in the end? These kinds of issues have plagued collective practices for decades. Artist Judy Chicago's taking full credit for *The Dinner Party* (1974–1979), despite the fact that over 400 volunteers worked on it, created quite a controversy back then. The lesson from Chicago's debacle is remuneration is key.

Of course, one simple answer is "the party who calls the meeting to order" wins the day, since he/she/they/it assembled all the accomplices for a particular reason. The question of "credit due whom" is especially crucial for design firms where minimally-paid, low-level designers and interns work long hours (salaried workers may be excluded from over-time pay) doing a lot of the technical work, but rarely receive extra compensation or formal credit (outside the firm) for their significant contribution. That is, they cannot sign the drawings (with their own names) even if they're licensed professional architects. Although I know of no case where a designer has expressed his/her feeling slighted for not having been permitted to sign the drawings that he/she actually produced, or for not having received public recognition for managing a design process from start to finish on behalf of his/her firm's principals, I wouldn't be surprised to one day meet such a disenchanted ex-designer.

Like Latour, Nelson and Stolterman consider designers responsible actors. Of course, most responsible are the (typically licensed) principals who sign off on the firm's drawings, which may explain why they inevitably win the awards, are often paid the big bucks, and receive credit for design achievements, even though they may have contributed very little to the actual design or its implementation. They are after all, the person(s) whose representation impacts the world (Case 2). With Case 1, the artwork's owner (the collector), as opposed to the artist, typically reaps the big bucks. Nelson and Stolterman particularly deplore the notion of designers attempting to escape responsibility for the consequences of their ill-conceived designs: "[T]hese attempts by designers to divorce themselves from responsibility for the ultimate outcomes of their designs cannot be justified and are unacceptable, given the accumulating effect of small designs on the emergent design of social reality" (Nelson and Stolterman, 204). They continue, "Design decisions are based on judgment and judgment is both personal and situational. In the end, design is always an act of faith in our abilities and ourselves" (204). For this reason alone, responsible designers deserve credit for building a winning team of experts that successfully implement and manage the design. The principal(s) who assume all of the risk receives the lion share of credit as a reward for having been the responsible party.

Related to the notion of the design principal being the responsible party (and thus the most rewarded) is the connection between a design firm's mission statement (a.k.a. design philosophy or design principles) and a particular design's values. The reason lower-level designers can do award-winning work with very little supervision is because they understand the design firm's design principles, as established by the principals in accordance with values earlier designs uphold. The principal(s) receive all the credit because the firm's designs convey their values. When performing their work on the firm's behalf, employees effectively enact their employer's design philosophy, not their own. This is yet another reason they don't expect to receive credit for their work…it's not necessarily their work! One imagines that those designers who routinely find their personal design principles diverging from those of their firm's principals will either soon search for a new firm whose design principles align with their own or start their own firm.

Even if designers adhere to principles that are expressed by prior designs' values, it doesn't mean that the principals could actually identify said principles, let alone articulate their designs' values. Principals are under no legal, ethical, or practical obligation to specify design principles and values up front. Some designers, especially architects, have clarified their principles, though mostly for marketing purposes, to attract clientele. Some principals eschew the idea of a principle specifically because they expect clients to establish the principles, so lacking a principle reflects their belief that their clients' briefs come first. Absent a stated principle, a firm's designs cannot be critiqued for veering from (or praised/blamed for steering) their firm's stated mission (Table 2).

Independent of a design firm's mission statement, every design action reflects particular values, and skilled observers not only recognize values in designed objects (see Table 2), but building enthusiasts can also infer the underlying design principle from implicit values. For example, a public building that integrates proportion with its sighting and accommodates solar angles and street views aims to relay that it connects with the community and is welcoming, ecological, sunny, and

Table 2 22 Design values culled during "Design Ethics" Course (2010)

		Roger	Royal Fine Art
General	Particular	Scruton	Commission
context	sustainable	mutability	order/unity
program	solar angles	taste	expression
practical	& views	façade	integrity
public		manners	plan and section
scale		street	detail
			integration
			sighting
			massing
			proportion
			rhythm
			materials

bright. To my lights, this is exactly what Latour has in mind when he attributes meaning to design. Nelson and Stolterman add, "Meaning as form, is revealed to us through the ordering and organizing of elements into systematic relationships and connections that have been created intentionally, in response to purpose, in fulfillment of an end. With this we mean that those unifying forces that cause things to stand together, in unity, provide comprehensible emergent qualities of presence, significance, and value, thus forming meaning for individuals who are part of the whole or served by the whole" (93–94). As already noted, designed objects (not design philosophies) make designers' intentions manifest. "It is intention that predisposes or directs us toward certain data and values" (121).

Moreover, skilled observers recognize when a designer's actions suddenly contradict his/her prior actions and thus either indicates some external influence such as a client's wish or some investor's demands, or the designer's changed principles. More worrisome than designers' contradictory actions veering from prior principles is their capacity to produce outright fakes, that is, solutions that some designer has promised his/her client, yet the delivered goods fail to fulfill the stated promise. Either the delivered goods were not designed to meet their specs or the design does not function as promised. As alluded to in the introduction, I worry that a lot of what passes as "green design" merely looks green, especially when it employs nonrenewable resources. Strangely, design theorists tend to ignore the preponderance of fakes! I guess they opt to leave this conundrum to courtroom judges (and philosophers), who can't ignore them.

5 Entwining Conception and Making

While Latour seems to sever design as conceiving from building as making, Nelson and Stolterman's notion of design entwines conception and making. They characterize designers as craftsmen, or practical actors who capably wield their "knowing hand," infusing their every move with "design wisdom," as briefly discussed already (Nelson and Stolterman, 18). From what I know about engineers, such expectations seem entirely unrealistic. It's hard to imagine that some designers, let alone most, fulfill their expressed craftsmen ideal, anymore than engineers are tinkerers, who build machines in their garages. Design attracts all types, from the brainy to the brawny. A handful of architects have built their own homes, yet I imagine many more who haven't a clue as to how to use an electric screw gun. That said, what seems more appropriate here is to convey the roles bestowed upon conception and making, while not pretending that any one designer necessarily balances both skill sets. In fact, collectives, whose members fulfill diverse roles, are purposely assembled to defeat inescapable deficiencies. After all, responsible designers recognize their shortcomings (210).

Because Nelson and Stolterman emphasize making, their notion of design, which involves bringing "things into existence" (127), departs from that of Latour, who sees design as an ongoing process of redesign. For them, "[t]his type of ideal process involves imagining and creating that-which-does-not-yet-exist, but which we desire to be in existence, in the service of humanity in general and specific people in particular" (Nelson and Stolterman, 132). Sharing Latour's concern about our Prometheus (or modernist) past, they concur: "This is quite different from a typical Western technological approach, which prescribes that something ought to be created, simply because it can be done. This assumed prescriptive reasoning is lifted from an economic frame of reference where money –as the measure of value and return on investment –stands in for deeper aspiration and intentions" (132).

The little takeaway here is that the "I can" (capability) is not reason enough to justify the "I will" (determination). As will be discussed in the next section, leveraging the "I can" to execute the "I will," while elevating the "I show" over the "I know," constitutes what Nelson and Stolterman term "evil design." For them, "design volition," what I call the "I will" (after Hannah Arendt), concerns the use of one's will to pursue desired ends and forms the "distinctive character of design judgment" (134). To my lights, "making" (design volition) drives every project, leaving conception (what they term "design interpretation') to keep making in check.

> In design, interpretation is not about determining a solution by closely and objectively analyzing reality in order to be informed of what action to take. Interpretation in design is not a search for the objective, true, and precise design imperatives, hidden somewhere in the richness of reality waiting to be observed. Instead, design interpretation is an act of judgment. A scientific assessment is an accounting of objective factors, while a design interpretation is an appreciative judgment –a picking and choosing of what is to be considered and in what way (121–122).

In design, conception doesn't stop at the factory door: it is invested throughout the design process from team building to fabrication, marketing, distribution, and eventual iconography. In contrast to Latour's view of design as continual reflection and rethinking, conception is no less material than intellectual. Conception thus requires the complete and total envisioning of every corner of Earth where the product could reach, and imagining how its function might be received, long before it ever materializes. This is how a creative concept is transformed into a "concrete particular addition to real life" (134). Nelson and Stolterman term this process *allopoiesis*, since it is "the making of something outside of one's self, with and on behalf of the other" (132). And they see design as "the act of creating something intentionally on behalf of another's desires and purposes" (132). Again, their all-encompassing reach may be unrealistic, but at least they convey how conception and making are entwined, not severed as Latour problematically claims.

6 Normative Constraints of Good and Bad (or Evil) Design

I now want to revisit Latour's positive conception of design, which for him engages human beings in collaborative action, while reconnecting them to their environment. As briefly mentioned, he characterizes design's five advantages over revolutionary practices as being: modest, attentive to details, semiotic, in constant transformation, and subject to the normative constraints of good and bad design. While I appreciate Latour's historicist account of design vis à vis modernist strategies of emancipation, detachment, progress, and mastery; I worry that his emphasizing design's positive capacities overlooks the way designs proposed by designers who lack experience making things can seem half-baked. That Latour considers design "low risk" seems myopic since "bad design" engenders unforeseen consequences, while "fantasy designs" risk dismissal. By contrast, environmental artists who propose ecoventions initially perform alpha tests (often on their own dime), so that they can analyze and revise their designs, long before proposing them as public commissions. As already noted, Latour severely under-estimates the significance of making, yet he rightly makes room for the "normative question of good and bad design" that invites praise or condemnation of the outcomes of design actions.

Keeping in mind the possibility of normative judgments that praise or condemn the outcomes of design actions, I finally return to the urgency awaiting environmental designers, whose most successful solutions, whether sustainable architecture, large-scale public works, or edible foodstuff; are the results of efforts either to restore "lost" practices or to replicate nature's processes. Relevant examples include bio-intensive farming, rammed-earth homes, phytoremediation, or land conservation (using herd-plant dependencies to minimize soil erosion). As well-documented elsewhere, our crisis on Earth is the result of climate change and the depletion of scarce-world resources, which capitalism's dependency on population explosions further exacerbates. Design that pursues a totalizing conception involves every step of the process from brainstorming to iconography, enabling designers to optimize resources expended by the array of handlers contracted. Much like Nike was in a position to demand that its manufacturers implement higher work standards, designers who conceive the entire chain of command from manufacturing to distribution and marketing, are in a position to demand that resources be conserved to the highest degree. Similarly, designers are in a position to ensure that the daily operation of designed products minimally expends resources. And quite frankly designed objects that expend more resources than their designers originally promised (as stipulated in design-spec contracts) are either fakes or failures. Take your pick.

I would imagine that the vast majority of bad design stems from over-confident designers who either promise more than they can deliver or don't know what is feasible to build with existing technologies or for contracted manufacturing fees. As Nelson and Stolterman point out, "[D]esign is based on a compound form of inquiry, composed of true [evidential], ideal [perfect world], and real [best practices] approaches to gaining knowledge" (34). "Design is a process of making close approximations, the closest possible, to these idealistic desires" (35). Bad design happens because designers failed to make close approximations. While bad design is inexcusable and is punishable to the full extent of the law, I see bad design for what it is: hubris, ignorance, lack of curiosity, poor connections, or an under-performing team. With bad design, the principal is blameworthy, even if the motives were good (or right). The intent to design was lacking, since the principal lacked the requisite skill set needed to carry out the contracted bid.

What interests me most is what Nelson and Stolterman term "evil design," which supersedes bad design (168). With evil design, the principals promise to design something that they know they cannot deliver. Even worse, they use the promise of delivering something that is either desirable or innovative as a way to gain attention from the press. Sometimes, principals know full well that their stated innovation not only adds little value, but unnecessarily expends additional resources. One wonders what inspires principals to attempt evil design, especially since they are most likely risking their professional accreditation. Most evil design takes the form of window-dressing, thus generating the false impression that the design is especially helpful, necessary and innovative. As Nelson and Stolterman point out, "[D]esigned artifacts are most commonly recognized by their most immediately accessible level of presence, their style or fashion. Style and fashion are characteristics of presence that appear across the compositions of the one designer, or school of design, or across eras of material culture. When particular design principles are used together, regularly and consistently implemented in multiple artifacts or system designs— a style is born" (170). Putting an end to evil design, especially as it infects "green design" is just a matter of demanding that designers account for every show-stopping embellishment and provide evidence that what they are recommending their clients do has undergone a time-tested evaluation.

7 Conclusion

With this paper, I have demonstrated that not all "green design" is desirable, since many examples fail to deliver promises the designer was contracted to deliver. I imagine designers establishing boundaries to ensure that the most important demands are met, while avoiding experimental options with only a small chance of working out. The most important point is that designers know their limits and don't incidentally (unbeknownst to either designers or clients) experiment on their clients. Although I don't share Latour's view that design is low risk, as in "harmless," I do think it ought to remain as low risk as possible. Leave the experimenting to environmental artists, whose budgets, commitments, and spheres of influence are comparatively infinitesimal. When an artist's design experiment goes awry, it seems charming. When a building collapses, a crane topples over, or streets flood, there are life-threatening consequences.

The differences between art and design are vast, so it is indeed odd that few philosophers have attempted to tease out their differences as I have tried to do here. Since I assign environmental art to be the experimental territory for environmental design, the distinction that I outline here is enormous, and demonstrates perhaps why architectural teams increasingly involve artists in their proposals.

References

Danto, A. (1979). Basic actions and basic concepts. *Review of Metaphysics, 32*.

Danto, A. (1981). *Transfiguration of the commonplace*. Cambridge: Harvard University Press.

Golonu, B. (2008). "Greening the revolution.." *Art Papers*. November/December.

Haeg, F. (2010). *Edible Estates*. New York: Metropolis Books.

Latour, B. (2008). A Cautious Prometheus? A Few Steps Toward a Philosophy of Design (with Special Attention to Peter Sloterdijk). In *Networks of Design*. http://www.bruno-latour.fr/sites/default/files/112-DESIGN-CORNWALL- GB.pdf. Retrieved 29 Dec 2015.

Margolin, V. (2006). In S. Smith & V. Margolin (Eds.), *Beyond green: Toward a sustainable art*. Chicago: Smart Museum of Art.

Nelson, H. G., & Stolterman, E. (2012). *The design way: Intentional change in an unpredictable world*. Cambridge: MIT Press.

Ridgely, C. (2008) http://www.fritzhaeg.com/garden/initiatives/edibleestates/baltimore.html.

Spaid, S. (2002). *Ecovention: Current art to transform ecologies*. Cincinnati: Contemporary Arts Center.

Spaid, S. (2012). *Green acres: Artists farming fields, greenhouses and abandoned lots*. Cincinnati: Contemporary Arts Center.

Spaid, S. (2016a). The future of environmental art or reimagining a sustainable art practice? Real-world problems. In *Sustainable art facing the need for regeneration, responsibility and relations*. Wrocłov: University of Wrocłov.

Spaid, S. (2016b). "Danto's Artworld: Where Nine Indiscernible Red Squares Yield Nine Distinct Contents." Proceedings from the 2014 Meeting of the Spanish and Portuguese Society for Aesthetics.

Wilson, E. O. (1984). *Biophilia*. Cambridge: Harvard University Press.

Zelenski, J. M., & Nisbet, E. K. (2014). Happiness and feeling connected the distinct role of nature relatedness. *Environment and Behavior, 46*(1), 3–23.

Scratching the Surface: "Appearance" as a Bridging Concept between Design Ontology and Design Aesthetics

Annina Schneller

Abstract What is design? Definitions range from design as a product or process of thinking, modelling or problem solving, to all-encompassing visions of design as the transformation of social environments. Some definitions of design stress the aspect of function, others the similarity with art. Even if we try to break down the definition to design objects in the sense of designed material artefacts such as chairs, books or buildings, defining their essential properties proves difficult. What is the special ingredient that makes an artefact a design object? Based on the philosophical method of conceptual analysis, the present chapter asserts that any definition of design objects necessarily includes their *appearance*. Since the creation of appearance and aesthetic experience is an essential task of design, the study of aesthetics should consider design among its paramount subjects. This argumentation leads to an astonishing conclusion for traditional philosophy: The philosophical divide between ontology and aesthetics is bridged when it comes to design objects.

Keywords Design definition · Design ontology · Design aesthetics · Appearance

1 Design Philosophy

Until now, design has not been a focus of philosophy, nor has design research hitherto devoted much attention to philosophical problems. The central issues of design philosophy – or the philosophy of design – are still to be determined. As both a philosopher and a design researcher, I see two possible questions that might help us to elucidate issues of design philosophy:

1. Of what concern can design be for philosophy?
2. Of what import can philosophy be for design?

A. Schneller (✉)
Bern University of the Arts, Bern, Switzerland
e-mail: annina.schneller@hkb.bfh.ch

© Springer International Publishing AG 2018
P. E. Vermaas, S. Vial (eds.), *Advancements in the Philosophy of Design*, Design Research Foundations, https://doi.org/10.1007/978-3-319-73302-9_3

I think that these two questions are interrelated. Looking at design from the perspective of philosophy might enable us to see design in a new light – and vice versa. Philosophy has a genuine interest in "what holds the world together in its inmost folds"[1] and in the clarification of fundamental terms of inquiry. For a start, philosophy could help to define the basic concepts of design: What is design? What does a designer do, when she designs? What meanings or implications does the concept of design involve? Finding an answer to these questions is of genuine import for any design researcher or reflecting design practitioner (cf. Schön 1983), because they apply directly to their subject of study and go to the very essence of their practice. On the other hand, designerly ways of knowing[2] and thinking can change the way we approach philosophical problems. Grasping what design is improves our philosophical understanding of the basic inventory and correlations of the world.

Philosophy is generally based on *theoretical* study, and design is considered to be an *applied* art. We should not deduce from this that philosophy is supposed to learn (only) about practice from design, while design will gain (only) theoretical knowledge from philosophy. I rather think that each field can provide and gain instructive insights on the level of theory, practice and method. Each field can fructify the other, either by providing new questions or methods for investigation – such as design thinking, user-centred or participatory design, rapid prototyping or brainstorming from design, conceptual analysis, dialectic or methodic doubt from philosophy – or by broaching previously unconsidered subject matters or drawing unfamiliar correlations between them. The latter might occur via the ideas of 'creativity', 'draft' or 'model' on the one hand, or the concepts of 'responsibility', 'value' or 'meaning' on the other. Learning from design could mean that philosophy asks for applicability of results, and gains an awareness of the creative aspect of knowledge generation and the importance of implicit knowing. Learning from philosophy could lead design research to adopt the exactness and scrutiny of reflection that is not always found in design theory today.[3]

In what follows, I will endeavour to begin bringing philosophy and design together by means of a philosophical investigation of the concept of design and by taking a designerly look at important philosophical issues. In a first step, a philosophical concept analysis will be carried out in order to gain a sharpened concept of design and to provide a useful argumentation basis for design theory, e.g. for distinguishing design from other forms of creation or problem-solving, such as crafts, do-it-yourself or fine arts. As will become apparent, a crucial criterion for design activity is the designer's skill of shaping the form, surface or *appearance* of objects. Taking a closer look at the phenomenon of appearance from a designerly point of view will then, in the remaining sections of this book chapter, reveal surprising insights into a handful of traditional philosophical problems, both in the field of ontology and aesthetics.

---

[1] Cf. Goethe's *Faust* (translation by George M. Priest)

[2] This is a reference to (Cross 2007)

[3] For a suggestion of how a philosophical approach to design could deepen our understanding, cf. (Vial 2015b).

2 Defining Design

As we have seen, philosophical examination can help us to determine with greater precision the central concepts of design theory. These concepts are often not well defined, partly due to design research being a young academic discipline, and partly due to its non-theoretical grounding. Both design research and design education have been conducted primarily by design practitioners. So what about the fundamental concept of design theory – namely 'design'? Existing definitions of 'design' range from design as a product or process of creating, drawing, thinking, modelling or problem-solving, to all-encompassing visions of design as a transformation of social situations. There is no consensus about what 'design' is.[4] In general, we can find a gap between positions that stress the aesthetic and form-giving aspect of design, where "the ultimate object of design is form" (Alexander 1971: 15), and positions that emphasize the aim of creating a desired state of affairs (following Simon 1969; cf. Kimbell 2011). On the one hand, the idea of design as giving form is linked to the physical making of things and craft, and necessarily involves materiality. It reflects a conception of design found in design movements and schools such as Arts and Crafts or the Bauhaus. Such a view might not easily cope with the complex structures that design faces today, e.g. in service or interface design, and, despite its technical interest, it runs the risk of perpetuating existing ways of making by just looking "at problems of modification of the surface of end products" instead of thinking things through (Fuller 1955: 61; cf. also Burckhardt 2012). On the other hand, a more effect-oriented or solution-oriented account opens up our focus on the general nature of "the artificial, made by human beings" (Bayazit 2004: 16), considering design in rather abstract terms as one stakeholder within multifaceted organizations or systems. With such a universal definition of design, we might lose sight of the specific techniques and material nuances involved in particular fields and practices of design, and start thinking of design ability as an anthropological feature possessed by all human beings. There have been attempts to preserve the uniqueness of design by postulating a special ability of designers to solve even ill-defined, underdetermined or 'wicked' problems (Cross 2007; Buchanan 1992) or by stressing the unorthodox interweavement of problem and solution in the design-thinking process (Rowe 1987; Dorst and Cross 2001).

Instead of further elaborating on the abovementioned conceptions of design or deciding in favour of one of them, I will investigate the concept of design using a philosophical method called *conceptual analysis*. Conceptual analysis does not consist in listing the etymological details of the word 'design', but rather tries to answer the following questions: What criteria must an artefact fulfil in order to count as a design object? What is involved in the act of designing? What are the necessary and sufficient conditions for the application of the concept? One test for attributing (or denying) specific properties to a concept is our actual use of the word. What do we

[4] This is not only a problem of design theory, but of most fields of study, at least within the humanities and social sciences.

mean when we talk of design? What things, processes or actions are named by our use of the word 'design'? Carrying out this test usually reveals that there is not just one use of the word, so that there is not just one possible definition. To gain clarity about the object of design theory, I will therefore have to exclude certain marginal uses and to differentiate between a narrow and a broad definition of design.

2.1 Prospect and Production

So what is design? What does the concept 'design' mean? How do we use the word 'design'? The expression 'design' seems to serve a double function: 'Design' can mean both an activity ('to design') and the product of this activity ('a design'). This double sense could be responsible for two different ways of differentiating design in design theory: there are definitions that focus more on the design process (as thinking, envisioning, planning, problem-solving, modelling, drafting, conceptualising or the like) and others that take into account the constitution and workings of the design artefact as a product. If we talk of 'design' as an activity, we refer to more than just doing, namely to an *action* involving intention. Although any action means to take a step into the future, design action seems to be directed towards the future course of events in a very special way. It explicitly takes into consideration how things should be in the future. Design is a *prospective* action.

Moreover, this prospective or envisaging endeavour has a very specific object: the design object, now in the second meaning of 'design'. Any design action seems to prepare and envision the *production* of an object or state. So envisaging the creation of a product is essential for any act of designing. This condition even holds true if the design product is not produced by the designer herself (e.g. in industrial design), or if it is not produced at all (e.g. due to technical or financial problems). Is the productivity of design already implied by the fact that the verb 'to design' is transitive, i.e. it necessarily involves an object? There are many other actions requiring an object, e.g. 'drinking' or 'throwing'. To drink means to drink *something*, and we can only perform an act of throwing if we throw *something*. The productive aspect of design differentiates designing from these other actions, as the transitive object of designing is not there *previous* to the action – indeed, it *may not* be there, in order to make it an act of designing.

So every design act aims at bringing something into being that has not previously been there. Design action plans to change or proliferate the inventory of the world. Let us thus tentatively characterize design as an action with a productive prospect.

2.2 *Effect and Deliberation*

There is, however, more to design than this. If we say that someone is designing something we do not mean that she is thinking about the production of *any* object or state with *any* properties. Rather, a design action must consider the specific *impact* that the design artefact is likely to have on people and their environment – the ways in which a design object will function and interact with the world. Design is an *effect-oriented* action. It can also be an act of communication, and design products can have meaning. Some theoretical approaches such as *product semantics* even think that *all* design products are semantic or semiotic entities with a communicative function. I prefer to speak of 'effect-orientation' instead of 'communication' in general, and of 'effect' instead of 'meaning'. This leaves design the option of adopting a communicative function without postulating meaningfulness in all tokens of design.[5]

So designers create highly specific situations by means of their designs. In order to do so, they need to think about the needs the design object is supposed to satisfy, about the use, purpose or function that it will fulfil, about the role it will play in social life, or about the meaning it will have for different people. This process can also involve considering the resources that are needed in order to produce the artefact (e.g. renewable resources, low-energy production) and what will happen when the design object becomes obsolete (e.g. recycling, avoidance of toxic substances). So designing is not simply thinking about means-ends relations, but involves a complex of prospective *deliberation*. Designers must consider how and to what extent the actions and emotions of beholders or users will be influenced, and how and in what ways the world will be changed by their design action. To be sure, the impact of design is not always great, neither is the depth of reasoning the same in all acts of designing. Some of the thought involved in the design process is even developed *implicitly* – as part of the designer's know-*how*. Taking into account the complexity of the correlation between design deliberation, know-how and possible effects on human living, we can easily see why design can become an interesting subject of philosophical (applied) *ethics* – but that is not the issue here.

We have now expanded our definition of 'design' by two conditions: Design is a deliberative, effect-oriented prospective productive action or the product of that action.

[5] For a design-rhetorical account of design effects, cf. (Schneller 2015); for a detailed phenomenology of the effects of design, cf. (Vial 2015a).

2.3 Form and Appearance

This definition is compatible with expansive definitions of design. Herbert Simon's famous definition of design, for example, says that design aims at "changing existing situations into preferred ones" (Simon 1969: 111). I find such a broad, overall concept of design dissatisfactory for two reasons. First, I think that many activities fit this definition that we would not normally call 'design'. Secondly, adding all these activities to the realm of 'design' would lead to a dilution of the term, with the consequence that the special skills are lost that actually make a designer. Think of someone doing housework – such as baking a cake or cleaning the bath – or of a politician carrying out legislative work. All these activities aim at improving existing situations. But are they really instantiations of 'design'? If baking, housekeeping and legislating are design tasks, this reduces the particularity and possibly also the worth of the work accomplished by professional designers.

We might rule out cleaning as a design action by means of the criterion of productivity, but even the cleaner creates a new state of the world in some sense. The criterion of deliberation about the specific outcome and intended effects should do the trick. Neither cleaning nor amateur baking are well-reasoned ways of creating products with clear-cut effects. In more deliberate and purposeful cases, however, the demarcation line becomes porous. Why not call sophisticated housekeeping or confectionery 'design'?

But what about legislating? Creating new laws seems to fulfil all four criteria: Laws are the product of prospective action, and they involve a great amount of deliberation. Laws are effect-oriented, as they are intended to organize social interaction and the community in sufficiently specific ways. Moreover, it is not unusual to talk of 'design' in connection with laws. The question "For what end was this law originally *designed*?" is perfectly meaningful. Nevertheless, there seems to be something missing in the work of the legislator that makes us refrain from calling her a 'designer' in the true sense of the word. What could this be?

We therefore need a final criterion to help us specify the particular skills of designers and to differentiate 'design' from other human activities or products that are similar to design in some ways, but are not 'design' in the more specific sense. The special know-how of designers, as I see it, lies in their ability to select or provide highly specific formal means relative to their field: means that are appropriate to implementing the intended effects in the design artefact or situation. Those formal means involve style elements, steps, procedures, patterns, figurations or ornamentations that are needed for the (re)production of the design object.

Let us call this the *formative* criterion of design. Only by fulfilling this further criterion will the prospective, productive, deliberative and effect-oriented design action become operative in a typically 'designerly' way. The formative criterion is also what contrasts (real) design from legislating. For the legislator, the specific forms and ways a law will be materialized are not his main focus. Designing a law surely involves formulating and structuring the legal text, but not its overall form or the material and technical details of representation and production. This would be the job of the (editorial) designer. The formative criterion thus requires more than just providing a scheme, a concept or an abstract idea. The work of the designer involves instructions about the overall form, the construction and the materiality as well as decisions about the colour and scale, composition and proportion, accentuation and contrast, and light and placement of the design object that is to be produced.

The conceptual analysis leads us to the following criteria for design:

1. **Prospect** (future oriented)
2. **Production** (directed towards the creation of a new thing or state)
3. **Effect** (effect-oriented, purposeful, functional)
4. **Deliberation** (awareness and control of effect).
5. **Form** (the specification of formal means for the implementation of intended effects)

This definition, with its five criteria, can also help to understand the difference between *amateur design* or *handicraft work* (that is not 'design' in a strict sense) and professional design. If we quickly sketch a note or fold a paper plane, this is clearly not an act of designing. We neither deliberate about the exact outcome, nor do we think or care much about formative matters – so points 4 and 5 do not obtain. Today, especially in visual communication, amateurs have access to the technical equipment of professionals and often use professional tools for creating visual media. If amateurs create artefacts with both diligence and formative skill, at some point we might start calling their work 'design'.

Different gradations can apply when discriminating between 'design' and 'non-design'. There might be amateur or self-taught designers who fulfil the criteria for 'real' design, but on a lower level. There is no need to draw a strict line here. In order to preserve a valid concept of 'design', however, we need to give an account of why not *everything* is 'design'. It seems to be the formative skill in particular that is not fully developed in amateurs. Even if an amateur designer is fully clear about the intended outcome of an envisaged product, she might fail to realize her aim because she is not able to choose the appropriate formal means or to apply them subtly enough. The formative criterion really seems to be the special ingredient that makes design 'real' design.

It is the combination of the prospective, productive, effect-oriented, deliberative criteria with the formative criterion that yields the specific capacity required from designers, namely a standard that is not mastered by everybody – at least not to a high degree. As we shall see, the formative endeavours of the designer are always directed towards the *appearance* of the design object. Only through the ways in which a design object appears to us can its intended effects really come into action. In what follows, the importance of the concept of appearance for design theory will be further investigated.

3 Scratching the Surface

My attempt to analyse the concept of design has led to the following definition: To design means prospectively and deliberately to envisage the production of artefacts or states that affect (and ideally improve) the emotions, actions and interactions of people. In order to demarcate design actions from other forms of prospective, deliberative, effect-oriented productive actions, the *formative* criterion has been added: Designing means to provide or select the *formal* means that are appropriate for creating the intended effects. To say that designing consists, *inter alia*, in determining the 'form' of the envisaged object, is not very precise. Grasping the idea of the 'formative' or 'form-giving' aspect of design becomes even more difficult when the design of complex states or situations is concerned instead of self-contained material objects.

A first problem is that our common concept of 'form' does not *prima facie* embrace all features that should be included in the formative criterion of design – aspects that are easily passed over when design is conceived in classical terms of 'form-giving'. Take for example illumination or lighting, which are crucial means in many areas of design, e.g. scenic, film, interior or wayfinding design. We would not normally consider light as part of the 'form' of a design object or of the 'form-giving' process. This is because light is added from the outside and is not a material or physical property of the design object in a strict sense. Similarly, the colour, weight, smell or sound of design objects should not be considered as extra elements alongside 'form', but as part of the formative endeavour of design. In my understanding, all these aspects of design add to the overall 'form' of a design object or state.

Another problem, quite contrary to the abovementioned drawback, is that not all aspects of 'form' are really important to design. There are formal properties that are negligible for the designer because they do not matter for the effect a design has on people and their environment. The details of technical construction of an industrial or architectural design product, for example a coffee machine or a waterpark, are formally vital for the designer only as far as they are *perceivable* by the user. Designers have a special interest in the *surface* of what they create. This is not to say that the technical construction or function does not matter for design. Designing a waterpark goes hand in hand with developing a functioning system of water pipelines, heating and a power supply. However, providing for such a supply system is not part of the design task, but rather of engineering. Its whole construction is usually completely invisible to the visitor, and if the pipes are made visible, then this is due to a design decision. It would become the designer's job to think about the form and effect of the perceivable part. The construction of the basins is a hybrid, as it involves a mix of visible, tangible, perceivable, and therefore designerly important formal features, and of formal parts hidden from the bathers. Constructing the latter is an issue of technical engineering and usually exceeds both the formative concern and the formative ability of the designer. Architecture is always hybrid in this sense: design and engineering join together in giving form.

If we grant that design connects the criterion of form with prospective, deliberative, effect-oriented production, the designer must consider matters of function, usability and sustainability that are linked to or will result from the chosen form. That is why the 'surface' of design artefacts cannot be split from their 'interior', their technical workings and overall material composition. If the basins of the waterpark or the coffee machine leak or if the water temperature is not appropriately regulated, then the design will be of no use, however elegant or striking it might be. Design goes beneath the surface because it thinks about the bigger picture. Yet the need to pay careful consideration to the technical construction and to enjoy a fruitful cooperation with engineers is not a consequence of the formative criterion alone, but rather of combining it with the criteria of purpose and deliberation. This is why certain technical issues of form-giving might lie beyond the formative skills and interest of designers, and why it can even happen that technical considerations spoil the primary formative endeavours of a designer.

To focus on the perceivable or tangible aspects of form does not mean stopping at the surface. It does not make design a purely *superficial* activity. Superficial design – design that is concerned only with the superficial effects of a design object, such as a smooth touch, nice colour, elegant shape or shiny surface but does not care about material quality, function, ergonomy, the user's needs or environmental protection – would not count as 'design' in its above, strict definition, but rather as 'styling'. Gilded plastic is not the same as pure gold, and someone will sooner or later scratch the surface and notice the difference. This does not imply that design always yields pure gold – for some design purposes, gilded plastic can be the appropriate material. But it means that design thinks things through. Design concerns itself with what lies *below* the surface, because designing the form necessarily involves questions of function, effect and use.

4 Design Ontology: Appearance and Reality

Placing our focus on the formative endeavours of design has culminated in the provocative claim that design is particularly occupied with the *surface* of designed objects. In philosophical terms, what distinguishes design from mere products of engineering or craft, such as a water pipe, a saw blade or a screw nut, is the special attention given to the *appearance* of the object, such as an attractive, futuristic or elegant look and feel. The formative features chosen and provided by the designer are intended to create this specific appearance. A knife is made for cutting, but only if its specific visual and haptic appearance was explicitly considered during the process of creation would we say that this knife is a *design* object. What designers do when they create design objects – above and beyond realizing a technical or communicative function – is to create a certain visual, haptic, acoustic, olfactory *appearance*, a specific manner in which the object comes across to the beholder or the user.

The special formative or form-giving ability of the designer could then be described as handling the ways her design will *appear* to the public. Choosing or providing the formal means of a design artefact thus means creating the specific appearance of this object. Or, to put it the other way round: By giving artefacts a particular appearance, designers manage to intervene in the lives of users or beholders in very specific ways. Creating or changing the appearance of things really seems to describe the unique designerly way in which prospective, deliberative, effect-oriented production is realized.

The concept of 'appearance' rings a bell in philosophers' ears. Usually, appearance is opposed to the 'real being' of the world, and ever since Plato's critique of the sophists, philosophy has aimed at grasping the truth instead of mere semblance. 'Appearance' is an important term of philosophical ontology and epistemology. John Locke's distinction between primary and secondary properties of objects (Locke 2012: II, viii), for example, is an attempt to distinguish the 'real' from the mere 'seeming' properties of things. While primary qualities are said to convey undeniable, objective facts about objects that are independent of any observer, secondary qualities are thought to be mediated by the observer's sensations, depending on the effects they have on human beings: "whatever reality we by mistake attribute to them, [they] are in truth nothing in the objects themselves, but powers to produce various sensations in us" (*ibid.* 14). While the shape, texture, extension or solidity of an object would be primary properties of this object according to Locke, colour, taste or smell only belong to its secondary properties. Primary qualities are about the *essence* of an object, while secondary qualities merely give hints about its *appearance*.

Bertrand Russell takes up Locke's distinction in his *Problems of Philosophy* (Russell 1998). By exploring the room he sits in and taking a closer look at the table he is sitting at, he tries to distinguish 'reality' from 'appearance' or 'what things are' from what they just 'seem to be' (Russell 1998: 2). Russell leads us to understand that not only the 'reality' of secondary properties such as the colour of the table is relative to the observer's sensations, but also the so-called 'primary' properties such as shape and texture. According to Russell, "the 'real' shape is not what we see; it is something we inferred from what we see. And what we see is constantly changing in shape as we move about the room" (*ibid.* 3). In the same vein, texture is something that changes with the distance of perception. From far away, the table looks smooth, up close we see the grain of the wood, and through a microscope we discern a rough texture with hills (*ibid.*). So what is the 'real' texture of the table – is it even or uneven?

In everyday life, it does not matter what the 'real' face of things is and what just belongs to their 'appearance'. So is this merely a problem created by philosophical sophistry, or could the distinction between appearance and reality also be of particular importance to design? Russell gives us an interesting hint: "For most practical purposes these differences are unimportant, but to the painter they are all-important: the painter has to unlearn the habit of thinking that things seem to have the colour which common sense says they 'really' have, and to learn the habit of seeing things as they appear" (*ibid.* 2). The habit of being aware of the way things appear, as described by Russell, is also a crucial habit of designers. Quite like the painter, the designer must be able to predict and control the ways her design is going to appear to other people. But unlike the painter, the designer does so by forming the 'real' properties of the object. Russell concludes that: "The painter wants to know what things seem to be, the practical man and the philosopher want to know what they are" (*ibid.*), but we should add that the designer needs both: control over the essence of objects as much as control over their appearance.

Of course we could say that, in a deeper philosophical understanding of the problem, the designer, just like the painter, never reaches the level of 'real' matter in her work, and that even handling materials such as print colour, paper, wood or metal means acting only on the level of 'appearance'. But why postulate a deeper level? At least for understanding 'design', something like Kant's *noumenon* is of no explanatory value if there is no possibility of reaching and experiencing it. We had better stay with the phenomena. For this reason, I do not want to go further into the debate about where exactly the distinction between reality and appearance might be drawn, or whether subjectivity is involved in any ascription of qualities.

My point is rather that the distinction between 'real' properties and mere 'appearance' becomes obsolete in the case of design objects. It is an essential part of the designer's task to take into account the specific appearance of design artefacts. From a designerly point of view, the appearance of a design object is by no means less real than any of its 'real' properties. In the end, matter and form, essence and appearance will melt together in the final design product.

Thinking about design enables us to see the philosophical problem of appearance and reality in another light. For the designer, and for an understanding of what 'design' is, the way things 'appear' to us is a 'reality' as worthy of study as their 'real being'. Designers essentially make use of the appearance of objects in their work. Appearance thus becomes part of the essence of objects qua *design* objects. In this sense, the work of designers creates a fundamental link between essence and appearance.

5 Design Aesthetics: Aesthetic Experience

The preoccupation of design with appearance brings us to another philosophical field, namely *aesthetics*. The way in which a design object appears to us is closely linked to the aesthetic experience we have when perceiving it. Traditionally, aesthetic experience was solely attributed to works of art or natural objects and was restricted to the aesthetic dimensions of the sublime and the beautiful. With the transformation of arts in the twentieth century, the dictates of beauty and representation were rejected and the distinction between 'high' and 'low' art (popular, folk or vulgar art) became porous. The study of aesthetics expanded and began to appreciate the multi-facetted sensual impact generated by the things surrounding us. We do not only enjoy (or suffer) aesthetic experience in museums or out in nature, but throughout our daily lives. We are confronted not just by the beautiful but also by the ugly, the frightening, the lustful and the cool, and not just through our eyes but through all our senses – whether we eat a meal or read the newspaper, watch television or observe people in the street. John Dewey already pointed in this direction in the 1930s in his aesthetic study *Art as Experience* (Dewey 2005).

In this cornucopia of things that provoke aesthetic experience in our daily life, design is special because it deliberately and purposefully does so and because of its specific form-giving skills. We have seen that design objects are not just the result of a productive, deliberative, effect-oriented endeavour, but that design tries to affect people by determining the form or appearance of these products. That is also how we can distinguish design from everyday actions that prompt an aesthetic experience, such as hanging the laundry in the garden, and from actions that even involve aesthetic intentions, such as planting flowers or preparing a meal. The food is supposed to smell, look and taste delicious, and the flowers are intended to make the garden attractive. But just as in the discussion of amateur design above, there must be a special relation between the consideration of effects and the formative endeavour to call these actions design. Everyday aesthetic actions show a lower degree of both deliberation and formative skill than would be the case with a professional garden designer or food designer, and they are less accurate in realizing intended effects in the act of giving form.

Opening the sphere of 'art' for other productive aesthetic practices is nothing genuinely new. In fact, in ancient times the fine arts were not yet set on a "remote pedestal" (Dewey 2005: 4), but considered as one *techne* among others, such as crafts and other productive technical activities. According to Dewey, the antique principle of *mimesis* made sure that the visual and performing arts remained closely connected with daily life: "the doctrine [of *mimesis*] did not signify that art was a literal copying of objects, but that it reflected the emotions and ideas that are associated with the chief institutions of social life" (*ibid.* 6). Art should be seen as a development of our "everyday enjoyment of scenes and situations", as an extension of the "everyday making of things" (*ibid.* 11). In this case, aesthetic quality would not be something invented by art or only found in art, but something adopted from daily experience and brought to mastery by different forms of art.

What Dewey described in his time still seems to possess validity in our own: "The arts which today have most vitality for the average person are things he does not take to be arts: for instance, the movie, jazzed music, the comic strip" (*ibid.* 4). To be sure, the list of the most 'vital' media would look different in the twenty-first century, but we still feel a bias when thinking about computer games, mobile apps, web shops, TV ads or the like in relation to the art found in museums. If we were to judge all these things by the aesthetic experience they skilfully produce in viewers and users, then they would all have to be counted as present-day forms of art.

Thinking along these lines, we can easily see that many of our daily enjoyments and aesthetic experiences are generated by design. Emerging as a special field in the early twentieth century, 'design' finds early consideration in Dewey's work: "Objects of industrial arts have form—that adapted to their special uses. These objects take on esthetic form, whether they are rugs, urns, or baskets, when the material is so arranged and adapted that it serves immediately the enrichment of the immediate experience of the one whose attentive perception is directed to it" (Dewey 2005: 121). He names two senses of the word 'design': purpose and arrangement, which fit the above criteria of deliberation and effect-orientation on the one hand, and the formative criterion on the other. For Dewey, giving form to a substance means to handle it in a way so "that it can enter into the experiences of others and enable them to have more intense and more fully rounded out experiences of their own" (*ibid.* 113). What was found to be the eminent criterion that distinguishes design from similar productive actions, namely providing for the *form* of an object and thereby creating its specific *appearance*, is mirrored in Dewey's aesthetic account.

6 Design and Art

Everything in the world might appear to us in one way or another. Prompting an aesthetic experience has turned out to be a trait common to things surrounding us. As we have seen, design objects are special because their specific appearance or aesthetic experience is the result of a deliberate, skilful formative process. However, the very same special quality seems to hold for the classical subject of aesthetics, namely works of art. It therefore remains for us to ask about the difference between design and the fine arts. Let us consider a thesis on the nature of art that was proposed by Emmanuel Alloa (2016). Referring to Gaston Bachelard's epistemological notion of 'phénoménotechnique', Alloa defines art as a technique of production that generates an *aesthetic spill-over* that cannot be fully functionalized (Alloa 2016: 180). This 'aesthetic spill-over' of art consists in a kind of self-referring loop. Every piece of art points out its own technical, material or formal making, e.g. the strokes of the brush (*ibid.* 181). According to Alloa, art objects are special in that they do not only *have* an appearance: They appear *in order to* appear (*ibid.*). Alloa thinks that the self-referential surplus of pointing to their own formal and technical making is a unique criterion of works of art, while being a characteristic that is unimportant to other functional processes of production. However, his idea of 'aesthetic spill-over' comes very close to what has been said about the formative criterion of design, namely the creation of appearance and aesthetic experience. Designers do not just create aesthetic experiences, but are highly aware of the connection between the specifics of making – the techniques, materials and other formative aspects – and the appearance they prompt as a result. Alloa's idea of art as making something 'emerge' or 'step out' (*etwas hervortreten lassen*) also reminds one of Gernot Böhme's aesthetic account of 'ekstasis', meaning the way a thing 'steps outside itself' (*Aus-sich-Heraustreten*) and thus creates an atmosphere (Böhme 2013: 107 f.). It is exactly this ability to emerge, to 'step out' or 'radiate', that is all-important not only for the fine arts, but especially for design (*ibid.* 108). In Böhme's view, it is a common trait of every *techne* not simply to *produce* things but rather to create the conditions that make something *appear* (*ibid.* 109).

The ideas of both 'aesthetic spill-over' and 'ekstasis' ultimately seem to underpin the similarities of the fine arts and design – but how can these two fields be distinguished from each other? Design is usually differentiated from the fine arts by its intrinsic purposefulness – a criterion that has been incorporated in the above definition of design. According to Richard Buchanan, design objects are "made for human use" (Buchanan 1985: 4) and have to serve the complex needs of a society (Buchanan 1992). In contrast, artwork is often said to be created for its own sake. But is a work of art really just *l'art pour l'art*? Even though *objets d'art* are not always directed at fulfilling clearly defined human needs or functions, we could still say that they are 'made for human use' in a wider sense. Art tries to make people think or see things in another light. It tries to affect people's feelings, or to leave them perplexed. So is art really free from any functional restraint? And what about the 'aesthetic spill-over' that is said to be devoid of any function? Does the generation of aesthetic experience not in itself serve a human need? The main difference is possibly that design *must* be very clear and deliberate about the aesthetic experience it wants to create and about the functions that are to be served. Art, however, may indeed aim to achieve specific effects and may use formal means in specific ways for a specific aesthetic impact, but it is *not pressed* to do so. In other words, neither serving a function nor creating an aesthetic experience by providing the productive formal means is necessary for a work of art, nor is it defined by them.

The first 'freedom' of art – its liberation from purposefulness – does not come as a surprise. But its freedom from aesthetic intention unveils an unexpected, provocative idea, namely that art, at least in its contemporary form, is no longer a prototypical candidate for aesthetics. The overall ideal of beauty that originally linked art with aesthetics has long become obsolete. We can still have an impression of beauty when enjoying works of contemporary art, and some artists might still strive for beauty in their works, but the creation of beauty can no longer count as a necessary condition of art. Nor is it a sufficient criterion. As we have seen, the products of any other *techne* can be scrutinized from an aesthetic point of view, as can all other things that surround us. If there is no specific aesthetic intention involved in the creation of art, then the unique status of the fine arts within the study of aesthetics is lost. Design might even take their place, since design seems to be the *techne* that is most eagerly and specifically directed towards the creation of aesthetic experience.

Let me illustrate how art can override its own aesthetic claim. Take Kazimir Malevich's *Black Square*, for example. To be sure, looking at this painting in its total blackness – or black totality – creates an aesthetic experience that must have been highly impressive at the time of its creation (1915), and it can still unleash some of its aesthetic power today. I argue, however, that *Black Square* is not supposed to act upon the viewer principally by means of its material, technical or formal qualities, but rather by the *idea* that is incorporated in it: the idea of absolute reduction. Although this idea could not be mediated without any act of materialization, and although the painting needs to have the form of a black square in order to promote exactly *this* idea, I still think that the *specific* formal elaboration with its *specific* aesthetic qualities does not really matter for this kind of art. The same idea could be transmitted by a black square painted in another technique, such as using acrylics instead of oil, or coloured with charcoal or even screen-printed. And it could be made manifest on a smaller or bigger scale.

Bazon Brock supports this line of thought in a magazine interview that pointedly argues the case of *Black Square*: "You don't see anything. But you think something – that is what counts [...], there is nothing to contemplate. You don't have to stand in front of the original in order to grasp it" (Behrisch and Brock 2015: 26).[6] If the main idea of *Black Square* can be grasped even without actually *seeing* it, this implies that the essence of the painting cannot be located in its material or technical constitution nor in any aesthetic properties resulting from it. The specific material execution and technique of painting is not essential to this piece of art: "It would be nonsense to consider the *Black Square* as part of the history of painting. It is very badly executed, the colour is flaking everywhere" (*ibid.* 25). It thus seems to be equally important in judging the quality of this kind of fine art to abandon the formative criterion. Today, the fine arts cannot be judged solely by the technical or aesthetic skills displayed: "This by no means makes Malevich a worse artist, since the gesture, the idea are indeed unique and epochal. Many great masters, think of René Magritte, were lousy painters. Just like Magritte, Malevich is a dead loss as a painter, but a brilliant artist" (*ibid.* 25 f.). Brock's statement, while strikingly simplistic, hits the mark. An artist today needs neither to be gifted with special formative skills, nor to work with special technical diligence, nor to create works of outstanding aesthetic quality. Many artists do still demonstrate special technical skills and create aesthetically remarkable and powerful works – but they are not compelled to do so. There are other possibilities for creating art, even for creating *great* art.

The same does *not* hold for design. A designer without any formative or technical skill, a designer who does not manage (or at least tries to manage) to implement specific aesthetic qualities in her work, will simply fail. Designers must create well-defined appearances and aesthetic experiences by virtue of mastering the principles of form in all its details. If they do not, then they will not be called designers. If they do particularly well, they can even become *great* designers. The creation of appearance and aesthetic experience is a principal task of design. The study of aesthetics is therefore well advised to consider design among its paramount subjects.

7 Conclusion

The aim of this chapter was to contribute to the emerging field of design philosophy by illustrating how philosophy can be of importance for design theory and – vice versa – what philosophy can learn from design. In the first place, the development of a design definition based on the philosophical method of conceptual analysis was meant to sharpen and scrutinize the fuzzy concept of design and to provide an

[6] All citations from Behrisch and Brock (2015): Translation from the German by the present writer.

argumentative basis for further design research. Five criteria proved to be essential for design: productivity, prospect, effect-orientation, deliberation and form, while the formative criterion seemed to be the special ingredient that distinguishes design from related activities such as legislating or amateur production.

Secondly, a deeper thought into the formative criterion of design activity revealed an intimate link between designing and shaping the *appearance* of design objects, whereby an *aesthetic experience* is created in users or spectators. Although the concept of appearance is eminent for our understanding of what design is, this does not make design a purely superficial activity: The overall construction and functioning of design objects is respected by the criteria of deliberation and effect-orientation.

Thirdly, the essential relation between design activity and appearance or aesthetic experience revealed design as an interesting subject both for philosophical *ontology* and *aesthetics*. As I have argued, in the case of design objects, the traditional divide between what things 'really are' and what they only 'seem to be' can be bridged. What is more, the proposed concept of design gives reason to turn the focus of aesthetical study from fine arts towards design – actually design might become the paradigm case for future aesthetics.

If we want to do more than just scratch the surface of design, then we have to recognize that in the case of design, appearance proves to be part of the essence.

References

Alexander, C. (1971). *Notes on the synthesis of form*. Cambridge, MA: Harvard University Press.

Alloa, E. (2016). Produktiver Schein: Phänomenotechnik zwischen Ästhetik und Wissenschaft. *Zeitschrift für Ästhetik und Allgemeine Kunstwissenschaft, 60*(2), 169–182.

Bayazit, N. (2004). Investigating design: A review of forty years of design research. *Design Issues, 20*(1), 16–29.

Behrisch, S., & Brock, B. (2015). Schwarzer Humor. *Das Magazin, 49*, 24–26.

Böhme, G. (2013). *Atmosphäre: Essays zur neuen Ästhetik*. Berlin: Suhrkamp.

Buchanan, R. (1985). Declaration by design: Rhetoric, argument, and demonstration in design practice. *Design Issues, 2*(1), 4–23.

Buchanan, R. (1992). Wicked problems in design thinking. *Design Issues, 8*(2), 5–21.

Burckhardt, L. (2012). *Design ist unsichtbar: Entwurf, Gesellschaft & Pädagogik*. Berlin: Martin Schmitz.

Cross, N. (2007). *Designerly ways of knowing*. Basel: Birkhäuser.

Dewey, J. (2005). *Art as experience*. New York: Perigee Books.

Dorst, K., & Cross, N. (2001). Creativity in the design process: Co-evolution of problem-solution. *Design Studies, 22*(5), 425–437.

Fuller, R.B. (1955). Influences on my work. Reprinted in J. Krausse & C. Lichtenstein (2001) (Eds.), *Your private sky: Discourse R. Buckminster Fuller* (pp. 48–61). Baden: Lars Müller.

Kimbell, L. (2011). Rethinking design thinking: Part I. *Design and Culture, 3*(3), 285–306.

Locke, J. (2012). *An essay concerning human understanding*. Oxford: Clarendon Press.

Rowe, P. G. (1987). *Design thinking*. Cambridge, MA: The M.I.T. Press.

Russell, B. (1998). *The problems of philosophy*. Oxford: Oxford University Press.

Schneller, A. (2015). Design rhetoric: Studying the effects of designed objects. *Nature and Culture, 10*(3), 333–356.

Schön, D. A. (1983). *The reflective practitioner: How professionals think in action*. New York: Basic Books.
Simon, H. A. (1969). *The science of the artificial. Cambridge, MA: The*. M.I.T. Press.
Vial, S. (2015a). The effect of design: A phenomenological contribution to the quiddity of design presented in geometrical order. *Art, 3*(4), 4.1–4.6.
Vial, S. (2015b). Philosophy applied to design: A design research teaching method. *Design Studies, 37*, 59–66.

The Varieties of Good Design

Salu Ylirisku and Mattias Arvola

Abstract This chapter explores the philosopher and logician Georg Henrik von Wright's metaethical treatise of the varieties of goodness in the context of design. von Wright investigated the use of the notion of 'good' in language, and he identified six kinds of goodness: namely *utilitarian goodness*, *instrumental goodness*, *technical goodness*, *medical goodness*, *hedonic goodness*, and *the good of man*. We discuss these different kinds of goodness in relation to six design traditions that we identify, namely *conceptual design*, *usability design*, *engineering design*, *ergonomics design*, *experience design* and *sustainability design*. We argue that the design traditions are grounded in different appreciations of goodness, and that designers and design researchers can benefit from a more precise discernment of values that underpin design processes and design critique in different traditions. von Wright's treatise serves as a point of departure for the appraisal of the multifaceted and relational character of the idea of good design and of the values of design.

Keywords Design · Goodness · Varieties of goodness · Values · Design traditions · Virtues

1 Introduction

The title of this chapter is a paraphrase of Georg Henrik von Wright's (1963) treatise The Varieties of Goodness. von Wright was a philosophical logician and analytic philosopher, and a student, close colleague, and successor of Ludwig Wittgenstein at the University of Cambridge. Wittgenstein's influence is visible in the way von Wright approached the metaethical treatise of goodness by analysing the plurality of ways in which the concept of good is displayed and used in *everyday*

S. Ylirisku (✉)
Aalto University, Denmark, UK
e-mail: salu.ylirisku@aalto.fi

M. Arvola
Linköping University, Sweden, UK
e-mail: mattias.arvola@liu.se

© Springer International Publishing AG 2018

P. E. Vermaas, S. Vial (eds.), *Advancements in the Philosophy of Design*, Design Research Foundations, https://doi.org/10.1007/978-3-319-73302-9_4

language, i.e. how people talk about what is good. His position was a teleological one treating the intended *ends of action* and the actual *consequences of action* as grounds for establishing what goodness is. According to his teleological view of goodness, von Wright set the perceived beneficial and harmful consequences of action as the conceptual frame for the assessment of goodness. His position hence lies close to utilitarian ethics. Utilitarianism and its descendants in different forms of consequentialism build on assumptions about the moral rightness of actions depending on the (judged) value of the consequences (Sinnott-Armstrong 2015). The varieties of goodness presented by von Wright represents a pluralistic consequentialism, taking multiple meanings of 'good' into account.

Herbert Simon (1996, p. 55), who was a prominent design theorist, defined designing in terms of "*devising action to transform existing situations into preferred ones.*" This definition depicts designing as action to intentionally change the prevailing conditions, and it relates to the study of goodness through *ends of action* and *consequences of action*. This bears a clear resemblance to von Wright's thinking on the notion of 'good'. Designers' action involves the introduction of designs, and it is essentially through the introduced designs that designers influence actual situations. The consequences that designs provoke when they are adopted into use can be evaluated through von Wright's framework. von Wright also explicitly considered an example of design in his treatise: the design of a knife. Knives are the result of design action, and hence, when evaluating a knife one is actually evaluating an extension of how designers intentionally act upon the world.

von Wright's approach to goodness relates to designing also in another way. Despite von Wright worked within analytic philosophy and studied the meaning and logic of value and goodness conceptually, his approach was unapologetically anthropocentric – the good of human resided at the centre. Human-centred design (or 'user-centred design') shares this premise, as it anchors the value of designs into how well they provide value to the people that become influenced by them. Hence, von Wright's treatise appears to be based on very much the same concerns about value as the field of human-centred design, where our own contributions lie.

Our interest in studying von Wright's work in connection with user-centred design stems from our own efforts in developing a framework for analysing the quality of designs (Arvola 2010; Arvola and Holmlid 2015) and analysis of how designers construct design concepts, i.e. expressions of possible ends of action (Ylirisku 2013; Ylirisku et al. 2015). We have also studied design processes in terms of the 'situated knowing' in conceptual designing (Ylirisku and Falin 2008) where designers anticipate the value that their designs bring to the people that will eventually use their designs. von Wright's work on the varieties of goodness enables us to see how partial and biased most evaluation and support frameworks for design are. The different frameworks, however, may serve the purposes for which they were intended, and the objective of this chapter is not to evaluate how well suited they are, but rather, to promote a more holistic view on goodness based on von Wright's work.

The chapter is organised so that the first part outlines von Wright's varieties of goodness and the second part highlights different design traditions that appear to be biased towards one of the six kinds of goodness that von Wright (1963) originally outlined. We have chosen key works amongst the traditions to draw the connections to von Wright's work, but we acknowledge that the traditions are more diverse than our portrayal of them in this chapter. We, nevertheless, argue that the different design traditions that we identify are grounded in different appreciations of goodness, and that designers and design researchers can benefit from a more precise discernment of values that underpin design processes and design critique in these different traditions. Designers trapped in a single tradition, with one way of appreciating design, run the risk of overlooking the consequences of their work with regard to different values.

2 The Varieties of Goodness

von Wright based his teleological study of the notion of 'good' on a conceptual analysis of the ways in which 'good' is used in language. This resulted in taxonomy of six 'varieties' of goodness, namely *instrumental*, *technical*, *utilitarian*, *medical*, *hedonic* goodness and *the good of human*[1]. In his treatise, von Wright related only some of the kinds of goodness explicitly to designed objects, especially to the design of knives. We shall continue this exploration further and deeper, relating each kind of goodness to design by covering different design practices, after we have summarised the varieties of goodness as presented by von Wright (1963).

2.1 Utilitarian Goodness

Utilitarian goodness, according to von Wright, is a synonym for the concept of 'useful'. To call something useful, von Wright (1963, p. 43) argued, is to say that it is "causally relevant to (the attainment of some) *end of action.*" In other words, when an object is said to be good for a purpose is to ascertain that the *object can be used to serve* this purpose.

An example that von Wright uses to illustrate usefulness is a knife that can be used as a hammer, because of its thick and heavy handle. It can be used for driving in nails, even though it is possibly not originally intended for the purpose. When people re-appropriate a design object, they are using the object for a purpose that was not likely associated for the object by its designers. Objects can be considered useful if they are 'causally relevant' for attaining the end of action, i.e. driving a nail down, when the consequences of the action taken with the design object serve the attainment of the desired end of action. The opposite of a useful design is one that

[1] We use the term 'good of human' to refer to what von Wright originally named the 'good of man,' for the use of the noun 'man' is old-fashioned and it is perceived too much gender-biased today.

cannot be used for a purpose. If that purpose is one that the design is intended to serve, it can be said to be *useless*.

2.2 Instrumental Goodness

According to von Wright (1963, p. 20) "to attribute instrumental goodness to some thing is *primarily* to say of this thing that *it serves some purpose well.*" Here we need to emphasise the last word – *'well'* – which marks the demarcation of instrumental goodness from utilitarian goodness. von Wright (ibid. p. 45) wrote, "*instrumental goodness* of the thing can be said to measure its *degree of usefulness.*" This difference between instrumental and utilitarian goodness may at first appear as a minute detail, but it has significant bearings on how the evaluation needs to be conducted. Assessing if something is useful (or useless) is different from assessing if it serves a purpose well or badly.

Like utilitarian goodness, instrumental goodness is relative to a *purpose*. Von Wright used the knife example again to draw a difference between a knife being useful and it being instrumentally good. He analysed it in relation to the purpose of cutting meat. A knife that is instrumentally good for cutting meat will be said to serve its purpose *well*. This may mean that the knife allows the user to cut smoother slices, cut them with less effort, feel safer while doing so, etc. The list of criteria can be expanded according to the practice of the user, which von Wright called the *subjective setting* of the purpose. When the criteria are made explicit, it is possible to 'objectively' measure the degree of instrumental goodness of a design object. von Wright also recognised that design objects, such as knives, which are created into a known product category, become *essentially associated* with a particular purpose (or set of purposes) that they are anticipated to serve in a certain subjective setting. For example, in order to be a knife the object needs to *function as a knife*. This connects instrumental goodness to technical goodness.

2.3 Technical Goodness

According to von Wright (1963, p. 33) technical goodness relates to the ability to "perform a certain *activity*". When a design object is evaluated regarding its performance in an activity, the evaluation implies a greater set of requirements than merely serving a purpose. The challenge becomes that of outlining the activity in terms of what constitutes *excellence* in it. von Wright's examples of technical goodness were all human examples, such as a good teacher or a good general, which is likely due to his non-technical background (von Wright 2001).

Technical goodness is often associated to the evaluation of an object (or agent) of some *kind*. For example, when a person is evaluated *as a teacher* the evaluation is done in regard to the requirements in the *constrained field of teaching*. In practice, however, the exact requirements are often left implicit, and instead of explicating all the necessary requirements and organising these into a balanced set of criteria, evaluations of technical goodness are conducted in two main ways, *competitive tests* and *achievement tests* (von Wright 1963). In competitive tests the objects, or agents, under evaluation are put to perform the same activity at the same time and then judged which one, or who, beats the other(s) in it. Achievement tests, such as running a marathon, can be conducted independently to evaluate the excellence in performance measured against particular criteria, such as the finishing time. Achievement tests are also common in the evaluation of the development of skills, in order to judge whether a person has attained a certain level of performance in an activity.

2.4 *Medical Goodness*

von Wright (1963) used the notion of medical goodness to refer to the effects that some thing has for the welfare of one's body and mind. What is considered good for the organs of the body, or good for the faculties of the mind can be said to be medically good. Medical goodness, when studied in terms of bodily organs and mental faculties, is relative to the *normal* functioning of the organs and faculties. Medicine is largely involved with the *failures* of bodily organs and mental faculties to serve their innate essential function. When studied in this narrowly framed way, medical goodness resembles technical goodness, i.e. the performance of bodily organs and mental faculties in the activities essentially associated with them. Medical goodness, however, involves also considerations of experiences, such as pain, suffering, misery, enjoyment, and happiness. This connects medical goodness to the next kind of goodness, that is, hedonic goodness.

2.5 *Hedonic Goodness*

von Wright (1963) admitted that his treatment of what he addresses with the name 'hedonic goodness' is superficial. Rather than speaking of a kind of goodness, von Wright used the term 'conceptual field' to address phenomena related to pleasure, as he appraised that the field related to sensations and emotions is highly heterogeneous. He identified three main forms of pleasure: passive pleasure, active pleasure, and the pleasure of satisfaction or contentedness.

With *passive pleasure* von Wright addressed the pleasantness attributed to sensations and states of consciousness that the world gives rise to. As a logician, he was stringent in not confusing pleasurable sensations with the concept of pleasure itself, and defined pleasure as an attribute, a characteristic, or a property, of sensations. When talking about passive pleasure, von Wright used the triadic relationship between a physical *object* that is sensed, the *sensations* of a sensing subject, and the *subject*. Hedonic goodness, when it functions on the level of passive pleasure, is based on a

> causal or dispositional characteristic of the physical object that it evokes or produces, under specific circumstances, [..] sensations in a sensing subject. These sensations are the primary logical subject of the hedonic value-judgment. The physical thing 'partakes', so to speak, in the goodness of the sensations only by being their cause. (von Wright 1963, p. 66).

von Wright admitted that when considering passive pleasure, and especially its opposite *pain*, there is a possibility to find "*intrinsic* connection between a section of the world of facts and a section of the realm of values" (von Wright 1963, p. 70), which a logician can find puzzling. It was, however, long after von Wright's treatise of goodness, that neuroscientists were able to dissect the multi-layered processes of the origins and constituents of the experience of pain, see e.g. (Damasio 2000), and discuss it in connection with life-sustaining emotive capacities of the organism, hence supporting von Wrights hunch. von Wright, however, recognised that pain can be induced on a more foundational level than pleasure, and be outbalanced by the positive responses to the pain. These pain sensations are not 'painful,' despite their ingredients, but instead they become "pain-sensations, which we happen to like" (von Wright 1963, p. 71).

Active pleasure refers to the pleasure of doing things a person is keen on doing, enjoys doing, or likes to do. von Wright presented three examples of active pleasure: watching cricket, playing chess, and getting up early in the morning. Despite watching a game can be considered a passive pleasure, it is often the case that the person watching has substantial knowledge of the rules of the game, and possibly of the history of the team and the players. Hence, the pleasure stems not from sheer sensations caused by external stimuli, but from how one responds to these on the basis of one's familiarity with the game. So, a passively appearing pleasure may actually be rather active, and the boundaries between active and passive pleasure is here very elastic, as von Wright also acknowledged.

Game playing is a more complex topic for the analysis of hedonic goodness. It may resemble passive pleasure in cases where someone is playing just for amusement, as a pass time. However, once a game is played not for the sake of amusement, but for one is *keen on* playing it, or is done *for its own sake*, the pleasure turns active.

Pleasure may also derive from a more complex relationship between activities that one needs to do. Some activities are such that a person *wants* to do. Some are practical *necessities* of life, and as such, they simply must be done if one is to survive. Some activities are such that *are simply done* to get things done. For example, getting up early in the morning may result from a complex set of anticipations by a person.

> The man who rises early may want to do so in order to avoid having to rush his day's work, which is an unpleasant thing. Or he may be anxious to finish his set work early in the day as possible, so that he can relax and do in the afternoon what he 'really likes,' i.e. that which affords him (active or passive) pleasure. (von Wright 1963, p. 79)

von Wright, however, warned that if we begin to see all activities as being prompted by a desire to avoid something unpleasant or secure something pleasant, we will subscribe to a thesis called 'psychological hedonism'. People do many things because they like or want to do them, but not everything happens this way. Many things are done because they are customary, and many things simply happen to people involuntarily, such as getting fat or falling asleep. Moreover, a great number of things are done in order to get things done irrespective of any related pleasure. Getting things done, i.e. the attaining of goals, is often accompanied by pleasure, and this points to the third kind of pleasure: *satisfaction*.

The pleasure of satisfaction refers to being content, i.e. to the feeling that arises when we get what we desire, need, or want. Satisfaction presupposes the existence craving, desire, or curiosity a priori to the existence of pleasure of satisfaction. Satisfaction always has an object, which we strive to attain to satisfy our desire.

2.6 The Good of Human

The most all-encompassing of the kinds of goodness is the *good of human*. Von Wright wrote:

> A being who, so to speak, 'has' or 'enjoys' its good, is also said to *be well* and, sometimes, to *do well*. (von Wright 1963, p. 86)

Someone who is said to be well is typically meant to be *healthy*. Moreover, a person who is doing well, who flourishes, thrives, and prospers, is often said to be *happy*. According to von Wright (ibid.), health and wellbeing are privative statements, i.e. they connote the absence of illness and suffering, whereas well-doing and happiness are positive statements, referring to an overflow and surplus of something desirable.

von Wright reserved the term welfare to be the best candidate as the synonym for the good of human. He identified several differences between happiness and welfare. According to von Wright, happiness is allied to pleasure, and thus to hedonic goodness, unlike welfare, which is better connected to wants and needs, and hence to utilitarian goodness. Moreover, happiness appears as a temporal 'state,' which can exist at one particular moment in time and be gone in the next. Happiness can thus be understood as an end that can be pursued and attained. Welfare, on the contrary, does not have similar relation to events in time, but nevertheless, is subject to causal considerations. For example, for the question "if smoking is good for you," the consideration is essentially about the negative or positive *consequences* of the activity of smoking. von Wright (1963, p. 88) wrote:

> Considerations of welfare are essentially considerations of how the doing and happening of various things will causally affect a being.

Happiness differs from welfare in that it can be considered in terms of consequences and antecedents apart. What may be bad for a person's welfare, such as

Table 1 Summary of the varieties of goodness

Goodness	Relative to	Context	What is 'good'
Utilitarian	Desired end of action	Task	Useful (yes/no)
Instrumental	Desired end of action	Task	Serving well
Technical	Requirements, competition	Activity	Excelling
Medical	Health, normalcy	Activity	Beneficial, not harmful
Hedonic	Pleasure, pain	Experience	Pleasure
Good of Human	Welfare	Life	Happiness and wellbeing

smoking, may sometimes contribute to a temporal happiness. The relationship between what happens and how it influences people's happiness and welfare is often complex, even to the extent of being impossible to accurately foresee. von Wright (ibid. p. 102) stated:

> the causes and consequences of things which happen, are often insufficiently known and therefore largely a matter of belief and conjecture.

Related to this, von Wright discussed the notion of a *wanted thing*, especially a thing that is *wanted in itself*. For the attainment of the thing the person who wants it needs, most often, to *pay* a *price*. Price, according to von Wright, is the sum total of those things, which are unwanted in themselves. A key challenge in designing is *prospectively* answering the question whether it pays off to pursue a wanted thing. This, nevertheless, may be extremely difficult, if not even possible in all cases, as von Wright admitted (ibid. p. 101):

> every event (change) 'strictly speaking' has an infinite number of consequences throughout the whole of subsequent time, and that for this reason we can never know for certain which all the consequences of a given event are.

He (ibid. p. 110) added:

> It should be difficult, or even humanly impossible, to judge confidently of many things which are known to affect our lives importantly, whether they are good or bad for us. I think that becoming *overwhelmed* by this fact is one of the things which can incline a man towards taking a religious view of life.

According to von Wright, *virtues* can provide a way for people to strive within the complexity of overwhelming influences. Virtues, or 'features of character,' enable people to escape being succumbed to immediate temptations at the cost of greater future good, and thus contribute to a morally greater conduct.

Table 1 provides a summary of the varieties of goodness. Utilitarian and instrumental goodness are relative to ends of action, and they are measured in how a designed object serves the attainment of a desired *end of action*. Utilitarian goodness is concerned with the question, whether a product serves an end of action or not, whereas, instrumental goodness is about serving the end of action well. Technical and medical goodness are both considered within a particular *activity*. A technically good product excels in an activity, and a medically good product does not cause harm or is beneficial for health. Hedonic goodness is related to the experience of pleasure, and a good product in terms of hedonic goodness is accordingly a pleasurable product. Finally, the good of human is related to welfare. A good product makes people happy or contributes to their wellbeing.

3 Varieties of Goodness in Design

In this section we adopt von Wright's (1963) varieties of goodness and turn attention to different traditions in design. von Wright's framework enables differentiation between design traditions based on their basic view of goodness. We have identified six design traditions without any claim of being exhaustive, and the distinction is based both on design research literature as well as on our own work as designers and design scholars. We refer to these traditions with the names *conceptual design, usability design, engineering design, ergonomics design, experience design, and sustainability design*. Goodness is appraised within these traditions in different ways, sometimes through explicit evaluation frameworks detailing out what makes a design good. The traditions are overlapping, and hence, the distinction between them is mainly analytic serving to sensitise design practitioners and researchers to qualities and aspects of features that makes a design good (what von Wright call 'goodmaking features') as well as to the different frameworks through which these features are perceived (what von Wright might call the 'kinds of goodness').

3.1 Conceptual Design

The tradition, amongst design traditions, which is the most concerned with von Wright's utilitarian goodness, is *conceptual design*. It can be viewed as a phase in a general design process, such as what Cagan and Vogel (2002) call the 'Fuzzy Front-End' of design, but also as an approach to innovation (Ylirisku et al. 2015). It is a process of iterative framing and re-framing that leads into relevant simplifications of what should be created, i.e. into design concepts (Ylirisku 2013). Conceptual designing results in radically new designs that are aimed at facilitating the *attainment of existing goals in new ways*, or serve the *pursuit of completely novel purposes*. For example, a new wrist-top computer could enable one to read e-mails, and hence, it would be useful in regard to the goal of receiving an e-mail message. The wrist-top computer could also be used to re-channel communications between internet-of-things applications, which is a new kind of a purpose. According to von Wright (1963), utilitarian designs are such that enable the attainment of a goal, or end of action. Conceptual designing is focussed on the explication of novel goals as well as attaining existing goals in radically new ways. This is the main connection between utilitarian goodness and conceptual design.

Conceptual design draws on different design approaches, processes, and methods, and the multi-disciplinary process of conceptual designing is typically very flexible. Conceptual design also employs collaborative multi-stakeholder activities in order to accommodate multiple voices and considerations of what is valued within particular domain that the design exploration addresses (Gottlieb et al. 2013). The thinking models, tools, and materials for facilitating collaboration vary from concrete *design materials*, e.g. (Sanders 2002; Ylirisku and Vaajakallio 2007), and *design games* (Brandt 2006) to abstract *design models*, e.g. personas (Blomquist and Arvola 2002; Cooper 1999; Grudin and Pruitt 2002), contextual models (Beyer and Holtzblatt 1998), customer journeys and service blueprints (Polaine et al. 2013), and *concepts*, such as design space (Sanders & Westerlund, 2011). A special category of design resources is *design formats*, such as sketches, mock-ups, posters, and prototypes (Agger Eriksen 2012), used to fuel collaborative design interactions and align these towards a coherent result.

The notion of a 'design concept' is central to conceptual design. In connection to von Wright's framework, *a design concept can be seen as an expression of a goal.* According to Keinonen (2006) a design concept refers to the description of a product (or service) that is anticipatory, well-founded, focused, and understandable. Minimally a design concept consists of a Name, Purpose, and Design Principles (Ylirisku 2013). In addition, a design concept can have descriptions of the 'character' of the concept, the 'actors' influenced by it, the 'scene' of its use, its 'form', and the consequences of its use 'act-scene relation' (Arvola 2014; Arvola and Walfridsson 2015).

During conceptual designing one of the main challenges is to distinguish the context into which the design object will be created. The process is typically highly iterative and involves the ideation of great varieties of ideas to work with. The key ideas may be re-framed multiple times over the process (Dorst 2015; Ylirisku et al. 2015). Ylirisku et al. (2009) described the problem of framing the design object in relation to context as the dilemma of relevance. According to Johnson and Henderson (2011) conceptual designing is crucial for attaining relevance, simplicity, and coherence in design. Designers often emphasise simplicity, relevance and desirability of a design concept in order to render the desired end, expressed as a design concept, easily communicated and actionable. As Winograd put it:

> The design itself cannot embody all of these complexities if it is to be constructible and understandable. The design must embody a simplification, leaving room for the texture of the world to be filled in by the interpretation and practices of those who use it. (Winograd 2006, p. 72).

A design concept expresses a new goal, or a particular way of attaining a goal. In von Wright's terms, the designed object is useful when it serves the attaining of the goal, and is useless if it fails to do so.

3.2 Usability Design

Usability design, also known as usability engineering, is the design tradition that is most purely focused on what von Wright (1963) called *instrumental goodness*. Formed during 1980s mostly in response to the spread of computing to a wide range of users, usability design began to focus on computer software applications that were too difficult to learn and use. Usability design is grounded in the identification of *users' goals*, which serve as the basis for the evaluation of the goodness of a design object. Usability design focuses on the optimisation of effort and resources in the attainment of the identified goals.

In addition to identifying the users behavioral goals, a usability design process involves specifying the user interface and collecting critical information about users (Gould and Lewis 1985). Understanding human cognition and perception is essential when developing optimal solutions for interaction, and usability design is rooted in psychology (Norman 1988) as well as in engineering (Nielsen 1993). Amongst the many tools for usability design are personas (Cooper 1999) and scenarios (Carroll 2000). These are representations of users and their activities, and serve to elicit users' goals for the design, and can they be utilised in usability tests when specifying tasks for test users.

The crux of usability design is the usability test, which provides measurable data on how well a design is serving users' goals. Usability tests typically measure the amount and characteristics of errors, the duration of time to complete each task (i.e. attaining a goal with the product), and the encountered usability problems on each task. Typical to usability tests is the study of individual products with individual users. (Nielsen 1993) Tight budgets and agile development processes have also given rise to "guerrilla" methods (Nielsen 1994), which are based on the idea that some user research and testing is better than none. They are also called 'discount methods'. They do not provide the same rigor as proper research-based design, and include for example remote testing.

Usability design focuses on designing for the instrumental goodness of a product and it incorporates practices of measuring how well the design serves the purpose within what von Wright (1963) calls the 'subjective setting' of individual users. In usability tests, products are studied in relation to particular goals of a user, and the performance of test subjects is measured against criteria of the subjective setting (e.g. time on task, errors, reported satisfaction) in order to decide the degree of usability of the studied product. Usability design addresses also hedonic goodness, however, only through the concept of 'satisfaction', which relates to the good feeling of attaining a goal.

3.3 Engineering Design

Engineering design is the design tradition most closely associated with von Wright's *technical goodness*. It is formed according to a systematic method of formulating a problem and then solving this with an optimal solution through a rigorous process (Pahl and Beitz 1996). Technical goodness is typically referred to as *product quality*, and for example, the systematic process by Pahl and Beitz (ibid.) for engineering design seeks to ensure product quality. Engineering design is similar to usability design in that the quality of a product is measured against recognised criteria, but it differs from both usability design and conceptual design in the breath and detail in which it addresses needs and technical criteria.

A deeply held belief within engineering design is that "*all design begins with a clearly defined need*" (Armstrong 2008, p. 12, see also Kamrani and Nasr 2010), and thus, the intended goodness of a product is typically grounded in a *requirements list*. This list outlines criteria (factors) against which the performance of the designed product is compared. The criteria may include aspects, such as required functions, working principles, embodiment, safety, ergonomics, production, quality control, assembly, transport, operations, maintenance, recycling and cost. According to Hofman (2000) 'good' criteria are such that are correct, unambiguous, complete, consistent, prioritized, verifiable and traceable. The requirements are defined in such great detail that the evaluation of the goodness, or the 'quality' of the product is accurately measurable. This typically quantitative approach is targeted at solving design problems in an *optimal* way (Kang et al. 2012). The criteria used for product evaluation vary depending on the phase of the product design process from the analysis of technical feasibility and performance towards analyses of customer satisfaction (Hart et al. 2003). In connection to von Wright's work it is apparent that in the evaluations of technical goodness considers a much wider variety of aspects than simply how well the product serves the attainment of a particular goal as is done in the case of instrumental goodness.

von Wright (1963) also outlined two kinds of tests for the evaluation of technical goodness: competitive tests and achievement tests. Explicitly defined requirements enable design engineers to evaluate products in achievement tests, where the evaluation is conducted independently from competing products. The excellence in performance is simply measured against the specified criteria. Competitive tests, nevertheless, are common in the comparison of different products within the same product category. For example, reviews of "the best printer," "best camera," "best SUV," etc. are common. In these tests, the details of the performance of the products are scrutinized regarding specified requirements.

3.4 Ergonomics Design

Ergonomics design (or Human Factors) is a design tradition, which is concerned with von Wright's *medical goodness*. It is a precursor of and close relative to usability design and both focus on users' performance. Sometimes usability design is considered being a part of ergonomics under the heading of cognitive ergonomics (Long and Whitefield 1989). The key difference between usability design and ergonomics design is their main focus. Ergonomics design focuses on human *health* whereas usability design focuses on pragmatic *goals*. Ergonomics originates in the mid-1900s after the World War II, at the time when human-focus became promoted especially by anatomists, physiologists, psychologists, industrial medical officers, and industrial hygienists (Murrell 1965).

Ergonomics design is also closely related to engineering design in the way it represents understandings of population wide anthropometric data on people's sizes, limits, movements, and functioning of limbs as design requirements. Human needs are expressed in terms of factors that influence health, and it is typical for ergonomics design to understand a human in terms of a biomechanical entity, which consists of bones, joints, muscles, nervous system, and features processes of metabolism and heat regulation. The practice of ergonomics design maps the potential sources of problems in the user's task allocation and work load, environment or equipment, and suggests ways to minimize the harmful consequences for the user's health and performance (Helander and Khalid 2012). The classic examples include considerations, such as the workspace layout (Das and Grady 1983), humidity and temperature (Chiles 1958; Hohnsbein et al. 1983), noise (Broadbent 1957; Edworthy 1997), illumination (Moore 1958) and vibration (Dieckmann 1958) as well as the formal properties of products.

According to von Wright (1963), medical goodness is predominantly concerned with the normal functioning of bodily organs and mind. Through an engineering-like approach ergonomics is targeted at creating solutions that will not impede the normal functioning of a healthy human being.

3.5 Experience Design

The design tradition that is most closely related to hedonic goodness we call 'experience design.' von Wright did not use the notion of 'experience' in his treatise (von Wright 1963), but the connection between hedonic goodness and experience is apparent, for example, in Jordan's (1999, 2000) writings on the design of pleasurable products. Norman's (2003) framework for emotional design is perhaps the closest to von Wright's three-fold scheme of active pleasure, passive pleasure, and pleasure of satisfaction. Norman (ibid.) divides product emotions into three levels: the *visceral* level of the senses, the *behavioural* level of activity, and the *reflective* level of ideas and conscious thought.

Norman's (2003) visceral level of experiencing resembles von Wright's (1963) characterisation of *passive pleasure* as the pleasantness of sensations and states of consciousness that the world gives rise to. According to Norman (2003), experiencing on the visceral level is highly automated and rapid. It is the initial impact of perceiving something. This appears similar to how Desmet and Hekkert (2007) outlined the 'aesthetic experience.' According to them (ibid.), this type of experience refers to "a product's capacity to delight one or more of our sensory modalities" (ibid. p. 3). Jordan (1999, 2000) conceptualized this kind of experience, the sensations from sensory organs as well as feelings of sensual pleasure, with the term physio-pleasure. According to Desmet and Hekkert (2002), these kinds of sensations are relevant when evaluating a *product as an object*, rather than as *an event*.

Norman's (2003) behavioural level of product emotions addresses the active engagement in using a product. In von Wright's (1963) terms we could say that when people actively enjoy using a product they are having *active pleasure* with it. These kinds of experiences also relate to how well the functionality is designed to serve the person, and hence, to the usability and utilitarian goodness. This connection can be also made with Jordan's (2000) characterisation of psycho-pleasure, which he associates to the ease of use and lack of cognitive burden. Hassenzahl et al. (2010) have shown a connection between positive product affect and the fulfilment of needs, such as stimulation, relatedness, competence and popularity. They (ibid.) draw a distinction between hedonic (pleasure) and pragmatic (utility) qualities of a product. In connection with product design, active pleasure is associated to perceiving products as events (Desmet and Hekkert 2002).

Norman's (2003) reflective experience relates to von Wright's (1963) *pleasure of satisfaction*, or contentedness. According to Norman (2003), reflective product emotions are mediated by conscious thought about a product. von Wright exemplified the conscious thinking that mediated these kinds of emotions with an example of where one can finish working early. The contemplation on the situation can lead to the pleasure of satisfaction, as one anticipates an opportunity to find time to relax. According to Kappas (2006), reflective processes may supplement and correct initial intuitive responses. An interesting aspect in von Wright's example is how experiencing at one moment involves satisfaction because of anticipation something pleasurable to happen. This is different to how reflective product emotions are typically considered, i.e. that they take place as a consequence of some experience, e.g. (Desmet 2002; Desmet and Hekkert 2002; Norman 2003).

3.6 Sustainability Design

According to von Wright (1963) the good of human encompasses welfare and happiness of a being. It addresses the complexity of human experiencing as well as the socio-material processes of thriving in a material world at once making it the broadest and most complex of the six different kinds of goodness. While sustainability design is a label for a broad set of design activities, the common denominator for these activities is the consideration of *longer time perspective* as well as the systemic or *holistic appraisal* of the *ecological* and *evolutionary* aspects of the *complexity* into which designers design their objects. Sandin Bülow (2007, p. 73, our translation) captures the agenda of sustainability design:

> Sustainable development means to, in tune with economical, ecological, social, and cultural conditions, consciously and thoughtfully take care of and develop things and environments.

Sustainability design is about broadening the scope and engaging with the complexity of design at a strategic level as the shaping the world we all live in (Fry 2009). von Wright acknowledged that it is practically impossible to know all the consequences of any given event exactly, and even less, to foresee the impact on the quality of experiences or wellbeing. He recognised that instead of becoming overwhelmed by the massive and dynamic complexity people may become inclined towards taking a religious life where *virtues*, which may provide sustained values, are taken as the foundation for making judgments and enacting in the world.

Typical to various approaches to sustainability design is the central role of various *principles*. For example, McLennan (2004) outlined six principles for sustainability design he named the Biomimicry, Human Vitality, Ecosystem/Bio-Region, "Seven Generations," Conservation and Renewable Resources, and Holistic principles. These principles aim to sensitise designers to the different sustainability-related aspects of design challenges in a manner that enables the shaping of the design process, and the making of pragmatic decisions while designing. For example, the Biomimicry principle urges designers to 'respect the wisdom of natural systems,' and involves the consideration of aspects, such as sources of energy, fit of form and function, diversity, recycling, and cooperation. There are numerous further sets of alike principles, such as the 'Hannover principles' (McDonough et al. 2003), the 'Ecological Design Principles' (Van der Ryn and Cowan 2010), and 'Principles of Ecological Design' (Todd et al. 1994).

In addition to becoming increasingly numerous, the principles have also become less intuitive than traditional virtues, such as kindness, diligence, humility, and patience. Despite the many principles and other rhetorical devices the real developments to achieve sustainability have hardly begun (Fry 2009). The Earth is facing unforeseen pressure due to the impact of humans (Steffen et al. 2005). Yet, we are dependent on the artificial world that we have designed and created. This means that design is, and will remain, a "*decisive factor in our future having a future*" (Fry 2009, p. 3). Hence, it is ever more important to find ways to support sustainability design. Perhaps by studying and developing *design virtues,* which could integrate wisdom about design processes, thinking about designing, and theories of sustainable change, we could serve this agenda.

To summarize, the varieties of goodness can be related to the identified design traditions as described in Table 2. It highlights the resemblances and differences between design traditions in terms of what forms of goodness that are in predominant focus of the design effort.

Table 2 Summary of the varieties of good design

Tradition	Dominant Goodness	Focus
Conceptual Design	Utilitarian Goodness	Finding new goals and exploring new ways to attain these
Usability Design	Instrumental Goodness	Optimising the cost and effort in attaining goals
Engineering Design	Technical Goodness	Specifying solutions that perform excellently
Ergonomics Design	Medical Goodness	Developing products that are not harmful for people
Experience Design	Hedonic Goodness	Creating pleasurable and meaningful products
Sustainability Design	Good of Human	Focus on a long perspective with a broad and ecological view

4 Conclusion

In this chapter we have described a set of different kinds of goodness, the varieties of goodness, which the philosopher and logician Georg Henrik von Wright developed on the basis of his analysis how the notion of 'good' is used in language. His treatise of goodness is teleological, which means that the intended *ends of action* and the actual *consequences of action* are employed as the grounds for investigating what goodness is. von Wright's approach to goodness is also unapologetically anthropocentric, considering the good of human to be the centre of consideration. His position is essentially similar to human-centred design.

We argued that the different kinds of goodness have been emphasised to different degree by different design traditions. *Utilitarian goodness*, which von Wright defined to be understood in terms of something being useful for a purpose, is expressed in the practice of conceptual design, which seeks to create radically new conceptions of ways to attain existing goals, conceive of radically new goals together with new design objects that could serve these goals. *Instrumental goodness* could be identified in the usability design practice, which seeks to develop optimal solutions with which users can reach the identified goals efficiently. Aspiration for *technical goodness* is best visible in the practices, processes and models of engineering design, which seeks to develop solutions that excel in performance of a particular activity, while *medical goodness* is reflected as the underlying orientation of many of the methods, models, and processes found in ergonomics design. It may be surprising that the practices of ergonomics design and usability design have different underlying appreciations of goodness, as the usability practice is considered as a form of 'cognitive ergonomics' (Long and Whitefield 1989). In terms of the underlying values, ergonomics is closer related to the practice of engineering design than usability design. *Hedonic goodness* could be identified as one of the distinguishing features that demarcated experience design from usability design. And finally, sustainability design appears as a label for the most all-encompassing practice of delivering results with the consideration of *good of human*. It involves processes, models, and practices that cover a larger timescale and an increasingly pluralistic set of viewpoints.

For a practicing designer the making of conceptual distinctions across the varieties of goodness may not be as relevant as the development of sensitivity to how different appreciations of goodness influence the design process. Considerations of goodness typically emphasise evaluation over construction. However, according to an old saying, "you get what you measure." It is apparent that the underlying values of the different design practices steer the design process towards different kinds of outcomes. Making the underlying values an explicit topic of discourse can lead to productive reframing of the design object. This kind of reframing, or reconsideration of what makes products good, is apparent in how the practice of usability design became challenged by the practice of experience design at the turn of the millennium. This has subsequently been further challenged by the practice of service design. The new practices tend to have a wider perspective in regard to what makes the result good.

The traditions are overlapping in practice, and separating them based on one of the kinds of goodness, does not do justice to the diversity of perspectives considered in practice. It, nevertheless, seems reasonable to claim that different design traditions, which are typically promoted by practitioners coming from different fields of expertise, appraise good in different ways. Moreover, the historical development of these practices has contributed to an increasingly multifaceted and nuanced knowledge base in all the traditions. Today, it would be difficult to find a practicing designer or design researcher that would argue for an objective, universal and absolute set of quality criteria, values, or goals to design for. Despite there is always the risk that a wider perspective is taken at the cost of the attention to detail embodied in the earlier practices, whenever there is someone proposing one single value (i.e. efficiency, security, or safety) to design for, it should raise suspicion. For example, in the name of safety and security, a case is often made without going into details of what it really means for people. A designer should not accept an account of what good design is at face value, but instead, look beyond it for a variety of desirable and undesirable consequences.

References

Agger Eriksen, M. (2012). *Material Matters in Co-designing: Formatting & Staging with Participating Materials in Co-design Projects, Events & Situations (Doctoral dissertation in interaction design)*. Malmö University, Malmö: Sweden.

Armstrong, J. (2008). *Design matters: the organisation and principles of engineering design*. London: Springer.

Arvola, M. (2010). *Interaction design qualities: theory and practice*. In Proceedings of NordiCHI 2010 the 6th Nordic Conference on Human-Computer Interaction: Extending Boundaries (pp. 595–598). Presented at the NordiCHI 2010, ACM Press. doi:https://doi.org/10.1145/1868914.1868982.

Arvola, M. (2014). *Interaktionsdesign och UX: Om att skapa en god användarupplevelse [Interaction design and UX: On creating a good user experience]*. Lund: Studentlitteratur.
Arvola, M., & Holmlid, S. (2015). User experience qualities and the use-quality prism. In Proceedings of the Fuzzy front end of experience design workshop. Presented at the The Fuzzy front end of experience design workshop, Espoo, Finland: VTT.
Arvola, M., & Walfridsson, A. (2015). The Mediated Action Sheets: Structuring the Fuzzy Front-End of UX. In E. Kaasinen, H. Karvonen, Y. Lu, J. Varsaluoma, & H. Väätäjä (Eds.), Workshop Proceedings – The Fuzzy Front End of Experience Design. Presented at the The fuzzy front-end of experience design, Espoo, Finland: Teknologian tutkimuskeskus VTT. http://www.vtt.fi/inf/pdf/technology/2015/T209.pdf
Beyer, H., & Holtzblatt, K. (1998). *Contextual Design: A customer-centered approach to systems designs*. Morgan Kaufmann.
Blomquist, Å., & Arvola, M. (2002). Personas in action: Ethnography in an interaction design team. (pp 197–200). ACM Press. doi:https://doi.org/10.1145/572020.572044.
Brandt, E. (2006). Designing exploratory design games: a framework for participation in Participatory Design? In PDC '06: Proceedings of the ninth conference on Participatory design (pp. 57–66). New York/Italy: ACM Press.
Broadbent, D. E. (1957). Effects of noises of high and low frequency on behaviour. *Ergonomics, 1*(1), 21–29. https://doi.org/10.1080/00140135708964568.
Cagan, J., & Vogel, C. (2002). *Creating breakthrough products: Innovation from product planning to program approval*. FT Press.
Carroll, J. M. (2000). In J. M. Carroll (Ed.), *Making use : scenario-based design of human-computer interactions*. Cambridge (Mass.): MIT Press, cop.
Chiles, W. D. (1958). Effects of elevated temperatures on performance of a complex mental task. *Ergonomics, 2*(1), 89–96. https://doi.org/10.1080/00140135808930404.
Cooper, A. (1999). *Inmates are running the Asylum: Why high-tech products drive us crazy and how to restore the sanity*. A Division of Macmillan Computer Publishing: SAMS.
Damasio, A. R. (2000). *The feeling of what happens: body and emotion in the making of consciousness* (1. Harvest ed.) . San Diego: Harcourt.
Das, B., & Grady, R. M. (1983). Industrial workplace layout design An application of engineering anthropometry. *Ergonomics, 26*(5), 433–447. https://doi.org/10.1080/00140138308963360.
Desmet, P. (2002). *Designing Emotions*. Delft: Delft University of Technology, Delft, The Netherlands.
Desmet, P., & Hekkert, P. (2002). The basis of product emotions. In W. Green & P. Jordan (Eds.), *Pleasure with Products, beyond usability* (pp. 60–68). London: Taylor & Francis.
Desmet, P., & Hekkert, P. (2007). Framework of product experience. *International Journal of Design, 1*(1), 13–23.
Dieckmann, D. (1958). A Study of the Influence of Vibration on Man. *Ergonomics, 1*(4), 347–355. https://doi.org/10.1080/00140135808964610.
Dorst, K. (2015). *Frame innovation: create new thinking by design*. Cambridge, MA: The MIT Press.
Edworthy, J. (1997). Noise and its effects on people: an overview. *International Journal of Environmental Studies, 51*(4), 335–344. https://doi.org/10.1080/00207239708711091.
Fry, T. (2009). *Design Futuring: Sustainability, Ethics and New Practice*. Oxford: Berg.
Gottlieb, F., Larsen, H., & Sørensen, V. (2013). Multi stakeholder innovation. In Participatory Innovation Conference.
Gould, J. D., & Lewis, C. (1985). Designing for usability: key principles and what designers think. *Communications of the ACM, 28*(3), 300–311. https://doi.org/10.1145/3166.3170.
Grudin, J., & Pruitt, J. (2002). Personas, participatory design and product development: An infrastructure for engagement. In Proceedings of participatory design conference PDC2002.
Hart, S., Jan Hultink, E., Tzokas, N., & Commandeur, H. R. (2003). Industrial Companies' Evaluation Criteria in New Product Development Gates. *Journal of Product Innovation Management, 20*(1), 22–36. https://doi.org/10.1111/1540-5885.201003.

Hassenzahl, M., Diefenbach, S., & Göritz, A. (2010). Needs, affect, and interactive products – Facets of user experience. *Interacting with Computers, 22*(5), 353–362. https://doi.org/10.1016/j.intcom.2010.04.002.
Hofman, H. F. (2000). *Requirements Engineering A Situated Discovery Process*. Springer Fachmedien Wiesbaden. http://nbn-resolving.de/urn:nbn:de:1111-20130220668. Accessed 9 March 2016.
Hohnsbein, J., Pekarski, C., & Kampmann, B. (1983). Influence of high ambient temperature and humidity on visual sensitivity. *Ergonomics, 26*(9), 905–911. https://doi.org/10.1080/00140138308963418.
Johnson, J., & Henderson, A. (2011). Conceptual models: Core to good design. *Synthesis Lectures on Human-Centered Informatics, 4*(2), 1–110. https://doi.org/10.2200/S00391ED1V01Y201111HCI012.
Jordan, P. W. (1999). Pleasure with products: Human factors for body, mind and soul. In W. S. Green & P. W. Jordan (Eds.), *Human factors in product design: current practice and future trends* (pp. 206–217). UK: Taylor & Francis.
Jordan, P. W. (2000). *Designing pleasurable products: An introduction to the new human factors*. Taylor & Francis.
Kamrani, A. K., & Nasr, E. A. (2010). *Engineering design and rapid prototyping*. New York: Springer.
Kang, J.-S., Lee, T.-Y., & Lee, D.-Y. (2012). Robust optimization for engineering design. *Engineering Optimization, 44*(2), 175–194. https://doi.org/10.1080/0305215X.2011.573852.
Kappas, A. (2006). Appraisals are direct, immediate, intuitive, and unwitting…and some are reflective…. *Cognition & Emotion, 20*(7), 952–975. https://doi.org/10.1080/02699930600616080.
Keinonen, T. (2006). Introduction to Concept Design. In T. Keinonen & R. Takala (Eds.), *Product concept design: A review of the conceptual design of products in industry* (pp. 2–31). New York: Springer.
Long, J., & Whitefield, A. (Eds.). (1989). *Cognitive Ergonomics and Human-computer nteraction: An introduction. Cambridge [England]*. New York: Cambridge University Press.
McDonough, W., Braungart, M., & Kerry, T. H. (2003). The hannover principles: Design for sustainability. W. McDonough Architects.
McLennan, J. F. (2004). *The philosophy of sustainable design: the future of architecture*. Kansas City: Ecotone.
Moore, R. L. (1958). Headlight design. *Ergonomics, 1*(2), 163–181. https://doi.org/10.1080/00140135808964583.
Murrell, H. (1965). *Ergonomics*. Dordrecht: Springer Netherlands.
Nielsen, J. (1993). *Usability engineering*. Boston: Academic Press.
Nielsen, J. (1994). Guerrilla HCI: Using discount usability engineering to penetrate the intimidation barrier. In R. G. Bias & D. J. Mayhew (Eds.), *Cost-justifying usability* (pp. 245–272). Orlando: Academic Press, Inc. http://dl.acm.org/citation.cfm?id=186524.186639.
Norman, D. (1988). *The psychology of everyday things*. Basic Books.
Norman, D. (2003). *Emotional design: Why we love (or hate) everyday things*. Basic Books.
Pahl, G., & Beitz, W. (1996). *Engineering design: A systematic approach* (2nd ed..) https://doi.org/10.1007/978-1-4471-3581-4. Accessed 27 Nov 2015.
Polaine, A., Løvlie, L., & Reason, B. (2013). *Service design: From insight to implementation*. Brooklyn: Rosenfeld Media, LLC.
Sanders, E. B.-N. (2002). Scaffolds for experiencing in the new design space. In *Information design*. http://maketools.com/pdfs/ScaffoldsforExperiencing_Sanders_03.pdf.
Sandin Bülow, K. (2007). Design = Kvalitet? [Design = Quality?]. In L. Strannegård (Ed.), *Den omätbara kvaliteten* (pp. 56–76). Norstedts Akademiska Förlag.
Simon, H. (1996). *The sciences of the artificial*. Cambridge, MA/London: The MIT Press.
Sinnott-Armstrong, W. (2015). Consequentialism. In E. N. Zalta (Ed.), *The stanford encyclopedia of philosophy (Winter 2015 Edition.). Stanford: The metaphysics research lab, center for*

the study of language and information (CSLI), Stanford University. http://plato.stanford.edu/archives/win2015/entries/consequentialism/. Accessed 13 Dec 2015.
Steffen, W., Sanderson, A., Tyson, P. D., Jäger, J., Matson, P. A., Moore, B., III, et al. (2005). *Global change and the earth system: a planet under pressure*. Berlin: Springer.
Todd, N. J., Todd, J., & Todd, N. J. (1994). *From eco-cities to living machines: principles of ecological design*. Berkeley: North Atlantic Books.
Van der Ryn, S., & Cowan, S. (2010). *Ecological design*. Washington: Island Press.
von Wright, G. H. (1963). *The varieties of goodness*. Routledge.
von Wright, G. H. (2001). *Mitt liv som jag minns det*. Stockholm: Albert Bonniers förlag.
Winograd, T. (2006). Designing a new foundation for design. *Communications of the ACM, 49*(5), 71–73.
Ylirisku, S. (2013). *Frame it simple! Towards a theory of conceptual designing* (Doctoral dissertation). Aalto University, Helsinki.
Ylirisku, S., & Falin, P. (2008). Knowing in situated design action. In T. Keinonen (Ed.), *Design connections: Knowledge, value, and involvement in design* (pp. 8–17). Helsinki: University of Art and Design Helsinki.
Ylirisku, S., Halttunen, V., Nuojua, J., & Juustila, A. (2009). Framing design in the third paradigm. In *CHI '09: Proceedings of the 27th international conference on Human factors in computing systems* (pp. 1131–1140). Presented at the CHI'09.
Ylirisku, S., Jacucci, G., Sellen, A., & Harper, R. (2015). Design research as conceptual designing: The manhattan design concept. *Interacting with Computers*. https://doi.org/10.1093/iwc/iwv040.
Ylirisku, S., & Vaajakallio, K. (2007). Situated make tools for envisioning ICTs with ageing workers. In *Proceedings of Include 2007: designing with people conference*. London: Royal College of Art.

Collisions, Design and the Swerve

Jamie Brassett and John O'Reilly

> *I had come to the conclusion that there was nothing sacred about myself or about any human being, that we were all machines, doomed to collide and collide and collide. For want of anything better to do, we became fans of collisions.*
>
> *Kurt Vonnegut, Breakfast of Champions*

Abstract If only everything were formed of neat laminar flows, with easy to understand conditions and determinable outcomes: there would be no risk to manage out, messy inconsistencies and uncertainties to disrupt well-laid out plans. Things are not so clear-cut however. Indeed, as scientists, poets and philosophers of science have pointed out it is under conditions of nondeterminism and complexity that everything comes into being. There is an issue, then, when creative disciplines in particular find such complexity problematic enough to design systems and models in which uncertainty, disruption and aleatory collisions are if not destroyed, then dampened. We wonder: what might become of a creative practice that championed its encounter with The Swerve, Lucretius's *clinamen*? This article examines the role, value and applicability of the concept of collision to design. It takes a philosophical approach to examining this concept and mapping the possibilities of its use in design. We will argue using concepts mainly from Lucretius and Serres—but also Deleuze and others—that collision is an important aspect of all creativity, and that there would be nothing were it not for collisions, disruptive deviation and swerves

---

J. Brassett (✉)
Reader in Philosophy, Design and Innovation; Subject Leader and MA Course Leader, Innovation Management, Central Saint Martins, University of the Arts London, London, UK
e-mail: j.brassett@csm.arts.ac.uk

J. O'Reilly
Associate Lecturer, MA Innovation Management, Central Saint Martins, University of the Arts London, London, UK
e-mail: john.oreilly@csm.arts.ac.uk

© Springer International Publishing AG 2018
P. E. Vermaas, S. Vial (eds.), *Advancements in the Philosophy of Design*, Design Research Foundations, https://doi.org/10.1007/978-3-319-73302-9_5

from equilibrium. The aim will be to articulate the conditions for the possibility of designing that is a 'fan of collisions'.

Keywords Collision · Deleuze · Design · Lucretius · Serres · The Swerve

1 Opening Remarks

To say that this paper is about design and philosophy is a little limiting, if true. In fact, it is about many things as it takes in science and literature too. This 'about' should be read not simply as 'focus' but as 'turning around', as a vortex rushes about its axis. These topics—design and philosophy, among the others—produce moments, affects, from the forces that turn around them. They are also the trajectories of these affects, these acts, fleeing in many directions at different speeds and slownesses. An exercise in what Félix Guattari calls 'transversality' (1984b)—which Gary Genosko explains as 'productively presentational and transdisciplinary' (Genosko 2002: 68)—this paper will take these different topics and push them slightly into collision with each other. This is done not only to see what happens, but also to shine a light on the concept of the collision. The language used so far here is infused with philosophical referents: Spinoza's affect, Lucretius's swerve and collision, Guattari's transversality; and each one of these has its own connections that bring them into contact with other philosophers, and concepts, as yet not mentioned but lurking: Deleuze and Serres, Foucault and Flusser, order and chaos, complexity and creativity. Any work is already many works pitched from places that while singular are connected, layered and manifold; and these complex *millefueilles* are themselves not only in the middle of current thinking, but also of millennia of historical thought and unfathomable æons of thought to come. Such is the way of all creative acts. And such it is that we will argue that they should be using some of the philosophical modes of creating already mentioned. We seek to position all of these not as transcendent, ideal forms towards which all creativity should point, but as expressions of ways of being creative that are immanent to all. These few particles of creative production we will let fall through this chapter, and introduce a swerve at a small angle of declension that will lead to collisions. This swerve is The Swerve, Lucretius's *Clinamen*. It is the point of this chapter and the condition of its existence. And design, what about design? We will see design as a collision, as well as in need of colliding. We will offer The Swerve as a principle of designing that ensures its collision, and a number of particles of thought and practice that we will set on collision course with each other in order to see where and how design's own ontologies might be constructed.

2 Lucretius, Serres and the *Clinamen*

Philosopher Michel Serres highlights, ceaselessly, in his work on Lucretius's *De Rerum Natura*,[1] the creative power of the *clinamen*. In fact, the swerve that the *clinamen* introduces to nature is the condition of all its (nature's) creativity. If the laminar fall of atoms—that describes the background condition of nature for Lucretius and some of the atomists who preceded him—has nothing to disturb it, there would be nothing more than this fall, this equilibrium, this *stasis*: the same for eternity. 'Nothing can happen,' Serres writes, 'nothing is produced, in a homogenous field' (Serres 2000: 33); and again: 'If we had only the principle of identity, we would be mute, motionless, passive, and the world would have no existence: nothing new under the sun of sameness' (Serres 2000: 21). It is only with the swerve in the fall of atoms—a movement introduced as a minimum angle of deviation from the norm by the *clinamen*—do we get things, stuff clumping in new ways, sometimes only momentary coagulations of turbulent, self-organising systems that dissipate almost as soon as they appear. 'For something to exist rather than nothing, there must be a fluctuation in this uniform flow, there must be a deviation from equilibrium. And this is the *clinamen*' (Serres 2000: 148). Deviation from uniformity and equilibrium leads to collisions, and collisions produce things. In an essay called '*Incerto tempore incertique locis*. The logic of the clinamen and the Birth of Physics', literature scholar Hanjo Berressem (2005) locates discussions of the *clinamen* 'in recent theory with the entry of chance into an ordered universe and the subsequent breakup of order and chaos into a universe lodged between the probable and the exceptional' (Berressem 2005: 61). Berressem's essay, which purports to establish an 'intelligent materialism' following the *clinamen* through Serres (2000) and Deleuze (2004), is itself a selection of atomistic moments[2] falling through intellectual space and knocked into creative clumps. The positioning of a universe between 'probable and exceptional'—in a region and attitude of complexity[3]—is key in realigning physics (and all science) as a practical *natural* philosophy even at its most speculative and metaphysical, not only for Berressem following Lucretius and Serres, but for Deleuze too (2004: 303).

[1] We have consulted two English translations of Lucretius's text, one as prose by R. E. Latham (a translation revised in 1994 of his original work of 1951) and another more recent translation into poetry by A. E. Stallings (2007).

[2] Some are named as 'Lacan', 'Derrida' and 'Foucault', along with the main protagonists; others include more general literary and scientific thoughts along with the philosophical.

[3] See the work of Stuart Kauffman (1993 and 2008) for a biologist's take on complexity of life, and Brassett (2013 and 2015) for a way of relating Kauffman and Deleuze to innovation and design. Serres's work is, of course, steeped in this complex space, with turbulence an 'intermittent state' between order and disorder (Serres 1995: 109). As is that of Nobel Prize for Chemistry winner Ilya Prigogine and philosopher of science Isabelle Stengers (see: Prigogine 1980; Prigogine and Stengers 1985; Stengers 1997a, b). Important philosophical work in this area includes that of philosopher of science David Webb (for example: 2006, 2010). We shall return to this issue below.

Even if there were nothing else (and there is, as we will discuss in a moment), Lucretius's *clinamen* gives those of us working in design—and other practices that can be brought under the auspices of creativity—a way of acting to maximise creative affect. As such, *to swerve* might act as an imperative that has both ethical and ontological import. This is because to consider whether, where or how we might be swerved from well-worn tracks of behaviour to have collisions that increase the possibilities of new creative clusters forming, necessitates the alignment of our systems (personal, organisational, and so on) as open and with increased opportunities for affecting and being affected. For Serres whether a system is open or closed is key for determining its creative or entropic nature. This is most explicitly discussed in his essay 'The Origin of Language. Biology, Information Theory, and Thermodynamics' (1982), but is conveyed with some marvellous poetic flourishes in *The Birth of Physics* (2000). For example, he writes: 'The laboratory, and every closed system, protects from turbulence' (Serres 2000: 68; translation modified)—and it is with turbulence, occasioned by the *clinamen*, that we create. 'The old closed systems,' he continues a few lines later, 'are abstractions or ideals. The time for openness has arrived' (Serres 2000: 68). 'Has arrived' with Lucretius, writing in Rome in the last century BCE, but also with Epicurus teaching much earlier (Greece, 341–270 BCE), and 'us' at the end of the twentieth, beginning of the twenty-first centuries. These times for the swerve to act, for openness to arrive and complexity to drift across as many disciplines, thoughts and practices as possible, have always been, and will always be. We are Greek, Roman, and whatever will exist when the stars go out, and all points in between, swerving and colliding. Serres in 'The Origin of Language' writes:

> And experience shows that there is no flux without eddy, no laminar flow which does not become turbulent. Now, and here is the crux of the matter, all times converge in this temporary knot: the drift of entropy or the irreversible thermal flow, wear and aging, the exhaustion of initial redundancy, time which turns back on feedback rings or the quasi-stability of eddies, the conservative invariance of genetic nuclei, the permanence of a form, the erratic blinking of aleatory mutations, the implacable filtering out of all non-viable elements, the local flow upstream toward negentropic islands—refuse, recycling, memory, increase in complexities. (Serres 1982: 75)

The emphasising of the negentropic upheaval of creative production from the eternal and universal drag of entropy is one of the most important aspects of Serres's work. The same is given a more artistic spin by philosopher Vilém Flusser, in a short essay on habit, which he gives as 'the aesthetic equivalent of "entropy"' (Flusser 2002b: 53).

2.1 *Æsthetics, Anæsthetics and Critical Decision-Making*

Here Flusser writes of habit as the tendency of the new to become probable, and that 'everything that is new is terrible, not because of what it is, but because it is new' (2002b: 51). An echo of Lucretius's lines:

> The roving stars, the moon, the sun's light, brilliant and sublime—.
> Imagine if these were shown to men now for the first time,

> Suddenly and with no warning. What could be declared
> More wondrous than these miracles no one had before dared
> Believe could even exist? Nothing. Nothing could be quite
> As remarkable as this, so wondrous would be the sight
> Now, however, people hardly bother to lift their eyes
> To the glittering heavens, they are so accustomed to the skies.
> That's why you should let go of any terror of the new. (Lucretius 2007: 67; 2:1031–1040).[4]

Flusser's aim in his short essay is to provide a way of considering aesthetic critique from the mixing of different types of philosophical, scientific and literary evaluation:

> everything aesthetic begins as a terrifying enormous noise ('big bang'), and as it grows habitual ('redundant') it ends in a quiet whisper (whimper). Thus one succeeds not only in making objectivity coincide with subjectivity, the sciences of nature with the sciences of culture, but even Eliot with Rilke. (Flusser 2002b: 53)

Habit anaesthetises and aesthetics terrifies with its newness. And so even in the inexorable flow of all things to habitual, probable, numbness there are opportunities for 'islands' of creativity to emerge, even if they are terrible. Lucretius's entreaty to 'let go of any terror of the new' (*desine qua propter novitate exterritus ipsa*) we will read not as requiring terror to be blunted, but that the terror of the new should not be feared. To be open to the new, even if it causes such strong sensations as to be feared, is to remove the constraints of habit, of closed-minded dogma, and to deliver a system up to the possibilities that there are ways out of entropic fall. But Lucretius also wants us not to remain numb to the wonder of the everyday, and that what might seem part of the normal flow of things has the opportunity to be affective.

For Lucretius as for Serres that there is something rather than nothing is not only worthy of record, but needs critical positioning. Critical, that is, in ways that both Serres (2014: x–xiii) and Flusser (2002a) explain comes from the Greek *krino/krinein*: to judge, decide. A critical action 'splits oneness, breaks it down, breaks in half: it casts doubt on oneness' (Flusser 2002a: 42). A doubt that Lucretius has no need of, so atomistic is his world already. The point of critical judgement is a moment of decision (de-cision, to cut in two), where paths fork (Serres 1995: 57; Serres 2014: xi) and the future superposition of possibilities urge us to critical creativity.[5] 'Normally,' Flusser says of crisis, 'this concept describes the point on a

[4] The prose translation is as follows: 'If all the sights were now displayed to mortal view for the first time by a swift unforeseen revelation, what miracle could be recounted greater than this? What would men before the revelation have been less prone to conceive as possible? Nothing, surely. So marvellous would be that sight—a sight which no one now, you will admit, thinks worthy of an upward glance into the luminous regions of the sky. So has satiety blunted the appetite of our eyes. Desist, therefore, from thrusting out reasoning from your mind because of its disconcerting novelty' (Lucretius 1994: 63; 2:1031–1040). We give book number and lines of the quotation in keeping with other work on Lucretius, in addition to the usual citation protocol.

[5] A dense nest of concepts is implied here. In *Genesis* (1995: 57) Serres relates the forking, bifurcating and dovetailing of paths and swallowtails as also an *instauration*. This word is heavy with resonance as it is used through æsthetician Étienne Souriau's work, where it relates 'inception' and 'establishment' (Souriau 2009: 108). Serres gives an etymology for instauration from the Greek

curve where it changes identity' (Flusser 2002a: 46). Criticality as crisis, as judgement and breaking open, decision and multiplicity brings us back to Lucretius, via Serres, and his proto-complexity. For complexity biologist and philosopher Stuart Kauffman also makes use of the term 'critical' to denote the complex region where chaos (supracritical) and order (subcritical) become each other, in which life is created and evolves (Kauffman 1993, 2008; see also: Brassett 2015). Such criticality as a spur to, and condition of, creativity is important for us, and we would like some more time on the complex aspects announced here.

'The origin of things and the beginning of order,' Serres tells us, 'consist simply in the narrow space between *turba* and *turbo*' (Serres 2000: 28). *Turba* is the chaos of the tumultuous crowd, and *turbo*, the spinning of the vortex, local order self-organising from never-ending chaos. We have the complexity of the vortex, and the disorder of the storm, and the silent background equilibrium of the multiplicity of atoms falling, with the nature of things created always in-between. Creativity for Serres is to be found in the in-between, the middle of all these, the *turba*, *turbo* and the fall, in the narrow spaces broken open by the swerve. But there is more: to be open to the swerve and the collisions it produces is as important for a rethinking of creative strategy as it is for creative ontology and ethics.

2.2 Beyond Strategy

As one of his consequences of reading Lucretius's *De Rerum Natura*, Serres provides us with not only collisions and creativity, repetition and the return, but also Mars and war, Venus and love. 'From Heraclytus to Hiroshima,' he writes, 'it [Western Science] has never known anything but martial nature' (Serres 2000: 108). Collisions are always lovely: markers of the processes of loving and its creative outcomes. Strategy is always martial—*strategos* in Greek was the head of the army—even if its primal warlike nature has been forgotten with its use in business contexts since the 1960s and others ever since. Lucretius opens onto Venus, placing at the forefront of the nature of things a poetic, philosophic and natural scientific call to creation, and closes abruptly with death, plague and pestilence. If his works, turbulent and swerving themselves, are ignored then the incessant fall of all things to death is all that is left. The promise of entropy must be creative declination. There is either swerve or death. Strategies demanding a clear road to goal—even the best, complex, topographic strategies—are martial acts nevertheless. So, keep the complex topography but remove the war and make the original mover Venus. What

fork as in bifurcation (Serres 1995: 57), but it is unclear where he gets this. The etymology of the French word *instauration* is the Latin *instauratio*: renewal especially after destruction, also restoration, that Serres acknowledges through referencing the Rennaisance Latin *instauratrix*, which has these meanings (Hoven 1993). The Proto-Indo-European root *sta* appears in many European words of control and stability; for example, English 'stand' and its Polish equivalent '*stać*', as well as the Greek '*stasis*'. *Stasis*, however, is interesting as it also relates to crisis in terms of 'civil strife' (Agamben 2015).

then? In one way, we are offering here an approach to designing that not only regards colliding as its model but is also, itself, a collision: a collision between philosophy and design. But this is not just about philosophy and designing. As it is also about creativity and *everything* the stakes are about as high as they can get. Serres and Lucretius recognise this: love and war, life and death, nature or otherwise. Collisions spun into fabulous turbulence by the swerve describe creative practice and the ethics of creative practice, as well as the creative possibilities of ethics. To be a 'fan of collisions' (Vonnegut 1992: 220) is to do Venus's equivalent of Mars's strategy, and its ripples will be felt politically and scientifically to the ends of the universe. To overcome Mars is to recognise the material atomisation of all things, their swerve off course, and their coming together in creation; that is, to recognise the collective constitution of all things as federations of nature. Serres explains that the 'natural constitution, in the last instance, is none other than the atomic constitution. Men, no less than things, are composed of atoms. Their soul and their conscience. Their collective is thus a composition of compositions' (Serres 2000: 121). Deleuze brings us to a similar position. 'With Epicurus and Lucretius,' he writes

> the real noble acts of philosophical pluralism begin. We shall find no contradiction between the hymn to Venus-Nature and to the pluralism which was essential to this philosophy of Nature. Nature, to be precise, is power. In the name of this power things exist *one by one*, without any possibility of their being gathered together *all at once*. (Deleuze 2004: 304. Original emphases)

Serres and Deleuze, philosophers of multiplicity both, find in these ancient atomists kindred souls: breaking open and asunder things that called themselves One or Whole. Nature, Deleuze writes, neither collects nor attributes nor totalises, but distributes, conjoins and disjoins. Nature *is* nothing but power, a relation of forces that themselves function according to the speeds and slownesses of their parts (Deleuze 1988a, 2004). The Whole, Deleuze will tell us in *Anti-Oedipus* written with Guattari and published a few years after *The Logic of Sense*, is itself a product (Deleuze and Guattari 1984: 42–44), a product of the machinic creation of multiplicities. The Whole is neither a lost original totality to be regained nor an ideal future one to be realised, but a product of every multiple, and which enters into the multiple from which a new whole might emerge. While Deleuze and Guattari here couch this discussion in terms of 'desiring-production'—'desiring-production is pure multiplicity,' they say, 'an affirmation that is irreducible to any sort of unity' (Deleuze and Guattari 1984: 42)—this is close to Deleuze's position on Lucretius: 'the multiple as multiple is the object of affirmation, just as the diverse as diverse is the object of joy' (Deleuze 2004: 315). Joyful affirmation of a multiplicity that occasionally comes together as things, which dissolves and recombines as principles and expressions of the *turbo* from the *turba*. 'We ourselves, born from the vortices, like naked Aphrodite in the foaming seas, are troublemakers full of troubles' (Serres 2000: 90). It is with us as 'troublemakers' that we will take this chapter to its next set of encounters: those that will bring us closer to particular creative practices, including—of course—design.

3 Colliding and Designing

Vilém Flusser, in his short essay 'About the Word *Design*' (1999), delves into design's etymology and unearths some gems; notably, that to design is also to trick, and designers are tricksters. 'The word [design],' Flusser writes, 'occurs in contexts associated with cunning and deceit. A designer is a cunning plotter laying his traps' (1999: 17). Troublemakers born from the turbulent seas, full of troubles, are also critically creative, we have seen. To this we add the designer as trickster. With the trick and the trap positioned as possible outcomes to troubling, even terrifying, complexity. And as we found that we should not try to dampen the terror of the new, the trickster may not need taming.

Philosopher Anne Sauvagnargues notes in *Deleuze et l'art* (2006: 146) that the 'creative posture reveals the blockage at the same time as its line of flight'. A trickster's 'creative posture' is one that sees the critical state of situations (its 'blockages'), as well as opportunities for novel outcomes ('lines of flight'). We noticed further above how Kauffman finds critical creativity happening at the moments and in the milieus where order and chaos become one another (Kauffman 1993, 2008). For us, the designer as trickster is both a collider and collided, always operating with faces turned to catastrophe and stagnation; critical in every sense of the term. Existing as troubled and troubling, terrified and terrifying, facilitating not only new production, but also the perpetuation of their own conditions for creativity. US architecture agency Studio Gang highlight just such a posture.

3.1 Polis Station: Deviating and Distributing

The 'Polis Station' (see Fig. 1) design concept was Studio Gang's entry to the 2015 Chicago Architecture Biennial (Chicago Architecture Biennial 2015, Studio Gang 2015). It is a project that emerged from a collision of two different, troubling, swerved and condensed processes: 'polis' as a coherent coagulation of parts, and 'station' as a moment for the production of such a coherence. Finding itself in a moment of crisis, Studio Gang has generated a particular, critical model for delivering different political, social and cultural entities. The conventional police station is designed to funnel citizens from the chaotic to the laminar through a process of arrest, judgement, criminalisation, incarceration and, maybe, rehabilitation. Such a police station is a closed building struggling in an open system, a laboratory for the production of tame results shielding itself from chaos as Serres says, organising a flow of people from a chaotic outside space towards the ultimate closed system: prison. Yet sometimes this has disastrous effects. The US police station—Studio Gang's reference point—while striving to produce material, social and psychological equilibrium from the chaos of criminality nevertheless seems to contribute its own

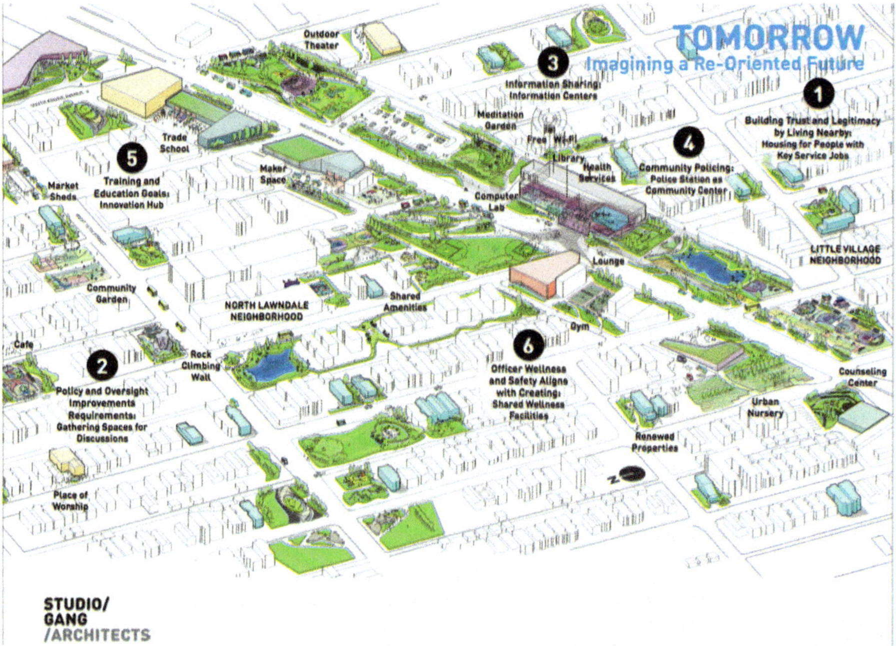

Fig. 1 Studio Gang (2015) **Polis Station.** The diagram of Studio Gang's entry to the 2015 Chicago Architecture Biennial, rethinking the place, people and Police Station. Image copyright Studio Gang

amount of strife.[6] It may not be surprising that in this climate Studio Gang approach the police station as an object for creative reconsideration, as the current brand of *stasis* produced by police stations appears closer to the word's meaning as 'civil strife' than equilibrium (Agamben 2015). Studio Gang's 'Polis Station' works in a different way.

Their company website describes the project thus: 'Polis Station proposes that police stations be reoriented toward their communities and become sites of social connection where officers and neighborhood residents can find many opportunities to interact' (Studio Gang 2015); thus enabling collision not only in the sense of permitting contingent encounters, but also in the sense that individuals are able to design relationships, to design what their communities might become. The research behind the vision of the project included a typology of police buildings ranging from: the Watch Box of the 1700s, with its technologies of the stove and extra lamp oil; to the Call Box, a box accommodating the new technology of the telephone;

---

[6] Recent US crime statistics published by the FBI show 1,165,383 violent crimes reported in 2014 from murder to rape and aggravated assault (FBI 2015). Further, 1,086 people killed by police in the US in 2015 (up to 16 December)—of which 27% had mental health issues, and according to *The Guardian* newspaper's project 'The Counted', 'Black Americans killed by police are twice as likely to be unarmed as white people' (Lartey et al. 2015).

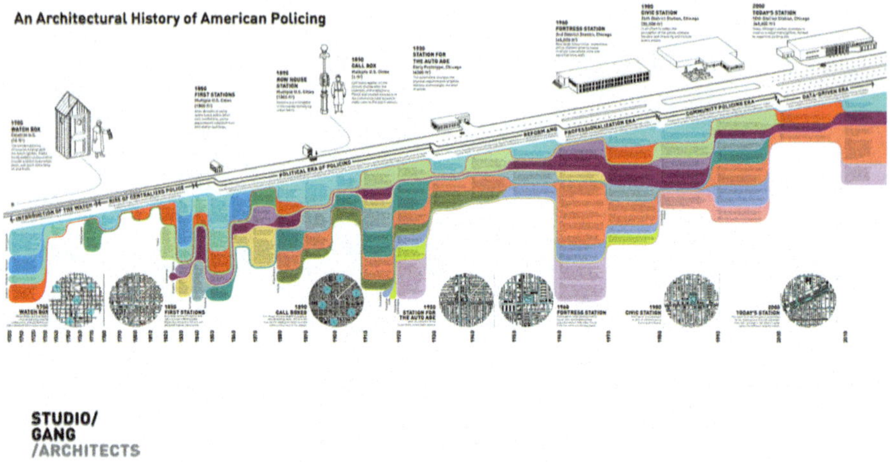

Fig. 2 Studio Gang's visual history of policing as the dynamic relationship between space, buildings and technologies. Image copyright Studio Gang

the fortress station of the 1960s and 70s, as police stations expanded to accommodate both growing bureaucracy and gang-related disorder[7] (see Fig. 2). 'Polis Station' is constructed as a series of encounters each posing the question of what it is to police, with the outcomes a series of decentred interfaces: the police station as community centre; police housing co-located with other public sector workers, such as teachers and health workers; workshops for shared maker spaces, with a trade school in an old industrial space to help those released from prison learn a trade; a police academy where local people can join up; a meditation garden; a counselling centre; and an urban nursery to help 'returning citizens' learn landscaping. The police station as static entity transforms into a dynamic and distributed 'Polis Station' through a series of collisions of people, practices, services, urban infrastructure and visibilities.

Dissolving the traditional experience of the police station as a site of disciplining and control, Studio Gang disorganise the fortress into a multiplicity of points of contact between citizens and police officers. In this way Studio Gang hopes to liquefy the blockages between police and community—and the troubles such divisions appear to intensify—and to reconfigure police stations as polis stations: emergent elements of polis-citizen-officer modalities in relation to each other under different conditions, and distributed across the neighbourhood in a more molecular fashion. In fact, Studio Gang's founder, Jeanne Gang, reveals the Epicureanism of this project on the business/design website Fast Company, describing 'the two prongs' of the 'Polis Station' project as: '"police station becomes community center" and police officers are "atomized" and become part of the community' (Budds 2015). We won-

[7] This period is expressed viscerally in John Carpenter's (1976) film *Assault on Precinct 13*, where a small group of stranded police officers, citizens and criminals in-transit are besieged in a local Los Angeles police station by a heavily armed mass of co-operating gang-members.

der whether the dissolution of the antagonistic, fortress-like structure into something more immanent will allow for better relationships between forces of order and those to be ordered. Nevertheless, we do see The Swerve at work.

While 'Polis Station' is admirably immanent to both the material and the structure of the milieu in which it operates, its reliance upon the concept of the 'station'—as a machine for producing equilibrium—might need readdressing. Through the heterogeneous elements of its codes, laws, uniforms, practices, regulations, rituals and stations,[8] the design of the US criminal justice system produces captive bodies and constrained bodies, as well as subjectivities. We have already noticed how police stations contribute to the design of a strategically effective, though dysfunctional, social and political equilibrium—a disorderly ordered disorder. But it might be worth considering further the ways in which the concept of deviation, *clinamen*, might critically unseal this unproductive lockdown of order and disorder.

Following Foucault, we might call the contemporary police station system a 'heterotopia of deviation' (Foucault 2008: 18): a system characterised by its contested spaces, its sites of difference, and exemplified for him by care homes, psychiatric hospitals and, of course, prisons. For us, and especially in relation to the example of 'Polis Station' that we have given already, the deviation announced here is already swerving from a heterotopia of crisis. Crisis heterotopias in, what Foucault calls, primitive societies were sacred or forbidden places, places for people in transition (we would also argue 'transformation') such as adolescents or menstruating women. For young men up to the twentieth century, the boarding school or military service was the critical space where virility was allowed to manifest; for young women up to the middle of the twentieth century, the 'honeymoon trip' where a young woman would be 'deflowered' in the honeymoon hotel—a place without a specific set of geographical markers. As with many concepts from Foucault, these (heterotopias of crisis and deviation) should not be seen in their purity alone, marking separate stages of progressive development. As should be the case with a concept such as 'heterotopia', crisis and deviation can exist as modes of any ontological space. This we have already noticed above, specifically with reference to Flusser and Lucretius. For Foucault here, deviation itself deviates from the moments of crisis, those events (spaces and times) of judgement production and power as domination, to enter a new trajectory generating the spaces of power as action (Dovey 2013); all the while carrying elements, atoms, of the other modes along for the ride. But as we have also seen, a crisis can be a moment of critical creativity at which all the possibilities of design can be superimposed. The heterotopic model can, immanently, exhibit the heteretopia that it allows for, with deviation and crisis providing key impetus for each other as well as opportu-

[8] We would like to draw attention to the relevance of Foucault's concept of the *dispositif* here as a way of thinking this heterogeneity of forms operating in accord to regulate the relationship of forces in the production of power. While a fuller encounter with this concept warrants much more than a footnote, we would like to highlight it as a point of collision with our chapter, and note that a future line of examination of *dispositif* and *clinamen* in terms of design and creativity would be fruitful. See Agamben (2009) for an exploration of *dispositif* that encounters more of the political and economic issues that emerge in this chapter.

nities where they can recoil and flee. 'Polis Station' already collides police and *polis*, deviates the relationship between community and law through buildings and spaces designed as a series of possibly troubling and troublesome encounters, ricochets between spaces enabling education, rehabilitation, mental health and policing, producing a heterotopia that has not only deviated but critically so.

3.2 Designing Heterotopias

In his examination of the concept of heterotopia, theorist Robert Topinka (2010) argues that scholars have focused on heterotopia solely as a site of resistance and not enough on the idea that such sites are where 'epistemes collide and overlap, creating an intensification of knowledge' (Topinka 2010: 55). Following this we would also say that through the relations enabled by this unusual space, knowledge becomes contested and multiple, ontologies become critical and creative, and from these new opportunities emerge. We would say, then, that Studio Gang offers such a critical moment of swerve and collision, and in doing so offer not only a foucauldian homeorrhesis[9] of epistemology and ontology and heterotopia, but also (to rework a quotation from Deleuze given above) a noble act of socio-political mulitplicity. Where 'Polis Station' works not to produce a culture of stasis (in all its problematic definitions), but one of open emergence, Topinka (2010: 56) tells that the term heterotopia originates in the field of medicine and refers to the displacement of an organ of the body into another place, a place it should not be. In the preface to *Order of Things* (1994) Foucault uses heterotopia in reference to an essay by Borges, in which Borges notes the classification of animals in a fictional Chinese encyclopaedia—'Celestial Empire of Benevolent Knowledge'—a whole whose multiplicity is affirmed through laughter. As Foucault quotes, the animals are classified like this:

> (a) belonging to the Emperor, (b) embalmed, (c) tame, (d) sucking pigs, (e) sirens, (f) fabulous, (g) stray dogs, (h) included in the present classification, (i) frenzied, (j) innumerable, (k) drawn with a very fine camelhair brush, (l) et cetera, (m) having just broken the water pitcher, (n) that from a long way off look like flies. (Foucault 1994: xvi)

Reminding us of Deleuze's (1988b) discussion of Spinoza's body and its definition along the lines of speeds and slownesses, and affective capacities (Spinoza 1996; see also: Deleuze 1990), Foucault provides storytelling as a critically designed taxonomy, a taxonomy that is swerved out of the norm, away—as Deleuze says following Spinoza—from issues of form, function and substance. Order and regula-

[9] We have used 'homeorrhesis' (the production of stability through movement) as it is important in Serres's work on Lucretius (2000). There is another philosopher, however, whose work is not only commensurate with the concepts we are putting together in this chapter, but to related issues of thermodynamics and creativity and homeorrhesis: Gilbert Simondon. See especially: Simondon (1989, 2009, 2012) where he mobilises the thermodynamic concepts of 'homeostasis' and 'metastability'; and Combes (1999), Chabot (2003), Hales (2015), Sauvagnargues (2012) and Brassett (2016). We will leave a fuller exploration of Simondon in relation to these concepts to another day.

tion in this taxonomy are subjected to a minimal deviation, and their new trajectories are swerved from their rational premise thereby creating an opening, a variation. Foucault's reading of Borges's redesign of narrative, so deviant yet appearing to perform its taxonomic function, is sent into a laughter that shatters 'all the familiar landmarks of my thought' (Foucault 1994: xvi). The redesign of the police station as 'Polis Station', shattering the familiar landmarks of power and control, and situating them across the manifold urban landscapes while not borne of laughter, may nevertheless construct new trajectories for socio-political narratives to be created.

Borges's classifications, 'Polis Station', Lucretius and so on, contain a number of equally weighted actions, characters and moments whose equality, equanimity and equilibrium are shaken out of their slumbers. This 'equal weightiness' is a decidedly Epicurean concept, and one that necessitates the swerve without which there would be nothing. And so Lucretius, with the *clinamen*, describes the ways in which something comes of the general fall of equally weighted things and becomes particular, new, locally combined into complexities in creative ways. Serres explains of the same concept:

> Equilibrium is global and distributed by chance in space and time. In uncertain places, and in unforeseeable times, another beginning takes place, somewhere else. There is no closed cycle on a local level. There are worlds which are scrap-heaps, there are worlds being born. Locally it is aleatory. Globally it is balanced. (Serres 2000: 173)

And Lucretius, who writes:

> [...] since this world is the product of Nature, the happenstance
> Of the seeds of things colliding into each other by pure chance
> In every possible way, no aim in view, as random, blind,
> Till sooner or later certain atoms suddenly combined.

So that they lay the warp to weave the cloth of mighty things:

> Of earth, of sea, of sky, of all the species of living beings. (Lucretius 2007: 68–69; 2:1057–1064)[10]

Deleuze (2004) makes similar points, highlighting the non-totalising, distributive and conjunctive character of nature and the relations between sums and parts. He notes well too that the swerve is not the movement that knocks the atoms off course, but the always present, original determination of direction and movement of atoms (Deleuze 2004: 306). The *clinamen* is not a secondary movement, he emphasises. This is an important consideration to make when thinking of The Swerve: things do not fall and then are hit by The Swerve; their swerving is part of their ontological milieu. We might do well, then, to highlight this in relation to design too: The Swerve does not hit already designed stuff; it is an important affective condition of the ontological milieu of designing. Design, all creativity, everything, is swerve and has been swerved; without the *clinamen* there would be nothing.

---

[10] The prose translation is as follows: 'This follows from the fact that our world has been made by nature through the spontaneous and casual collision and the multifarious, accidental, random and purposeless combinations could serve as the starting-point of substantial fabrics—earth and sea and sky and the races of living creatures.' (Lucretius 1994: 64; 2:1057–1064)

3.3 *The Swerve and the Design Process*

For all the differences relating to local contexts and particularities of upbringing and education, most designing operates as a linear and goal-oriented process. Designing usually starts with a brief from a client, progresses into a research phase, thence with conclusions from research helping to generate a range of relevant design concepts, which are discussed with the client to help formulate a particular design to develop and resolve, until a final outcome is reached. Notwithstanding that there are many iterative instances where data is folded-back into earlier stages of the process (the concepts generated can be taken back to users or markets to test before being developed as designs; more research can be demanded by various trajectories taken in resolving designs, and so on), this process is a teleological one. But there is no necessity here: linear, teleological and identity-driven design is not the only option. There are ways in which the *clinamen* that has constructed design and the collisions that design can produce can be emphasised in any process. We shall now give four examples expressing a Lucretian swerve of the design process.

Example 1. Inspiration Found in the Random
One can extract oneself from the everyday, commercially driven teleology of production and wander; become a *flâneur* (Baudelaire 1964, Benjamin 2002), adrift in the complex flows of the world and open to the contingent delivery of unexpected outcomes (O'Reilly and Linkson 2009: 76–79). Again, we meet the heterotopic and the clinamen, with organs of creation swerved from their proper places into chance collisions; with the once normal, normalised practices and products disturbed from their orbits and open to the possibilities that being dissolved and distributed brings. O'Reilly and Linkson (2009) note the way that graphic designer Nick Clark wanders, *flâneur*-like, collecting ephemera (tins of beans, old toys and sticker albums) as if they were drawn into his strange orbit, to accrete with him as he drifts through his universe. Clark's creative system obviously benefits from such additions, and his inspiration—O'Reilly and Linkson (2009: 76) note—'is to be found in the random'; as Lucretius writes (and we quote above) Clark and his stuff becomes 'combined/So that they lay the warp to weave the cloth of mighty things' (Lucretius 2007: 69; 2:1062–1063). Serres places Lucretius's text in relation to more recent sciences of nonlinear dynamics, chaos and complexity, especially with his (Lucretius's) discussions of meteorology. For Serres *turba* and *turbo*—disorder, confusion and tumult, and vortices and spinning tops respectively (2000: 27–31) as we note above—help him position Lucretius at the birth of a physics of which we are still part, as opposed to an aberrant, early mistake. The aleatory and stochastic, chance and randomness, of the social norms, cultural forms and everyday actualities of life that design engages in all their complexity, therefore announces that design is always already entangled with a world that is complex, distributed in entropic and creative clusters, in critical moments between chaos and order. That is, design is already Lucretian.

Example 2. The Welcome Openness of Research

Another way designers develop negentropic processes is by placing importance on the Research Phase—for investigating cultural contexts, socio-political and historical issues, user/consumer behaviours—such that the focal point of creative agency can be extracted from of the closed-system of the designer/design team. In his preface to Brenda Laurel's influential *Design Research* (2003), new media theorist Peter Lunenfeld writes:

> In the twenty-first Century, the linear narratives of research progress are dissolving into decentered threadings, less branches off a main root than tide pools by the shore, or the rhizomatic growth of peanuts in the soil (Lunenfeld 2000). As information and data about everything explode in a frenzy of rhizomatic connectivity the very search for what to research becomes its own research issue. The research model becomes a design problem that can also serve as its own solution. (Lunenfeld 2003: 14).

Referring to his own book *Snap to Grid* (2000)—rather than Deleuze and Guattari (1988)[11] whose concept of the rhizome has been so influential in creative thinking (see Wilson 2003, Coyne 2005, 2008, Teal 2010, van der Beek 2015 for just a few examples)—Lunenfeld recognises the distributed, nonlinear and complex act of research as it is involved in designing. Lunenfeld calls this preface 'The Design Cluster', deliberately referencing Marshall McLuhan's *The Gutenberg Galaxy* (1962), in order to emphasise the varieties of clumping of (designed) matter, work and cultures in a vast expanse. The image here is decidedly atomist with nature the result of the tiniest elements of matter swerved off course to collide and coagulate. Design, in Lunenfeld's cosmological image, seems to be both equated with the *clinamen* and that which results. Referring back to Laurel's book, he praises its 'welcome openness' in its 'understanding that no single research methodology could possibly account for the diversity of inputs and outputs to contemporary design practice and process' (Lunenfeld 2000: 10). For Lunenfeld here, as well as for Laurel, design research allows a space (albeit vast) for the complexities with which designing has to deal to be experienced and accessed. The intersection of designing with complexity theory is strongly evident here, and something that has been attracting attention (recent examples include: Findeli 2001, Kearnes 2006, Alexiou and Zamenopoulos 2008, Johnson 2008, Zamenopoulos 2012, Zamenopoulos and Alexiou 2012, Brassett 2013, 2015).

Example 3. Diagramming Forces

Engineering and liberal arts academic Kenneth Knoespel argues (2001) that a diagram for Deleuze is a drawing to think with but never simply a visual representation (though visual representations are never simple). A diagram is always connected to usage/function/context (as we will see Guattari's elaboration of the diagram as machine). A diagram may take a specific visual form with semiotic, social and political functions, and be ephemeral (the doodle), professional or scientific. In Foucauldian terms, it is

[11] In *The Digital Dialectic* (Lunenfeld 1999: 236 n. 1), a collection he edits, Lunenfeld notes the influence of Deleuze and Guattari on theoretical engagements with digital culture in general, as well as in this particular collection. Lunenfeld also acknowledges the paradox here, given the title of the book and Deleuze's detestation of dialectic (Deleuze 1995: 6).

technical practice (for example by hand, or in geometrical form, or by advanced imaging techniques) producing, configuring and distributing knowledge. For Deleuze the diagram relates to the complexity of forces arranged in different assemblages (see: Deleuze 1988a, Deleuze and Guattari 1988; also: De Landa 2000, Teyssot 2012, Vellodi 2014), and as such this concept has direct bearing on what we have been discussing regarding design. In his book on Foucault (1988a) Deleuze gives one of his best characterisations of the diagram, according to four 'definitions':

> [the diagram] is the presentation of relations between forces unique to a particular formation; it is the distribution of the power to affect and the power to be affected; it is the mixing of non-formalized pure functions and unformed pure matter [... and] it is a transmission or distribution of particular features. (Deleuze 1988a: 61–62)

Differently to stratified and segmented knowledge, power is diagrammatic: it mixes, folds, distributes and relates.[12] Diagram production is also one of Deleuze and Guattari's four markers of pragmatism (along with tracing, mapping and programming), they explain in *A Thousand Plateaus* (1988: 139, 145–146). Diagrams, design theorist Betti Marenko (2015: 118) explains in a paper on design and divination, 'articulate the conditions that make possible conceptual creation and the manifestation of new expressions – but do not determine directly the outcome'. The diagram is, thus, a reworking of the semiotic process into a machinic one, metamorphosing any idealisation of significations of meaning into actual, material, creative, future-focussed work. It is for this that Guattari first uses the term 'diagrammitisation' (in a lecture course in 1975, published later in *La Révolution moléculaire* in 1977, with English translation in 1984)—identifying it with a quote from pragmatic philosopher Charles Sanders Peirce. He writes:

> this work of diagrammitization, has become the necessary condition for the deterritorializing mutations that affect the fluxes of reality; no longer is there representation, but simulation, pre-production, or what one might call 'transduction'. The stratum of signification disappears; no longer are there two levels and a system of double-articulation; there is only a constant return to the continuum of machinic intensities based upon a pluralism of articulations. (Guattari 1984a: 95)

For Guattari the diagram materialises flows of reality as well as 'deterritorialises' them from any formal systematisation to which they may have been subjected. Meaning is neither imposed nor unearthed in and with the diagram, but connections made leading to what Marenko describes as the manifestation of possible future expressions (see also: Brassett 2016, Brassett and Marenko 2015, Brassett and O'Reilly 2015, O'Reilly 2015). Indeed, as Knoespel argues, the Greek etymology of the verb of *diagramma* means something figured, mapped, planned – marked out by lines – but also, 'carries the secondary connotation of marking or crossing out' (2001: 147). So intrinsic is this mutability to the practice and experience of the dia-

[12] Deleuze also marks the diagram as strategic rather than stratified (Deleuze 1988a: 62). Deleuze's use of 'strategic' has none of the negative connotations that it has for Serres, and aligns with the ability/need to 'think otherwise' (Deleuze 1988a: 98): thought set free to head to the future along a strategic line that is developed as becoming. We will leave to another piece of work a more critical look at design strategy in relation to Deleuze and Serres.

gram that its very cognition is swerved/swerving, carrying with it the sense that it may be redrawn, changed, re-arranged. It is why Knoespel diagrams the diagram (the diagram as example) as a 'relay'. He writes:

> While a diagram may have been used to visually enforce an idea one moment, the next it may provide a means of seeing something never seen before. Because diagrams mark a gesture or momentum toward recognition, they function as vehicles that invite elaboration through narrative. (Knoespel 2001: 147)

In other words, diagrams trace the genesis and direction of travel of their own forming as social, cultural and political forces. Take for example the diagrammatic narrative of John Cook (see Fig. 3).

Research around the Camdeboo National Park in South Africa where a number of oil companies had been given drilling licences to explore fracking, led University of Westminster (London, UK) architecture, digital art and landscape design student John Cook (2015) to design an alternative scenario, where entropic processes become regenerative. Combining tourism and solar farming, the initial construction of 'Camdeboo Solar Estate', seeks to design through collision a Stonehenge of the future from a working, sustainable solar farm. The spatial arrangement of this technology is narrative designed around the movement of the earth around the sun, the changing equinoxes and the mythological relationships that have been generated over time.

In an article for *Architectural Digest* peppered with references to Deleuze and Guattari and Foucault, and many of the concepts we have been expressing in this chapter, architect François Roche writes that 'science fictional architecture is a space of confrontation [...] By necessity, it confronts its emergence, its Gestalt, and can only be negotiated in the visible spectrum: that is its political and operational condition' (Roche 2010: 66–68). Like Lucretius, Cook's design collides science and fiction, inventing mythologies to collide on different plateaus across millennia. Over time, Cook's original solar farm transforms into vineyards, which then degrade and disintegrate with the environment (see Fig. 4), so that by the year 3000 there is revealed a series of landscape-scaled astronomical instruments that enable the park-solar farm-vineyard to be used as an astronomical observatory (see Fig. 5).

Cook opens the conventional architect's master plan to change, where the entropy of decay delivers moments of negentropic creation and wonder. But it is also the production of a diagram; an act of diagramming that is always already Lucretian in its swerves and collisions. Which is at least one of the directions that Cook, Deleuze, Studio Gang et al. deliberately open up for us. There are, of course, many others who follow different lines from these and other diagrams that cross philosophical, design and literary constructs.

Example 4. Externality

The final example that we wish to address of the design's *clinamen* coming out in the open is more organisational: the use by in-house design teams of outside consultancies and pushing 'normal' practices outside of the regulations of company equilibrium by accessing energy from the outside. Deleuze emphasises throughout his work (alone and with Guattari) the power of exteriority. At the end of *Foucault*

Fig. 3 Camdeboo Solar Estate. John Cook (2015) Camdeboo Solar Estate. Landscape, architecture and speculative design project by John Cook, for Camdeboo National Park, South Africa. "Opening in 2050, The Camdeboo Solar Estate looks to address both the agricultural and energy difficulties faced by South Africa and the Karoo region—the proposal combines the ancient practices of terrace agriculture, astronomy and solar observance with the modern day technologies of solar energy harvesting. The masterplan arrangement, its axial pathways and internal orientations are calibrated to the positions of the celestial objects within our solar system at the time of opening." Image and caption copyright John Cook

for example, he tells that 'forces always come from outside' (Deleuze 1988a: 100): when the outside folds over, becoming inside. A few pages earlier Deleuze provides a source-free open-ended sentence set-off in quotation marks, as follows: '"I have never written anything but fictions . . ."' (Deleuze 1988a: 98). We imagine that these words are from Foucault, but they could just as well be from Deleuze, or any of the characters he mentions in the two preceding paragraphs—Blanchot, Nietzsche, Heidegger, Gogol, Checkov and Bely. These outsides of Foucault fold effortlessly into Deleuze's presentation of Foucault's work, as well as his own. He continues, not quoting now: 'But never has fiction produced such truth and reality' (Deleuze 1988a: 98). The folding of philosophy's exteriority—fiction in this particular case—back into philosophy is a sure way of producing 'truth and reality', albeit in a fictional sense: Lucretius is a fine example of this, fulling[13] the different fibres of philosophy, poetry and physics into a many folded and entwined felt.[14] Design's outside folds inside in many different ways too, as we have been seeing, and may be its only way of avoiding the impositions of heroic (Julier 2013) or paternalistic (Thorpe and Gamman 2011) practices and ways of thinking.[15] In the same vein, and for quite some time, design companies have been introducing into their teams people with both multidisciplinary or specifically non-design skills: psychologists, anthropologists and sociologists, literature specialists and even philosophers (The Design Council 2005, Kimbell 2011, 2012). Like the Möbius strip so often associated with Deleuze and Guattari's work there may be times when the exterior is, can or needs to be, located inside.

---

[13] To 'full' is to mash up fibres—of wool, for example—into felt. Deleuze and Guattari (1988: 474–500) remark on the difference between systems of weaving (striated) and making felt or patchwork in their section of the book, '1440: The Smooth and the Striated'. 'Felt is a supple product,' they write, 'that proceeds altogether differently [to weaving], as an anti-fabric. It implies no separation of threads, no intertwining, only an entanglement of fibers obtained by fulling (for example, rolling the block of fibers back and forth' (Deleuze and Guattari 1988: 475). See also: Brassett (2005) for a discussion of this in terms relating to the current chapter.

[14] It is worth mentioning here the practice of Design Fictions (Sterling 2009, Hales 2013). Often linked to Critical Design (Dunne 1999, Malpass 2015) and Speculative Design (Dunne and Raby 2013), Design Fictions operates too as critical response to culture, as well as providing speculative and designerly expressions of the future-oriented modal question 'what if?' (Booth et al. 2009, Hales 2013, Brassett and O'Reilly 2015). Hales (2013) notes that in their speculative capacity Design Fictions 'take their cue from science fiction', and further that 'the notion of design fictions opens design to theoretical and artistic methodologies that can be used to excavate past, present and future media through its fictions' (Hales 2013: 2, 4). This is different to our presentation of the colliding of fiction/literature, philosophy and design, but not radically so.

[15] Such 'top-down' practices are not universally frowned upon of course. Management scholar Roberto Verganti (2009) sees a top-down, meaning-centred approach necessary for any radical innovation. See also Norman and Verganti (2014).

Fig. 4 CSP Plant Deconstruction: 2100-2105. "After the 60-year life expectancy of the CSP apparatus expires, the energy infrastructure is de-constructed, and the estate remains operating as a series of vineyards." Image and caption copyright John Cook

Fig. 5 CSP Plant Decomposition: 2200+ "By 2200, the plants building materials begin a sequence of planned and choreographed decomposition—as the buildings ruins reconfigure and settle, the celestial alignments and orientations of the CSP landscape are unveiled." Image and caption copyright John Cook

3.4 Designing and Colliding

The benefit of collisions that force the design process out of equilibrium is therefore not such big news. Nevertheless, as we have seen, what constitutes the design process can itself be forced into colliding with *outsiders* to that process. It has been a tenet of user-centred design that the locus of authority for designing things for people should be the people who will be the end-users. This is further complexified by the manifold practices, researches and related design activities that go by the name: 'co-creation' (see, for example, Sanders and Stappers 2008). Such processes are still teleological, however, with the inputs from users helping only to iterate a design towards a more ideal (or relevant, or appropriate) final outcome (Brassett and Booth 2007a, b). We wonder whether these are ever more than gestures serving to salve some designers' bad consciences at participating in 'phoney' practices (Papanek 1984)? A different model of an open, complex and collision-inducing design process, then, might be to dispense with the notion of a final outcome altogether. Where impacts from each moment of collision with users, with other designers, with clients, with any actors across the landscape in which design is practised, are valued and promoted, leaving only an evolving, dynamic and symbiotic process with no end but with myriad, non-privileged outcomes that can emerge at any moment.

4 Last Words

In the epigram to this chapter Vonnegut suggests that far from being privileged (spiritual) beings, we are just machines. Machines that in the normal run of things have not much else to do—that is, no real purpose or aim or meaning—other than to collide and collide and collide (Vonnegut 1992: 219–220). And so creativity emerges, just as Lucretius shows. We offer the same for design: it too needs to be a 'fan of collisions'.

But is it enough, really, to be a 'fan of collisions'? This phrase serves well to emphasise a direction to take and the rationale for taking it. The journey we have been on however takes it further still. It is not enough simply to say that designing should develop a collision-loving attitude, true though this may be. Neither should we imply that it is adequate simply to uncover the collisions that lurk behind or beneath every creative act, while this may be necessary also. Designing and colliding are co-extensive, they are immanent. They are proof that the swerve has knocked atoms off their inexorable course to the stagnant death of equilibrium. Designing emerges from the collisions—and the ensuing coagulations and ricochets—that the *clinamen* delivers, it is proof of a negentropic eruption that develops 'crystals sunk in ash' (Serres 1982). Designing also produces collisions: it is a collision-inducing machine.

The trouble is that design—as a set of practices and processes, and too often identified as objects—often forgets or actively negates the swerves, collisions and the rest with which it is implicated.[16] While we hope that this chapter provides one way of rekindling design's Lucretian nature, we are loath to position the ways in which it has been forgotten etc.—rationalisation, meaning, modelling and thinking—in opposition to collisions and The Swerve. It is obvious that for some such activities are themselves full of creative opportunity and can lead to much. What we would like to emphasise is that these moments (rationalisation, and so on) have a tendency to dictate and overcode; that is, to offer interesting lines of investigation and practice only to fold them back onto well-worn structures of power that strive for domination and control.

The ways in which a practice and process of designing might proceed such that their swerves and the collisions that ensue are championed, we announce earlier in this chapter, especially through the examples we have given where colliding/designing occur: random; openness; diagramming; and externalising. We are wary of introducing these as axioms of good design however. That is, as we have noted above, 'The Swerve does not hit already designed stuff; it is an important affective condition of the ontological milieu of designing'. With characteristics such as random, openness, diagrams and externality, then, it is not so much an issue of forcing these onto practice but uncovering where these are already taking place, where they were blocked, or where they might possibly erupt in the middle of our creative practices. We might ask designers, philosophers, any creative practitioners, then: 'where have you swerved?' Because it has happened. 'Map your collisions!' Because you may have forgotten how well they served you. 'How have you diagrammed your practice?' Because this will show how you relate, fold, distribute and mix as active verbs, rather than identifiable traits. 'Where do the most internal parts of your practice open up to the most external?' Because you know that those internal parts, the most protected and defended are also the most ossified. 'When, how and why have you blocked any engagement with randomness?' Because those ruts you have worn are the ways that you anæsthetise yourself against the randomness of creative collisions through habit. 'How might you make yourselves fans of collisions?' Because then you are The Swerve of all things.

Acknowledgements Our thanks go: to the editors for incisive and creative feedback on earlier versions of this chapter. To Professor Victor Margolin for comments on a *much* earlier version of this paper. To Professor David Webb, Staffordshire University for sharing thoughts on Serres, as

[16] We note that there are many ways in which design's Lucretian nature has been either ignored, forgotten or actively negated, as follows: it has striven to be serious, proper and mature (Whicher et al. 2015); or sought to emphasise its axiomatisation in rationalisation (Simon 1969), meaning (Krippendorf 2006, Verganti 2009), or thinking alone (for example: Brown 2009, Martin 2009, Neumeier 2009). There have been many efforts to counteract such axiomatising of design along the lines noted: by thwarting its rationalisation through bringing wicked problems closer to designing (for example, Hatchuel 2001, Coyne 2005); its domination by meaning-production by highlighting the role of affect in design (for example: Marenko 2010, Brassett and Marenko 2015, Brassett and O'Reilly 2015); its linear modelling by opening designing, again, onto the chaotic and complex (for example: Brassett 2015); and its overcoding as a form of thinking only by showing where design's value can be developed along different lines (for example: McCullagh 2010, Kimbell 2011, 2012, Tonkinwise 2011).

well as some of his pre-published papers. To Derek Hales, whose swerves align with ours but are different. To colleagues and students on MA Innovation Management at Central Saint Martins, University of the Arts London, for discussing some of these issues with us over the years.

References

Agamben, G. [2006] (2009). *What is an Apparatus? And other essays* (trans. Kishik, D. aand Pedatella, S.). Stanford: Stanford University Press.

Agamben, G. (2015). Stasis. Civil War as a Political Paradigm (trans. Heron, N.). Encounters in law and philosophy series. Edinburgh: Edinburgh University Press.

Alexiou, K., & Zamenopoulos, T. (2008). Design as a social process: A complex systems prespective. *Futures Special Issue: 'Design out of Complexity', 40*(6), 586–595.

Baudelaire, C. [1863] (1964). The painter of modern life. New York: Da Capo Press.

Benjamin, W. [1927–1940] (2002). *The Arcades Project* (trans. Eiland, H., ed. McLaughlin, K.). Cambridge, MA: Belknap Press.

Berressem, H. (2005). *Incerto tempore incertique locis*. The logic of the clinamen and the birth of physics. In N. Abbas (Ed.), *Mapping Michel Serres* (pp. 51–71). Ann Arbor: University of Michigan Press.

Booth, C., Rowlinson, M., Clark, P., Delahaye, A., & Procter, S. (2009). Scenarios and counterfactuals as modal narratives. *Futures, 41*(2), 87–95.

Brassett, J. (2005). Entropy (fashion) and emergence (fashioning). In C. Breward & C. Evans (Eds.), *Fashion and modernity* (pp. 197–209). Oxford/New York: Berg.

Brassett, J. (2013). Networks: Open, closed or complex. Connecting philosophy, design and innovation. In J. Cai, T. Lockwood, C. Wang, G. Tong and J. Liu (Eds.), Design-driven business innovation. 2013 IEEE-Tsinghua international design management symposium proceedings (pp. 1–11). Beijing: IEEE.

Brassett, J. (2015). Poised and complex: The becoming each other of philosophy, design and innovation. In B. Marenko & J. Brassett (Eds.), *Deleuze and Design* (pp. 31–57). Edinburgh: Edinburgh University Press.

Brassett, J. (2016). Speculative machines and technical mentalities: A philosophical approach to designing the future. *Digital Creativity Special Section: Speculative Hardware, 27*(2), 163–176.

Brassett, J., & Booth, P. (2007a). Design digestion. A work in progress. *Design Principles and Practices: An International Journal, 2*(3), 75–82.

Brassett, J., & Booth, P. (2007b). Ecstatic innovation. Digesting, designing and democracy. In A. Yagou (Ed.), *Re/Public, Special issue: 'Distributed creativity and design'*. http://www.republic.gr/?p=320. Accessed 30 July 2015.

Brassett, J., & Marenko, B. (2015). Introduction. In B. Marenko & J. Brassett (Eds.), *Deleuze and Design* (pp. 1–30). Edinburgh: Edinburgh University Press.

Brassett, J., & O'Reilly, J. (2015). Styling the future. A philosophical approach to design and scenarios. In *Futures 74, Special issue: 'Scenarios and design'* (pp. 37–48).

Brown, T. (2009). *Change by design. How design thinking transforms organizations and inspires innovation*. New York: Harper Collins Publishers.

Budds, D. (2015). .5 Big Ideas From The Chicago Architecture Biennial. http://www.fastcodesign.com/3051919/5-big-ideas-from-the-chicago-architecture-biennial. Accessed 19 Dec 2015.

Carpenter, J. (Dir.). (1976). Assault on precinct 13. The CKK Corporation/Overseas Film Group.

Chabot, P. (2003). *La Philosophie de Simondon*. Paris: Librarie Philosophique J. Vrin.

Chicago Architecture Biennial (2015). Polis Station. http://chicagoarchitecturebiennial.org/public-program/calendar/polis-station/. Accessed 19 Dec 2015.

Combes, M. (1999). *Simondon. Individu et collectivité*. Paris: Presses Universitaires de France.

Cook, J. (2015). The Camdeboo Solar Estate. https://www.behance.net/gallery/27063941/The-Camdeboo-Solar-Estate. Accessed 19 Dec 2015.
Coyne, R. (2005). Wicked problems revisited. *Design Studies, 26*(1), 5–17.
Coyne, R. (2008). The net effect: Design, the rhizome, and complex philosophy. *Futures, 40*(6), 552–561.
De Landa, M. (2000). Deleuze, diagrams, and the genesis of form. *Amerikastudien/American Studies, 45*(1), 33–41.
Deleuze, G. [1986] (1988a). *Foucault* (trans. Hand, S.). London: The Athlone Press.
Deleuze, G. [1981] (1988b). *Spinoza: Practical Philosophy* (trans. Hurley, R.). San Francisco: City Lights Books.
Deleuze, G. [1968] (1990). *Expressionism in Philosophy: Spinoza*. New York: Zone Books.
Deleuze, G. [1973] (1995). Letter to a harsh critic. In G. Deleuze, *Negotiations. 1972–1990* (trans. Joughin, M.), (pp. 3–12). New York: Columbia University Press.
Deleuze, G. [1969] (2004). Lucretius and the simulacrum. In G. Deleuze, *The Logic of Sense* (trans. Lester, M. with Stivale, C.), (pp. 303–320). London/New York: Bloomsbury.
Deleuze, G. and Guattari, F. [1972] (1984). *Anti-Oedipus. Capitalism and Shizophrenia 1* (trans. Hurley, R., Seem, M. and Lane, H.R.). London: The Athlone Press.
Deleuze, G. and Guattari, F. [1980] (1988). *A Thousand Plateaus. Capitalism and Shizophrenia 2* (trans. Massumi, B.). London: The Athlone Press.
Dovey, K. (2013). Assembling architecture. In H. Frichot & S. Loo (Eds.), *Deleuze and architecture* (pp. 131–148). Edinburgh: Edinburgh University Press.
Dunne, A. (1999). *Hertzian Tales: Eelectronic products, aesthetic experience and critical design*. London: RCA/Computer Aided Design Research Publications.
Dunne, A., & Raby, F. (2013). *Speculative Everything: Design, Fiction, and Social Dreaming*. Cambridge, MA/London: The MIT Press.
FBI (2015). FBI Releases 2014 Crime Statistics. https://www.fbi.gov/news/pressrel/press-releases/fbi-releases-2014-crime-statistics. Accessed 19 Dece 2015.
Findeli, A. (2001). Rethinking design education for the 21st Century: Theoretical, methodological, and ethical discussion. *Design Issues, 17*(1), 5–17.
Flusser, V. (1999). About the word *Design*. In V. Flusser, *The Shape of Things: a philosophy of design* (trans. Mathews, A.), (pp. 17–22). London: Reaktion Books.
Flusser, V. [1986] (2002a). Criteria—Crisis—Criticism. In V. Flusser, *Writings* (ed. Ströhl, A. and trans. Eisel, E.), (pp. 42–50). Minneapolis: University of Minnesota Press.
Flusser, V. [1991] (2002b). Habit: The true aesthetic criterion. In V. Flusser, *Writings* (ed. Ströhl, A. And trans. Eisel, E.), (pp. 51–57). Minneapolis: University of Minnesota Press.
Foucault, M. [1967] (2008). Of other spaces. In M. Dehaene and L. De Cauter (Eds.), Heterotopia and the City. Public space in a postcivil society (pp. 13–29). Oxford/New York: Routledge.
Studio Gang (2015). Polis Station. http://studiogang.com/researchproject/chicago-architecture-biennial-polis-station. Accessed 19 Dec 2015.
Genosko, G. (2002). *Félix Guattari: An Aberrant Introduction*. London/New York: Continuum.
Guattari, F. [1975] (1984a). Towards a micro-politics of desire. In F. Guattari, *Molecular Revolution. Psychiatry and Politics* (trans. R. Sheen), (pp. 82–107). Harmondsworth: Penguin.
Guattari, F. [1964] (1984b). Transversality. In F. Guattari, *Molecular Revolution. Psychiatry and Politics* (trans. R. Sheen), (pp. 11–23). Harmondsworth: Penguin.
Hales, D. (2013). Design fictions an introduction and provisional taxonomy. *Digital Creativity, 24*(1), 1–10.
Hales, D. (2015). Re-designing the Objectile. In B. Marenko & J. Brassett (Eds.), *Deleuze and Design* (pp. 139–172). Edinburgh: Edinburgh University Press.
Hatchuel, A. (2001). Towards design theory and expandable rationality: The unfinished programme of Herbert Simon. *Journal of Management and Governance, 5*(3), 260–273.
Hoven, R. (1993). *Lexique de la prose latine de la renaissance*. Leyde/New York/Cologne: E. J. Brill.

Johnson, J. (2008). Science and policy in designing complex futures. *Futures*. Special Issue: 'Design out of Complexity' 40(6), 520–536.
Julier, G. (2013). *The culture of design* (3rd ed.). London: Sage.
Kauffman, S. A. (1993). *The Origins of Order: Self-organization and Selection in Evolution*. New York/Oxford: Oxford University Press.
Kauffman, S. A. (2008). *Reinventing the sacred*. New York: Basic Books.
Kearnes, M. (2006). Chaos and control: Nanotechnology and the politics or emergence. *Paragraph, 29*(2), 57–80.
Kimbell, L. (2011). Rethinking design thinking: Part 1. *Design and Culture, 3*(3), 285–306.
Kimbell, L. (2012). Rethinking design thinking: Part 2. *Design and Culture, 4*(2), 129–148.
Knoespel, K. J. (2001). K. Diagrams as Piloting Devices in the Philosophy of Gilles Deleuze. In *Litérature, Théorie, Enseignement* (Vol. 19, pp. 145–165).
Krippendorf, K. (2006). *The semantic turn*. London: Taylor and Francis.
Lartey, J., Laughland, O., & Swaine, J. (2015). The counted. Black Americans killed by police twice as likely to be unarmed as white people. In *The Guardian*. http://www.theguardian.com/us-news/2015/jun/01/black-americans-killed-by-police-analysis.
Lucretius [55BCE] (1994). *On the Nature of the Universe* (trans. Latham R.E., revised Goodwin, J.). London: Penguin Books.
Lucretius [55BCE] (2007). *The Nature of Things* (trans. Stallings, A.E.). London: Penguin Books.
Lunenfeld, P. (Ed.). (1999). *The digital dialectic: New essays on new media*. Cambridge, MA/London: The MIT Press.
Lunenfeld, P. (2000). *Snap to Grid: a user's guide to digital arts, media and cultures*. Cambridge, MA/London: The MIT Press.
Lunenfeld, P. (2003). Preface. The design cluster. In B. Laurel (Ed.), *Design research. Methods and perspectives* (pp. 10–15). Cambridge, MA/London: The MIT Press.
Malpass, M. (2015). Criticism and function in critical design practice. *Design Issues, 31*(2), 59–71.
Marenko, B. (2010) Contagious affectivity – The management of emotions in late capitalist design. Proceedings of the Swiss Design Network Conference 2010: 134–149.
Marenko, B. (2015). When making becomes divination: Uncertainty and contingency in computational *glitch-events*. *Design Studies, 41*, 110–125.
Martin, R. (2009). *The Design of Business: Why design thinking is the next competitive advantage*. Boston: Harvard Business Press.
McCullagh, K. (2010). Design thinking: everywhere and nowhere, reflections on the big re-think. http://www.core77.com/blog/featured_items/design_thinkingeverywhere_and_nowhere_reflections_on_the_big_re-think__16277.asp. Accessed 19 Dec 2015.
McLuhan, M. (1962). *The Gutenberg galaxy: The making of typographic man*. Toronto: University of Toronto Press.
Neumeier, M. (2009). *The Designful Company. How to build a culture of nonstop innovation*. Berkeley: New Riders.
Norman, D. A., & Verganti, R. (2014). Incremental and radical innovation: Design research vs. technology and meaning change. *Design Issues, 30*(1), 78–96.
O'Reilly, J. (2015). Milieu and the creation of the illustrator: Chris Ware and Saul Steinberg. In B. Marenko & J. Brassett (Eds.), *Deleuze and Design* (pp. 191–218). Edinburgh: Edinburgh University Press.
O'Reilly, J., & Linkson, T. (2009). *Recharge your design batteries. Creative challenges to stretch your imagination*. Cincinnati: HOW Books.
Papanek, V. [1971] (1984)*Design for the Real World*. 2nd edition. London: Thames and Hudson.
Prigogine, I. (1980). *From being to becoming. Time and complexity in the physical sciences*. San Francisco: W. H. Freeman and Co.
Prigogine, I., & Stengers, I. (1985). *Order out of chaos. Man's new dialogue with nature*. London: Flamingo.

Roche, F. (2010). (Science) fiction, Ecosophical apparatus and Skizoid machines. Animism, Vitalism and Machinism as a way to rearticulate the need to confront the unknown in a contradictory manner. *Architectural Digest, 80*(6), 64–71.
Sanders, E. B.-N., & Stappers, P. J. (2008). Co-creation and the new landscapes of design. *CoDesign: International Journal of CoCreation in Design and the Arts, 4*(1), 5–18.
Sauvagnargues, A. (2006). *Deleuze et l'art*. Paris: Presses Universitaires de France.
Sauvagnargues, A. (2012). Crystals and membranes: Individuation and temporality (trans. Roffe, J.). In A. De Boever, A. Murray, J. Roffe and A. Woodward (Eds.). Gilbert Simondon. Being and technology (pp. 57–70). Edinburgh: Edinburgh University Press.
Serres, M. (1982). The origin of language: Biology, information theory and thermodynamics. In M. Serres, *Hermes. Literature, Science, Philosophy* (ed. Harari, J.V. and Bell, D.F.), (pp. 71–83). London/Baltimore: Johns Hopkins University Press,
Serres, M. [1982] (1995) *Genesis* (trans. James, G. And Nielson, J.). Ann Arbor: University of Michigan Press.
Serres, M. (1977). *La Naissance de la physique dans le texte de Lucrèce. Fleuves et turbulences*. Collection « Critique ». Paris: Les Éditions de Minuit. English edition: Serres, M. (2000). *The Birth of Physics* (ed. Webb, D., trans. Hawkes, J.). Manchester: Clinamen Press.
Serres, M. [2009] (2014). *Times of Crisis. What the financial crisis revealed and how to reinvent our lives in the future*. New York/London/New Delhi/Sydney: Bloomsbury.
Simon, H. (1969). *The sciences of the artificial*. Cambridge, MA: The MIT Press.
Simondon, G. [1958] (1989). Du modes d'existence des objets techniques. Paris: Aubier.
Simondon, G. (2009). The position of the problem of ontogenesis. *Parrhesia, 7*, 4–16.
Simondon, G. (2012). Technical mentality. In A. De Boever, A. Murray, J. Roffe, & A. Woodward (Eds.), *Gilbert Simondon. Being and technology* (pp. 1–15). Edinburgh: Edinburgh University Press.
Souriau, E. [1943] (2009). *Les différents modes d'existence*. Paris: Presses Universitaires de France.
Spinoza, B. (1996). *Ethics* (trans. Curley, E.). London: Penguin.
Stengers, I. (1997a). *La vie et l'artifice: visages de l'émergence*. Paris: Editions La Découverte.
Stengers, I. (1997b). *Power and Invention. Situating Science. Theory out of bounds 10*. Minneapolis/London: University of Minnesota Press.
Sterling, B. (2009). COVER STORY: Design fiction. *Interactions, 16*(3), 20–24.
Teyssot, G. (2012). The diagram as abstract machine. *V!RUS* 7. http://www.nomads.usp.br/virus/virus07/?sec=3&item=1&lang=en. Accessed 10 Nov 2015.
The Design Council. (2005). Eleven Lessons: managing design in eleven global brands. In *A study of the design process*. London: The Design Council.
Thorpe, A., & Gamman, L. (2011). Design with society: Why socially responsive design is good enough. *CoDesign: International Journal of CoCreation in Design and the Arts, 7*(3–4), 217–230.
Tonkinwise, C. (2011). A taste for practices: Unrepressing style in design thinking. *Design Studies, 32*(6), 533–545.
Topinka, R. (2010). Foucault, Borges, heterotopia: Producing knowledge in other spaces. *Foucault Studies, 9*, 54–70.
van der Beek, S. (2015). Rhizome as a toolkit for fluid design. From a paper given at *Spaces of Learning Conference*, Toronto (18 April 2015). http://www.sannevanderbeek.nl/2015/04/rhizome-as-a-toolkit-for-fluid-design/ Accessed 10 Nov 2015.
Vellodi, K. (2014). Diagrammatic thought: Two forms of constructivism in C.S. Peirce and Gilles Deleuze. *Parrhesia, 19*, 79–95.
Verganti, R. (2009). Design-Driven Innovation. In *Changing the rules of competition by radically innovating what things mean*. Boston: Harvard Business Press.
Vonnegut, K. [1973] (1992). *Breakfast of Champions*. London: Vintage Books.

Webb, D. (2006). Michel Serres on Lucretius. Atomism, science, and ethics. *Angelaki. Journal of the Theoretical Humanities*, 11(3) December, 125–136.
Webb, D. (2010). Penser le multiple sans le concept: Vers un intellect démocratique. In F. L'Yvonnet and C. Frémont (Eds.) *Cahier Michel Serres* (pp. 87–94). Paris: Éditions de l'Herne. English version: Thinking multiplicity without the concept: Towards a democratic intellect. Paper given at the *Joint Conference: Society for European Philosophy/Forum for European Philosophy*, University of Dundee (September 2006).
Whicher, A., Swiatek, P., & Cawood, G. (2015). *Design Policy Monitor 2015*. In *Reviewing innovation and design policies across Europe*. Cardiff: PDR/Cardiff Metropolitan University.
Wilson, B. (2003). Of diagrams and rhizomes: Visual culture, contemporary art, and the impossibility of mapping the content of art education. *Studies in Art Education, 44*(3), 214–229.
Zamenopoulos, T. (2012). A complexity theory of design intentionality. *Artificial Intelligence for Engineering Design, Analysis and Manufacturing, 26*(1), 63–83.
Zamenopoulos, T., & Alexiou, K. (2012). Complexity: What designers need to know. In S. Garner & C. Evans (Eds.), *Design and designing: A critical introduction* (pp. 411–428). London: Berg.

Part II
Design Thinking

Arational Design

Thomas Wendt

Abstract Too often, designers rely on rationalist notions of their work: from planning to execution, from theory to practice, from strategy to execution, from problems to solutions, from thinking to making, etc. While these sharp distinctions can serve to hyper-focus individual designers on their unique role and responsibilities, the confusion they create outweighs any potential benefits. Rigid distinctions between modes of practice often create confusion and illusions of certainty, especially when two poles come together, even rely on and co-construct one another. Much of the rationalist sentiment in contemporary design stems from a bias in Western philosophy that introduces a hierarchical relationship between mind and body – the mind dictates and the body executes. But there is no designer equivalent to cogito ergo sum – no positivist statement we can make to delineate and prioritize mental functions over bodily engagement. As an alternative to these dualisms, this paper will take a phenomenological and arational perspective on the components of design, with the end goal of articulating an arational understanding of design. It will examine the emergence of design from a rationalist epistemology and contemporary practices that are attempting to break the boundaries of reason-based methods.

Keywords Design philosophy · Design practice · Rationality · Critical design

1 Introduction

I will argue here that design has limited itself as a field of study by failing to recognize the detrimental effects of rationalist thinking. Just as design is working on shedding the baggage of its perception as a purely aesthetic practice, so too should it ask itself to further engage with its philosophical roots, with the end goal being to interrogate that nature of design as a cultural/social/political force and to question its underlying assumptions. Rational thought is in need of a similar reframe. The

T. Wendt (✉)
Surrounding Signifiers, Design Research and Strategy, New York, NY, USA
e-mail: thomas@srsg.co

© Springer International Publishing AG 2018

P. E. Vermaas, S. Vial (eds.), *Advancements in the Philosophy of Design*, Design Research Foundations, https://doi.org/10.1007/978-3-319-73302-9_6

extent to which rational thinking is useful for designers should be called in to question. This paper will argue that similar to shifting common notions of constraints, design can and should also shift popular notions of rationality. The prioritization of rational thinking has become an unnecessary constraint.

The following essay attempts to articulate how the rationalist bias in western philosophy has influenced design, beginning with the emergence of Cartesian thinking during the Enlightenment, how early design theory (consciously or unconsciously) incorporated Cartesian rationalism and dualism into its articulation of design, and why now is the time to start thinking about an arational theory of design. Arationality is neither rational nor irrational. It is a-rational, or simply "not rational," uninterested in questions of rationality because such questions are unnecessarily limiting. Arationality lies somewhere between the rigid functionalism of mass quantification and the nonsense of the Cheshire Cat. This in-between space will be our focus.

The main philosophical lens used here is phenomenology, particularly the relationship between phenomenology and design, but we will not provide a deep analysis of the phenomenological tradition here. (See Wendt 2015) The aim of phenomenology from Heidegger onward has been to highlight the limits of Cartesian rationality and introduce a philosophical practice that does not rely on strict dualisms. We will apply certain phenomenological notions to the theory of design, in attempt to argue for an arational approach to design.

2 Enlightenment

Design as we know it emerged from a rationalist epistemology. From Cartesian rationality during the Enlightenment up to the Design Methods movement in the twentieth century, our conception of design has been shaped by (overly) rational notions of knowledge. Modern concepts of design as "problem solving" stem directly from this early formulation of design and have shaped design practice for decades. Designers are encouraged to adopt a sort of predator/prey model, in which they seek out problems and eliminate them via solutions. More recently, however, we are beginning to see practices that break rational problem solving in favor of a praxis-based mode. We will examine some of these practices in more detail, but first a more detailed look at rationalism and its effects on design.

Our focus here is on Cartesian rationalism in particular, as opposed to rationalism as a broader school of thought. The rationalist philosopher strives for a pseudo-scientific truth, a set of knowledge based on reason rather than sensual information, believing that human senses are not to be trusted, and humans must take advantage of their capacity to reason to understand the world. René Descartes took these ideas even further by relying on doubt as the primary driver for getting to Truth and a sort of paranoid introspection as a means of exploring the nature of Truth, as shown by his famous mistrust of a potential evil genius: "I shall consider that the heavens, the earth, colours, figures, sound, and all other external things are nought but the illusions and dreams of which this [evil] genius has availed himself in order to lay traps for my credulity; I shall consider myself as having no hands, no eyes, no flesh, no blood, nor any senses." (Descartes 1955) Cartesian thinking is defined by a disavowal of all things bodily, as the corporeal can and should be doubted. If there is an evil genius pulling the strings, we should use our reason to outsmart him.

Taken to its logical conclusion, Descartes's prioritization of thought over sense implies that we can cleanly separate mind and body. The mind, for Descartes, is the executive seat of reason, and the body is a bundle of nerve endings that processes ambiguous and potentially deceptive information from the external world. This dualist notion implies that the primary means of action begins with mental activity and intention, followed by bodily execution of mental "strategy," and that knowledge acquisition is predicated on finding an ultimate Truth through a sort of paranoid/narcissistic process of elimination. All action becomes a product of omnipotent reason, instilling an illusion of control and insistence on a hard categorical line between thought and action.

In short, Cartesian rationalism holds that we think and then act. This simple formulation of thought and action shaped much of the western approach to scientific inquiry up until the present. The past century, however, has seen a significant backlash against rationalist thinking, from phenomenology questioning the primacy of the mind, to critical theory pointing out flaws in hyper-technical capitalist systems, and most recently, design practices beginning to think beyond the bounds of problems and solutions.

Martin Heidegger famously rejected Cartesian subjectivity as a purely mental activity, instead positioning phenomenology as the practical study of Dasein, the contextualized being characteristic of human life. Dasein avoids the harsh dualism of mind and body, the transcendentalism of *res cogitans*, and the reduction of existence to the capacity to formulate thoughts. Dasein is defined by the objects with which it interacts, and knowledge acquisition is associated with engaged interaction with those objects rather than pure reason.

Critical theorists took up a critique of Enlightenment rationalism and its focus on hyper-objective scientific thinking, from the perspective of political inquiry. Critical theorists argue that the radical Enlightenment adoption of science over mysticism has resulted in an over-emphasis on scientific objectivity over all other forms of inquiry, a domination of nature in attempt to institute reason throughout all existence, and the creation of a system in which modern capitalism can thrive. They took issue with most Enlightenment-Capitalist topics, especially the imposition of rationalist work processes, the domination of nature, and commercialization. We will return to critical theory in the section on critical design.

Finally, contemporary design theory has begun to articulate the Enlightenment's effects on design thinking. Most relevant for us is how Enlightenment rationalism foreclosed on craft knowledge and cultural wisdom that shaped social practice, as rational thinking tends to dismiss non-reason-based forms of inquiry. For design, however, cultural knowledge is highly important as a driver for inspiration and models of practice and preserving it is necessary.

3 Design's Emergence Out of Rationalism

What we commonly think of as design emerged largely as a product of the Industrial Revolution, when the rationalization and standardization of production introduced greater complexity into prior modes of craft, and design came to be as a means of planning before acting. The core components of pre-industrial craft – cultural knowledge, local material, and unified maker and user – were shuttered by design's violent separation of planning and making. Rational-industrial modes of production introduced planning as a project task, and design emerged as a field that planned, sketched, modeled, and attempted to predict future effects of products. Klaus Eder explains this phenomenon in terms of architecture and engineering:

> The craftsman's work is in turn the starting point for an additional natural division. The knowledge acquired in craft work is systemized logically; the observable effects of this work become the object of explanation attempts. A new type of interaction with nature arises, supported first by architects and finally by engineers. They interact with nature by calculating it. The architect plans on the drawing board with the aid of assumptions on statics and material properties. The engineer recombines nature and calculates the energetic effects that result from it for people. By expanding this activity, society enters a new state of nature. (Eder and Ritter 1996)

Craft making shifted from a mostly individual act of creation based on cultural knowledge to a standardized, logical process of production. The craftsperson handled natural materials by working *with* them, interacting with the materials by allowing the materials themselves guide the making process, while the architect and engineer attempt to manipulate nature for the benefit of humans. This new relationship set the context for modern design: an act of domination and manipulation to benefit human and (or?) commercial endeavors.

Herbert Simon's work represents one of the first serious inquiries into the nature of design as the production of artifice. Simon was largely successful in framing design as an inherent process in all industries, not just "capital D Design," thus articulating a design theory that moved away from pure aesthetics. However, his work also attempted – less successfully – to establish a science of design. In his famous *Sciences of the Artificial*, Simon sought to define design as a scientific activity, which can be broken down into discrete parts and formulated as a standardized procedure. His well known definition of design as the movement from "existing situations into preferred ones" (Simon 1996) is an easy way to understand design in an everyday sense – it evokes a sense of improvement, a movement toward future states that improve upon our current state. It also, however, positions design as a logical movement from one point to another; it refuses to come to terms with the complexity of design, the limits of designer intention, and non-linear designerly activities. His conception of design is a primary example of how rationalist thinking can often serve as the default framework for theoretical work.

Another key aspect of Simon's work on design is his reliance on a hard dualism between "inner" and "outer," or the sense that the "self" has some quality of being removed from the rest of the world. This of course stems directly from Cartesian thinking, in which the mind separates itself off as the executive, reason-based function, and the body takes the position of a secondary, practical component of mental activity. Supporting his insistence on binary opposition, Simon's notion of the artifact sets up a way of thinking about artifice and the artificial world in a neutral way, as opposed to sticking to the purely negative connotations of artificiality. At the same time, his overly rationalistic conceptualization of an artifact as an "interface [...] between an "inner" environment, the substance and organization of the artifact itself, and an "outer" environment, the surroundings in which it operates" (Simon 1996) is entirely too simplistic to account for the complexity of object relations. He states further that artifacts "serve their intended purpose" and are thus successful "if the inner environment is appropriate to the outer environment, or vice versa". (Simon 1996) Again, this rationalist, dualist view is accurate but not sufficient. What is the difference between "inner" and "outer" in this context, and why is the artifact the point of interface between the two?

These dualist and rationalist influences gained major significance in design communities with the Design Methods movement, which aimed to introduce scientific thinking into design work. Theorists such as John Chris Jones, Horst Rittel, and Christopher Alexander devoted their time to working out methods designers can and should follow in order to achieve certain goals and solve problems. The technical rationalism of Enlightenment thinking had finally made its way to design via the Methods movement, causing palpable anxiety among designers for their lack of scientific rigor. The Methods movement was an attempt to bring this sense of rigor and objectivity to design, mostly for noble reasons: if design is actively shaping the environment, and the types of activities designers perform have definite political and ethical implications, then we should be as rigorous as possible when articulating the nature of design work. And while many authors of the Methods movement explicitly avoided step-by-step processes, or insinuations that following a method will always result in a certain outcome, standardizing design as a set of phases will always be read sequentially and technically as a means to an end. The issue is that design includes technical activities, which have discrete goals and usually a set of steps to achieve them, as well as praxical activities, which do not necessarily have an end point. We will look closer at this difference in the following sections.

The common thread that runs through this emergence of rationality in design is the assumption that science is the preferred model by which we measure design. The following sections will question this assumption via a discussion on contemporary methods that resist being measured against science.

4 Arational Turn

The rationalist methods introduced to design discourse in the post-war period served commercial purposes: if we can position design as a scientific activity, which potential clients already understand and trust, then it will be easier to sell design work. There was (and still is) a strong desire to standardize design work into a nice, clean framework that "non-designers" can easily understand and ultimately pay for. This impulse to make design understandable to non-designers has resulted in a few outcomes: it has decreased the esoteric aura that tends to surround design, ultimately making it easier for clients to understand and pay for it; but this effect has also begun to revert back on itself, causing designers to internalize the overly-simplified frameworks meant for clients and the public. It did not take long, however, for designers to realize that the complexity involved in most design projects does not lend itself to clean, linear frameworks. Some projects go so far as to rupture the entire rationalist mode and call for a more flexible, adaptive, and one might even say spontaneous model.

The concept of wicked problems emerged as a way to explain this rupture. Rationalist models of design often rely on the notion of problem solving – specifically, that designers use reason to fully understand a problem, and only when that problem is understood do they apply creativity to envision solutions. This model relies on many assumptions: 1) problems lend themselves to complete understanding; 2) designers have access to information needed to understand a problem and possess the skills to interpret that problem in its entirety; 3) the problem space is confined enough to set non-porous boundaries around it; 4) once the problem is understood and the design moves to solutioning, the original problem space remains static; 5) we have valid means of assessing whether solutions actually fit the problem space. Seeing the many difficulties with relying on these types of assumptions, Horst Rittel and Melvin Webber articulated their theory of wicked problems to help deal with the inherent complexity and fluidity of design projects. In a certain sense, wicked problems mark what some theorists think of as the failure of Design Methods; thinkers who were involved in the Methods movement (Rittel and Jones, especially) began to rethink their earlier formulations of design in the light of this wickedness.

Wicked problems have a number of characteristics according to Rittel (1972), including some that are relevant for our current work: 1) There is no expertise. Wicked problems are complex, and complex fields do not have experts. They call for unique, emergent approaches (Snowden 2000); 2) They have no definitive formulation. It is impossible to understand the problem and then solve it. There are simply too many inputs and outputs, all of which are constantly changing; 3) Problems and solutions are interconnected. Each solution is a reframe of the problem; 4) There is no end point. Each solution creates new problems, which must be dealt with; 5) Conceptions of truth or falsity are not relevant. A solution is simply better or worse; 6) Each problem has multiple possible causes, and it is often impossible to trace a problem back to its single root cause; 7) Every problem is a symptom of another problem. The retroactive line of problems is infinite; 8) Every problem is unique and context-dependent. "Best practices" are irrelevant, as each problem exists with in a unique context.

We get the sense from these eight criteria that wicked problems are massively complex; they exist in systems with multiple inputs and outputs, they resist understanding, and even when "solved," they replicate themselves as new problems. This is not to say wicked problems are beyond understanding and are thus not worth our time and effort. Quite the contrary: wicked problems have a huge potential to affect our future as a species, and breaking our obsession with rational thought is necessary to deal with them. We need to reframe how we think about the nature of problems and solutions from an arational and non-positivist perspective. Design must move beyond a simple problem solving perspective. I have argued elsewhere (Wendt 2015) that there exists a paradox in design, which I have called the problem-solution paradox (for lack of a better name), which follows Rittel and Webber's anxiety over the hyper-rationalist conception of design planning (Rittel and Webber 1973) and Dorst's (2006) continuation of design paradoxes. In short, the problem-solution paradox states that we cannot design solutions until we understand the nature of the problem, but it is also true that we cannot understand a problem until we explore solution possibilities. Given this paradoxical relationship between problems and solutions, it is necessary to rethink the categories themselves – not to simply dismiss them as anachronistic Enlightenment relics, but rather to understand them beyond simply stating that one comes before the other.

Rittel (1972) goes on to explain two more components of wickedness that are relevant here. First, any attempt to work with wicked problems involves a rhetorical method. If expertise is irrelevant – that is, the system of a wicked problem is complex and involves emergent, novel solutions – then the potential for understanding cannot reside in the mind of a single person. There is no omnipotent and omniscient designer-god who can claim specialized knowledge. Design in this context is rhetorical and argumentative; it is a truly collaborative effort, in which any potential solution is distributed among many contributors. Second, every act of design involves a sense of what ought to be, which is of course quite different from what *is*. Processes based on scientific rationalism often guide toward a specific end goal, as when one applies the scientific method to obtain a sense of clarity around the current state of things. When concerned with what ought to be, however, the designer's focus is not necessarily on the technical aspects of how to reach the end of the process, but rather the end goal of what ought to be often unfolds in the process itself.

The work around wicked problems represents a larger arational turn within design theory. Following the Methods movement, it quickly became evident that not all design work calls for a reason-based approach to thinking about what ought to be. This is not to say design does not contain components of problem solving in the traditional, positivist sense, but simply that the hyper-rationalist assertion that design *is* problem solving is myopic and insufficient. While not all design problems are inherently wicked, I would argue that much commercial design attempts to oversimplify problems to fit existing processes. Contemporary design methods, however, are attempting to account for complexity by balancing the desire for process and the variability of design work.

4.1 Design Thinking

One such contemporary method falls under the umbrella of Design Thinking, a popular and somewhat contentious method that has emerged over the last few decades as an attempt to analyze the cognitive activities in design. It is disconcerting how many practicing designers seem to believe that the history of design thinking began in the 1990's with large firms like IDEO adopting design thinking as a process-driven way of engaging with client work. Of course, while IDEO was highly influential in popularizing design thinking, its roots go much deeper than IDEO's brand of neo-methods. Indeed, we can locate many of the principles of design thinking in Simon's *Sciences of the Artificial* and perhaps even earlier in the early writings on industrial design, Rittel and Webber's theory of wicked problems and Buchanan's (1992) continuation, Donald Schön's (1983) *The Reflective Practitioner*, and Nigel Cross's (2006a) *Designerly Ways of Knowing*. What IDEO and subsequently the Stanford Design School did so well was to take the previous academic work and reframe it for a business audience and introduce human centricity.

We have seen already how the notion of wicked problems introduces major issues for the problem solving vision of design, despite Rittel and Webber's attempt to reconcile the two. Schön's work on reflective practice and critique of sequential thinking also played a part in the non-rational articulation of design thinking. Schön explored how professional activity often does not follow a predefined plan but rather the course of action emerges out of engaged activity with the environment. The chef's work, for example, is not defined by following recipes; the interesting and productive aspects of cooking are when the chef diverges from the plan, or has no plan at all, adding "a little of this and a little of that" tasting it, and then adjusting. For design, we might conclude that the argument for design as planning is insufficient, as the most interesting, and perhaps the most important, aspects of design work occur when decisions are made in the moment. Science and design diverge at this crucial point: while science attempts to factor out variability through rigorous experimentation, design embraces the unknown by leaving room for the emergent properties of the creative process.

These approaches focus on the process of design, whereas work from Nigel Cross and others has focused on the cognitive components of design, or the "thinking" part of design thinking. Cross showed how different cognitive modes of thinking play in to the cognitive activity of designers, including inductive and abductive reasoning. Induction deals with claims to truth based on experience. The scientific method is an obvious example; it attempts to build empirical evidence to make claims about how the world works, thus articulating the current state of things. Design research methods certainly have a strong inductive component, as they aim to uncover explanations of how people interact with their environment. What we commonly think of as the "creative" components of design – what allows us to move from an understanding of the current state of things to a preferable future state – includes aspects of abductive thinking, the style associated with futuring. Abductive thinking is a unique design skill, one that sets it apart from the sciences as an active shaping of future worlds, and a style that differentiates design from rational technique. Cross holds that "designing is not a search for the optimum solution to the given problem, but that it is an exploratory process" and that "in the process of designing, the problem and the solution develop together." (Cross 2006b) This convergence of problem and solution in the 'exploratory process' of design indicates a certain non-rationality. Design is not always a technical process that positions itself toward an end goal but can also be a praxical movement in which the goal reveals itself through action.

Kees Dorst (2004) goes so far as to say designers evolve out of rationalistic thinking through experience. He argues that what he calls novice, advanced beginner, competent, and proficient designers rely on rules and reason to solve problems. Novices need structure in their process, following "best practices" to complete tasks. Beginner and competent designers work to pick out situational aspects on the design environment to choose which rules are most relevant and then reasons through them. Proficient designers can immediately see these situational elements. Even expert designers, for Dorst, rely on planning, although it tends to be an intuitive rather than conscious planning. Then there is a shift in his hierarchy of design expertise in which reliance on rules and reason dissolve. Master and visionary designers no longer need rules and reason, instead relying on intuition to fuel innovative ways of combining elements of a design problem, paying attention to how strange combinations create different perspectives. While we might quibble with Dorst's choice of words to describe these levels ("visionary" is a bit over the top), it is significant how rationality and reason never go away, they simply become less important for the designer. In the evolution of a designer, s/he becomes less and less reliant on the guardrails of rationality and begins to experiment with novel, emerging ways of understanding and shaping worlds.

Moving from the more academic origins of design thinking, we can start to see how this early work shaped the mindsets and processes involved in modern design. Contemporary design thinking extracted key mindsets from academic design thinking work in attempt to form a theoretical foundation on which to build process. Some of these mindsets might include:

Thinking = Making The traditional binary opposition between strategy and tactics, thinking and doing, etc. are no longer valid within design thinking. Thought activities are associated with creation, and vice versa. The act of sketching or building prototypes is not a purely "making" process; the creative action serves to assist the designer in better understanding – specifically, understanding through experiencing. This is perhaps the most important example of the rupture of Cartesian mind-body dualism in design practice; the connection between thinking and making demonstrates how mind and body work together, not only supporting one another but necessitating the other's involvement.

Context and Experiential Understanding Design thinking emphasizes the need to experience in order to understand. Designers are never able to separate themselves off from the human needs, behaviors, and thoughts they seek to understand, as a scientist might do, leading to an inherent acceptance of bias in design research. What is lost in scientific rigor is ideally made up in depth of insight. This play between breadth and depth is characteristic of design thinking.

Divergence and Convergence One of the greatest skills in design is to know when to diverge and when to converge. Divergence can allow for breadth of thought and possibility exploration without the hindrance of artificial constraints. But it can also result in lack of focus and getting lost in seemingly infinite options if not used properly. Convergence helps design teams make decisions and focus their energy, assisting designers in making choices and preventing them from spending too much time on inconsequential tasks. But it can also hinder creative thought when introduced too early.

These mindsets are then translated into a process, which on the surface looks sequential, phased, and limiting. These phases go by many names, but usually include:

Discovery Design problems are sought out. Researchers will explore an area of interest and gather as much information as they can within time and budget constraints, diverging in focus to take in as many observations as possible.

Synthesis Once information is gathered, designers work to make sense of it and converge on needs, problem frames, or insights.

Idea Generation Based on the products of Synthesis, designers diverge again to generate many solution options.

Testing Finally, designers converge again to test, refine, and potentially abandon ideas.

These phases, if properly planned (i.e., flexibly and adaptively), tend to overlap and allow for backward movement. In other words, unlike most phased approaches that require forward movement from step 1 to 2 to 3, etc., this generalized process of design thinking tends to account for the inherent connection between "problems" and "solutions," and that no problem or solution space emerge separately from one another. The arationality we can read into design thinking begins with its espoused mindsets and flows through its process. Of course, this formulation is quite idealistic, and not many design teams work strictly within this process. But this is not necessarily a detriment, as the arational qualities would begin to dissolve if a process is followed too dogmatically.

4.2 Lean

To a certain extent, Lean was popularized in design communities on the heels of design thinking. Originating as a manufacturing process for Toyota, Lean emphasizes waste reduction to maximize output and worker time. From a design perspective, it attempts to eradicate the "theoretical" or "speculative" aspects of design, opting instead to focus on the so-called "making" or "doing" activities. In practice, this often takes form in rapid cycles of building and testing prototypes that help gather evidence for or against pre-determined hypotheses. Many design thinking mindsets mentioned above are also embodied in Lean design, particularly the role of context and the relationship between thinking and making.

Most contemporary versions of Lean, for better or worse, are based on the Lean Startup method, a process popularized around 2011 (Ries 2011) as a way for startup companies to embody lean principles of waste reduction and continuous customer discovery. In its ideal form, Lean design minimizes time spent on activities that do not contribute to greater customer understanding, especially in the more ambiguous phases of design research. It tends to see upfront research as speculative and thus wasteful. On the other hand, in its most surface level form, Lean design tends to fetishize action over thought, even when thoughtfulness is sacrificed for a false sense of productivity.

One might argue that Lean tends to be a larger product and service development process rather than focusing strictly on design, but it can be difficult to determine where design stops and product/service development begins, especially when working in a cyclical process. The Lean Startup process is commonly composed of three modes – Build, Measure, Learn – which are best thought of as a circle rather than a line. It is uncertain where the process begins and ends, lending to the complexity of planning Lean projects but also the advantages of the practice. The Build mode involves the creation of a prototype, a materialization of a key question emerged in the act of design. Lean commonly opts out of large research phases at the beginning of a project, instead positing that any design is contingent on assumptions, so it is best to identify those assumptions and build prototypes to gather evidence for or against them. Once prototypes are made, the Measure mode involves testing the prototypes with potential users of the product or service, soliciting feedback, and most importantly, experiencing real world use scenarios (or as close to them as possible). Finally, the Learn mode involves taking the evidence gathered in Measure and working it into another Build cycle.

We can see how this process explicitly avoids the linearity of most rationalist design methods. There is no predetermined end point toward which designers are working. Instead, the end point reveals itself almost intuitively through the process of customer discovery and prototype creation, allowing designers to break free from the technical impulses that commercial design tends to impose. At least, this is the ideal. In reality, Lean Startup (as distinct from Lean in general) lends itself to amateurish adoption by untrained designers fueled with investment money and dreams of neoliberal fame. Lean Startup is so attractive because it feeds an illusion of shortcutting to the end – i.e., the inversion of what Lean does so well in avoiding rationalist impulses. Profit motive trumps design rigor. The "talk to your customers" and "design things people want" mantras disguise the complexity of uncovering unmet, often tacit needs, resulting in a surface level understanding of how to perform design research. As opposed to applying methods and principles from design research to discover unarticulated needs, it is much easier to simply ask people what they want and then go build it, actively ignoring or simply being ignorant of the idea that most needs, wants, and desires are unconscious and therefore not easily discovered in a 15 minute interview.

Despite these potential shortcomings, Lean Startup takes much of its arational inspiration from design thinking insofar as it resists an exclusive focus on strategically moving toward a predetermined end goal.

4.3 Frame Innovation

A third method worth noting is the Frame Creation Model developed by Kees Dorst (2015) in his book *Frame Innovation*. The Frame Creation Model adds a bit more rigor to traditional/commercial design thinking. This is not to say design thinking is unrigorous in and of itself, but that its incorporation into commercial design practice has the tendency to be somewhat "dumbed down" to fit within budgets and timelines. Dorst calls for a greater sense of depth in the design process, a need to take time to actually think through complexity, to make bold leaps in attempt to surface real needs, and consider cultural/social context. He also makes it clear that that rationalist approach to design is insufficient, and that design must also promote a sense of irreverence: "The "self-made box"of received wisdom and conventional practices is often considered the very core of the culture of our societies, and eagerly reinforced by popular media. The 'rational high ground' that is often implied in this claim to authority sparks another archetype: the clever outsider who runs circles around accepted behavior." (Dorst, 2015) These 'clever outsiders' are commonly known as tricksters, an archetype historically used as a rhetorical device for authors to convey a popular, yet socially unacceptable opinion. We will have more to say on this archetype later, especially on how design can benefit from the trickster's arational approach to existence.

Dorst develops his Frame Creation Model from the perspective of breaking the self-imposed box of rational limits. His process is laid out in nine steps:

1) Archaeology

Designers begin by examining the nature of what Dorst calls the "apparent" problem, acknowledging that many design projects begin from a place of ignorance. A design problem does not exist simply because someone says it does, and even if it does, it can easily take form as something other than originally believed. The archaeological phase is one in which designers attempt to reveal contextual elements of the design problem that will be useful later – elements that would otherwise surface at inopportune times if left buried. This is a historical practice, in which designers look at the history of a problem space, including why previous attempts at solutions have failed.

2) Paradox

Dorst then advocates for identifying what he calls the "clash of rationality." This is the point at which designers begin to realize the complexity of the design problem, and within its complexity, there are likely paradoxical relationships among the design problem's elements. Paradoxes might include conflicting stakeholder views, requirements that cannot exist simultaneously with one another, or even the general paradox of rationalistic conceptions of design problems and solutions. (see Wendt 2015) An abundance of paradoxes usually means the original design problem might need to be reframed and examined from a different perspective. For Dorst, the best way to proceed through paradoxical situations is to learn as much as possible about the context in which they exist.

3) Context

Learning about context involves design research methods such as stakeholder interviews, customer interviewing and observation, and a variety of ethnographic methods that aim to gather a wide variety of contextual information. The idea here is to resist a narrow focus on the problem space and to understand it from a systems perspective. No design problem exists in isolation from social, cultural, economic, ethical, etc. influences. In this phase, designers focus on the influence, or potential influence, of current stakeholders.

4) Field

Field and Context are inherently interwoven insofar as observations about the context of a design problem will likely also uncover observations about the cultural and social field of the problem. Combined together, Context and Field provide the necessary human-focused background information needed to form key insights through deep sensemaking.

5) Themes

Finding themes is an exercise in pattern recognition and hermeneutics. The development of insights requires both deep knowledge of context and field, and a willingness to take risks associated with making inferences. Stakeholders work together to develop these themes.

6) Frames

Themes allow designers to form Frames, or perspectives that can launch the more "creative" aspects of design. Frames are described as "as if" statements that provoke designers to consider multiple perspectives and look at problems through many lenses to consider possibilities for future states. Designers always have a point of view, which Framing helps to articulate, as opposed to designing from nowhere. It also helps align design teams on a shared perspective, instead of leaving it unarticulated and working against one another.

7) Futures

In the creation of Futures, designers finally transition to envisioning possible futures based on the Frames created in the previous step. They use generative abductive reasoning to expand the areas of possibility as wide as possible before narrowing down to the preferable. This phase most closely aligns to colloquial notions of "creativity," in which designers push past the present moment into possible ways of being.

8) Transformation

Critique and prototyping come in to play in the Transformation step, in which designers interrogate their decisions and make choices about which Future to pursue. Much critique happens in this stage. Designers will often create prototypes—whether of tangible products or intangible services and experiences—that manifest intended Futures, and then use these prototypes to critique whether they address the Themes and whether they present preferable Futures.

9) Integration

Finally, Integration accounts for fitting Futures into the larger organizational context of the design problem.

The Frame Creation Model is significant for the current study in its expansion of design thinking into more rigorous territory that can potentially break the bounds of rationality even further. Even from a surface level examination, the Frame Creation Model resists the teleological (profit) motives of commercial design to the last responsible moment. It forces designers to become comfortable with the inherent discomfort of not knowing "the answer." It acknowledges paradoxes instead of ignoring them, promoting an active engagement with the ambiguity of design work, which often runs counter to more scientifically grounded design methods. Finally, it refuses to decontextualize design decisions from their complex environments, choosing instead to take the necessary time and effort to understand both internal and external contexts in effort to integrate design solutions into them ethically and responsibly.

4.4 Critical Design

Perhaps the most radical break with rationality in design is an approach somewhat redundantly called critical design, or a design philosophy and corresponding practice expressly aimed at evoking critical discourse around a topic, as opposed to a "solution" introduced to address a discrete problem. If we take it seriously, critical design is inherently impractical: it produces objects and experiences that serve no concrete purpose other than influencing opinionated debate. In its impracticality, however, it tends to reproduce the conditions it wishes to overthrow. Insofar as qualifying design with adjectives like "critical" reinforces the relegation of design to style (Tonkinwise 2015) and aesthetics, critical design easily becomes obsessed with the form of its outputs rather than the critical debate it sparks.

Nonetheless, critical design is highly representative of the arational movement within design in that it attempts to avoid design practice being designated as a technical skill for producing commercial ends. Anthony Dunne and Fiona Raby, originators of critical design as an articulated practice, hold that the core objective of critical design is to resist the status quo, or in other words, to practice design in a non-commercial sense by creating prototypical objects that challenge common, uncritical ways of viewing the world. It attempts to explore possible futures in a way that highlights the dystopian aspects of uncritical thinking and hopefully to encourage the creation of more preferable scenarios. It is an exploration in utopias via their extreme opposites and rooted in arationality: "Driven by poetry, imagination, and intuition rather than reason and logic, [critical design projects] have their own sense, an alternative to our everyday scientific-industrial one. These are tales about the space between rationality and reality, which in an industrial society have come to be synonymous." (Dunne 2008) In their ideal form, critical designs embrace this difference between rationality and reality by playing with the irrationality that industrial production attempts to suppress.

There are several aspects of critical design that are important for the current study, but we will look at two for the sake of being concise. First, critical design projects depend on defamiliarization and the uncanny. Making the familiar unfamiliar is a key rhetorical technique that displaces any sense of historical comfort with familiar objects. The real power of defamiliarization comes not from how an object can be presented as the diametric opposite from its normal interpretation but rather how from how close the two poles actually are. Sigmund Freud referred to this phenomenon as the uncanny, (Freud 2003) or the resulting feeling of encountering the "familiar unfamiliar," the thing that is so defamiliarized specifically because it is so "close to home." The word "uncanny" is a translation of the original German *Unheimlich,* or the sense of eerie familiarity of "homeliness." Take for example a project entitled *Scary Beautiful* (Fig. 1), which depicts an exaggerated woman's high heel shoe, with the heel and toe in their opposite position. The resulting image is of the model's body contorted, knees bent, struggling to keep her balance, looking both strangely elegant and in pain, as (supposedly) opposed to normally designed high heels, which frame a woman's body into a pleasing shape, lengthening the legs, pushing out breasts, and arching the back. The image is so disturbing not because it creates such an unnatural shape in the human body but because, upon encountering the image and interpreting it, the viewer realizes that the familiar image of a woman in stiletto heels is not radically different than the image intended to be disturbing. The lasting effects of the image now begin to seep into everyday life as the viewer sees high heels in mundane settings. This is the uncanny: the recognition that the everyday is just as disturbing as what is meant to be outrageous. It is important to point out, however, that the designer must walk a line between familiarity and strangeness. If critically designed objects are too familiar, they become status quo; if too unfamiliar, they become art (Kjærsgaard and Boer 2014).

Fig. 1. *Scary Beautiful.* Artist: Leanie van der Vyver. (Source: dezeen.com)

Second, critical design relies on the grotesque. These projects tend to be comically absurd and irregularly formed; their physical imagery is exaggerated to show just how far ideas and interpretations can be pushed before breaking. "Grotesque" originates from French and Italian words associated with "emerging from the cave," (an apparent reference to the excavation of cave paintings) indicating that the grotesque tends to reveal that which has been hidden from everyday life. We can see this phenomenon in action with projects such as *Tender* (Fig. 2), which aims to critique the mobile dating app Tinder by imitating the main interaction of swiping left or right. Tinder users are presented with profile pictures of other users and must swipe to the left or right to indicate whether they are interested in learning more about that person. *Tender* pokes grotesque fun at the interaction by fixing a piece of meat to a rotating fork, slapping the meat against a mobile device with Tinder loaded, swiping the screen with each rotation. The video (Tender 2014) allows viewers to see the full extent of the project, particularly the characteristically grotesque sight of raw meat slapping a surface and leaving residue, perhaps the symbolic residue of conventional dating practices. The grotesque elements of this project highlight the anxiety around the convenience of modern dating and the transition of the proverbial "meat market" mentality into digital spaces.

Insofar as critical design practices exist, they refuse to follow rational progression of problem to solution, opportunity to execution, or strategic vision to tactical execution. There is a sense of play in these projects, a vision of possible (mostly dystopian) futures that can only be expressed through playful exploration rather than strategic visioning. This haphazard way of designing, however, is the basis for

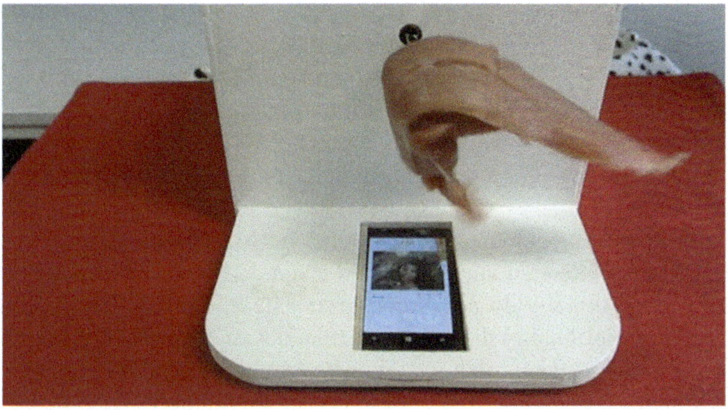

Fig. 2. *Tender*. Artists: Cors Brinkman, Jeroen van Oorschot, Marcello Maureira, and Matei Szabo. (Source: boingboing.net)

legitimate critique against critical design – that it fails, or actively ignores, current problems in the world that affect real people, instead focusing on hypothetical ways of being that have little to do with current material existence. The irony here is critical design's still largely unexplored relationship with critical theory. Attempting to take Marx seriously on his point that philosophers have merely interpreted the world but have largely failed to do anything about it, (Marx 1845) critical theorists highlighted the paradoxes, contradictions, and ethical imbalances of modern capitalism in hopes that other enlightened citizens would act on their insight. Critical theory took consumerism, ideology, and alienation as their targets of criticism, whereas critical design seems to take specific instantiations of these forces as its target. Especially pertinent here is both critical theory and critical design's discomfort with hyper-rationality. For critical theorists, capitalism's rationalist conception of work lead to profound worker alienation; for critical designers, rationalist design processes lead to designer alienation. Further exploration of this relationship is needed to address the tension between speculative futures and material realities.

5 Toward Arationality

The discussion of design methods is as much about *who the designer is* as it is about *what the designer does*. What rationalist design methods forget is that activities and practices are intimately connected to identity. While design is something that everyone does, the identity of "designer" indicates an intense focus on designing as a practice that defines oneself. Rationalizing design tends to decontextualize practice, standardizing methods into identity-free activities, which then results in designers becoming alienated from the work that defines them. In other words, rationalization positions design as a technical output and thus limits the potential benefits of design labor to the end result; instead of the work acting as a potential source of value, technicalization works to ensure that designers can only measure the quality of their work based on the final output, and too often, whether it can be bought and sold. An arational design approach holds that technical aspects of design are important, but we should not forget about the phronetic and praxical components of design work.

The act of designing can teach the designer how to live well through adaptation and coping with unintended and unplanned events. The problem is that living well and acting unintentionally do not directly translate into profit. Commercial design inserts a hyper-intentionality that mandates that designers predict the unpredictable. The so-called data-driven methods, rising popularity of A/B testing, analytics benchmarking, etc. are celebrated for their ability to create a (false) sense of security in inherently complex environments, and the illusion of certainty pervades the methods by which we design. Rationalist design cannot account for the pleasure of unintentional action or the idea that the success of design hinges on adaptive skill that can only be acquired when one is forced to adapt to the unplanned.

In *Strategy Without Design*, Robert Chia and Robin Holt argue that the success of a given activity is not dependent on a singular plan of action, an individual intention, or standardized activity. Practical success comes from the "phronetic capacity to continuously make timely and ongoing adjustments and adaptations to local circumstances." (Chia and Holt 2009) This perspective has lead to some of the more arational design methods outlined in the previous section. It is my hypothesis that the "more arational" the design method – that is, the closer it is to the middle ground between rationality and irrationality, the space of flexible structure that is neither bound by scientific rigidity nor subject to random whims of an individual – the more satisfied the designer is with his or her work, and the stronger connection s/he feels with the identity of "designer." Humans are egotistical animals. The connection between self and activity must be highlighted in order to promote successful and fulfilling work. If one cannot project oneself into work, it becomes simple pantomime and fails to contribute to a greater purpose.

The imposition of rational strategy has stripped away the individual designer's personal connection to work. While we should hesitate to simply reverse this affect and advocate a return to craft, it is worth considering how inherently non-strategic craft activities might be incorporated back in to design work. One way to do this is beginning to incorporate some of the arational components of the methods discussed in this paper, keeping in mind that design methods are never plug-and-play. Instead, designers must take the preferable aspects of methods and shape them to their context rather than reading a book, attending a workshop, or taking a class and attempting to adopt a method wholesale. Design thinking introduces a sense of non-linearity to traditionally forward-moving commercial design. Instead of setting goals in an unforeseeable future and working sequentially toward them without adapting to change, design thinking promotes the use of designer intuition, experience, and "gut feeling" to know when to move forward, when to back up, and when to dwell in one place. Lean is attractive in its orientation toward action.

Designers in Lean settings are able to see the empirical results of their work and can course-correct based on real time feedback loops. If balanced with thoughtful planning, Lean promises to cut wasted time and effort that has become so normalized in commercial design. Frame Innovation has the benefit of honesty, especially in its direct interrogation of paradox. Many designers and design methods attempt to ignore, sublimate, and deny paradoxes, allowing them to fester in the background. Frame Innovation takes care to acknowledge and deal with paradoxes before they appear later as symptoms. Finally, critical design offers the idea of provocation and playfulness. Rational design methods leave little room for play, as their focus is on meeting strategic goals and attempting to materialize a pre-determined intent; and they leave less room for political engagement, as they attempt to neutralize any sense of individuality the designer might bring, especially if employed by a for-profit company. Critical design, however, embraces political action through play and provocation, allowing for design to consciously take the form of social commentary.

I have tried to argue throughout this paper that rationalist design methods serve commercial purposes but fail to account for the complexity of design and for the role of individual perspective. Rationality continues to impose itself as an artificial constraint. A movement toward arational design methods would need to combine the best parts of existing methods along with inventing new ones to balance business goals with the individual perspective of designers. Apart from new methods, however, there lies the more basic need to combat rational thinking as the default cognitive style for all professionals; this concern is certainly much more complex than promoting one method over another, involving a radical shift in epistemological approach. Nonetheless, if design is an arational practice, epistemological support is not necessary for change in praxis.

References

Buchanan, R. (1992). Wicked problems in design thinking. *Design Issues, 8*(2), 5–21.

Chia, R. C. H., & Holt, R. (2009). *Strategy without design: The silent efficacy of indirect action*. Cambridge University Press, New York.

Cross, N. (2006a). *Designerly Ways of Knowing*. New York: Springer.

Cross, N. (2006b). *Design thinking: Understanding how designers think and work*. Oxford: Berg.

Descartes, R. (1955). *The philosophical works of Descartes (2 Vols.)*. Oxford: Dover Publications.

Dorst, K. (2004) *Investigating the nature of design thinking*. Future Ground: DRS Conference

Dorst, K. (2006). Design problems and design paradoxes. *Design issues, 22*(3), 4–17.

Dorst, K. (2015). *Frame innovation: Create new thinking by design*. Cambridge: MIT Press.

Dunne, A. (2008). *Hertzian Tales: Electronic products, aesthetic experience, and critical design*. Cambridge: MIT Press.

Eder, K., & Ritter, M. T. (1996). *The social construction of nature: A sociology of ecological enlightenment*. Thousand Oaks: Sage Publications.

Freud, S. (2003). *The uncanny*. New York: Penguin.

Fry, T. (2009). *Design futuring*. Oxford: Berg.

Kjærsgaard, M., and L. Boer. (2014). The speculative and the mundane in practices of future–making–exploring relations between design anthropology and critical design. *Research Network for Design Anthropology Seminar.*
Marx, K. (1845). Theses on Feuerbach. https://www.marxists.org/archive/marx/works/1845/theses/theses.htm
Ries, E. (2011). *The lean Startup: How today's entrepreneurs use continuous innovation to create radically successful businesses.* New York: Random House.
Rittel, H. W. J. (1972). On the planning crisis: Systems analysis of the first and second generations. *Institute of Urban and Regional Development.*, 390–396.
Rittel, H. W. J., & Webber, M. M. (1973). Dilemmas in a general theory of planning. *Policy Sciences, 4*(2), 155–169.
Schön, D. A. (1983). *The reflective practitioner: How professionals think in action.* New York: Basic Books.
Simon, H. A. (1996). *The sciences of the artificial.* Cambridge: MIT Press.
Snowden, D. (2000). Cynefin: A sense of time and space, the Social Ecology of Knowledge Management. http://citeseerx.ist.psu.edu/viewdoc/download?doi=10.1.1.196.3058&rep=rep1&type=pdf Accessed 28 Sept 2015.
Tender - It's How People Meat. (2014). https://vimeo.com/111997940. Accessed 28 Sept. 2015.
Tonkinwise, C. (2015). Just Design. *Medium,* https://medium.com/@camerontw/just-design-b1f97cb3996f. Accessed 28 Sept 2015.
Wendt, T. (2015). *Design for Dasein: Understanding the design of experiences.* USA: Crreatespace.

A Case for Graphic Design Thinking

Katherine Gillieson and Stephan Garneau

Abstract This chapter draws on a range of historical, philosophical and contemporary design references to argue for a broad view of graphic design thinking as a distinct approach to problem-solving. The history of epistemology is linked to the history of modes of communication, and this in turn is intimately tied to the development of textual organization, typographic and diagrammatic representation. Modes of thought and the way we acquire and share knowledge are shown to exist in symbiosis with technological development. In order to help define this mode of thinking, we consider the distinctions and similarities between forms of academic research in the humanities, the sciences, and in design. We also present seven characteristics pertaining to graphic design thinking in particular, making reference to the history and theory of graphic communication as well as to standards of professional practice. The final outcome of this mode of thinking implies action; it is inherently synthetic, generative and future-conscious. Defining and making explicit these seven characteristics puts into sharp relief the potential for graphic design thinking to be a tool of practical and ethical engagement with the world.

Keywords Philosophy of design · Graphic epistemology · Design research · Design theory

1 Introduction

In this chapter we will argue for a particular definition of graphic design thinking as a problem-solving and knowledge-building process that is distinct in its influence on many areas of human activity. To do this, we start with a curated historical view of visual communication (Sect. 2), to show the development of forms and how they relate to prevailing cultural norms and conceptions of knowledge. This developmental history of graphic forms serves to underpin a discussion of graphic design

K. Gillieson (✉) · S. Garneau
Faculty of Design and Dynamic Media, Emily Carr University of Art + Design, Vancouver, BC, Canada
e-mail: kgillieson@ecuad.ca; sgarneau@ecuad.ca

© Springer International Publishing AG 2018

P. E. Vermaas, S. Vial (eds.), *Advancements in the Philosophy of Design*, Design Research Foundations, https://doi.org/10.1007/978-3-319-73302-9_7

thinking in Sect. 3, which begins with an overview of the main cultures of knowledge in academic discourse: the sciences, the humanities, and that of design, proposed as a 'third culture of knowledge' by theorists such as Nigel Cross (2006).

The unique place of visual communication within the spectrum of designed forms is central to this proposal; building on the concept of 'design thinking', broadly conceived, we elaborate a series of features that pertain specifically to graphic design (Sect. 3.2). With reference to the historical examples as well as professional standards and conventions, we propose that this series of seven characteristics sets graphic communication apart from other design practices and methods.

Graphic design thinking, in contrast to other forms of design, aligns itself as an epistemic practice through its dealings with language, information, and communication. Rather than a materially-oriented 'science of design' (after Horst Rittel in Rittel and Webber 1972), we suggest an *epistemology of design* that involves a dialogical dimension that is also one of social responsibility.

When asked in an interview to comment on the state of modern philosophy, Michel Foucault described the fractured state of the discipline in the twentieth century:

> It seems to me that philosophy no longer exists; not that it has disappeared, but it has been disseminated into a great number of diverse activities. Thus the activities of the axiomatician, the linguist, the anthropologist, the historian, the revolutionary, the man of politics can be forms of philosophical activity. (Foucault 1996: 29)

Philosophy has become *applied* and taken form through new disciplines, and we believe that design is due to be added to the list. Graphic design in particular has often been linked to linguistics, an area already well influenced by philosophy. Through a closer discussion of some of the central elements of graphic design methods for problem-solving, it becomes apparent that this field is already engaged with substantial epistemological themes. Through greater awareness of its nature, parameters, processes, and outcomes, it can be capable of an even deeper level of social, cultural and ethical engagement.

2 Historical Roots of Graphic Design Thinking

Modern design methods are largely considered to have emerged with the socio-political upheavals of the twentieth century, with mechanized modes of production and the modern nation-state. These practices have lead to what we think of as the objects of design, products such as posters, chairs, books, or more recently, apps. However, the epistemological foundation of graphic design thinking in particular reaches much farther back, with the development of human communication systems. The history of epistemology is linked to the history of modes of communication (Ong 1983, Olson 1994); this in turn is intimately tied to the development of textual organization, typographic and diagrammatic representation.

To illustrate this, we can turn to examples of graphic formats for communicating information developed far back in human history. The very term 'history' is a reflection of our inherent tendency to take notes, distinguished as it is from 'pre-history' through the presence of written records, themselves a technological advance. As an extension of memory and as a physical form for shared understanding, our history can be seen to reside in these objects of communication, which are also evidence of the graphic nature of epistemology. This becomes increasingly relevant in new media forms that engage the user more explicitly in curating and selecting content. Johanna Drucker (2004) argues that the visual forms we conceive of as empirical representations of reality are in fact rhetorically charged products of interpretation; she argues for a humanities-based 'observer-dependant model of knowledge' in graphical information, in which digital formats are acknowledged to reflect the co-dependency between the user (observer) and the resulting interface, reflecting the entanglement of thought and visualization.

An early and critical example of graphic abstraction that relates to the development of modes of thought is the Sumerian King's List, also called the Weld-Blundell Prism, which dates back to 3000 BC (Fig. 1). This list of kings' names is an administrative document of ancient Sumeria that the anthropologist Jack Goody (1977) describes as the earliest sophisticated form of graphic abstraction. The list developed out of a need to record events, people or objects, to outline future plans (the 'shopping list') and to keep track of inventories; this deceptively simple format is highly sophisticated as it embodies a spatial and generative mode of thought independent of the linearity of the spoken sentence. It is an everyday, publicly accessible visual representation that lays the groundwork for more complex organisations of knowledge through the possibilities of further operations, hierarchies, subdivisions, and other manipulations of its constituent relationships.

Another important form of graphic abstraction is that of the map, a representation of various kinds of geographic space. A prototypic example of this form is the Dunhuang star map, thought to date from the reign of Emperor Zhongzong of Tang, around 705–710 AD. It represents the North Polar section of the sky, providing a visual account to back written astronomical knowledge of the time (Bonnet-Bidaud et al. 2009). As an example of an abstraction of spatial forms (the actual distances between stars), it employs a simplification of detail for the purposes of orientation. The further the elements are abstracted, however, the greater the need for supporting information, as indicated on the map (Fig. 2).

In medieval Europe, the constraints of handwriting and the hand-reproduction of images helped shape the manuscript culture of reason; the intellectual approach of medieval Europeans fortunate enough to be a part of the economy of writing worked in a metaphorical mode of thought in which "written texts are meant to be open to interpretation, to create a meditative string of concepts and allusions in the reader." (Ong 1982: 104, 110).

In the medieval mind, written texts combine into series of signs to compose arguments. The standard practice of writing notes or glosses in the margins of books reflects the cumulative quality of oral argumentation. At its heart, the text is open for interpretation and the reading experience is itself a process of creating meaning. The

Fig. 1 The King's List (Weld-Blundell Prism), a four-sided clay tablet inscribed in cuneiform with lists of Sumerian kings, dating from around 3000 BC. (Ashmolean Museum)

medieval mode of thought, one of 'metaphor and metonymy', is reflected in this approach to communication; here signs are seen to stand for things as they are, in a bipartite relationship (Olson 1994).

Eventually a more structured form of Latin and refinements in argumentation led to a scholastic turn towards a more rational mode of thought: "The scholastics thus saw the emergence of the three-part distinction of signs, whereby signs, the ideas they represent, and underlying reality were seen as separate categories" (Olson 1994: 165). This development allowed for increased standardization in the communication of ideas and a reduction in the ambiguity of metaphorical images. Logical argumentation through many forms of communication became normative.

Over this period, graphic formats evolved alongside developments in technologies of reproduction, and the advent of moveable type in Europe encouraged considerable cognitive leaps through visualizations that were facilitated and shared

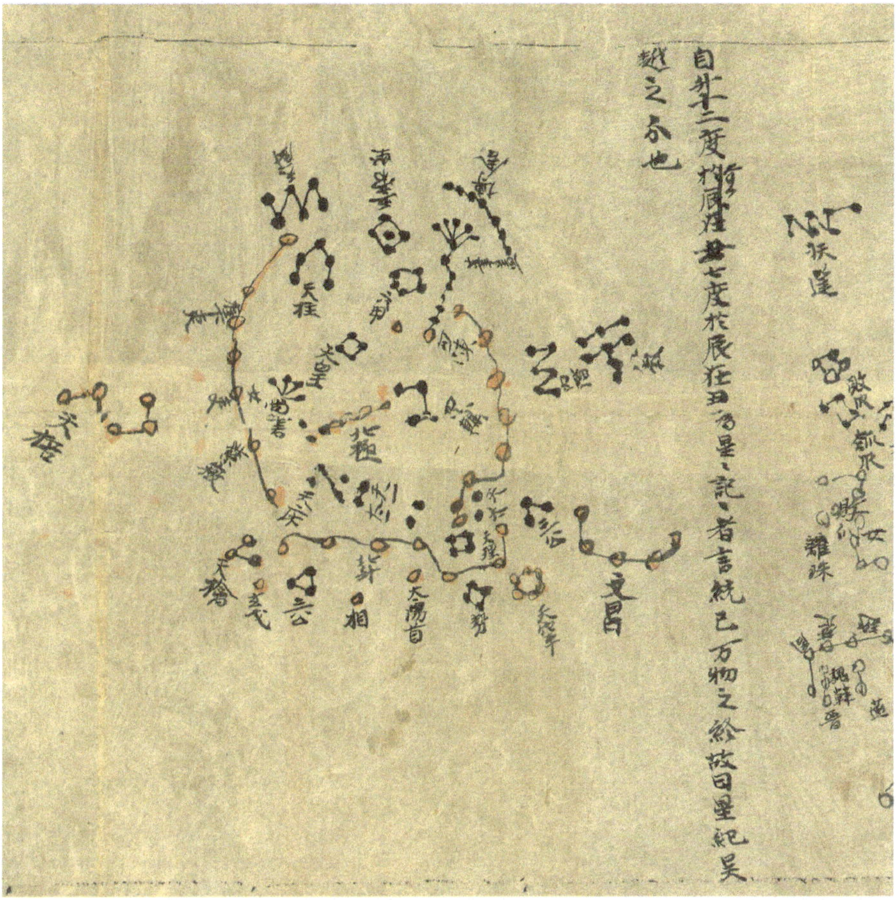

Fig. 2 A detail of the Dunhaung Star Map, one of the earliest known graphical representations of the night sky, dating from the Tang Dynasty in China (approximately 705–710 AD). (British Library)

through print. Early Renaissance culture was marked by a sudden profusion of printed matter; not only did the introduction of the press allow much more affordable books, it also prompted innovations in graphic forms from its very inception.

For example, the publishing house of Aldus Manutius, active in Venice in the late fifteenth century, developed several typographic formats and conventions still in use today, including italic type (Fig. 3), punctuation marks such as the semicolon and the comma, and a standard format for printing books, the octavo, which is commonly recognized today as the paperback book size (Bajetta 2000). The standardization of physical formats led to more efficient production methods, which in turn helped to transform these new forms into conventionalized genres.

The High Renaissance mode of thought led naturally to the intellectual modes characteristic of the early modern era. Increasingly concerned with logic and

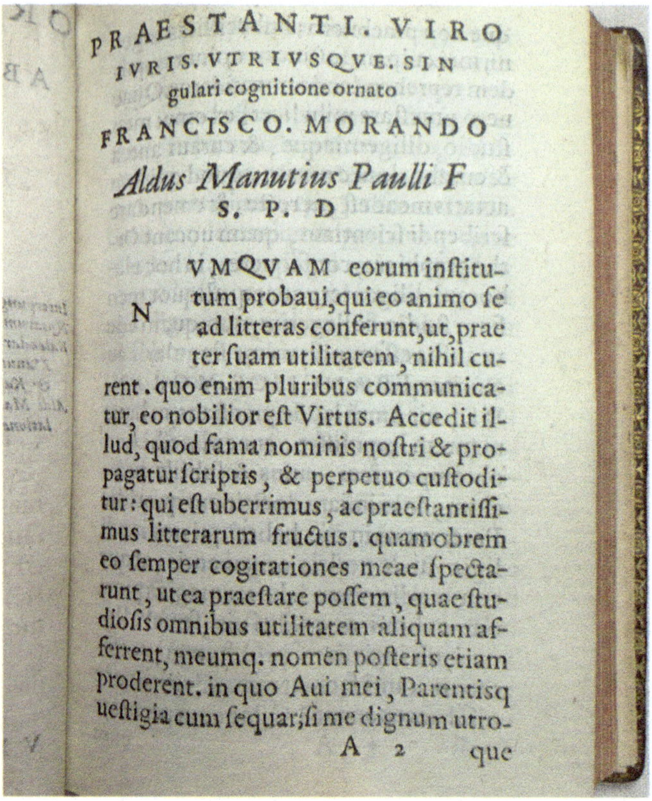
PRAESTANTI. VIRO
IVRIS. VTRIVSQVE. SIN
gulari cognitione ornato
FRANCISCO. MORANDO
Aldus Manutius Paulli F
S. P. D

NVMQVAM eorum inſtitu-
tum probaui, qui eo animo ſe
ad litteras conferunt, ut, prae
ter ſuam utilitatem, nihil cu-
rent. quo enim pluribus communica-
tur, eo nobilior eſt Virtus. Accedit il-
lud, quod fama nominis noſtri & pro-
pagatur ſcriptis, & perpetuo cuſtodi-
tur: qui eſt uberrimus, ac praeſtantiſſi-
mus litterarum fructus. quamobrem
eo ſemper cogitationes meae ſpecta-
runt, ut ea praeſtare poſſem, quae ſtu-
dioſis omnibus utilitatem aliquam af-
ferrent, meumq. nomen poſteris etiam
proderent. in quo Aui mei, Parentiſq
ueſtigia cum ſequar; ſi me dignum utro-
A 2 que

Fig. 3 *Orthographiae*, a book on spelling published by Aldus Manutius in 1449, demonstrates several innovations of Manutius' press, including its practical octavo or 'paperback' format and its use of italic type. (Wosk-McDonald Aldine Collection, Special Collections and Rare Books, Simon Fraser University)

organized around the rationalist notion of the dichotomy, early modern thought progressively lost medieval notions of metaphorical allusion in texts. The rise of rationalist epistemologies saw the proliferation of visual representations of mathematical and scientific principles and schemas, such as Copernicus' *De revolutionibus orbium coelestium libri V* (1543) (Fig. 4).

The work of the Humanist logician Petrus Ramus (1515–1572) is pivotal in the history of visualization for its graphic classifications of knowledge. A central innovation of his was popularized as the Ramist method, a technique he developed to represent all human knowledge in the form of branching tree diagrams. Figure 5 shows a Ramist diagram included in Chambers' *Cyclopaedia* (1728), which is also mentioned further below.

For Ramus and others writing at this time, a visual representation is meant to be taken literally, and presumes an objective rather than a subjective interpretation. The spatial representation of thought results from a logical process in which larger

Fig. 4 A spread from *De revolutionibus orbium coelestium libri V*, part of the seminal work by the astronomer Nicolaus Copernicus (1566) shows a diagrammatic representation of his heliocentric model of the solar system. (Reading University Library)

concepts are graduated sums of individual elements. This work reflects what Walter Ong refers to as a 'hypervisual noetic world' (1982).

Ramus' innovation stands as a good example of how visualizations of knowledge in the sciences developed through the technology of print, which in turn helps to explain how book design has evolved to the present day. Since the advent of the press, books were able to contain complex visual representations in large quantities that brought about not only new ways of reading, but what historian Elizabeth Eisenstein argues was a re-presentation and re-conception of scientific knowledge. She argues that print culture "actually increased the functions performed by images while reducing those performed by words" (1979: 69) due to the repeatability of visual forms allowed by print.

In addition to the role of images, the communicative quality of diagrams and of layout itself must be considered as a vehicle of meaning. Various types of schematization, or visualization of forms of knowledge, have been essential to the representation of epistemologies through print culture. Though readers think of themselves as reading words, they are very much 'reading design' (Gillieson 2008). We will expand on the notion of schematization further below.

As printing techniques improved, visualizations of intellectual space progressed from more rationalistic modes (exemplified by diagrams of mathematics, logic structures such as Ramus', etc.) to more empirical ones. At this point we can refer to sophisticated visualizations based on sense perception that maintain information

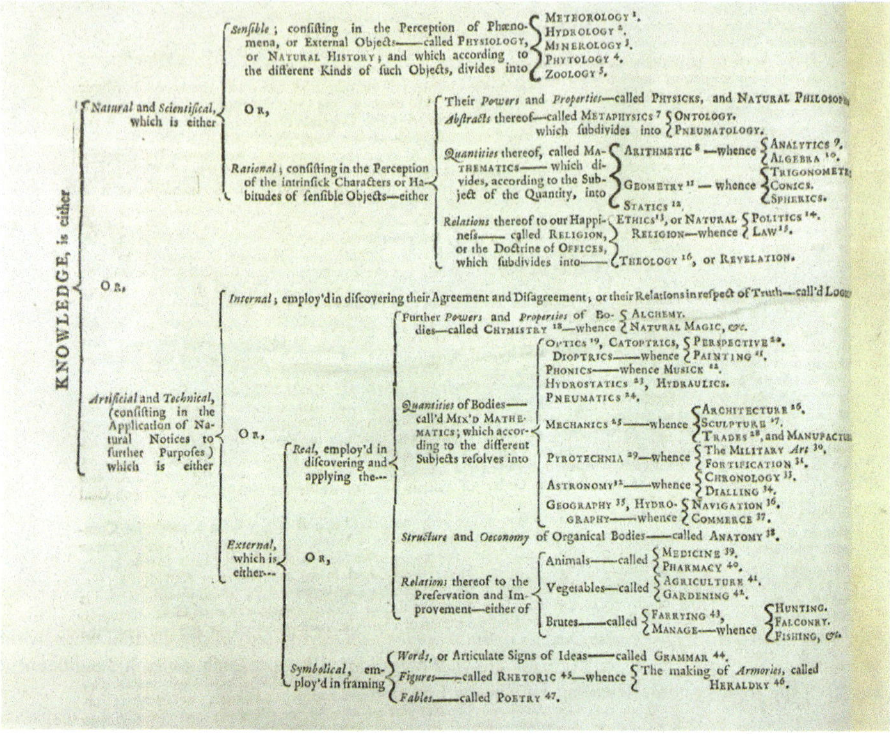

Fig. 5 A Ramist diagram reproduced from the *Cyclopaedia*, a general English-language encyclopaedia first published by Ephraim Chambers in 1728. This branching-tree structure was an innovation of the Renaissance logician Petrus Ramus and became a standardized form for showing classifications of knowledge. (Rare Books and Special Collections, University of British Columbia Library)

value through their use of selective visual syntax; for example, Vesalius' *De humani corporis fabrica* (1543) with its ground-breaking anatomical illustration (Fig. 6).

In the seventeenth century, the epistemology of early modern science is reflected in visualizations employing a quintessentially empirical approach, as demonstrated in Robert Hooke's *Micrographia* of 1665 (Fig. 7). The volume is dedicated to reproducing a variety of objects and creatures as seen through a microscope in vivid detail, a world until then inaccessible to the vast majority of readers. Here the 'visual truth' of the text is most directly related to demonstrable sense perception. It is a visual account of experience that can only exist through the extension allowed by tools—the microscope and the reproducibility of the drawing.

The epistemological approach of the modern era also prescribed ways of making sense of the universe of knowledge through the conception of complex structures in which knowledge could be classified and accessed. In publishing, the most prescient example of this is the format of the encyclopaedia, the contemporary form of which evolved mainly in the eighteenth century; key examples are Ephraim

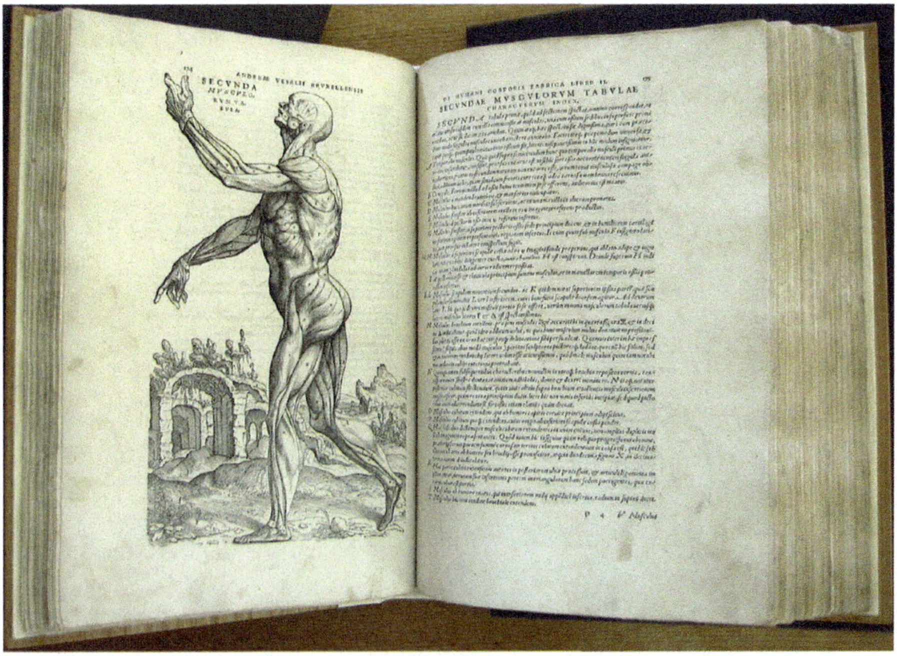

Fig. 6 *De humani corporis fabrica,* a milestone in the history of medical publishing produced by the physician Andreas Vesalius in 1543; the publication is remarkable for the very high quality and scientific accuracy of its numerous woodcut engravings. (Reading University Library)

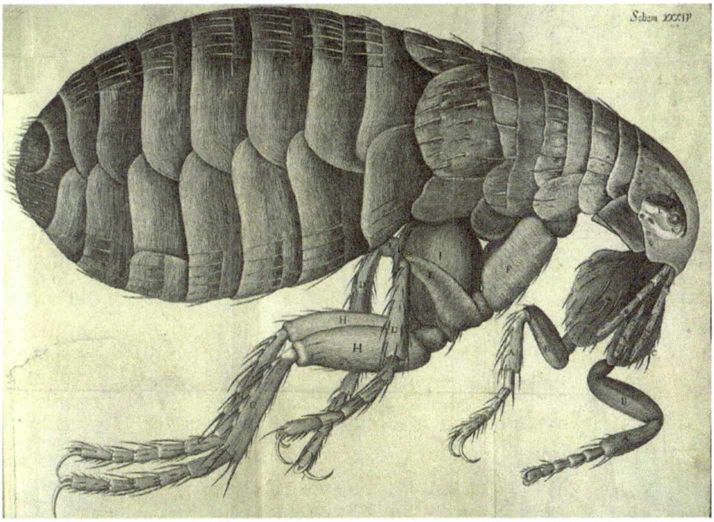

Fig. 7 A copperplate engraving based on the drawing of a flea in *Micrographia* by Robert Hooke (1665). Hooke's visualizations and observation of the microscopic world presented the British public with views of the natural world never before seen. (British Library)

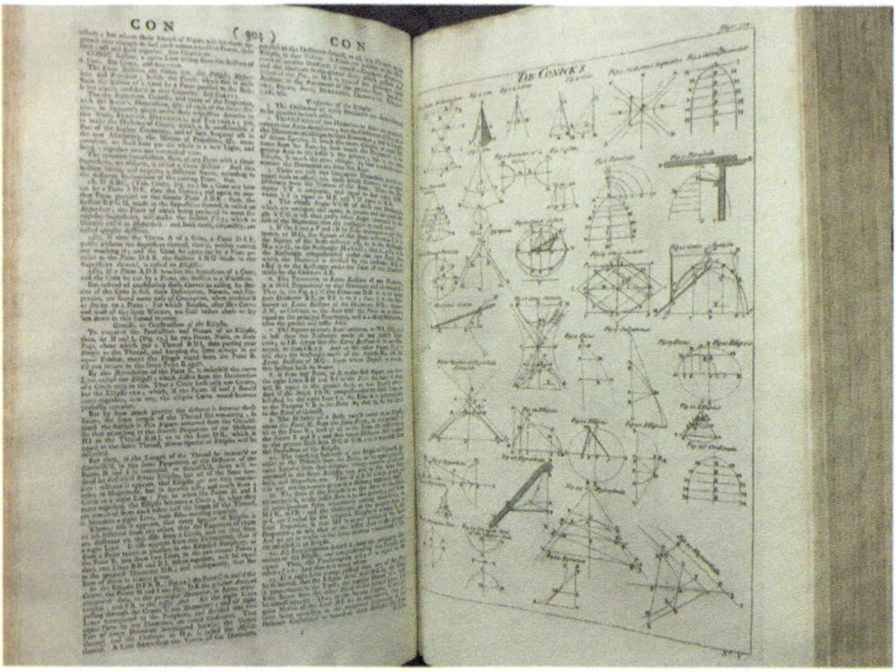

Fig. 8 A spread from the *Cyclopaedia* published by Ephraim Chambers in 1728. Here the structural elements of the work as a whole are evident, including the use of alphabetical ordering through headers and the systematic annotated illustration of various topics. (Rare Books and Special Collections, University of British Columbia Library)

Chambers *Cyclopaedia* (1728, Fig. 8), and the *Encyclopédie* of Diderot and d'Alembert, from 1751 onward. The physical form of these publications embodies their aims, presenting what was intended to be complete accounts of current human knowledge in a systematic and particularly logical framework that could be extended over multiple volumes as needed.

The rise of the encyclopaedia, a highly structured and spatial organization of information, is contemporary to the visualizations of scientific knowledge developed in the eighteenth century, such as Linneaus' taxonomies for classifying plants, as well as the contemporary format of the dictionary, such as Samuel Johnson's canonical example of 1755.

Another lineage that can be traced in the history of graphic communication relates to the dissemination of news. News formats are characterized by their large scale, the speed and immediacy of their production, and the frequency and reach of editions. News outputs are intentionally short-term and ephemeral in nature, qualities that are also the source of their influence and power. As representations of and reflections on current events, news formats act as socio-political frames for their time.

News forms have existed since shortly after the advent of the printing press in Europe, as broadsides and other street literature. In the nineteenth century, ever-increasing population and literacy levels helped fuel a rising 'flood of print' which included periodicals, tracts, pamphlets, almanacs and more, much of which related to social and political reformist movements. The dissemination of newsworthy information was supported by infrastructures of production that were commensurate with their use value. In terms of content, our expectation of the news has not altered as much as the technologies surrounding its dissemination; readers understand news as an aggregation of information in which many contributions are brought together.

Digital news media today break down spatial and temporal boundaries and allow new levels of reader agency in terms of selecting and viewing content. News aggregators (first launched in 1999 by Netscape), also called RSS aggregators, pool web content from blogs, newspapers and more into a single interface. Two examples of these provide a good contrast in their approach to the display of information: Google News and Flipboard.

The simple web-based news aggregator Google News (launched in 2002) combines material from various sources using XML to represent the news items in a single list format. The paradigmatic claim of Google News UI is that of a raw, unedited and automated interface. The user is presented with content as 'data' to be interrogated in a functional, neutral experience.

In contrast, Flipboard (launched in 2010) presents social news feeds through a highly modulated magazine-like format that can be accessed via the web or the standalone app. The graphic and interactive reconceptualization of web content subsumes the feeds to the distinctive style of the Flipboard UI. Where Google News presents news to be explored directly and functionally, Flipboard offers a rhetorically charged, customized subscription interface. As with the older historical examples of visual communication, these are designed forms developed through decisions made by their creators; the degree of user customization and input is heightened, however, to the degree that it is contributing to the visualization of the content itself.

If the history of human communication, in particular that dealing with the acquisition and sharing of knowledge, is dependant on technology, it must also be said that the formats allowed by these technologies live in a symbiotic relationship with these epistemologies. Our modes of thought are linked to the ways we visualize and share knowledge. This can be considered through any form of technology, from the pencil to the browser window. Consistent in the use of these technologies are the inherent motivation for sharing experience, their support in externalizing expression, and the inevitable tautology created by the link between the normative formats we use and the limits of our worldviews.

This brings up the notion of linguistic relativity, and serves as an introduction to the discussion of contemporary academic thinking in design in Sect. 3.1. This notion, also referred to as the Sapir-Whorf hypothesis, posits that language is intimately tied to our worldview. If we include a definition of graphic formats in our understanding of the term 'language', the study of the evolution in the way we make meaning of text and images should tell us something about how we 'read design' today, as well as how this language might serve to mediate between the environment and our cognitive processes (Gillieson 2008). Julia Kristeva argued that semiotics should study the constraints or limits of the system, to specify what falls outside it (1986); similarly the language of graphic presentation describes the limits of our intellectual universe.

The art historian Ernst Gombrich describes this type of 'visual relativity' through a perceptual lens that highlights the process of learning and differentiation in acquiring knowledge. Viewers decode images by drawing on an internal dictionary of representations about the world, or schema, to make sense of unfamiliar forms. We possess an internal tendency to organize what we know into systems or bodies of knowledge. This is a feature of adaptation that involves two processes: that of *assimilation*, in which new experiences are interpreted through incorporation into existing schemas, and *accommodation*, where existing schemas are modified to adapt new experiences. New forms are internalized as 'relational models' (Gombrich 1962: 78). Critically, these limits to perception are necessary; where anything is possible, communication breaks down.

Thus the mental set of a particular group, historical period or culture is created by and directly informs its conception of knowledge. The nature of the learning process is such that our understandings are shared and communal; these are cultural building blocks which designers today use to construct a broad variety of messages.

It can be argued that, just as there are histories of formal representation in fine art, there are genres and vocabularies of design; the creation of something entirely new is impossible and in any case would be incomprehensible. It is the very essence of graphic representation to refer to established genres and standards that are shared in common. The modern viewer reads a branching tree diagram in a biology textbook just as a Renaissance scholar might examine one of Ramus' trees of knowledge; the contemporary anatomy text contains carefully composed linear illustrations in the tradition of medical illustration, the lineage of which dates back to works such as Vesalius. More recently developed formats for showing information can be seen in the same light as longstanding formats such as the list and the diagram. In this sense, the 'language of design', like verbal language, is the product of a social process and therefore a social object, a set of shared conventions.

If graphic conventions can be considered a constituent and indeed necessary element in the construction of knowledge, we might characterize them as the *noesis graphikos* (to use the Greek root words for 'of the mind' and 'graphic'); thought as graphic. The generative form of this is what we refer to as *graphic design thinking*.

3 Cultures of Knowledge and Graphic Design Thinking

In the previous segment we examined how vocabularies of graphic design can be traced back through time; here we take a closer look at the array of approaches to research in academia, a cross-section of present 'cultures of knowledge' that prevail in the humanities, the sciences, and in design. These definitions help lead us to define more specific characteristics of graphic design thinking, with reference to the examples presented above.

3.1 Cultures of Knowledge

By 'cultures of knowledge' we mean the distinctions that exist among the tools for thinking and conducting research in the humanities, the sciences, and in design. These 'ways of knowing', to quote the design researcher Nigel Cross (2006), are naturally adapted to their subject areas, and align with particular approaches to logic, argumentation and discovery. The scholarly alignment of design research, and whether it is more naturally attuned to scientific or humanistic modes of inquiry, continues to be a subject of debate amongst design scholars. However, in its prevalence in all aspects of living, its unique methodologies and constraints, and its progressive stance, design research is very much distinct from the central approaches of traditional academic disciplines, and it is from this distinction that we will further characterize graphic design thinking in Sect. 3.2.

Research models in traditional areas and in design research can be distinguished according to the mode of logic employed. One mode of traditional academic thought is based on *deductive* logic; this is an argument that logically guarantees the truth of its conclusion, if its premises are true. The sequence of this mode of thought entails pursuing a general known rule or truth, creating a hypothesis based on the rule, noting specific observations, analysing the data, and confirming (or proving) the original truth (Kolko 2010). In the sciences, this is demonstrated in the format of the experiment, in which scientists aim to control all environmental and procedural variables in order to reveal potentially repeatable truths. According to the philosopher Karl Popper, the hallmark of a scientific proposition is that it can be proven wrong; this possibility of fallibility is the key distinction between scientific theories and the 'unprovable' ideas of pseudo-science (Miller 1985).

The other normative mode of academic thought is based on *inductive* logic; a process that offers sound evidence that something might be true, based on structured experience. Its sequence entails making observations (of possibly multiple occurrences) to infer patterns leading to the formulation of a tentative hypothesis, so that an analysis of the data may allow the inference of truth of the hypothesis (Kolko 2010). In humanities-based research, this can entail the discovery of subtexts or links that reframe our understandings of established values, beliefs or opinions, as has been the case with the development of feminist and post-structuralist theory in the latter twentieth century.

Although research in the sciences and humanities can employ either of these modes of thought, conceptions of knowledge can be contrasted. In the sciences, knowledge tends to be apparent and its findings and products are results of experiments that are intended to be repeatable and fallible. These endeavours strive to find the limits of nature and are largely, though not exclusively, involved with the physical world and natural processes. In the humanities, knowledge is built on previous knowledge in a way that dialogue is continual, potentially argumentative, and infinite; it tends to satisfy human desire for knowledge and an understanding of the world from numerous dimensions such as the historical, cultural, and social. These contrasting approaches are unified in their generation of arguments or propositions that exist to create new knowledge and reflect the meaning of human nature.

In response to these modes, Cross proposes that design embodies a 'third culture of knowledge' (2006) that possesses its own tools of logic. He draws a contrast between the scientific method and the design method:

> The scientific method is a pattern of problem-solving behaviour employed in finding out the nature of what exists, whereas the design method is a pattern of behaviour employed in inventing things of value which do not yet exist. Science is analytic; design is constructive. (Cross 2006: 7)

From outside the field, design processes and ways of thinking can seem to be highly subjective and what Jon Kolko describes as 'magically derived' (2010). Yet design research relies on distinct and comparable approaches to logic and its activities result in knowledge production of the same consequence as those undertaken in other academic disciplines.

Design research as an academic discipline is increasingly concerned with defining what can otherwise appear to be a nebulous process. Kolko argues that "[design] principles and methods are teachable, repeatable, and understandable." (2010: 17) Its development has been reliant on a continual critical evolution from within and outside the discipline. What sets the culture of design apart from others is its relative youth, its emergence from a merging of traditions in crafts and arts, its reliance on industrial production, and perhaps most importantly, its mode of discourse embodying not only texts, but products and processes.

A central tool of logic used in the knowledge culture of design is that of *abduction*, as defined by Charles Sanders Peirce, the process of hypothesizing through inference. Distinct from deduction and induction, abductive thinking 'creates new meaning' and embodies an 'argument to the best explanation' (Kolko 2010). The conclusions from an abductive argument might turn out to be false, even if the premises are true. Abductive thinking is a key resource in the process of design problem-solving, which can be characterized as *synthesis*, "an abductive sensemaking process" (Kolko 2010) where sensemaking refers to the process by which we give meaning to experience.

The logical leaps characteristic of abductive thought are also reliant on the background and knowledge of the individual designer, when faced with partial or inconclusive data. The designer's personal knowledge, developed both through previous life experience and work history as well as larger cultural narratives in which they are embedded, bridges the gap between the given requirements of the solution and the established methods of design that in themselves do not provide a solution.

This brings us back to the key element of design thinking which involves its engagement with complex and chaotic problem spaces, which can often require the collection and compilation of vast amounts of data that may not be coherently presented. The mechanism of synthesis, that of finding resolutions or cohesive proposals to make sense within these problem spaces, forces the designer to push for organization, reduction and clarity.

Because of the vast quantity of data under consideration, designers employ various techniques to visualize or spatialize through sketching, mind maps, and related formats (Kolko 2010). These techniques can be compared to some historical forms for recalling and making sense of information using spatial metaphors, most principally the mnemonic technique of *loci* (Latin for 'place') in use since the time of the Greeks, which entails the use of structured visualization to aid in the recall of information. In addition to these characteristics is the technique of 'reframing', in which a design problem is recast to create a new point of view. This shifting of semantic perspectives can help suggest ways of solving the design problem.

Looking broadly at the central modes of research it is apparent that general attributes of design research and knowledge are applicable to all design practices, and are thus relevant to graphic design or visual communication, our own area of specialization. However, we propose that the unique position of visual communication across cultures, with its considerable connection to media, language and thought, requires a more precise notion of design problem-solving. Graphic design exists at an intersection of design, language and literacy; it is a conduit of dialogue, consensus and cooperative meaning-making. Its products can be incorporeal, anchored in our shared graphic epistemology or *noesis graphikos*, as argued in Sect. 2. Considering all of this, we outline a series of characteristics pertaining to graphic design thinking that distinguishes it from design thinking more broadly, and its capacity for a particular form of knowledge production.

3.2 Seven Characteristics of Graphic Design Thinking

Based on the historical overview of various graphic formats and the characteristics of different cultures of knowledge discussed above, this segment presents the central proposal of this chapter: an elaboration of particular ways in which graphic design thinking may be distinguished from the broader design cultures of knowledge. The characteristics we discuss are not meant to be exhaustive and comprehensive, but rather a compilation of some of the basic features of visual communication, a starting point to an inquiry that we hope will be built upon in the future. And, while there are aspects of these that surface in other forms of design, they will necessarily operate in ways that are specific to each discipline; in graphic design they operate in a particular way as part of a web of interrelated features.

This series of characteristics particular to graphic design thinking include: its inherent language element, its nature as a constituent of mass communication, its implication in cultural narratives, its covert power, its particular ways of organizing complexity and schematizing knowledge, and its approach to reframing.

3.2.1 Language Element

A central distinction in how graphic design thinking differs from design thinking in general is that of the *language element* inherent in visual communication. This simple phrase implies considerable ramifications if we consider the vast contribution and entanglement of language with thought, philosophy and culture. Graphic communication itself has been assessed and analysed as a language in various ways by theorists in linguistics, information design and more (amongst others Kress and van Leeuwen 1996, Engelhardt 2002, Richards 1984, 1998).

Here we can speak of the 'language of design', which historically is evidenced in products of graphic design embodying systematic forms of communication, such as the language of charts and diagrams, or the development of information design techniques such as ISOTYPE (International System of Typographic Picture Education). Originally developed by the social scientist Otto Neurath in interwar Vienna, the ISOTYPE system of public education seeks to clarify information through the use of standard, legible and repeatable symbols (Fig. 9).

Alongside the use of design as a language of information, there is also the affective and expressive dimension of visual communication. Theories of visual rhetoric and proposals to map out and codify varieties of visual metaphor were developed especially at the Ulm School of Design in the 1960s, by theorists including Tomas Maldonado and Gui Bonsiepe.

The idea of the 'language of design' extends to the sub-disciplines of graphic design, such as that of typography, with its rules and principles of legibility, readability, and good form. These sub-languages allow an infinite number of solutions through the selection of finite compositional elements. One key characteristic of complex forms of visual communication, described by Gillieson (2008), is the possibility of recursivity in graphic language; this suggests that graphic design can be seen to possess characteristics of *generative grammar*, because it proposes a system of rules that can be used to produce meaningful statements in the form of nested graphic groupings at various levels of discourse, from smaller clusters to larger compositions such as a double-page spread. As a common and natural system, visual communication can also be seen as a form of universal grammar, one that is shared without being explicitly taught and that requires its own form of literacy:

> Literacy, as will be seen, is absolutely necessary for the development not only of science but also of history, philosophy, explicative understanding of literature and of any art, and indeed for the explanation of language (including oral speech) itself. (Ong 1982: 15)

3.2.2 Graphic Communication as a Constituent of Mass Communication

In addition to (and as a result of) its heavily linguistic nature, graphic design thinking can be described as *a central ingredient in systems of mass communication*; it is a key constituent of the media networks that exist in the world today, including not

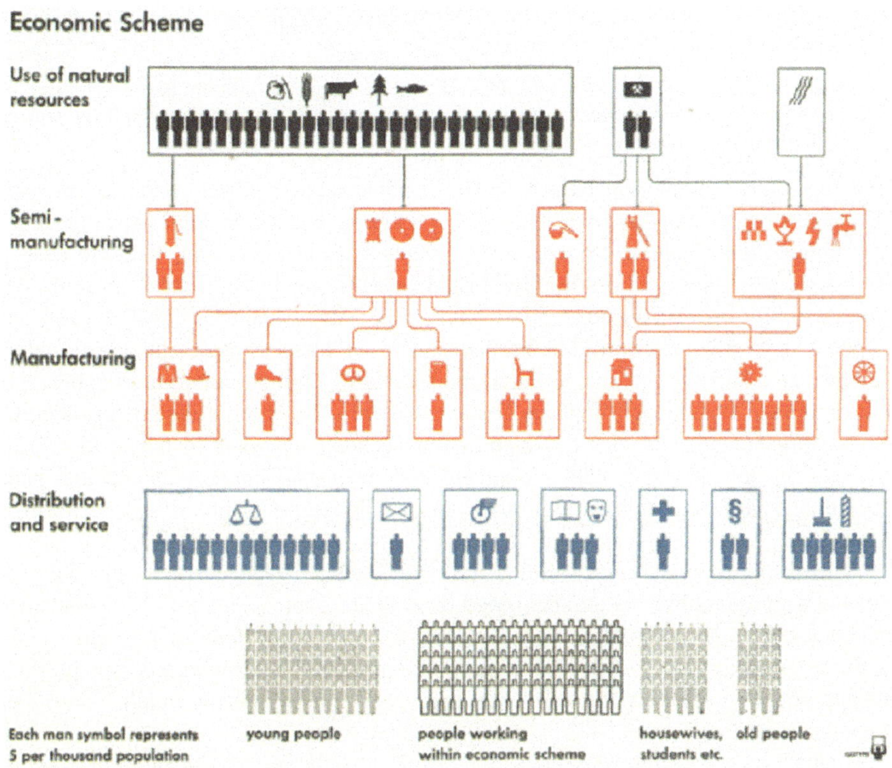

Fig. 9 An ISOTYPE diagram drawn from the volume *Modern Man in the Making* by Otto Neurath and published by Knopf in 1939. ISOTYPE (International System of TYpographic Picture Education) sought to represent social and political data to the general public through a particular and highly principled approach to information design. (University of Pittsburgh)

only traditional news media and publishing, but digital media, social networks, and visual languages of mass culture such as signage and the packaging employed in the full range of commercial products.

Although many design practices are implicated in contemporary mass communication systems (which involve the development of particular technologies and infrastructures), mass communication per se can be traced through its capacity to convey normative power structures and paradigms. The communication strategies employed in mass media rely on the language we use to think and align closely with our conceptions of reality. The way we conceive of ourselves and our environment is reinforced through this alignment, so that the constructed world of mass communication becomes naturalized as part of the background of our experience. Our shared visual language supports our sense of personal, tribal, and community belonging, fulfilling a need for social identification as part of a group. This feature of communication confirms not only social identity but the web of meanings intertwined with it. Designers become acutely aware of these interrelated systems of meaning in developing new messages, which carry new meanings and can embody new knowledge:

> Designers have the ability to 'read' and 'write' in the (material) culture: they understand what messages objects communicate, and they can create new objects which embody new messages. (Cross 2006: 9)

The Frankfurt School theorist Jürgen Habermas discusses communicative rationality as a form of reason concerned with clarifying the norms and procedures by which agreement can be reached, via a social form of reasoning. Graphic design conventions, including those highly normative ones employed in mass communication, also rely on norms and standard processes. After Habermas (1989), we might suggest that graphic communication embodies a form of procedural rationality rather than a substantive one, and that the social engagement inherent in communication design is culturally loaded. Made evident in the graphic face of mass communication systems, graphic design practice has been described by many schools of thought as one that is ethically engaged and politically motivated and that sees visual language as ideology, as in the work of Bonsiepe (1968, 1999), mentioned above,.

The design theorist Horst Rittel (Rittel and Webber 1972) views design as a process of argumentation, so that designed objects are expressions that form part of a broader cultural dialogue. In visual communication, this feature is explicit and reflected in the history and uses of graphic formats. The constraints surrounding design decisions allow patterns of argumentation employing the many systems of visual semiotics.

Today graphic design remains a pervasive force, functioning through highly tailored messages often formulated according to user analytics and leveraging the power of visual rhetoric and metaphor. Graphic communication is a meta-language of commercial culture, one that is utilized in every form to provide context to the message, either through description, amplification or as the visual object of 'text' itself. In this sense mass culture texts can be understood from the point of view of their identity value, as a historical actor in society. Examples of this type of group identity can be found in ancient coins, political pamphlets printed after the advent of the press, and in modern graphic design language through posters, magazines, newspapers and television and digital media formats. Throughout history, the role of graphic communication as an instrument of social cohesion has given it a dubious reputation as a tool of propaganda; the Ulm School's view of visual communication as ideological imbues its power of persuasion with an ethical dimension.

3.2.3 The Implication of Cultural Narratives

Because of the immediacy and power of visual forms, graphic design thinking is heavily concerned with understanding *cultural narratives*, not only propagating but re-presenting them for various audiences. We use graphic design to express our culture and ourselves in various ways. The products of visual culture can therefore be traced as anthropological evidence.

For instance, variances in styles of Eastern and Western comics reflect differences in worldviews and conceptions of time and being. The cartoonist and theorist Scott McCloud demonstrates how standardized techniques of framing and the presentation of the passage of time in Manga reflects fundamentally Eastern notions of time and space that stand in contrast to those of Western-style comics (1993).

Cultural narratives in graphic design can also be identified in their self-consciousness. The re-use of certain visual forms in pop and postmodernist visual culture is undertaken in a way that expects the viewer to be conscious of the appropriation; the effect is to give the audience capable of reading this referentiality a feeling of association to a subculture. As an example, we can cite the contemporary trend of deskilled graphics, intentionally poorly executed type and graphics, as a form of humour or irony in contemporary subcultures of art, design, fashion and music. Cultural manipulation is exercised through the use of naive typography and aesthetics to disrupt notions of beauty and the expectation of perfect mechanical reproduction. The Vancouver-based record label Mood Hut's event flyers and record stamps provide an example of this intentionally flawed yet compelling aesthetic (Fig. 10).

3.2.4 The Covert Power of Graphic Communication

Another central feature of graphic design is what we refer to as the *highly covert nature* of the contribution of visual communication to the organization of human experience. This is an element related to verbal language which itself possesses many levels of covert meaning, reflecting and legitimizing social and cultural norms and structures at a deep level. The graphic face of a culture becomes powerful precisely because it is a self-organizing system that ensures its own hegemony. The covert power of graphic design is pervasive and hinges on its use of and ability to generate metaphors to convey messages. We understand the world through conceptual analogies that are so deep-seated they are invisible. Visual communication is a generative and constituent element of thought, yet its power and potential is not fully understood outside of certain fields related to visual forms, such as communication design, illustration, and visual art. Art theorists such as Michael Baxandall (1985) and Ernst Gombrich (1962) have pointed to the historical and cultural anchors informing the rhetoric embedded in a wide assortment of visual forms. The way that we read is bound by psychological and environmental factors and is a natural product of historical processes; in this way language, thought and typographic communication may be linked.

All objects of design (graphic, industrial, etc.) constitute applications of knowledge that they both embody and transmit. Graphic design specifically is inextricably tied to surrounding cultural trends and movements in such a way that its products can be particularly slow to evolve, and embody what we might term 'everyday truths' in addition to grand cultural narratives. The knowledge residing in these products can reflect their environment, the knowledge of their creators, the history of their genres, and their contexts of use. Argument or discourse through graphic

Fig. 10 Two record stamps, Jack J *(top)* and Cloudface *(bottom),* and event flyer *(last)* demonstrating the visual language of deskilled, 'naïve' aesthetics that are visual hallmarks of certain contemporary cultural and musical subcultures. (Mood Hut Records)

Fig. 10 (continued)

messages is facilitated through conventions, expectations, standards, and the everyday truths necessary for readers to engage in semiotic assimilation and accommodation. They allow for many messages to exist simultaneously, and for these to be classified (e.g. in paradigmatic groups of *style* or *genre*) and deployed. It is this hidden or covert feature of graphic design that allows it to be embedded in cultural and social discourses.

3.2.5 Particular Ways of Organizing Complexity

The next characteristic of graphic design thinking presented here is a particular approach to what we term *organizing complexity*. In order to be precise in our meaning, we must first mention that a range of design practices engage in 'information synthesis', which is a form of organizing complexity. This typically occurs in the early stages of the design process, during which sketching and visualization of the problem space can also be seen as a nascent form of applying knowledge. This engages Cross' concept of the 'dialectics of sketching', a dialogue between 'seeing that' (a form of reflective criticism) and 'seeing as' (a form of analogical reasoning) (2006, 86). In this phase, making sense of the collected information is of crucial importance:

> Because of the vastness of data gathered in even a simple design problem, the quantity of data that must be analysed is often too large to hold in attentive memory at one time, so a designer will externalize data through a process of spatialization. (Kolko 2010: 18)

In graphic design practice, however, there is a further extension of the direct application of this form of knowledge in the design object itself. The crucial difference in graphic design is that this feature is both a *product* as well as a *process*. The outcomes of graphic design that organize complexity are apparent in the historical examples for classifying knowledge in all fields of human endeavour, cited in Sect. 2. Extensive bodies of knowledge related to science, philosophy, and other fields are organized into elaborate and sophisticated forms such as the encyclopaedia, the catalogue, or the dictionary. These provide navigable and extendable systems for large bodies of information at a high level, whether linguistic, diagrammatic or pictorial, in a way that identifies and reveals connections and groupings to create ordered and searchable collections.

The encyclopaedia, discussed above, is perhaps the prime example of this feature, its epistemological foundations reflecting the empirical mode of thought prevalent at the time of Chambers' and the Encyclopaedists' in the mid-eighteenth century:

> In its preface, Ephraim Chambers' influential *Cyclopedia* (1728) – the first encyclopaedia with an elaborate system of cross-references between alphabetically ordered entries – opens with a knowledge map as well as an analysis of its forty-seven interrelated divisions of knowledge, in a preface that is intended to serve as a table of contents and a directory to the articles collated in the rest of the book. (Beltran 2012: 7)

A further example of the particular capacity of graphic design thinking to organize complexity is demonstrated in the organizing principles of book design more generally. The page plan of a publication (a visual map that shows all its double-page spreads in order, as on a photographic contact sheet), exposes the characteristic rather well; seeing all the pages simultaneously shows the overarching design features and principles, headers and footers, contents list, indexes and other components that serve to frame, cross-reference, and provide easy access to the 'main text'. The various levels of organizing elements are metadata that act as supporting access structures to the content. The graphical and spatial organization of the content, including its access structures, is the literal embodiment of the design solution. The book is itself an outcome of the graphic design process, rather than a metaphor or technique for managing large quantities of information, as are the phase of sketching and organizing in the early stages of general design processes.

3.2.6 Particular Ways of Schematizing Knowledge

The next feature of graphic design thinking highlighted here we refer to as *schematizing knowledge*. Though there are many conventions and approaches related to visualizing knowledge across all design practices, these effectively conform to what Kolko refers to as spatialization techniques, the formative sketching and other methods used to externalize information and help make connections in the data as part of the phase of information synthesis cited in our discussion of organizing complexity (Sect. 3.2.5).

In graphic design however, this feature is not only present in early problem-solving stages but in its *products*, which are literally spatializations and visualizations of knowledge. To illustrate, we can refer to examples in the history of communication; the form of the list, one of the earliest abstractions of linear writing, served to clarify information schematically. Subsequently the development of diagrams over time reveals a wealth of schematization techniques for efficiently showing various kinds of information, including forms such as pie charts, bar charts, tree diagrams, and custom formats such as the periodic table. Figure 11 shows a table used to compare characters of assorted alphabets. The table as a format acts as a structure for simultaneous contrasting and paradigmatic linking of variables. It simplifies the comparison by highlighting formal commonalities—they are all alphabetic units within a system, created according to the constraints of varying production technologies (chisels, pen widths, etc.) A more sophisticated form of the list, the table schematizes knowledge in a way that reveals connections between its representations along 2 axes.

There are implications to the ends and the means of the process being one and the same; graphic design schematizations are standardized and universal in a way that is not possible in the formative spatialization techniques used in other design disciplines. The process is a generative activity that imbues knowledge to the original 'raw data' through the complex operation of bringing it into visual form, so that the schematization that is the end result at once realizes and augments the content.

3.2.7 Reframing

The last characteristic we would like to put forward for its unique application in graphic design is that of *reframing*. As with previous features, the process of reframing can be used in a broad range of disciplines; it involves changing one's perspective on a problem to produce a semantic shift that can shed new light on it. This can be used as a heuristic device to develop lateral thinking in design generally; in

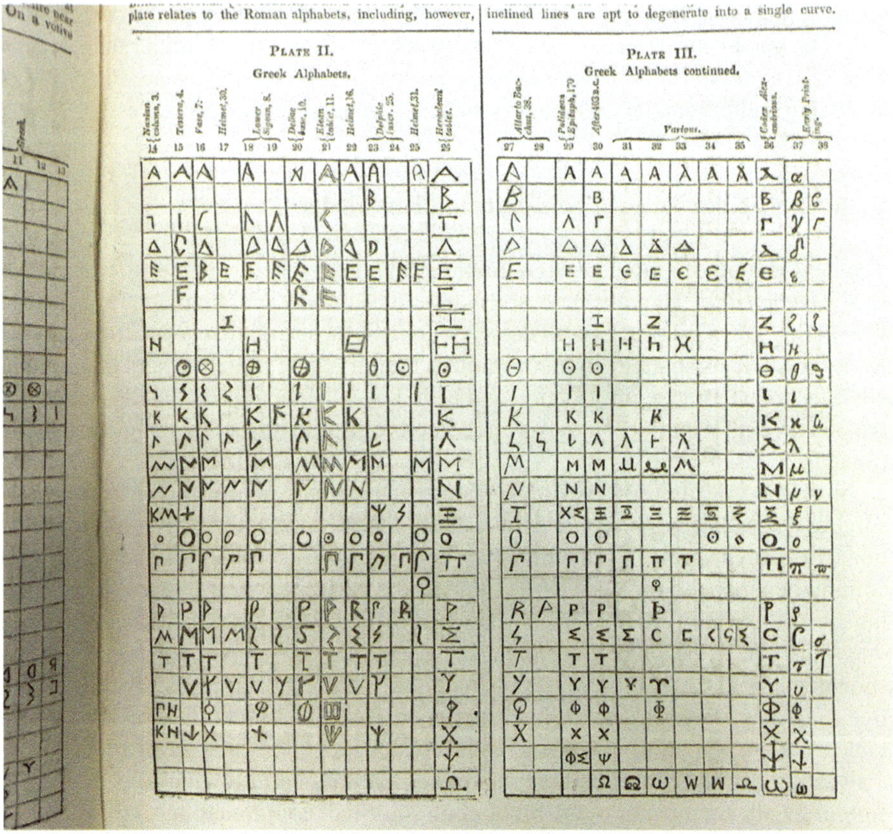

Fig. 11 A Greek alphabet table drawn from *The Penny Cyclopaedia of the Society for the Diffusion of Useful Knowledge* published by Charles Knight in London (1833). The matrix structure allows a density of information and a very efficient use of space in comparison to simpler list formats or the use of running text. (Rare Books and Special Collections, University of British Columbia Library)

graphic communication, however, reframing takes on a particularly focussed and linguistic character. To illustrate this, reference can be made to Barthes' discussion of the *commutation test*, which is a technique in linguistics and semiotics that involves identifying signifiers (whether in a word, a layout, etc.) and then substituting these for equivalent signifiers as a way of testing the variants in meaning. This can occur in the early stages of the design process, where features such as colour, typeface, and images are selected, manipulated and combined to form the larger whole. The combination of these features are what give the project a 'frame' and altering individual elements can create new frames or arguments.

The commutation test lends itself best to forms that follow historically-derived, explicit principles such as those of verbal language, typography, and symbol systems. As an example of commutation in graphic design, we can look at a series of book covers for a novel, Aldous Huxley's *Brave New World,* which has been published many times since 1932, for a demonstration of how small changes in design choices produce very different styles or moods (Fig. 12). The book cover itself is a 'frame' in graphic design that can be manipulated and thus modify the reception of the entirety of the book to its audience. Isolating and replacing individual elements of the typography, colour, visual metaphor, illustrative or photographic technique may produce variants that can be used to test the limits of recognizable styles in these covers. The 'limits of style' are reached when a substitution results in nonsense or discontinuity in the graphic language.

To summarize, reframing is a process of creating coherent and refined micro-cultural references through the modification of signifiers. The cultural frequency of these signifiers means they can be traced with reference to colour psychology or the historical connotations of typography, for instance. Consequently, they also exist in continuity with the previous characteristics of graphic design thinking we have presented above. Thus, a piece of graphic design must make semantic sense (the language element), so that it can be easily be understood by many (as mass communication); it needs to exist within our conception of reality (cultural narrative), in a way that is consistent with our deepest levels of understanding (covert power), links all complex elements, including the previous characteristics, appropriately (organizing complexity), and finally puts it in an optimal format (schematizing knowledge). As a logical sequence the 'tested' arguments within a frame must to conform to sets of principles and rules (reframing). A product of graphic design must conform to expectations and conventions of its language, a fact that is reflected in this interdependence of the characteristics of graphic design thinking.

In this section we began with a consideration of the knowledge cultures prevalent in the very high-level categories of academic discourse; those of the sciences, humanities, and also more recently that of the domain of design. The seven characteristics of graphic design proposed above are not meant to be an exclusive or comprehensive list, nor do they form the necessary conditions of visual communication; rather they are the beginning of a vein of inquiry into graphic design as a 'way of knowing', a notional checklist that, as we have mentioned, may be edited and added to in the future.

These features, together with the brief history of graphic communication presented in Sect. 2, contribute to what we propose is an epistemology of graphic perception. If writing is metalinguistics, used in reflecting about language, then design can be seen as meta-perception, reflective not only of verbal language but of all the components of visual communication, including pictorial, spatial and structural systems of signification. The features of graphic design thinking described earlier provide the conditions that allow us to expand on the notion of graphic design as an ethically conscious practice in the following segment.

Fig. 12 Book covers of *Brave New World* by Aldous Huxley ranging in publication dates from 1932 *(top left)*, 1946 *(top right and bottom left)*, and 1958 *(bottom right)*. Here the various approaches to layout reflect contemporary aims to cast the book in a certain light through the use of the 'design languages' of commercial publishing of the various time periods

3.3 *Expanding the Remit of Graphic Design Practice*

Having outlined the seven characteristics above, it is fitting at this point to draw out some thoughts about the ramifications of a more conscious approach to visual communication. Graphic design is an activity that is involved in an abstract continuum of cultural standards and patterns, and that engages in speculation as a future-facing activity while simultaneously being anchored in real-world concerns. Through a heightened understanding of graphic design thinking, both in the scope of its application and its consequences, we suggest that graphic design, as a social practice, needs to be both responsible and ethically engaged.

The form of intelligence that can be associated with graphic design thinking possesses a particular quality of abstraction. As we have argued, graphic design process is inherently synthetic, generative and future-conscious; its products are embedded in the continuum of design history and propel it forward. Because of their strong communicative qualities, graphic design outcomes merge in a rhizome of interrelated statements, so that individual projects need to be seen in their larger context. Christopher Alexander calls for a new conception of urban planning in *A Pattern Language* (1977), in which he proposes that we should not deal with individual objects in design practice, but that we should consider 'integral composites' of these, such as a street corner, as the significant units for discussion (Burckhardt in Gretzinger 2012: 116). Patterns in the forms and placements of composite statements can be identified and developed; this approach to understanding design is highly synthetic and aligns well with graphic design in particular.

Because of its composite quality, visual communication acts to abstract and sublimate thought; every design project, every application of graphic communication in the cultural sphere acts as part of a larger discourse that includes other work. It follows then that each design problem is a conversation within a larger conversation, a phrase within a paragraph. The process of development of universal standards in design, "the repertoire of shared and embodied norms, techniques and interpretive frameworks that all societies build up" (Beltran 2012: 8), involves a very slow-moving process of abstraction, because messages are classified through cultural production into signifying categories. These amount to the standards of everyday truths used by graphic designers in developing messages of mass communication, as discussed earlier. According to Michael Gardener (in Beltran 2012: 7), everyday knowledge possesses specific features: it is pragmatic, it makes use of analogical forms of reasoning, it tends to generalize, and it forms an essentially practical engagement with the world. Visual communication propels everyday knowledge in a generative process and through a use of analogy that remains anchored in a real-world dialogue; it employs a functional abstraction.

As a form of cultural production that sets out to improve (or at least influence) the human condition, meaning-making and public interaction, it is essential for design to operate ethically within its particular circumstances. We have suggested that the 'epistemology of design' involves a dialogical dimension that is also one of social responsibility. When contributing to this dialogue, in any form of graphic design, the role of the designer is one of a knowing participant operating according to the implied consent of the audience or receiver. Designers are already aware, to varying extents and in some cases only implicitly, of the principles of graphic design thinking presented above. To some degree, this knowledge comes through in their work. In the direct impact of graphic design to the public sphere, it could be argued that it is a professional responsibility to explicitly understand the impact and potential of their work on a broad cultural level.

As a highly consequential practice and one that is capable of a high degree of abstraction, graphic design can be characterized further as aspiring to possible worlds; the act of design is a future-facing activity which involves an element of speculation. In this sense, it can also be seen as an inescapably ethical practice; designers collect, analyse and synthesize to problem-solve. In communication design especially, this implies an engagement in philosophical and political discourse. Design solutions are driven by the expectation that they may lead to an improved future state.

A conundrum in this view of graphic design thinking is that, although our shared sources of knowledge also provide us social continuity, they are also the sources of constraint. Despite a running myth of design as a creative and autonomous practice, designers work within various cultural constructs; they may look to lauded historical examples and to high-profile contemporary projects to inform their own future-facing work. Their choices of medium are as important as, and indeed indistinguishable from the message; together these elements form an ideological sum that assimilates the decisions of the designers that have come before. Errors, inefficiencies and oddities of past forms are perpetuated in new work, and the socio-political sphere in which designers may feel able to act upon in turn constrains them. A designer may produce work that they may feel is not optimal, but that is the best possible response to the problem as it has been presented, given the circumstances. As a result, the ideological truths that we expect in design are confounded; the work of designers is inevitably reliant on the commonly understood standards and languages built up before them, a repertoire which is not necessarily optimal or even effective in some circumstances. For this reason, we argue that graphic designers need to be more explicitly conscious of the nature of graphic design thinking, such as through the characteristics we have outlined in this chapter. The constraints and cultural standards by which designers work need to be knowingly questioned and tested, and an awareness of their ethical responsibility is needed to ensure that designers, as social agents, employ design as a tool of practical and ethical engagement with the world.

4 Conclusions

Through an exploration of various aspects of the history of graphic communication and considerations of prevalent notions of design thinking in comparison to other forms of knowledge production, we have sought to demonstrate that the tools and techniques of *graphic design thinking* are particularly distinct and set it apart from methods used in other disciplines and fields of design. The series of seven proposed characteristics of this mode of problem-solving makes reference to historical and theoretical dimensions of the field, contemporary heuristics and standards of professional practice. These include: the inherent language element of visual communication, its nature as a constituent of mass communication, its implication in cultural narratives, the covert nature of its power, particular ways of organizing complexity and schematizing knowledge, and its particular approach to 'reframing'.

A greater awareness of the capacity of graphic design and the tools it offers as a mode of thinking-through-problems can contribute to an understanding of the embedded and universal nature of much of the graphic resources we draw on. Further, graphic design thinking proposes that although there are no purely unique and original expressions (in the sense that it has all been done before), the key difference is that the expression is powerful through its differentiation with what surrounds it; how it acts in a specific and unique setting against other objects and ideas, in conjunction or confluence with a particular set of circumstances, in relationship to the reaction of particular people in that setting.

We also suggest that this form of design thinking can be considered a particular type of contextualized intelligence; it proposes a dialogical epistemology that is rooted foremost in its outward-facing and social concerns. We have mentioned the capacity for design, and communication design in particular, to engage in dialogue; while theorists such as Rittel see an element of dialogue in the design of material objects, we have proposed that the discursive function inherent in graphic design defines it as a political form, suggesting an ethical and responsible dimension to practice. With the perspective of this ethical approach, graphic design thinking can encourage reasoned debate and socio-political engagement,—an actualization of 'dialogical reasoning' at a broad cultural level.

This becomes ever more important due to the relatively limited general knowledge of design that exists in those who do not practice it. The designer has the responsibility to act with privileged knowledge and an understanding of their contributions to the designed world. Another proposal that extends from this notion is that a base level of design literacy in the general population is needed, because methods of design thinking can be thought of not only as a fundamental requirement in our media- and information-saturated society, but as a form of critical thinking that can contribute to democratic decision-making.

Nigel Cross (2006) draws a distinction between the 'science of design' and the 'philosophy of design' with regard to industrial design practices. He refers to the domain of the epistemology of design as that of the logic of creativity and innovation. In considering graphic design thinking in particular, we argue that the logic involved is deeply rooted in the structure of channels of communication, language and other semiotic systems, and therefore in thought and knowledge. This particular domain knowledge sets it apart, and we have attempted to specify at least part of it above, through the series of characteristics. With a clearer understanding of the ways in which it is distinct, its full potential may be harnessed not only in developing work in other disciplines of design, but in broader social and cultural contexts.

References

Bajetta, C. M. (2000). *Some notes on printing & publishing in renaissance venice*. New York: Typophiles.
Baxandall, M. (1985). *Patterns of intention: On the historical explanation of pictures*. New Haven: Yale University Press.
Beltrán, E. (2012). *The world explained: A microhistorical encyclopaedia*. Amsterdam: Roma Publications.
Bonnet-Bidaud, J.-M., Praderie, F., & Whitfield, S. (2009). The Dunhaung Chinese sky: A comprehensive study of the oldest known star atlas. http://idp.bl.uk/education/astronomy_researchers/index.a4d. Accessed 14 Aug 2015.
Bonsiepe, G. (1968). A method of quantifying order in typographic design. *Ulm, 21*, 24–31.
Bonsiepe, G. (1999). *Interface: An approach to design*. Maastricht: Jan van Eyck Akademie.
Cross, N. (2006). *Designerly ways of knowing*. London: Springer.
Drucker, J. (2004). Speculative computing: Aesthetic provocations in humanities computing. In S. Schreibman, R. Siemens, & J. Unsworth (Eds.), *A companion to digital humanities*. Oxford: Blackwell.
Eisenstein, E. (1979). *The printing press as an agent of change: Communications and cultural transformations in early modern Europe*. Cambridge: Cambridge University Press.
Foucault, M. (1996). Interview with Michel Foucault. In S. Lotringer (Ed.), *Foucault live: (interviews)* (pp. 1961–1984). New York: Semiotext(e).
Gillieson, K. (2008) A framework for graphic description in book design. PhD thesis. Department of Typography & Graphic Communication, University of Reading, UK.
Gombrich, E. (1962). *Art and illusion: A study in the psychology of pictorial representation*. New York: Pantheon Books.
Goody, J. (1977). *The domestication of the savage mind*. Cambridge: Cambridge University Press.
Gretzinger, K. (2012). *In a manner of reading design*. Berlin: Sternberg Press.
Habermas, J. (1989). Technology and science as "ideology". In S. Seidman (Ed.), *Jürgen Habermas on society and politics: A reader*. Boston: Beacon Press.
Intelligence [Def. 1, 2]. (n.a.) (n.d.). *Merriam-Webster Online*. http://www.merriam-webster.com/dictionary/intelligence. Accessed 25 Nov 2015.
Intelligence [Def. 1]. (n.a.) (n.d.). *Oxford Dictionaries*. http://www.oxforddictionaries.com/definition/english/intelligence. Accessed 25 Nov 2015
Kolko, J. (2010). Abductive thinking and sensemaking: The drivers of design synthesis. *Design Issues, 26*(1), 15–28.
Kress, G., & Leeuwen, T. (1996). *Reading images: The grammar of visual design*. London: Routledge.
Kristeva, J. (1986). In T. Moi (Ed.), *The Kristeva reader*. Oxford: Basil Blackwell.
Lindinger, H. (1990). *Ulm design: The morality of objects*. Cambridge, MA: MIT Press.
McCloud, S. (1993). *Understanding comics: The invisible art*. New York: Harper Collins.
Miller, D. (Ed.). (1985). *Popper selections*. Princeton: Princeton University Press.
Olson, D. (1994). *The world on paper: The conceptual and cognitive implications of writing and reading*. Cambridge, MA: Cambridge University Press.
Ong, W. (1982). *Orality and literacy: The technologizing of the word*. London: Methuen.
Ong, W. (1983). *Ramus, method, and the decay of dialogue: From the art of discourse to the art of reason*. Cambridge, MA: Harvard University Press.
Peirce, C. (1991). On the nature of signs. In J. Hoopes (Ed.), *Peirce on signs: Writings on semiotic* (pp. 141–143). Chapel Hill: University of North Carolina Press.
Richards, C. (1984). *Diagrammatics*. London: School of Graphic Arts, Department of Graphic Design, Royal College of Art.
Richards, C. (1998). Getting the picture: Diagram design and the information revolution. *Information Design Journal, 9*, 87–110.

Rittel, H., & Webber, M. (1972). *Dilemmas in a general theory of planning*. Berkeley: University of California Press.
Smith, F. (2004). *Understanding reading: A psycholinguistic analysis of reading and learning to read*. London: Routledge.
Taylor, C. (1991). *The malaise of modernity*. Concord: House of Anansi Press.

The Role of Abduction in Production of New Ideas in Design

Lauri Koskela, Sami Paavola, and Ehud Kroll

Abstract The pragmatist philosopher Peirce insisted that besides deduction and induction there is a third main form of inference, abduction, which is the only type of inference capable of producing new ideas. Also he defined abduction as a stage of the methodological process in science, where hypotheses are formed to explain anomalies. Basing on these seminal ideas, scholars have proposed modified, widened or alternative definitions of abduction and devised taxonomies of abductive inferences. Influenced by Peirce's seminal writings and subsequent treatments on abduction in philosophy of science, design scholars have in the last 40 years endeavoured to shed light on design by means of the concept of abduction. The first treatment was provided by March in 1976. He viewed that abduction, which he called "productive reasoning", is the key mode of reasoning in design. He also presented a three-step cyclic design process, similar to Peirce's methodological process in science. Among the many other later treatments of design abduction, Roozenburg's definition of explanatory and innovative abduction is noteworthy. However, an evaluation of the related literature suggests that research into abduction in design is still in an undeveloped stage. This research shows gaps in coverage, lack of depth and diverging outcomes. By focusing on the differences between science and design as well as on empirical knowledge of different phenomena comprising design, new conceptions of abduction in design are derived. Given the differences of context, abduction in design shows characteristics not yet found or identified in science. For example, abduction can occur in connection to practically all inference types in design; it is a property of an inference besides an inference itself. A number of the most important abductive inference types as they occur in design are identified and discussed in more detail.

L. Koskela (✉)
University of Huddersfield, School of Art, Design and Architecture, Huddersfield, UK
e-mail: L.Koskela@hud.ac.uk

S. Paavola
Faculty of Educational Sciences, University of Helsinki, Helsinki, Finland
e-mail: sami.paavola@helsinki.fi

E. Kroll
Department of Mechanical Engineering, ORT Braude College, Karmiel, Israel
e-mail: kroll@braude.ac.il

© Springer International Publishing AG 2018

P. E. Vermaas, S. Vial (eds.), *Advancements in the Philosophy of Design*, Design Research Foundations, https://doi.org/10.1007/978-3-319-73302-9_8

Keywords Types of abduction · Design · Design reasoning

1 Introduction

The American pragmatist philosopher C. S. Peirce (1839–1914) developed understanding of an inference type he called abduction throughout his creative research career lasting around 50 years. He used different names in addition to abduction, for example, retroduction and hypothesis, for this inference, and changed his conception of abduction as his research progressed. He viewed abduction as a type of inference that is ubiquitously used in everyday life by humans – sometimes comparing abductive innate tendency of humans to instincts by animals, for example, to chickens having an innate tendency of finding proper food and eating (Peirce CP 5.591, 1903). However, it is the role of abduction in science that Peirce's interest was especially focused on. Abduction was conceived by him as the type of inference through which all new ideas, still hypothetical, are produced in science. It starts when something surprising has been observed, requiring explanation. It was characterized by Peirce as the third type of inference, besides deduction and induction.

Although the concept of abduction stems from the latter part of the nineteenth century, it was only in the latter part of the twentieth century that it started to be discussed again, especially in philosophy of science but also in other fields. This discussion has taken different directions. Based on the original ideas of Peirce, scholars have proposed modified, widened or alternative definitions of abduction and devised taxonomies of abductive inferences. Also, later authors have endeavoured to reach a better understanding of Peirce's original ideas. Nevertheless, Hintikka (1999) holds the problem of abduction as the most important question in epistemology.

Inspired by treatments of abduction especially in philosophy of science, there is a broadening interest in analyzing design in terms of abduction. However, in critical examination, it can be asked whether this work has led to conceptual gains in design science or to useful advances in design practice.

Why has the progress of clarifying abduction in design been slow? Three main problems can be recognized, namely lacking maturity of the concept of abduction in philosophy of science, the differences in context between the two fields, and the embryonic state of the science of design. As different authors in the design field pick up different conceptions of abduction from the wider literature, the discussion has tended to be fragmented and has hardly led to cumulative increase in understanding. The concept of abduction is still deeply embedded in specific interpretations within the context of science, where the question is about suggesting a cause (i.e. explanation) for a surprising phenomenon. In design, similarly a "cause" for the required behavior is searched, but design may comprise also other types of inferences through which a new idea can emerge. Unfortunately, there has been little scholarly work on the different types of design inferences. Indeed, many theories of design conflate types and chains of creative inferences under the term "synthesis".

Thus the question remains: how do new ideas emerge in design, and how, if at all, abduction can help in analyzing these processes? In view of this, this paper attempts to clarify the phenomenon of abduction in design. The main starting point is that the concept of abduction, as it has been discussed in philosophy of science, carries implicit contextual assumptions, which are not compatible with the context of design. We contend that the focus should be directed, besides the concept of abduction, to the underlying phenomena and their context as they occur in design. In this way, it will be possible to define and characterise abductive inferences in design.

The paper is structured as follows. In the next section, an overview on the prior research on abduction is given. It should be noted that the conceptions of abduction have developed over time, and are about to be developed further. We aim at giving a short picture of a variety of interpretations on abduction. Then, a survey of prior research on abduction in design is presented as well as a critical evaluation of it. In the following section, the newly-developed understanding of abduction in design is presented, and a number of main types of abductive inferences in design are introduced. A section on conclusions, including further research needs, completes the paper.

2 Abduction in Philosophy of Science

2.1 Origin of the Concept of Abduction: Peirce

Throughout his career Peirce insisted that besides deduction and induction there is a third main form of inference (see e.g. Peirce W 1:180, 1865; CP 8:385–388, 1913). But there were changes in his notions and formulations of abduction over the years while he interpreted it in relationship to various questions concerning reasoning, inquiry and human cognition (Paavola 2012, 46–47). It is customary to discern two main phases in Peirce's treatment of abduction (Fann 1970). In his early works Peirce defined abduction as an *evidencing process* by using syllogistic formulations. Typical examples of abduction and induction can be formulated by inverting a deductive syllogism in two different ways. Abduction is an inference of a cause from its effects (W 1:180, 1865), or "reasoning from consequent to antecedent (CP 5.276, 1868), or "making an hypothesis" (CP 2.623, 1878). This is different than induction, which is typically about generalising and inferring a rule on the basis of cases (CP 2.622–624, 1878). In these early formulations, the strength of abductive arguments was presented quite vaguely. Peirce said that abduction (or 'hypothesis' as he then called it) is a "weak kind of argument", and about surmising (CP 2.625, 1878), but in his early works he often presented it also as a form of probable reasoning (e.g. CP 2.511, 1867).

In his later works Peirce emphasised abduction as a part of a *methodological process* (Fann 1970). A basic formulation on abduction is often cited (see Peirce, CP 5.189; Hanson 1958, 86):

> The surprising fact, C, is observed;
> But if A [an explanatory hypothesis] were true, C would be a matter of course,
> Hence, there is reason to suspect that A is true.

Here abduction is a part of a process of inquiry and central in that stage where hypotheses are invented on the basis of anomalies or clues ("the surprising fact"). Abduction provides only tentative conclusions ("there is reason to suspect"). Besides abduction, the process of inquiry essentially needs deduction (which is important when hypotheses are explicated) and induction (which is central for testing these hypotheses). These three (abduction, deduction, induction) provide the basic phases of the process of inquiry (CP 6.469–473, 1908), and the role of induction is then different compared to Peirce's early formulations (this difference will be further discussed in Section 3.3).

Peirce emphasised in his later works more the weakness of abduction: "Deduction proves that something *must be*; Induction shows that something *actually is* operative; Abduction merely suggests that something *may be*" (CP 5.171, 1903). Peirce also insisted now that human beings must have a *guessing instinct* for finding fertile hypotheses (CP 7.220, 1901). In his early formulations Peirce had rejected the connection of abduction (or reasoning in general) to instinct (CP 2.749–754, 1883), but in his later works a guessing instinct is precisely the root of abduction (CP 8.234, c. 1910), perhaps somewhat paradoxically, because he emphasised now that abduction is a weak form of inference. Peirce had, however, different kinds of formulations on abduction, maintaining, for example, that it is close to perceptual judgment and an "act of insight" (CP 5.181, 1903), an "inference through an icon" (Peirce CP 2.96, c. 1902), or a first starting of a hypothesis as a simple interrogation (Peirce CP 6.524, 1901). One central feature of abduction is that it is a way of arranging facts in a new way (Peirce, PPM 282–283, 1903):

> A mass of facts is before us. We go through them. We examine them. We find them a confused snarl, an impenetrable jungle. We are unable to hold them in our minds. We endeavor to set them down upon paper; but they seem so multiplex intricate that we can neither satisfy ourselves that what we have set down represents the facts, nor can we get any clear idea of what it is that we have set down. But suddenly, while we are poring over our digest of the facts and are endeavoring to set them into order, it occurs to us that if we were to assume something to be true that we do not know to be true, these facts would arrange themselves luminously. That is *abduction*.

There are many predecessors and earlier, close formulations to Peirce's conception of abduction (see Paavola 2012, 21–22). Peirce himself referred to earlier writers who had used the term "hypothesis" in a sense of "the conclusion of an argument from consequence and consequent to antecedent" (see EP 1:34–35, note, 1868). Clear influences were also Aristotelean syllogisms and Kantian philosophy. Especially in his early lectures Peirce also analysed the "logic of science" by his contemporaries like Whewell, Mill and Comte (see Peirce W 1: 205–223, 1865).

Peirce appreciated especially Whewell's work (W 1:211 1865) and there are interesting, later discussions on similarities and differences between Peirce's and Whewell's conceptions of the logic of science (Snyder 1997; Niiniluoto 1999a). Regarding Aristotelean syllogisms, Peirce even speculated that Aristotle had been formulating the basics of abduction in *Prior Analytics* with the notion of '*apagoge*', but this sense had been lost because the text of Aristotle had been misunderstood and mistranslated as a consequence of having been corrupted (CP 1.65, c. 1896). Later he emphasised that this theory on the (mis)interpretation of Aristotle was doubtful and only conjectural (CP 8.209, c. 1905). There are, however, interesting recent interpretations maintaining that there are abductive syllogisms present in Aristotle's work but in places that Peirce neglected. Abduction comes close to '*anchinoia*', which is for Aristotle a skill of conjecture or sagacity discovering or inferring a cause (Florez 2014). One interesting parallel to Peirce's abduction (especially when interpreted as a regressive inference of a cause from its effects) is the method of analysis (where regressive inferences also play a central part) and synthesis by Greek geometers, which is often referred to and applied as heuristics in the history of science (Niiniluoto 1999b). This topic will be taken up in Section 3.

2.2 Later Understandings of Abduction

Abduction was for long quite marginally treated (Paavola 2012, 31–45). There were a number of reasons for that. Peirce's philosophy in general was not much studied. The twentieth-century philosophers of science (like Popper and Hempel) were mostly against any logical treatments of the area of discovery. Abduction was then seen as conflating discovery and justification, as defined by Reichenbach (1938), and either being a form of induction, or close to the hypothetico-deductive model of science. A notable exception was Hanson (1958) who defended Peirce's "logic of discovery" as a means of conceptualising what is the key issue in science, that is, the search for new hypotheses. But it was not until the late 1970s and early 1980s that abduction started to raise a broader interest, especially in philosophy of science and in methodology (Nickles 1980; Eco and Sebeok 1983). Besides Peircean tradition, abduction is nowadays influenced also by the "Inference to the Best Explanation" (IBE) model that was formulated by Harman in the late 1960s (Harman 1965, 1968). IBE is close to Peirce's formulations on abduction although there are also clear differences (Minnameier 2004; Paavola 2006).

Nowadays the uses and discussions on abduction have expanded on various fields of research, such as logic, philosophy of science, research methodology, semiotics, cognitive sciences, artificial intelligence and diagnostics. It is then no wonder that there are different emphases and interpretations of abduction. This might cause confusion but is also a strength when formulations of abduction are developed further.

Different kinds of distinctions and taxonomies on various forms of abduction have been suggested. One distinction is if abduction is supposed to be a way of generating new hypotheses ("creative abduction"), or a way of evaluating and choosing a candidate from given options ("selective abduction") (Schurz 2008; cf. Magnani 2001; Eco 1983). Another basic distinction is if the abduced conclusion is a singular fact, law, or theory (or a theoretical model) (Schurz 2008). Still another dimension is what triggers abductive reasoning: a novelty or an anomaly, or perhaps just a phenomenon to be explained (see Aliseda 2006, 45). Schurz (2008) presents a thorough classification of abduction patterns, all of which are "special patterns of inference to the best explanation". Based on Schurz' classification, Hoffman (2010) develops a taxonomy of 15 forms of abductive inference.

Abduction is usually connected to searching for *explanatory* hypotheses. However, according to Gabbay and Woods (2005) abduction is not necessarily tied to explanation, but the characteristic of abduction is that it is "ignorance-preserving" reasoning, in contrast to deduction as "truth-preserving" and induction as "probability-enhancing". Furthermore, abduction can also be interpreted: more clearly as a form of reasoning; or more closely related to such cognitive issues as perception, guessing and insights; or as a part of a broader process of inquiry (Paavola and Hakkarainen 2005).

3 Abduction in Design

3.1 Prior Treatments of Abduction in Design

March (1976) suggests that abduction, which he calls "productive reasoning", is the key mode of reasoning in design. He also points to the confusion and misunderstanding created by not distinguishing between scientific and design hypotheses, and between logical propositions and design proposals. Whereas the goal of science is to establish general laws, he says, design is concerned with realising a particular outcome. The pattern of abduction proposed by March is: from certain characteristics that are sought, and on the basis of previous knowledge and models of possibilities, a design proposal is put forward. March presents a three-step cyclic design process that is similar to Peirce's three modes of reasoning (abduction-deduction-induction), and says that rational designing has three tasks:

1. Creating a novel composition (i.e., the artefact) as the outcome (the 'case') of productive (=*abductive*) reasoning,
2. Predicting the performance characteristics of the artefact by *deduction*,
3. Accumulating habitual notions and established values by *induction*.

Induction, therefore, may have two related roles: a background activity that represents ongoing acquisition of experience and expertise, and an evaluative step in the design cycles.

Goel (1988) proposes to extend and complicate March's production-deduction-induction model if we wish to use it in knowledge-based systems. He maintains that it is too simplistic to divide the phases of design activity into three categories. His argument is based on the fact that the laws (also called rules or knowledge) can have different logical natures: some are universal, others statistical, and a third category is the quasi-laws that govern the behaviour of complex adaptive systems such as humans and organisations. Depending on the type of law involved, the prediction of performance may no longer be a deductive inference, and the inference of design descriptions, while still abductive, poses new problems related to affirming the consequent and to instantiating an individual.

Takeda et al. (1990) propose a cognitive model of the design process based on general design theory (GDT), which contains three types of reasoning: deduction, abduction and circumscription. Circumscription is used to find exceptions that cause a contradiction, so the incomplete knowledge of object properties and behaviours at any time during the design process can be modified. The design process consists of a 5-step cycle: (1) identification of problems by comparing the designed object with the required specifications, (2) suggestion of key concepts to solve the problems, (3) development of candidate solutions from the key concepts and design knowledge, (4) evaluation to confirm the candidate solutions, and (5) conclusion to decide which candidate to adopt. Step (1) is carried out by circumscription, step (2) by abduction, and steps (3) and (4) by deduction. Step (5) is a decision-making process that is not analysed with the logical framework. Step (2), the abductive step, basically follows the paradigm: from the properties of the current design candidate and the knowledge of object properties and behaviour available at the current state, obtain a description of the current design candidate. In this sense, it is similar to the "standard" syllogism of "given fact + major premise ➔ conclusion".

Takeda (1994) continues this work, but emphasises the role of abduction as the ampliative process of making integrated hypotheses and theories. He claims that Peirce's abduction is unattractive in design because it only enumerates many hypotheses. In contrast, design abduction is different because it not only generates hypotheses (descriptions of objects), but also considers the background theory and uses part of it, the "explanatory theory", by selecting relevant rules and laws and even generating new ones. Takeda's view on design abduction is therefore a step in the direction of Roozenburg's innovative abduction (see below).

The design support system proposed by Takeda et al. (2003) is based on a model of the design process that consists of iterations of abduction (to create a new idea or artifact) and deduction (to validate the design). They claim that creative design must contain two aspects: generating a new product and expansion of knowledge. Abduction should include more types of reasoning then Peirce's "abduction as inversed deduction", which is merely one category of factual abduction. It should include also discovery of new laws. Factual abduction is used to create a new design while law abduction creates a new theory. Both abductions take place concurrently to achieve "integration of knowledge". Among the several methods that can be used for law abduction (some of which are mentioned by Schurz (2008)), they focus on analogical inference. The procedure for carrying out the analogical abduction starts by choosing a candidate theory (i.e., knowledge base), preferably one with low similarity to the original knowledge in order to increase the likelihood of a surprising result. Next, correspondences are created among concepts in the chosen theories, followed by generation of candidate design knowledge, which are the hypotheses found appropriate to solve the current design problem. Finally, candidate design solutions are created from the new integrated knowledge by a process of hypothesis verification.

Tomiyama et al. (2003) connect the above work even more strongly with Schurz's classification. They claim that Schurz's *first order existential abduction* (one type of his factual abductions) can generate a design that performs given requirements, but because both the requirements and rules are known, no creative design can result. Another important type is *theoretical-fact abduction* that generates new initial or boundary conditions that apply to the yet-unknown design solution. This abduction does not generate solutions, but conditions that the solution should satisfy. These conditions can become new design sub-problems or additional requirements. A third type of abduction is a combination of modes of Schurz's *second order existential abduction*, and plays a role in integrating multiple theories. This "abduction for integration" consists of identifying the applicability and domain of seemingly irrelevant theories (as in analogical reasoning) and merging them with the existing knowledge.

Roozenburg (1993) discusses in depth the question whether the reasoning towards a tentative description of a design with plausible reasoning follows the conventional view on abduction, or whether it should be defined differently. He argues that the commonly presented view, especially in artificial intelligence literature, deals with "explanatory" abductions, which are good for diagnosis or troubleshooting, but that the core of design reasoning follows another type of abduction, for which he proposes the terms "innovative" abduction and "innoduction" (Roozenburg and Eekels 1995). In fact, says Roozenburg (1993), Habermas (1978) distinguished between *explanatory* abduction and *innovative* abduction, and it was March who did not make that distinction. In the case of innovative abduction, says Roosenburg, "Starting from a surprising, not yet explainable fact (the result), we try to conceive of a new rule (a principle, law, or theory) that allows us to infer the cause (the case); the rule itself, therefore, is not yet assumed to be true". He goes on explaining that the conclusion of this inference is a hypothesis that still needs to be tested by deduction and induction before it becomes a new rule with explaining power. Roozenburg's pattern of innovative abduction is therefore:

q	(q is a given fact, a desired result)
--	
$p \rightarrow q$	(a rule to be inferred first, IF p THEN q)
p	(p is the conclusion, the cause, that immediately follows)

He even says that the above pattern is Peirce's original intention, because p cannot be part of the premise and needs to be part of the conclusion of the inference (cf. CP 5.189, 1903). This means that both $p \rightarrow q$ and p 'present' themselves together, at the same moment.

Roozenburg's innovative abduction is claimed to represent the kernel of the design process. The desired result is the *function* to be accomplished, his rule follows the formula "if *form* + *way of use* then *function*", and the conclusion is *form* + *way of use*.

Dorst (2011) proposes another view on design abduction that revolves around the following formula:

what (the artefact) + *how* (the working principle) ➔ *value* (aspired)

in which the (aspired) *value* is always given. If the *how* is also given, then the *what* is generated by a so-called *abduction-1*, which is precisely "explanatory" abduction. Dorst calls this case "conventional ('closed') problem-solving that designers often do". If, however, the *how* is not given, then we have a more 'open' problem in which we need to decide on both the working principle and the artefact. This is accomplished by *abduction-2*, which is the same as Roozenburg's innovative abduction. *Abduction-2* is carried out by first developing or adopting a 'frame' (after Schön), which is a "general implication that by applying a certain working principle we will create a specific value". The framing activity is characterised by Dorst as being "a form of induction", because it is reasoning back from consequences (this is in conflict with Peirce to whom that kind of reasoning represents abduction). With the help of framing, *abduction-2* takes place according to the following pattern:

q	(q is the given desired *value*)
--	
$p \rightarrow q$	(IF *how* THEN *value*, the first conclusion)
p	(*how*, the second conclusion)

When a possible or promising frame has been proposed and the *how* is known, says Dorst, *abduction-1* can take place to design the *what*, the artefact.

Kolko (2010) defines design synthesis as an abductive sensemaking process of manipulating, organizing, pruning and filtering data related to the design problem in order to produce information and knowledge. Three methods of formalising the synthesis process are proposed: reframing, concept mapping and insight combination. Each of the methods emphasises actions of prioritising, judging and forging connections, which are claimed to have been derived directly from the logical processes of abduction and the cognitive psychology theory of sensemaking. However, the connection of these actions to abductive reasoning seems rather loose, and is mostly based on Kolko's claim that abduction (defined as IBE) is the only type of inference that can generate something new, and that it involves intuition and the designer's own life experiences.

Lu and Liu (2012) refer to Peirce's description of abduction as "intelligent guessing" and show how abductive reasoning can be applied in three different ways to design synthesis: (1) inferring functional requirements from the customer's need, (2) deriving design concepts from the requirements, and (3) diagnosing faults within design concepts to facilitate the selection among them. Using Schurz's classification,

Lu and Liu associate these three types with second order existential abduction, factual or law abduction, and observable-fact abduction, respectively.

Ullah et al. (2012) attempt to connect the notion of "classical abduction" to the C–K theory of design. They conclude that conceiving a creative ("undecided" relative to existing knowledge) concept is more complex than abduction, being a motivation-driven process. Motivation here consists of a "compelling reason"—why a certain concept is pursued, and an "epistemic challenge"—seeking new knowledge.

Pauwels and Bod (2014) adopt Peirce's model of scientific inquiry, claiming that it (and design) consists of repetitive cycles of abduction (to make hypotheses and interpretations of "the world" or "the situation"), deduction (to make predictions of anticipated consequences), and induction (to devise experiments to test the anticipated consequences and to learn new knowledge). Abduction is identified here with analogical thinking and the design cycle is demonstrated through an example of a kitchen sketching episode. It starts with the designer looking at a partial sketch and trying to find an explanation (analogy, interpretation, idea) to it by abduction. The abduction results in a hypothesis that a specific other configuration might improve the design. Now, from this hypothesis, the designer deduces a specific prediction about what to do next, resulting in consequences: re-arranging the sketch lines so that the kitchen will improve. Finally, an induction step is used to verify that the new design has indeed been improved, and this step constitutes learning.

More recently, a need-function-principle-system (NFPS) model has been proposed for conceptual design (Chen et al. 2015a, b). A clarification stage converts subjective needs into objective functions, followed by a synthesis stage to find abstract principles for satisfying the functions. The abstract principles consist of combinations of action classes and behaviour classes, which are generalised actions and behaviours, respectively. Next comes an embodiment stage, where action classes and behaviour classes are instantiated as a system (i.e., structure) having corresponding specific actions and behaviours. The actions and behaviours are verified in an analysis stage, followed by a prediction stage to identify unintended side effects and possibly generate new functions for the next design cycle. Both synthesis and development stages are claimed to be "implicit abductions" or innoductions, because they lack sufficient premises for generating only one result.

An empirical approach to study abduction in design is demonstrated by Cramer-Petersen and Ahmed-Kristensen (2015). They define abduction as the use of a known principle, law or theory for the purpose of a causal explanation, and investigate aspects of idea-generation sessions by protocol analyses. Abductive reasoning is identified by the use of the following "indicator words": *could, maybe, think, could be, imagine, probably, likely;* deductive by *so, then, therefore, that is, must be, as, can;* and induction is correlated with *I, me, you, they, we, them.* They found that abduction was the least frequent type of reasoning in the protocols, although most ideas were initiated by it. The explanation was that abduction only requires a few statements to hypothesise and introduce new frames of understanding the problem. Similarly, the high proportion of deductive reasoning was explained by observing that deduction often comes in series of several statements about the structure of an idea. Inductive reasoning was found to occur more frequently than abductive reasoning, but its occurrences rarely happened at the beginning of idea-generation episodes. They also found that abductive reasoning led to more radical ideas, whereas deductive reasoning led to ideas related to project requirements (especially cost reduction). The latter type had a higher proportion being rejected as not valuable.

3.2 Critical Evaluation of Prior Treatments of Abduction in Design

The previous section has made it clear that there is a variety of interpretations of abduction in design. They do not provide a unified picture of abduction. In this section, the prior treatments of abduction in design are analyzed and critically evaluated. The discussion is structured according to the main topics arising from the materials examined.

3.2.1 Lacking Interest and Contribution by Philosophers and Logicians

It is striking that understanding on design abduction has exclusively been advanced by scholars in the design domain. Professional philosophers have shown little interest in the clarification of abduction in design. This is further exemplified by the fact that the authoritative and thorough handbook *Philosophy of technology and engineering sciences* (Gabbay et al. 2009) has "deductive reasoning" in its index, but nothing on abduction. Unfortunately, the inevitable academic division of work has implied that the studies of design scholars on abduction tend to be thin on the side of philosophy of science and logic. There have been basic misunderstandings in the interpretation of literature in philosophy of science.

3.2.2 Fragmentation and Divergence

The existence of various somewhat different characterizations of abduction already by Peirce and the lack of unity in philosophy of science regarding abduction has led to a situation where design scholars have picked up somewhat varying concepts and interpretations of abduction and often developed them into new directions. It seems that this reflects partially the breadth of discussion within the area of design, and also of abduction. Arguably, however, this situation has tended to lead to fragmentation of discussion, lacking accumulation of understanding and divergence on views on abduction.

3.2.3 Methodological Issues

Most research into design abduction is conceptual, driven by the pursuit of adoption and adaption of abduction concepts from science to design. The developed design abduction concepts are illustrated—or perhaps initially justified—through schematic or worked examples. Empirical research on abduction in general has been scarce (cf. Cramer-Petersen and Ahmed-Kristensen 2015; Dunne and Dougherty 2016).

3.2.4 Contextual Differences Between Science and Design

Although the contextual differences between science and design were discussed already in the seminal contribution by March (1976), it is doubtful whether they have been fully covered in design abduction research. In the following, the topics arising from such differences are discussed.

3.2.4.1 Starting Point for Abduction

In the accounts on science, there is usually a focal point, surprising observation or anomaly, that forms a starting point for abduction. Based on this, singular (and often celebrated) acts of abduction have been addressed in discussions on abduction in science (for example, Hanson 1958). However, as Hanson himself pointed out, discovery processes often extend over longer time periods (ibid.) and clearly involve several mixed forms of reasoning. In treatments on abduction in design, the focus similarly has been on singular abduction from the problem to the solution, usually from function to structure. However, the design counterpart for surprising observations in science has hardly been discussed and characterized.

Another question hardly discussed is related to the location of abduction in the respective process. In science, abduction is located next to the surprising observation, in the beginning of the research cycle. This is in connection to the situation where in science typically one aspect or part of a phenomenon is explained (Eekels and Roozenburg 1991). However, in design, the totality of the targeted artefact is designed. This implies that in design, a problematic situation (requiring an abductive solution) may occur in any part of the design process, not only at its start, and that there may be many different problems in the framework of a design cycle.

3.2.4.2 The Abductive Inference

The approaches to design abduction have not always been sensitive to differences in the type of inference in science and design. March (1976) explained that abduction in science is about generalisation, whereas in design it is about particularisation. However, the situation in design (and as a matter of fact, also in science) is more nuanced. Often also in design, the first step is to find an (abstract) natural law or fundamental concept, from which the form and structure of the artefact is abducted. Thus, abductions in both ways, generalisation and particularisation, occur in design.

Another question is that in the classical treatment of abduction in science, the inference is a regressive one, from effect to cause. This idea has largely been accepted also for design abduction. However, there are other types of problems in design, say decomposition of functions, physical composition of the artefact or finding the most suitable framing for a design problem. Whether ampliative solutions to these should be viewed as types of abduction has not been discussed.

3.2.4.3 Relation of Abduction to Other Forms of Inferences in Science and Design

The Peircean definition of abduction as a third form of reasoning, besides deduction and induction, is adopted by authors on design abduction. However, the relation of abduction to other forms of reasoning and mixed forms of reasoning in design tend to remain vague as there is little scholarly consensus on reasoning types in design. Also, that Peirce, in his later works, defines induction in a non-conventional manner has not been discussed in design abduction literature. The question whether all regressive inferences in design are identified as abduction remains without substantive discussion.

3.2.4.4 Outcome of Abduction

In science, the outcome of abduction is hypothetical. In design, the same applies as more or less all intermediate outcomes are hypothetical until the final validation. However, in design there are additional criteria. The embodiment design stage can embrace only such forms, materials and assemblies that can be realised in the purported context. These have not been discussed in prior treatments.

3.2.4.5 Context for Novelty

In science the context of abduction, at least in the end, is the whole scientific community interested in the topic in question. Thus, novelty depends on what is known by the community. In design, it is rather an individual designer or a design team, who provide the context for abduction. Novelty is thus relative to what is known by the designer or a design team. This issue has hardly been discussed in the literature on design abduction.

3.2.5 Design Creativity

The literature on abduction in design makes little reference to the literature on design creativity, although the subject matter in these two fields seems highly overlapping, namely emergence of novelty.

3.3 Reflections on Abduction: Potential Problems in Peirce's Conception of Abduction

The difficulties of conceiving the form and role of abduction in design, as discussed above, seem to derive, at least partly, from the specific features of Peirce's treatment of abduction and gaps in its subsequent interpretations.[1] Such problems are especially related to using only the logical scheme of syllogism as the starting point, the position of induction in relation to abduction, and the relation of abduction to regressive inferences as they occur in the method of analysis.

Peirce started his research on abduction by using the syllogistic approach to logic, originated by Aristotle, to illustrate the differences between deduction, induction and abduction. Syllogisms had been the dominant approach in logic for several centuries till the mid nineteenth century, when new approaches started to be developed (actually Peirce was among the initiators of these). A syllogism is an inference from two premises, both containing a "middle term" that thus connects them, towards a conclusion (Lagerlund 2016). The problem with the syllogistic conception of an abductive inference is that the new idea seems already to be in the premises of the inference instead of being created through that inference[2] (see Sect. 3.1). The syllogistic conception thus seems to compromise the central notion of abduction engendering a new idea. It might be one reason for Peirce to adopt the broader view of abduction as a stage in the research process (Psillos 2009).

The existence of these two Peircean conceptions of abduction (as an inference and as a stage in a research process), and also the discussions on the nature of the syllogistic presentation of abduction have been confusing. It seems this was the

[1] Recent developments of abduction in philosophy of science highlight broader and more dynamic understanding of this concept but they have not yet had impact on the research on abduction in design.

[2] Psillos (2009) states on this: "This creates a certain tension in Peirce's account. Hypothesis is ampliative and the sole generator of new ideas or content. And yet, in the syllogistic conception of hypothetic inference, the new ideas or content must already be there before they are accepted as the conclusion of the inference." However, this commonly presented view can be debated. Verene (2008) has argued that Aristotle's theory of syllogisms was actually twofold: a syllogism serves both as an instrument of demonstration and as a means for the generation of new ideas. Thus (Verene 1981): "The invention of an argument requires the invention of the middle term of the syllogism. The creation of the middle term and the needed premises are aspects of a common process; they come into being at the same time." If we subscribe to the conception of syllogisms as advanced by Verene, the tension alleged by Psillos disappears (see also Paavola 2004).

reason for Habermas (1978) coining the terms explanatory abduction and innovative abduction, based on which Roozenburg (1993) saw it necessary to invent the term "innoduction", innovative abduction. However, as he himself admits, this kind of abduction is what Peirce may have had in mind. Thus, the necessity of a new name for it can be questioned.

Another source of confusion is related to induction. In the early writings of Peirce, induction was about generalisation of a rule from a sample, corresponding to the still common usage of this term. However, in his later work, the meaning of induction switched to confirmation of an abductive hypothesis through experimental verification.[3] Both in the earlier and later conception by Peirce, abduction was related to deduction and induction; however, the meaning of induction is very different in these two cases.

Based on the conception of Peirce of the scientific process as abduction-deduction-induction, March (1976) presents a tripartite model of design, consisting of the corresponding stages of production, deduction and induction. However, when trying to fit design into the mold provided by science, he mistakenly—so it seems to us—describes the last stage as follows: "the design and its expected characteristics are used to infer new generalisations". In our view, the evaluation stage in design is occupied by the question whether the design fulfills the requirements of the particular situation; it is not primarily about the generalisation. Also Pauwels and Bod (2014) similarly adopt the tripartite model of Peirce, and call the evaluation stage in the architectural process induction.

The third source of confusion is the question how abduction relates to regressive inferences as they occur in the stage of analysis in the context of the geometrical method of analysis (Hintikka and Remes 1974). This method has been influential in science; no less figure than Newton (2003) writes on analysis in Opticks: "By this way of Analysis we may proceed from Compounds to Ingredients, from Motions to the Forces producing them; in general, from Effects to their Causes, ...". During the active years of Peirce, the method of analysis was adopted as the scientific model by several leading scientists and philosophers of science, such as Duhamel, Whewell and Mach. As argued by Koskela et al. (2014), analysis proceeds through regressive inferences (but also through decomposition and transformation), whereas synthesis, being an inverse of analysis, proceeds through deduction and composition. Peirce[4]

---

[3] It is noteworthy that Hintikka (2007, p. 55) advises not to use the word induction in the case of hypothesis testing: "[...] I do not think that it is instructive to call such reasoning inductive, but this is a merely terminological matter". Indeed, if the meaning of the term induction is changed in the way Peirce does, the question emerges how should the types of reasoning traditionally referred to as induction be called.

[4] Actually it is not clear whether he was aware of the method of analysis in the first place. In one instance, he wrote about analysis and synthesis as used in science: "This method of procedure is that Analytic Method to which modern physics owes all its triumphs." (Feibleman 1969). However, from the context it emerges that he is not discussing the method of analysis from geometry but differential calculus (also called analysis). Even more strangely, he compares analysis, in this same sense of differential calculus, to Hegel's method of analysis and synthesis (Feibleman 1969). These failures to discuss the method of analysis, when it would have been apt and deserved, would be understandable if he was not aware of the method of analysis.

did not relate abduction to the method of analysis. Intriguingly, the method of analysis has been proposed also as a model of design since Aristotle (Koskela et al. 2014).

Now, analysis would seem to serve the same function as abduction—proceeding from effects to causes—but the question arises whether we should equate all regressive inferences[5] with abductive inferences, especially in the case that a regressive inference is not providing a new idea. Note that as Peirce's viewpoint is the syllogism, he is typically discussing one inference, whereas in analysis it is customary to acknowledge a chain of regressive inferences (for example, in Aristotle's account in Nicomachean Ethics), some of which may be habitual, some selective and some creative. This boils down to the question whether abduction should be understood as a type of inference (that is, regressive inference as in the method of analysis) or a property of an inference (that is, any inference producing a new idea). Our view is that the distinguishing characteristic of abduction is the generation of new ideas.[6] In this way, it will be possible to enrich the conceptualisation of design with a new distinction. However, as evident from the presentation on prior views on abduction especially in the design domain, the term abduction is sometimes understood to also cover regressive inferences that do not yield novelty.[7] This terminological issue cannot be solved in this presentation; for communicative purposes, we use the noun "abduction" as it is generally understood in the Peircean sense, to denote a regressive inference producing a new idea, while the adjective "abductive" is used to refer to a property of an inference, namely that a new idea has been engendered.

4 Towards Defining Abduction in Design

4.1 How Should We Define Abduction in Design?

Why to define and clarify abduction in design? While a major part of designing is based on habitual solutions and generally available knowledge, there is also a need for novelty, new ideas, through which better design outcomes can be targeted. The existing theories of design do not provide a systematic and detailed account on how new ideas emerge in the design process. Fundamental understanding on this crucial part of design has thus been lacking. Here, we aim at a descriptive account of

[5] Peckhaus (2002) characterises regressive analysis in terms that could be used for abduction as well:"...regressive analysis is not completely logically determined, but has elements of contingency, creativity and intuition".

[6] Thus, we tend to agree with Pietarinen (2014): "Another way of putting a related point across is to observe that, taking retroduction only as a converse of deduction, or simply as reasoning from effects to causes, or from the major premiss and the conclusion to the minor premiss, is a limiting view of retroduction."

[7] For example, *abduction-1* as defined by Dorst (2011).

abduction in design. Often, prescriptive methods evolve based on descriptive theories (Kroll 2013), and thus it may be possible, as a next step, to start developing a method providing a prescription for innovative design.

In the following, a framework for conceptualising abduction in design is presented. It is based on two main starting points discussed in the preceding sections: (1) abduction is about introducing new ideas into the situation[8]; (2) abduction is a property of an inference besides an inference type as such. The first feature is of course directly coming from Peirce. Regarding the second, the common interpretation of Peirce is that it is an inference type. However, through accepting abduction (or abduction-like inference) happening by a mixture with induction (in the case of analogy) or deduction (in the case of theorematic reasoning), he *de facto* gave support to the idea that it is a property of an inference. The implications of this second starting point are summarized in Section 4.1.3.

We recognise the following important dimensions of abduction in design: triggering factor, position in the process, nature of the abductive inference, outcome of the abductive inference, and psychological character of abduction. In the following, these are presented in more detail and also in relation to understanding of abduction in scientific research. Based on these considerations, we then present the most common types of abduction in design.

4.1.1 Triggering Factor: Abductive Problem

Taking lead from Velázquez-Quesada et al. (2013), we propose that abduction is typically triggered by an abductive problem. In science, the search for fertile problems and their formulations often start the process. A surprising observation (Peirce) or an anomaly (Kuhn) provide a basis for such a problem. In design, we define abductive problem as one that the designer is not capable of solving using habitual or (generally) known solutions. This notion of abductive problem is relative, depending on the knowledge and experience of the designer or design team in question. An abductive problem does not necessarily lead to abduction; for example, if a solution can be found directly, in a non-abductive way from literature or from more experienced colleagues.[9] On the other hand, abduction may occur without a preceding abductive problem (this will be discussed below).

It is noteworthy that an abductive problem may be deliberately created through stretch targets (Thompson et al. 1997): create a new product that is, say, 10% better in performance than the best current corresponding products.

---

[8] Here we follow the intepretation of Suwa et al. (2000), according to which it is situated invention (interpreted here as situational novelty) that occurs in design, in contrast to historical invention (first time in history) and psychological invention (first time for the person in question), concepts defined by Boden (1996). Note that in science, historical inventions are targeted; nevertheless, these will emerge only through psychological invention.

[9] However, of course information acquired from literature or colleagues may also trigger a creative abduction.

4.1.2 Position in the Process: Anywhere

In typical formulations of abduction in science, a surprising observation motivates and initiates the inquiry process. Thus, abduction is positioned at the start of the process. A design process is often complex and covers a multitude of different tasks. *A priori*, that an abductive problem may emerge in any part of the design process can be expected.[10]

4.1.3 Abductive Inference: Property of the Inference, Not Just a Type

In science, the main type of abduction is the inference of the cause and/or rule from effect. This represents regressive inference, reasoning backwards. Reasoning backwards is similarly a common context for abduction in design. In design parlance, the question is about moving in the direction of means in an ends-means chain. However, it is suggested that other types of inference in design may be abductive, especially transformation (of the problem) as well as decomposition (of the problem and solution) and composition (of the solution). Further, there may be manipulative abduction such as sketching (Magnani 2004). All such inferences may be abductive or non-abductive, depending on whether they create a new idea or stay in the domain of the habitual and known (in the given context).

4.1.4 Outcome of Abduction

The outcome of a scientific abduction is a hypothetical statement about nature—"something may be". The other criterion Peirce attaches to the outcome of abduction is "uberty", fertility. In design, the outcome of abduction is a hypothetical solution candidate (conceptual or detailed, partial or total) or a design step facilitating the solution. Especially in the latter case, uberty is thus the evaluation criterion. Uberty can also be related to the strategic role (Paavola 2004) of abductive inferences in the design process. The mood of an abductive conclusion is "investigand", it points out how to continue the (tentative) process of finding a solution (Ma and Pietarinen 2016).

The result of design abduction is hypothetical in the sense that it is novel and it is not yet certain whether the move leads to a successful design. Thus the hypothesis is not about validity of a natural law or theory, but rather the utility of the outcome of the abduction as a part of the design solution or for the design process.

---

[10] Dorst's (2006) critical remarks regarding the customary idea of a specific design problem providing a starting point for design emphasise this point; "If the "design problem" in general is not knowable at any specific point in the design process; and if it is evolving in the design process—at least until the creation of the design concept, and possibly beyond that point; and if the connotations of the very concepts that are used to describe a "design problem" are shifting as a part of the design effort; then we need to radically reconsider our use of the term "design problem.""

Additional criteria can be set to design abduction, depending on the situation. Thus, the outcome of design abduction may be abstract and general or concrete and particular. That the thing abduced is possible, doable, is an important criterion especially in embodiment design. But in other cases, the uberty of the outcome, its ability to propel the process forward and to inspire further steps, may accentuate.

4.1.5 Cognitive/Psychological Character

Peirce connects abduction to a psychological process, not dissimilar to what Poincaré has suggested for creativity. Incubation of the problem, the solution emerging in a flash (CP 5.181, 1903), and the subjectively felt certainty connected to the outcome, are the hallmarks of this description. Although such a process seems to imply an abductive inference, it is not clear whether abduction always emerges through such a process.

4.2 Types of Abductive Inference in Design

4.2.1 Abductive Regressive Inference

This is the original type of inference named by Peirce as abduction in science: regressive (backwards) inference from effects to a (hypothetical) cause. In design, the concept of regressive inference is similarly time honoured; already Aristotle (n.d.) described deliberation in the following way:

> They assume the end and consider how and by what means it is to be attained; and if it seems to be produced by several means they consider by which it is most easily and best produced, while if it is achieved by one only they consider how it will be achieved by this and by what means this will be achieved, till they come to the first cause, which in the order of discovery is last.

It has been customary to call this reasoning chain the means-ends chain (Hughes 2009). Usually, neither inferences nor ends or means are characterised in any more detailed way; for example, regarding creativity.

However, as presented above, under influence by Peirce, design theorists have focused on regressive inference as abduction, starting from March (1976), with follow-up especially by Roozenburg (1993), Dorst (2011) and others. With their starting point in Peirce (rather than the traditional view on means-ends), these authors depict design abduction as one singular inference, especially from desired behaviour or function to structure. Note that this inference from behaviour to structure has usually been called synthesis in design literature; for example, by Hubka and Eder (1992) and by Gero (1999).

This single-inference view has been criticized by Kroll and Koskela (2015, 2016), who have pointed out that this should rather be a two inference chain, covering also the stage of conceptual solution. In other terms, a generalising inference creates the conceptual solution, and a particularising inference, the embodied solution.

However, it can be argued that there is in design a wider generic chain of reasoning backward and forward. For example Vermaas (2013) defines it as follows: (1) Goal of the device; (2) Action with the device; (3) Function of the device; (4) Behaviour of the device; (5) Structure of the device. Further, rarely is there one single chain of inferences in design but rather both the functions and structures have to be decomposed into their constituent parts, with their own means-ends chains.

The regressive inferences in design show a multitude of characteristics; some are habitual, some are selective and some are creative. It is these creative regressive inferences that we call abductive. It can be asked whether abductive regressive inferences in design are located only between "behaviour and structure", as implied by the former treatments of design abduction. We contend that abductive regressive inferences may occur in all parts of the means-ends chain of design.

4.2.2 Abductive Composition

The term composition refers here to a spatial or relational (abstract) arrangement of component parts of a system. Although Peirce's canonical definition of abduction does not explicitly capture this kind of mental move, he discussed it similarly to abduction, as something leading to a new idea[11] (Peirce, CP 7.498 1898):

> Suppose I have long been puzzling over some problem, — say how to construct a really good typewriter. Now there are *several ideas* dimly in my mind from time, none of which taken by itself has any particular analogy with my grand problem. But someday these ideas, all present in *consciousness* together but yet all very dim deep in the depths of *subconscious* thought, chance to get joined together in a particular way such that the combination does present a close analogy to my difficulty. That combination almost instantly *flashes out* into vividness. Now it cannot be contiguity; for *the combination is altogether a new idea.* It never occurred to me before; and consequently cannot be subject to any acquired habit. It must be, as it appears to be, its analogy, or resemblance in form, to the nodus of my problem which brings it into vividness. Now what can that be but pure fundamental association by resemblance?

That Peirce did not explicitly consider creative composition as abduction (in science) may be due to his focus especially on physics where causality relations accentuate as major scientific problems. Instead, in disciplines like chemistry and biology, finding the composition of an entity, the component ingredients of which are known, is a frequently encountered problem type. The characterisation of DNA

[11] Indeed, Paavola's (2004) characterization of strategic abduction approaches the understanding of abduction as composition: "This is strategic thinking: the constraints and hints that help in hypothesis finding are taken into account. And the goal in abductive inference (at least in most cases) is to find an overall *pattern* into which all evidence and clues fit … and this phase especially requires that various inferential moves be put together skillfully and by taking various clues and constraints into account (a paradigmatic case is detective stories, but this is in itself a very general model)."

as a double helix (Pray 2008) and the discovery by von Kekulé of certain organic compounds being closed chains, rings, regarding their molecule structure (Koestler 1975), provide examples of abductive composition.

Composition has a time honoured place in literature on design and its antecedents. In rhetoric, arrangement was the second of the five canons, referring to the positioning of topics into a speech. This linear, one-dimensional, understanding of composition was already in Antiquity generalised into two-dimensional cases (painting) and three-dimensional cases (sculpture, architecture). Still at the brink of the modern period, the last great scholar in rhetoric, Vico, held that *ingegno* (invention) gives things "a new turn or puts them into proper arrangement and relationships" (Verene 1981). The position and influence of rhetoric waned, but the idea of composition held its position, for example in architecture. In fact, the seminal author on design abduction, March (1976), named the outcome of abduction in architectural design "composition".

A vivid and illustrative example of an abductive composition is provided by the architect Aalto, who describes his working method as follows (Wilson 1979):

> The large number of different demands and sub-problems form an obstacle that is difficult for the architectural concept to break through. In such cases I work – sometimes totally on instinct – in the following manner. For a moment I forget all the maze of problems. After I have developed a feel for the program and its innumerable demands have been engraved in my subconscious, I begin to draw in a manner rather like that of abstract art. Led only by my instincts I draw, not architectural syntheses, but sometimes even childish composition, and via this route I eventually arrive at an abstract basis to the main concept, a kind of universal substance with whose help the numerous quarrelling sub-problems can be brought into harmony.

The many similarities of this description to Peirce's account are noteworthy. However, there is a new element, namely turning to an external medium, sketching, as support to composition. This is another type of abduction to be discussed next.

To sum up, we define abductive composition as such a spatial or relational (abstract) arrangement of component parts of a system, which can be interpreted as a new idea in the context.

4.2.3 Manipulative Abduction

This is not a type of abduction discussed by Peirce. Magnani (2005) characterises manipulative abduction: "In this kind of action-based abduction the suggested hypotheses are inherently ambiguous until articulated into configurations of real or imagined entities (images, models or concrete apparatus and instruments)." One of the most common forms of manipulative abduction in design may be sketching. According to Goldschmidt (1991), "in sketching, the designer creates visual displays which help induce images of the entity that is being designed". Kroll et al. (2001) describe a conceptual design method called "parameter analysis", in which the concepts and ideas ('parameters') proposed while designing cannot be evaluated directly; rather they have to be implemented as hardware representations ('configurations') first. The configurations themselves are only temporal and mostly serve to facilitate an evaluation.

In his book "The reflective practitioner", Schön has many examples of this type of design operation. He describes an academic supervisor in architecture (Schön 1993, 95):

> Quist makes his moves in a language of designing which combines drawing and speaking.

Such moves may lead to abductive insights (Schön 1993, 102):

> He discovers in the situation's back–talk a whole new idea which generates a system of implications for further moves.

The outcome may also be another type of abduction, to be discussed below, namely reframing (Schön 1993, 166):

> But their on-the-spot experiments, conducted in the virtual worlds of sketch-pad or storytelling, also function as transforming moves and exploratory probes. Hypothesis testing has the limited function of enabling them to achieve satisfactory moves or to surface phenomena which cause them to reframe the situation.

4.2.4 Abductive Transformation

Peirce (1907) discusses this type of inference in connection to Desargues' theorem in geometry. In this particular problem, the two-dimensional case is best proven by seeing it as a projection of the three-dimensional case. For explaining this move, Peirce introduces theorematic or theôric reasoning, which is deductive and undisputable (in contrast to the Peircean abduction that is non-deductive and hypothetical), but otherwise similar to abduction: "…theôric reasoning. It is very plainly allied to retroduction, …". The core of theôric reasoning is "in the transformation of the problem – or its statement – due to viewing it from another point of view" (Peirce, MS 318:68, 1907). Peirce translates "theôric", coming from the Greek word for theory, as "the power of looking at facts from a novel point of view" (Peirce, MS 318:68, 1907). Noteworthily, Hoffman (2010) classifies theoric reasoning as one type of abduction.

However, what Peirce describes has been a well-known step, namely drawing of auxiliary figures, towards a solution in geometry already since Antiquity. Aristotle's suggestion of the similarity of deliberation (into which design arguably falls) and geometric analysis has been interpreted as covering also such transformation of the problem (Koskela et al. 2014). Unfortunately, Aristotle's design theory was hardly followed up.

In the current design literature, Schön (1993) seems to have seminally described abductive transformation (although not connecting it to abduction explicitly):

> When he finds himself stuck in a problematic situation which he cannot readily convert to a manageable problem, he may construct a new way of setting the problem—a new frame which, in what I shall call a "frame experiment", he tries to impose on the situation.

Inspired by Schön, design theorists have adopted the term of frame (Dorst 2011) to refer to transformation.

An abductive transformation in design does not create a hypothetical solution, rather it creates a new problem that hypothetically is easier to solve than the original one. It is thus an example of strategic abduction (Paavola 2004).

4.2.5 Abductive Decomposition

Decomposition[12] refers to division of a whole into constituent parts. As such, this type of mental move has been known in philosophy at least since Plato, for whom division was an important operation for defining things.[13] However, as far as we know neither Peirce nor later abduction scholars have related abductive insights to decomposition.

In the domain of design, decomposition is a recognized mental move, referring especially to the decomposition of functions and structures (Smith and Browne 1993). In addition, there is the important step of allocation of functions to different parts of the structure, which we subsume under the term decomposition.

Our justification for the existence of abductive decomposition is empirical. Dorst and Cross (2001) describe an experiment with experienced designers, where all took the opportunity of separating the functions of the targeted artefact, a litter bin in a train, into two parts, realising them separately when earlier approaches had ended up at one integral artefact. All designers, working alone, considered this to be a creative, new idea.

Suh's (1990) method of axiomatic design has increased interest in creative and novel decomposing by pinpointing the importance of independence in the realisation of functional requirements.

---

[12] It may be asked whether it is necessary to discuss abductive composition and decomposition separately when they logically seem to be the two sides of one coin: the former starts from parts and creates a whole, the latter starts from a whole and creates parts. Both are applicable to the design problem and to its solution, and they often occur sequentially. A problem is often decomposed into subproblems (sub-functions) because it is easier to handle smaller problems. Once sub-solutions (solutions to sub-functions or subproblems) are found, they need to be composed into a whole solution. However, decomposition can also be applied to a solution (structure), if we need to allocate its components to different manufacturers or distinguish between parts that need to be made vs. those to purchase off-the-shelf. Likewise, composition can also be applied to functions, if we identify sub-functions that are realisable as a single entity. For example, if we need to convert electrical energy to rotational motion (realisable as an electric motor), and we need to reduce the speed of rotation (a gearbox), we may combine the two sub-functions because we recognise the possibility to use a gearmotor (a combination of motor and gearbox). Thus, the problems triggering abductive composition, on one hand, and abductive decomposition, on the other hand, are different and various, and it seems safe to treat them separately.

[13] In more recent times, Bergson has had ideas aligning to Plato's method of collection and division (Lawlor & Moulard Leonard 2013).

4.2.6 Abductive Analogical Reasoning

Analogical reasoning can be characterised as an inference from one particular to another particular. Thus it is opposed to conventional notions on induction, deduction and abduction, where at least one premise or conclusion is general.[14] In other words, it is reasoning from case to case - hence also the term case-based reasoning.

Analogies and analogical inferences have been considered as a valid way of reasoning since Antiquity (Bartha 2013). They have been commonly recognised as important elements of scientific discovery (Schickore 2014). For Peirce, an analogical inference appeared as a mixture of abduction and induction (possibly with some elements of deduction) (McJohn 1993; also see Peirce CP 5.277, 1868; CP 2.787, 1902). Commonsensically, the abductive element of an analogy is the recognition of the similarity of the cases, whereas the projection from the one case to the other provides for the inductive element.

In the recent design literature, analogical reasoning has been analysed from many angles, and methods have been devised for supporting "design-by-analogy" (Goel 1997, McAdams and Wood 2000). The whole area of biomimicry (several other terms, like biomimetics and bionics, are also used in roughly the same meaning) is based on utilising analogies from the biological world in product design. A well-known example is provided by the type of fabric hook and loop fastener colloquially known as Velcro, the design of which was inspired by burs from the plant burdock.

The use of analogies in design takes many forms, not yet much explored in the scholarly literature:

- analogy based on form (the Velcro example)
- visual analogy (Casakin and Goldschmidt 2000)
- analogy based on common physical principles (for example, taking inspiration from maple seeds for a single-wing air vehicle (Fregene and Bolden 2010) or looking at a pendulum as a spring because both generate a restoring force when their equilibrium is disturbed (Kroll 2013)).

As a general concept, analogy has been characterised as transfer of information from a source situation to the target situation (Casakin and Goldschmidt 2000). Thus, it seems to cater to many specific design inference types with insights, such as regressive, compositional or transformational inferences. Obviously, analogies in design may be abductive, resulting in new ideas, or non-abductive. The common approach of using precedent cases known to the designer, may generally fall into the category of non-abductive analogy.

[14] This way of contrasting analogy to other well-known types of reasoning, presented in ("Analogy" 2015), deserves to be justified. Of course, the very purpose of induction is to generalize. Regarding deduction, Peirce (CP 2.620, 1878) states: "All deduction is of this character; it is merely the application of general rules to particular cases." And regarding abduction (Peirce CP 7.218,1901): "Abduction seeks theory".

4.2.7 Abductive Invention of Requirements

The design problem, in the sense of constraints and functional requirements, is hardly ever given to the designer in its totality; rather, adding understanding of the problem is one integral part of design, usually called need analysis or task clarification. The first approaches to deal with this issue can be traced back to the ancient discipline of rhetoric, which emerged from the need to prepare – indeed design – speeches for defense or accusation at a court of law. The stage of finding ingredients and arguments (called topics) for the speech was called *inventio,* and authorities on rhetoric, such as Aristotle and Cicero, created guidelines for this activity. The main underlying idea was that it is easier to find something if one knows where to search, and thus pinpointing generic places (*topoi*) for arguments emerged as the popular methodological approach.

As far as it is known, the issue of adding understanding on the problem has not been identified and discussed as an individual stage in science, perhaps for the simple reason that scientific research as such equates to it. Thus, it has not been discussed in terms of involved inference types, neither by Peirce nor by later scholars.

In the early design literature, the "analysis-synthesis-evaluation" model of design contained an explicit stage, analysis, for extending the understanding of the problem (Braha and Maimon 1997). Subsequently, the focus has shifted to considering analysis (in this specific sense) as an intertwined aspect of the design effort, as expressed in the slogan[15] "analysis by synthesis", made popular by Lawson (1980). The underlying idea is that hidden requirements are found through the attempt to solve the problem rather than preceding it. The same idea is also forwarded through the claim that the problem space and the solution space are co-developed in design (Maher and Poon 1996).

The capture of hidden requirements is illuminatingly described by Suwa et al. (2000):

> There are at least two distinct ways in which this architect invented design issues or requirements. One way is to retrieve explicit knowledge or past cases and generate issues or requirements as the knowledge or the cases prescribe. The other way is to invent design issues or requirements by some justifications or reasons which are spontaneously constructed at the moment; those justifications or reasons are constructed on the fly by being mediated by a tacit component of the designer's knowledge.

---

[15] The concept seems to originate from research on perception in the 1950s (Halle and Stevens 1959). The idea is that "speech perception involves reconstructing the production plan" (Bever and Poeppel 2010), where speech perception refers to analysis and (internal) reconstruction of the production plan of speech equates to synthesis. This mechanism has been argued to exist also in reading and visual recognition. An up-to-date description of analysis by synthesis is provided in (Poeppel et al. 2008): "In particular, analysis-by-synthesis, or perception driven by predictive coding based on internal forward models, is a decidedly active stance towards perception that has been characterized as a 'hypothesize-and-test' approach. A minimal amount of signal triggers internal guesses about the perceptual target representation; the guesses (hypotheses) are recoded, or synthesized, into a format that permits comparison with the input signal." This understanding of analysis by synthesis seems to be applicable also to the use of this approach in design.

It is the latter way that arguably is abductive. But how is abduction triggered? Through solutions attempts (Suwa et al. 2000). Note that it is not only requirements relating to the original design problem but also conflicts between different design elements that get retrieved in this way:

> The architect's encounter of unintended visuo-spatial features in his sketches somehow activated parts of the tacit component of his knowledge, and thus the resulting interaction between the parts of the tacit knowledge and the unintended visuo-spatial features led to the construction of justifications or reasons for an inventive idea.

Abductive invention of requirements leads to unexpected findings, and there is no well-articulated problem preceding it, only the diffuse understanding about incompleteness of the requirements at hand. Research relates this type of abduction to sketching but it has to be noted that sketching also operates in and expands the solution space, having thus a wider scope.

4.2.8 Other Forms of Abduction in Design

The preceding discussion on types of abduction in design is by no means exhaustive. There are rarer types of abduction that cannot be discussed in detail due to space limitations. Two examples will suffice for illustration.

Abduction for integrating scientific theories in design has been discussed by Tomiyama et al. (2003), using the integration of knowledge on cooling and access in refrigerator design as an illustrative case. They identify it as second order existential abduction in Schurz' (2008) classification. In turn, the 1977 design of the Gossamer Condor, the first human-powered aircraft capable of sustained flight, was inspired by manipulating the equation representing the forces affecting an aircraft into such a form that emphasises what should be prioritised as design targets, especially a long wing span, low weight and low speed (Kroll and Farbman, 2016). In the mentioned Schurz' classification, this arguably represents theoretical model abduction.

5 Conclusion

By focusing on the differences between science and design as well as on empirical knowledge of different phenomena comprising design, we have derived new conceptions of abduction in design. Similarly to seminal and many subsequent treatments of abduction in science, we hold that an abductive design insight leads to a new idea, still hypothetical, by means of often subconscious, uncontrolled mental processes (implying thus an incubation period and often the flash of insight when it surfaces to the consciousness).

However, given the differences of context, abduction in design shows characteristics not found (or at least not identified clearly up to now) in science. Design abduction may emerge in any part of the design process (not just in the beginning as in typical reconstructions on science). Abduction can occur in connection to practically all inference types in design (rather than just through regressive inferences as in typical formulations of science); it is a property of an inference besides an inference itself. Abduction usually leads to an idea new in the context (rather than to entirely new ideas as in science). The main criterion of an abduced insight in design is its utility (rather than its truth as in typical examples of science).

What is the importance of these new conceptions of abduction for the design domain? For the first time, the different mental moves, which lead to new ideas in design, have been at least initially identified and described. The new conceptions invite empirical validation and verification. Based on this description, hopefully gaining validity but also perhaps evolving, new prescriptions for creative design may be devised in the future. Especially, the different categories of abduction might serve as places (*topoi*) in which to search for novel solution ideas when facing a design task, similar to existing creativity techniques but at a higher level of abstraction. Empirical research could then be applied to better understand the process of discovery within design and offer new insights on different forms or kinds of creativity. Generally speaking, expanded understanding on abduction offers new means and models to discuss and teach various phenomena in design.

However, we also humbly offer an unintended side outcome for the philosophy of science. The rich variety of abductive inferences found in design easily creates the suspicion that Peirce's seminal discussion on abduction in science has been interpreted in too schematic and narrow a way, failing to capture types of abduction not following the canonical form of "given the result, find the rule and the cause". The question arises whether an overhaul of the conception of abduction is also needed on the side of science.

Acknowledgement The authors are grateful for the insightful comments and helpful suggestions by the editors and by Dr. Glenn Ballard.

References

Aliseda, A. (2006). *Abductive reasoning. Logical investigations into discovery and explanation, Synthese Library, vol. 330*. Dordrecht: Springer.

Analogy. (2015, November 15). In Wikipedia, The Free Encyclopedia. Retrieved on Nov 28, 2015, from https://en.wikipedia.org/w/index.php?title=Analogy&oldid=690822710

Aristotle (n.d.) Nicomachean ethics [electronic resource]. http://classics.mit.edu/Aristotle/nicomachaen.html

Bartha, P. (2013). Analogy and analogical reasoning. The Stanford Encyclopedia of Philosophy. http://plato.stanford.edu/archives/fall2013/entries/reasoning-analogy/. Accessed 11 Dec 2015.

Bever, T. G., & Poeppel, D. (2010). Analysis by synthesis: A (re-) emerging program of research for language and vision. *Biolinguistics, 4*(2–3), 174–200.

Boden, M. A. (1996). *Dimensions of creativity*. Cambridge, MA: MIT Press.

Braha, D., & Maimon, O. (1997). The design process: Properties, paradigms, and structure. *IEEE Transactions on Systems, Man and Cybernetics, Part A: Systems and Humans, 27*(2), 146–166.

Casakin, H. P., & Goldschmidt, G. (2000). Reasoning by visual analogy in design problem-solving: The role of guidance. *Environment and Planning B: Planning and Design, 27*, 105–119.
Chen, Y., Zhang, Z., Xie, Y., & Zhao, M. (2015a). A new model of conceptual design based on scientific ontology and intentionality theory. Part I: The conceptual foundation. *Design Studies, 37*, 12–36.
Chen, Y., Zhao, M., Xie, Y., & Zhang, Z. (2015b). A new model of conceptual design based on scientific ontology and intentionality theory. Part II: The process model. *Design Studies, 38*, 139–160.
Cramer-Petersen, C.L., & Ahmed-Kristensen, S. (2015). *Reasoning in design: Idea generation condition effects on reasoning processes and evaluation of ideas*. Proceedings of the 22nd Innovation and Product Development Management Conference, June 14–16, Copenhagen, European Institute for Advanced Studies in Management (EIASM).
Dorst, K. (2006). Design problems and design paradoxes. *Design Issues, 22*(3), 4–17.
Dorst, K. (2011). The core of 'design thinking' and its application. *Design Studies, 32*, 521–532.
Dorst, K., & Cross, N. (2001). Creativity in the design process: Co-evolution of problem–solution. *Design Studies, 22*(5), 425–437.
Dunne, D. D., & Dougherty, D. (2016). Abductive reasoning: How innovators navigate in the labyrinth of complex product innovation. *Organization Studies, 37*(2), 131–159.
Eco, U. (1983). Horns, hooves, insteps: Some hypotheses on three types of abduction. In U. Eco & T. A. Sebeok (Eds.), *The sign of three. Dupin, Holmes, Peirce* (pp. 198–220). Bloomington: Indiana University Press.
Eco, U., & Sebeok, T. A. (Eds.). (1983). *The sign of three. Dupin, Holmes, Peirce*. Bloomington: Indiana University Press.
Eekels, J., & Roozenburg, N. F. (1991). A methodological comparison of the structures of scientific research and engineering design: Their similarities and differences. *Design Studies, 12*(4), 197–203.
Fann, K. T. (1970). *Peirce's theory of abduction*. The Hague: Martinus Nijhoff.
Feibleman, J. K. (1969). *An introduction to the philosophy of Charles S. Peirce: Interpreted as a system*. Cambridge, MA: M.I.T. Press.
Fl6rez, J. A. (2014). Peirce's theory of the origin of abduction in Aristotle. *Transactions of the Charles S. Peirce Society, 50*(2), 265.
Fregene, K., & Bolden, C. L. (2010). Dynamics and control of a biomimetic single-wing nano air vehicle. *Proceedings of IEEE American Control Conference (ACC)*, pp. 51–56.
Gabbay, D. M., & Woods, J. (2005). *The reach of abduction. Insight and trial. A practical logic of cognitive systems* (Vol. 2). Amsterdam: Elsevier.
Gabbay, D. M., Thagard, P., Woods, J., & Meijers, A. W. (2009). *Philosophy of technology and engineering sciences*. Burlington, MA: Elsevier.
Gero, J. S. (1999). Constructive memory in design thinking. In *Design thinking research symposium: Design representation* (pp. 29–35). Cambridge: MIT.
Goel, V. (1988). Complicating the 'logic of design'. *Design Studies, 9*(4), 229–234.
Goel, A. K. (1997). Design, analogy, and creativity. *IEEE Expert, 12*(3), 62–70.
Goldschmidt, G. (1991). The dialectics of sketching. *Creativity Research Journal, 4*(2), 123–143.
Habermas, J. (1978). *Knowledge and human interests* (2nd ed.pp. 147–148). London: Heinemann.
Halle, M., & Stevens, K. (1959). Analysis by synthesis. In W. Wathen-Dunn & L. E. Woods (Eds.), *Proceeding of the seminar on speech compression and processing* (Vol. II.) Paper D7.
Hanson, N. R. (1958). *Patterns of discovery*. Cambridge: Cambridge University Press.
Harman, G. H. (1965). The inference to the best explanation. *The Philosophical Review, 74*(1), 88–95.
Harman, G. (1968). Knowledge, inference, and explanation. *American Philosophical Quarterly, 5*(3), 164–173.
Hintikka, J. (1999). What is abduction? The fundamental problem of contemporary epistemology. In *Inquiry as inquiry: A logic of scientific discovery* (pp. 91–113). Dordrecht: Springer.

Hintikka, J. (2007). *Socratic epistemology: Explorations of knowledge-seeking by questioning*. Cambridge: Cambridge University Press.
Hintikka, J., & Remes, U. (1974). *The method of analysis: Its geometrical origin and its general significance*. Boston: D. Reidel.
Hoffmann, M. H. (2010). "Theoric transformations"and a new classification of abductive inferences. *Transactions of the Charles S. Peirce Society: A Quarterly Journal in American Philosophy, 46*(4), 570–590.
Hubka, V., & Eder, W. E. (1992). *Engineering design*. Zürich: Heurista.
Hughes, J. (2009). Practical reasoning and engineering. In D. M. Gabbay, A. Meijers, P. Thagard, & J. Woods (Eds.), *Philosophy of technology and engineering sciences* (pp. 375–402).
Koestler, A. (1975). *The act of creation*. London: Picador.
Kolko, J. (2010). Abductive thinking and sensemaking: The drivers of design synthesis. *Design Issues, 26*, 15–28.
Koskela, L., Codinhoto, R., Tzortzopoulos, P., & Kagioglou, M. (2014). The Aristotelian proto-theory of design. In A. Chakrabarti & L. T. M. Blessing (Eds.), *An anthology of theories and models of design: Philosophy, approaches and empirical explorations* (pp. 285–304). London: Springer.
Kroll, E. (2013). Design theory and conceptual design: Contrasting functional decomposition and morphology with parameter analysis. *Research in Engineering Design, 24*(2), 165–183.
Kroll, E., & Farbman, I. (2016). Casting innovative aerospace design case studies in the parameter analysis framework to uncover the design process of experts. *Design Science, 2*(e2).
Kroll, E., & Koskela, L. (2015). On abduction in design. In J. S. Gero & S. Hanna (Eds.), *Design computing and cognition '14*. Cham: Springer.
Kroll, E., & Koskela, L. (2016). Explicating concepts in reasoning from function to form by two-step innovative abductions. *Artificial Intelligence for Engineering Design, Analysis and Manufacturing (AIEDAM), 30*, 125–137.
Kroll, E., Condoor, S. S., & Jansson, D. G. (2001). *Innovative conceptual design: Theory and application of parameter analysis*. Cambridge: Cambridge University Press.
Lagerlund, H. (2016). Medieval theories of the syllogism. In E. N. Zalta (Ed.), *The Stanford Encyclopedia of Philosophy* (Spring 2016 ed..) URL = <http://plato.stanford.edu/archives/spr2016/entries/medieval-syllogism/>. Accesses 20 Apr 2016.
Lawlor, L., & Moulard Leonard, V. (2013). Henri Bergson. In E. N. Zalta (Ed.), *The Stanford Encyclopedia of Philosophy* (Winter 2013 ed..) URL = <http://plato.stanford.edu/archives/win2013/entries/bergson/>. Accessed 20 Dec 2015.
Lawson, B. (1980). *How designers think*. London: Architectural Press.
Lu, C.-Y., & Liu, A. (2012). Abductive reasoning for design synthesis. *CIRP Annals – Manufacturing Technology, 61*, 143–146.
Ma, M., & Pietarinen, A. V. (2016). A dynamic approach to Peirce's interrogative construal of abductive logic. *If CoLog Journal of Logics and their Applications, 3*(3), 73–104.
Magnani, L. (2001). *Abduction, reason, and science. Processes of discovery and explanation*. New York: Kluwer Academic/Plenum.
Magnani, L. (2004). Model-based and manipulative abduction in science. *Foundations of Science, 9*(3), 219–247.
Magnani, L. (2005). *Inductive generalizations and manipulative abduction*. Department of Philosophy and Computational Philosophy Laboratory, University of Pavia. http://www.cs.bris.ac.uk/~oray/AIAI05/paper_magnani.pdf. Accessed 11 Dec 2015.
Maher, M. L., & Poon, J. (1996). Modeling design exploration as co-evolution. *Computer-Aided Civil and Infrastructure Engineering, 11*(3), 195–209.
March, L. (1976). The logic of design and the question of value. In L. March (Ed.), *The architecture of form* (pp. 1–40). Cambridge: Cambridge University Press.
McAdams, D. A., & Wood, K. L. (2000). Quantitative measures for design by analogy. *Proceedings of ASME Design Engineering Technical Conferences (DETC'00)*, September 10–13, Baltimore.

McJohn, S. M. (1993). On uberty: Legal reasoning by analogy and Peirce's theory of abduction. *Willamette Law Review, 29*, 191–235.

Minnameier, G. (2004). Peirce-suit of truth – Why inference to the best explanation and abduction ought not to be confused. *Erkenntnis, 60*, 75–105.

Newton I. (2003) *Opticks.* (great mind series), Prometheus books.

Nickles, T. (Ed.). (1980). *Scientific discovery, logic, and rationality*. Dordrecht: D. Reidel.

Niiniluoto, I. (1999a). Defending abduction. *Philosophy of Science*, S436–S451.

Niiniluoto, I. (1999b). Abduction and geometrical analysis. Notes on Charles S. Peirce and Edgar Allan Poe. In L. Magnani et al. (Eds.), *Model-based reasoning in scientific discovery*. New York: Kluwer Academic/Plenum.

Paavola, S. (2004). Abduction as a logic and methodology of discovery: The importance of strategies. *Foundations of Science, 9*(3), 267–283.

Paavola, S. (2006). Hansonian and Harmanian abduction as models of discovery. *International Studies in the Philosophy of Science, 20*(1), 93–108.

Paavola, S. (2012). *On the origin of ideas. An abductivist approach to discovery.* Revised and enlarged edition of a dissertation (2006). Saarbrücken: Lap Lambert academic Publishing.

Paavola, S., & Hakkarainen, K. (2005). Three abductive solutions to the Meno Paradox – With instinct, inference, and distributed cognition. *Studies in Philosophy and Education, 24*(3–4), 235–253.

Pauwels, P., & Bod, R. (2014). Architectural design thinking as a form of model-based reasoning. In L. Magnani (Ed.), *Model-based reasoning in science and technology, Studies in applied philosophy, Epistemology and rational ethics* (Vol. 8). Berlin: Springer-Verlag.

Peckhaus, V. (2002). Regressive analysis. *Philosophiegeschichte und logische Analyse (Logical Analysis and History of Philosophy), 5*, 97–110.

Peirce, Charles S. [CP (volume.paragraph, year] (1931–1958). Collected papers of Charles Sanders Peirce, Vol. 1–6, Hartshorne, C. and Weiss, P., (eds.), Vol. 7–8, Burks, A. W., (ed.). Cambridge, Mass: Harvard University Press.

Peirce, C. S. (1967). *Manuscripts in the Houghton library of Harvard University, as identified by Richard Robin, Annotated catalogue of the papers of Charles S Peirce*. Amherst: University of Massachusetts Press.

Peirce, Charles S. [W (volume: page numbers, year)] (1982–) Writings of Charles S. Peirce: A Chronological Edition, 6 vols. (to date), the Peirce Edition Project (eds). Bloomington: Indiana University Press.

Peirce, Charles S. [EP (volume: page numbers, year)] (1992–8) The Essential Peirce: Selected Philosophical Writings, 2 vols., the Peirce Edition Project (eds). Bloomington & Indianapolis: Indiana University Press.

Peirce, Charles S. [PPM (page numbers, year)] (1997) Pragmatism as a Principle and Method of Right Thinking. The 1903 Harvard Lectures on Pragmatism, Ann Turrisi (ed.). Albany: State University of New York Press.

Pietarinen, A. V. J. (2014). The science to save us from philosophy of science. *Axiomathes*, 1–18.

Poeppel, D., Idsardi, W. J., & van Wassenhove, V. (2008). Speech perception at the interface of neurobiology and linguistics. *Philosophical Transactions of the Royal Society B: Biological Sciences, 363*(1493), 1071–1086.

Pray, L. (2008). Discovery of DNA structure and function: Watson and Crick. *Nature Education, 1*(1), 100.

Psillos, S. (2009). An explorer upon untrodden ground: Peirce on abduction. *Handbook of the history of logic: Inductive logic, 10*, 117–151.

Reichenbach, H. (1938). *Experience and prediction*. Chicago: University of Chicago Press.

Roozenburg, N. F. M. (1993). On the pattern of reasoning in innovative design. *Design Studies, 14*, 4–18.

Roozenburg, N. F. M., & Eekels, J. (1995). *Product design: fundamentals and methods. Chapter 4*. Chichester: Wiley.

Schickore, J. (2014). Scientific discovery. *The Stanford Encyclopedia of Philosophy*. http://plato.stanford.edu/archives/spr2014/entries/scientific-discovery/. Accessed 11 Dec 2015.
Schön, D. A. (1993). *The reflective practitioner: How professionals think in action*. New York: Basic Books.
Schurz, G. (2008). Patterns of abduction. *Synthese, 164*(2), 201–234.
Smith, G. F., & Browne, G. J. (1993). Conceptual foundations of design problem solving. *IEEE Transactions on Systems, Man and Cybernetics, 23*(5), 1209–1219.
Snyder, L. J. (1997). Discoverers' induction. *Philosophy of Science, 64*(4), 580–604.
Suh, N. P. (1990). *The principles of design*. New York: Oxford University Press.
Suwa, M., Gero, J., & Purcell, T. (2000). Unexpected discoveries and S-invention of design requirements: Important vehicles for a design process. *Design Studies, 21*(6), 539–567.
Takeda, H. (1994). Abduction for design. In J. S. Gero & E. Tyugu (Eds.), *Formal design methods for CAD* (pp. 221–244). Amsterdam.
Takeda, H., Veerkamp, P., Tomiyama, T., & Yoshikawa, H. (1990). Modeling design processes. *AI Magazine, 11*(4), 37–48.
Takeda, H., Sasaki, H., Nomaguchi, Y., Yoshioka, M., Shimomura, Y., Tomiyama, T. (2003). Universal abduction studio–proposal of a design support environment for creative thinking in design. *Proceedings of the 14th Int. Conf. Engineering Design (ICED 03),* Stockholm, Aug. 19–21.
Thompson, K. R., Hochwarter, W. A., & Mathys, N. J. (1997). Stretch targets: What makes them effective? *The Academy of Management Executive, 11*(3), 48–60.
Tomiyama, T., Takeda, H., Yoshioka, M., Shimomura, Y. (2003). Abduction for creative design. *Proceedings of ASME Design Engineering Technical Conferences (DETC'03),* (pp. 543–552), Sept 2–6, Chicago.
Ullah, A. M. M. S., Rashid, M. M., & Tamaki, J. (2012). On some unique features of C–K theory of design. *CIRP Journal of Manufacturing Science and Technology, 5*, 55–66.
Velázquez-Quesada, F. R., Soler-Toscano, F., & Nepomuceno-Fernández, Á. (2013). An epistemic and dynamic approach to abductive reasoning: Abductive problem and abductive solution. *Journal of Applied Logic, 11*(4), 505–522.
Verene, D. P. (1981). *Vico's science of imagination*. London: Cornell University Press.
Verene, D. P. (2008). *The history of philosophy: A reader's guide*. Evanston: Northwestern University Press.
Vermaas, P. E. (2013). The coexistence of engineering meanings of function: Four responses and their methodological implications. *Artificial Intelligence for Engineering Design, Analysis and Manufacturing, 27*(3), 191–202.
Wilson, C. S. J. (1979). Alvar Aalto and the state of modernism. *International Architect, 1*(2), 27–32.

Perennial Prototypes: Designing Science Exhibits with John Dewey

Kim Kullman

Abstract This chapter evokes the writings of John Dewey to investigate his pragmatist design philosophy through a study on exhibit development at the Exploratorium science museum in San Francisco. The chapter begins by describing the main features of Dewey's thinking, concentrating specifically on two notions: *experience* and *experiment*. It then transposes these into the area of exhibit development to explore their potential for contemporary design practice and their largely unexamined implications for design theory.

Keywords John Dewey · Pragmatism · Design · Science museums · Exhibits

1 Introduction

This chapter traces a distinctive philosophy of design through the pragmatist writings of John Dewey (1859–1952). For him, the purpose of designed objects and spaces is to facilitate a shared exploration of novel experiences that encourage involvement with others and the world. This makes design an ethical endeavour that reorganises materials to induce affective and perceptual transformations and uses such transformations to cultivate new modes of engagement.

Contributing to nascent work on design and pragmatism (see Dalgaard 2014; Melles 2008; Rylander 2012), the chapter brings these ideas into contact with a specific empirical setting, the Exploratorium in San Francisco, which is a museum known for its exhibitions that combine art, design and science to foster collaborative enquiry among visitors. Applying Dewey to analyse exhibition development, the chapter outlines the workings of a design philosophy that is committed to experimenting with the conditions of experience and their reconfiguration. It demonstrates how exhibits on motion, weather phenomena and nonhuman life-forms are crafted to attune bodies to the countless entities and forces that connect them with

K. Kullman (✉)
Department of Geography, The Open University, Milton Keynes, UK
e-mail: kim.kullman@open.ac.uk

© Springer International Publishing AG 2018

P. E. Vermaas, S. Vial (eds.), *Advancements in the Philosophy of Design*, Design Research Foundations, https://doi.org/10.1007/978-3-319-73302-9_9

broader earthly processes. As a consequence, the exhibits have been assembled into prototypical arrangements that invite completion by visitors, who become co-experimenters alongside designers, learning together about the shifting foundation of material environments.

The Exploratorium, which was studied in 2015 and 2016 through ethnographic methods (see Gunn et al. 2013; Koskinen et al. 2011; Yaneva 2009) as part of a European Reasearch Council project (323777), serves as a productive test site for Dewey's design philosophy due to its grounding in constructivist learning theories that stem in part from the pedagogical writings of Dewey and related thinkers (Allen and Gutwill 2004; Gutwill 2008; Hein 1990; Hein 2016). As the full potential of his vocabulary has not been examined in actual exhibit development, it is the purpose of this chapter to mobilise his concepts to investigate whether pragmatism and exhibit design might accentuate previously unarticulated features in each other and therefore advance design practice and theory.

2 Dewey as a Design Philosopher

Although it is beyond the scope of this chapter to present an exhaustive discussion on the work of Dewey and its relationship to American pragmatism, some introductory remarks will ease our way into his thinking. Dewey is often credited as having systematised and popularised the seminal writings of Charles Sanders Pierce (1839–1914) and William James (1842–1910), which he developed across a variety of fields, including aesthetics, education, ethics and science. While differing over their understanding of pragmatism, the three are united in their views of philosophy as a mode of enquiry that grounds knowledge in experience and tests ideas in concrete situations to bring theory and practice into a closer conversation. Pragmatism presupposes a "radical empiricism" (James 2003, 22–23), where the world is seen as constituted in ever-diversifying relations between entities, rather than as possessing fixed properties to be established through prior reasoning. Stressing the experiential character of philosophy, pragmatists suggest that attempts to know the world are entangled with it and thinking is a practice among other practices:

> the object of knowledge is not something with which thinking sets out, but something with which it ends: something which the processes of inquiry and testing, that constitute thinking, themselves produce. Thus the object of knowledge is practical in the sense that it depends upon a specific kind of practice for its existence. (Dewey 2004a, 212)

Thinking involves arranging and shaping materials in such a way that they produce desired effects, which alter both theories and practices. Dewey (2004a, 264) likens thinking to boat building, where wood is given "a form which it did not have, in order that it may serve the purposes to which it is to be put." Whether designing boats or thoughts, the process is precarious, because variegated materials need to be balanced against each other and because the "usefulness" of the emerging object is "whatever, from infinity to zero, experience may subsequently determine it to be" (Dewey 2004a, 213). Such a line of argument, as will be demonstrated in the remainder of this section, opens a space for a distinctively Deweyan version of pragmatist design philosophy.

One entry-point into the work of Dewey are his writings on aesthetics, where he develops an argument about art and its material qualities, embodied effects and cultural embeddedness. Significantly, Dewey (2005, 222–254) demonstrates that architecture, design, music and related practices, strive for an intensification of experience. Art objects, Dewey (2005, 204) claims, operate as mediators between humans and the wider world by channelling energies and forces so that those who come in touch with them have transformative experiences—works of art are "the middle, the intervening, things through which something now remote is brought to pass." Art, furthermore, has the quality of "expressiveness" (Dewey 2005, 80–81) that stems from the ability of the practitioner to assemble materials into a medium that generates "a concrete situation," which evokes a new "emotional response" in people (Dewey 2005, 70). Fashioning materials to facilitate expression is a process that Dewey (2005, 121) refers to as *design*: "an ordered relation of many constituent elements" that is characterised by "the intimacy of the relations that hold the parts together". Dewey's views of design are permeated by a relational ontology, which is based on the idea that relations are always prior to entities—a relation "denotes something direct and active, something dynamic and energetic. It fixes attention upon the way [elements] fulfill and frustrate, promote and retard, excite and inhibit one another" (Dewey 2005, 139). Design does not "superimpose" form on matter but brings disparate elements into novel arrangements that address bodies in their "unified vitality" (Dewey 2005, 122). The distinctiveness of designed objects, whether these are paintings or buildings, can be found in their aliveness and potential for growth that "carry forward" (Dewey 2005, 172–173) experience in unexpected ways:

> A work of art elicits and accentuates this quality of being a whole and of belonging to the larger, all-inclusive, whole which is the universe in which we live [...] We are carried out beyond ourselves to find ourselves [...] the work of art operates to deepen and to raise to great clarity that sense of an enveloping undefined whole (Dewey 2005, 202-203).

The notion of *experience* forms the centrepiece of Dewey's empiricism, where it refers to the pervasive quality of existence (Johnson 2007, 71–78). Humans do not passively receive affects and sensations from the world, but, as embodied beings, engage with their environments in co-producing experiences through ongoing adjustments in response to other human and nonhuman entities (Dewey 2005, 184). Through its dynamism, experience is constantly evolving, which means that it "extends much further than that which at any time is *known*. [...] there are always potentialities which are not explicit" (Dewey 2015a, 20–21; original emphasis). For

Dewey, experience serves as a growing ground for experimentation, where bodies and materials become involved in testing novel types of relationality.

Dewey is advancing a form of experimentalism that is based on discovering the conditions under which new experiences might emerge and how these conditions could be kept productive (Kullman 2013). Societies abound with experimental sites, from science, where laboratories have become places for reordering relations among culture, technology and nature (Dewey 2004a), to politics, where collectives explore alternative ways of living together by trying out new modes of governance and organisation (Dewey 1991). Dewey devotes considerable attention to educational experiments, which incorporate material "arrangements that will permit and encourage freedom of investigation" (Dewey 2015b, 59) among teachers and pupils alike. This is a key feature of Dewey's constructivist pedagogy, where people learn through active engagement and by "keeping alive the ordinary bonds" (Dewey 2015b, 47) to their worlds. Another aspect of this pedagogy is acquiring an "instinct of making" that "seeks outlet in shaping materials into tangible forms" (Dewey 2015b, 30), from cooking and sewing to building, carpentry and painting. Dewey's notion of making is indistinguishable from his concept of design, as both involve a process of testing materials and their permutations to learn about how they could be combined to create specific effects. Making does not simply entail producing isolated objects, but exploring new relations through emerging artefacts. Experimenting with materials, Dewey explains, one learns about the broader economic, physical and social processes that shape them—for example, "carpentry" and "textile shops" can connect pupils "with the country, as the source of their materials, with physics, as the science of applying energy, with commerce and distribution, with art in the development of architecture and decoration" (Dewey 2015b, 51). Making, much like design, can gradually expose a wider geography of relations that both uphold an object and transport it beyond its immediate setting (Massey 2005; Murdoch 2005).

Taken together, the discussion so far has outlined the main ingredients of Dewey's design philosophy. For him, there are, strictly speaking, no separate entities in the world, because everything is folded into the "pervasive qualitative whole" of experience (Johnson 2007, 75–76). Similar to other entities populating the earth, designed objects are better understood as relational phenomena that need to be studied as part of a wider meshwork of bodies and materials on which they necessarily depend. A challenge for design is to produce objects that make this relationality more apparent and, like artwork, has the potential to modulate experiences in such a way that our attachments to human and nonhuman others are both revealed and creatively reorganised. As people shape and share objects, they come to trace out the profound interconnectedness of all entities and cultivate an awareness of the "common world" in which they participate (Dewey 2015b, 55).

Above all, Dewey suggests that the purpose of design is to create objects and environments that facilitate experiments with experience. He invites us to consider design as a speculative undertaking that may give rise to more responsive ways of being and relating. A consequence of such an approach is that "design is not a thing but a process," which "can be thought, articulated, embodied and practised" in myriad ways (Brassett and Marenko 2015, 12). Instead of seeking to establish an essence for design, pragmatism brings into relief its open-endedness and the possibilities it offers for a continuous reconfiguration of relations that make up the world. Let us now turn to the Exploratorium in order to investigate this argument further.

3 Designing Science Exhibits with Dewey

The Exploratorium is a 31,000m² science museum located on Pier 15 by the eastern waterfront of San Francisco.[1] The institution was founded in 1969 by educator Frank Oppenheimer (Shapin 2010) and occupied the Palace of Fine Arts in the Marina District until 2013, when it moved into the present building. Since opening, the primary goal of the Exploratorium has been to make "natural phenomena and the world around us both exciting and understandable" through carefully crafted and curated exhibits that invite collaborative and explorative forms of learning.[2] There are currently around 670 exhibits on display in the museum, which in 2016 was visited by some 850,000 people of all ages.[3]

Writing about the original location of the Exploratorium, Hein (1990, 86) points out that exhibit design "was never hidden away in the bowels of the museum but was itself an exhibit on display." Placed at the centre of the Palace of Fine Arts, the studio showed developers at work on exhibits, thereby exposing the material labour behind the museum experience. The architecture of the new building on Pier 15 is arranged in a similar manner with the workshop occupying a prominent place in the middle of the museum, divided from the passing visitors by a low fence, over which people can look into the exhibit development space and invariably see pieces at diverse stages of completion (Fig. 1).

For exhibit developers at the Exploratorium, science is less about "objectively pure empiricism" or "certainty and a sense of mastery over the forces of nature" (Toon 2005, 34) than about tinkering with materials in a collaborative spirit. Working on exhibits in front of onlookers, developers, as Hein (1990, 112) notes, "help demystify [...] equipment by displaying it in the context if its origin and its current use and operation, at the same time reconciling science with ordinary experience". They also involve visitors as experimenters in the shaping of exhibits in line with the constructivist learning theories that have been influential for the Exploratorium as an institution and could be crystallised into three components: "open-ended exploration with gentle guidance", "self-driven discovery" and a "shift of the visitor's role from that of recipient [...] to that of participant [...] in the generation of activities, questions, and explanations" (Humphrey and Gutwill 2005, 3). While this pedagogy has been broadly inspired by Dewey (Allen and Gutwill 2004;

[1] https://www.exploratorium.edu/press-office/press-releases/exploratorium-visitor-overview-fact-sheet
[2] https://www.exploratorium.edu/press-office/press-releases/exploratorium-2015-overview-fact-sheet
[3] https://www.exploratorium.edu/about/fact-sheet

Fig. 1 The workshop in the Exploratorium, image by author

Gutwill 2008), among others, the full conceptual potential of his design philosophy remains to be explored as part of actual exhibit production work. The ensuing discussion will therefore use two notions from Dewey, *experience* and *experiment*, as a way to create dialogue between his thinking and exhibit development, with the aim of allowing each to bring out new features in the other and offer insight into a pragmatist design philosophy.

3.1 Experience: Perceptual Phenomena

Let us begin with Dewey's concept of experience, which directs attention to the conditions under which new affects, sensations and thoughts emerge (Dewey 1997a, 33–50). Dewey is not referring to any experience, but to "*an* experience"—a qualitative perceptual transformation that is "demarcated in the general stream of experience from other experiences" (Dewey 2005, 36–37; original italics). This orientation brings to the fore the diverse perceptual effects sought by exhibit developers at the Exploratorium. As an example, we may take Charles Sowers, who produces installations that address the aesthetics of weather phenomena to involve visitors in multisensory ways with natural processes. These include *Illuminated Fog*, a round table containing a tray with patches of fog floating over a dark and damp surface (Fig. 2). Instead of covering the device in glass, Sowers realised that the exhibit would be more engaging were the visitors allowed to use their hands to feel the dampness of

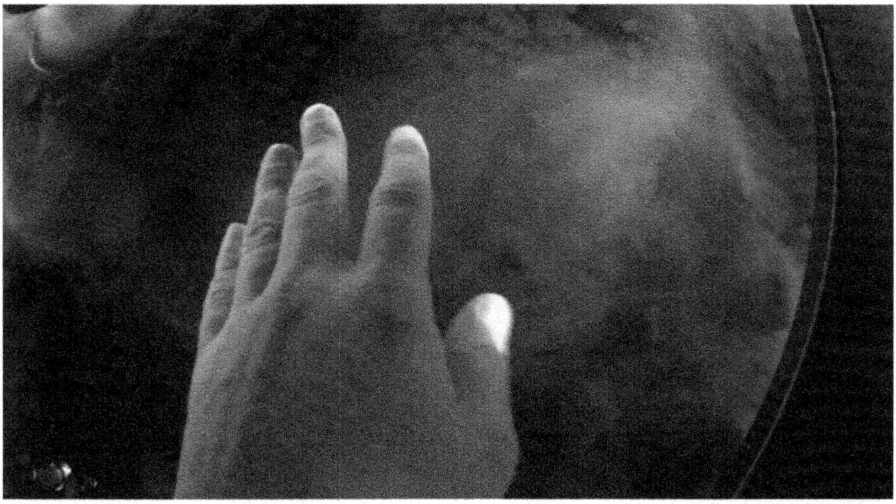

Fig. 2 *Illuminated Fog* by Charles Sowers, image by author

the fog and create turbulent patterns. The exhibit therefore illuminates what Oppenheimer, the founder of the museum, called "perceptual phenomena",[4] which is based on the idea that if "everything we know is filtered through our perceptual apparatus" then "all of science hinges on it one way or the other" (Cole 2009, 157). The work of Sowers demonstrates how developers take nonhuman forces and, similar to the artists described by Dewey (2005, 57), render them "perceptible as part of the common world".

Another fog-related exhibit offers further insight into how developers summon perceptions that are transformative and speak of the inseparability of humans and environments. *Fog Bridge* by artist Fujiko Nakaya is an outdoors installation that envelops a bridge between Piers 15 and 17 in a foggy microclimate. The exhibit, based on decades of experimentation by Nakaya, was designed to produce fog patterns in response to the changing weather in the surrounding bay. The success of *Fog Bridge* has been due to the variability of the experiences it creates, depending on the weather and the time of the day. The fog can assume an endless diversity of shapes, from billows to blankets, with shifting degrees of density and permanence (Markopoulos and McDougall 2013). As one commentator notes, the *Fog Bridge* begins to alter perceptions of the relationship between weather, space and bodies:

> The fog momentarily de-solidifies the hard materials of the bridge, obscuring its structure and passengers. It changes not only our physical sensations, but also our social and emotional expectations, as structure and people disappear from view and our sense of place is disrupted by an improbable curtain (Rockwell 2013, 5).

[4] https://www.exploratorium.edu/files/about/our_story/history/frank/pdfs/playful_museum.pdf

Fig. 3 *Glass Settling Plate* by Denise King, image by author

Skilfully blurring the boundaries between subjects and objects, experience and phenomena, humans and nonhumans, the *Fog Bridge* ties in with other exhibits in the Exploratorium, which are rarely seen as self-contained pieces, but temporary conduits that all kinds of energies, forces and entities pass through. The work of Denise King, for instance, "recycles", as she terms the process, living organisms from the bay. One of her exhibits is called *Glass Settling Plate* and consists of a large display cabinet with a digital microscope directed towards a transparent plate, where organisms from the bay are growing (Fig. 3). It is removed every week and lowered into the sea from the pier, so that marine organisms can feed in their natural environment and new colonies settle on the plate. Although the glass plate appears unspectacular to the bare eye, the microscope reveals an infinite complexity of colours, shapes, textures and movements that offer visitors access to a barely perceptible micro-world of nonhuman entities.

A distinct feature of the above exhibits is that they address visitors as multisensory beings by expanding their embodied interactions with phenomena. The body is central to all forms of enquiry, because people, as Dewey (1997b, 190) writes, use their "eyes, ears, and sense of touch as guides to action" when acquiring knowledge about the world. At the same time, visitors can sense the forces that are channelled through the exhibits, as suggested by pieces such as the *Motor Effect*, which invites people to push down an electrified wire suspended between two magnets and feel the forces acting on the wire. Likewise, the *Finger Tingler* asks visitors to operate a manual generator to produce electricity, which is allowed to pass through their fingers via a brass plate. As the exhibit plaque explains:

> The intensity of an electric shock depends on the amount of current flowing through your body. The current in turn depends on both the voltage and your resistance. Moist skin has a lower resistance than dry skin, so the same voltage will give you a bigger shock if you wet your fingertips. You can also lower your resistance by pressing more skin into contact with the plates.

Dewey describes in a similar vein how a group of school children, as they experimented with "simple material things", such as rocks and metals, "felt the connection" (Dewey 2015b, 37) between earthly processes and thus gained a grounded understanding of their workings to supplement textbook knowledge. The exhibits in the Exploratorium function in a comparable manner, because visitors are encouraged to become moved by embodied encounters with phenomena and actively look for an explanation for their variegated effects. As the following section indicates, the fact that exhibits function as temporary relays for wider transformations makes it difficult to impose a sense of object-ness on them. Instead, they are open to constant reconfiguration and hence resonate with Dewey's idea of artefacts as emerging only momentarily against the background of experience (Johnson 2007, 75–76).

3.2 Experimentation: Perennial Prototypes

The notion of experimentation developed by Dewey provides further purchase on exhibit design, particularly on the unique approach to prototyping cultivated in the Exploratorium. Prototyping is often seen as a short-lived phase in design, but the developers expand this practice and turn it into a permanent feature of their pieces. For Dewey, experimentation is similarly an ongoing process of testing new ideas in everyday settings. As Dewey (2004b, 264; original emphasis) argues, experimentation is based on the realisation that people "have to *do* something to the things when they wish to find out something; they have to alter conditions."

Full-scale prototyping begins early and is central to every project at the Exploratorium, which means that developers have extensive building skills and the workshop is filled with an array of tools, machinery and materials that enable various forms of making and mending. The developers therefore have the ability to "build up simultaneously the idea and its objective embodiment" (Dewey 2005, 53). According to Dewey, this is increasingly rare in the world of architecture and design, where ideas and objects are not allowed to "modify one another" over a longer period, which was the case, for example, in the construction of medieval cathedrals—a process that was not "controlled by plans and specifications made in advance [...] Plans grew as the building grew" (Dewey 2005, 53–54). Prototyping could be understood as a source of such intimate design knowledge by constituting a "way to understand touch, materials, shapes, and the style and feel of interaction" (Koskinen et al. 2011, 134). Likewise, prototyping involves designers in embodied practices, where "hands are in the middle of the process, offering nonverbal feedback based on texture, ergonomics and proportion" (Adams 2013, 106).

Fig. 4 An incomplete exhibit in the workshop, image by author

In the Exploratorium, early prototyping enables developers to save time and resources by creating an immediate understanding of the viability of a design, including its workings, the cost of its parts and its appeal to visitors. While developers produce sketches, technical drawings and other representations, prototyping serves as the main vehicle for design work (Fig. 4). The prototype is a reminder that designs are required to undergo a series of difficult trials to become exhibits, from tests by colleagues to interactions with visitors, whose responses are key from the outset, as one developer, David Torgersen, explains:

> You are just trying to get the phenomenon working as cheaply and as easily as possible [...] You build it to a point where visitors can use it, but perhaps it is not completely stand-alone and I have to make it a mediated experience, so I will [...] put it on the floor and I will stay back a few feet [...], but in case something goes wrong I can step right in.

Torgersen is describing a recurrent practice among developers, where prototypes, if deemed to be in workable order, are promptly placed on the museum floor for trial. Designers then spend time alongside the prototype, "mediating", as Torgersen calls it, between the exhibit and the visitors to make the encounter as engaging as possible. Almost without exception, these moments result in new insights about exhibits that can be incorporated into subsequent iterations. A case in point is *Water Drop Photography* by Erik Thogersen, which consists of a camera set-up and a screen that visitors can employ to create and view images of water dripping into a vessel. When Thogersen brought an early prototype to the floor, he and a visitor started to experiment with the set-up, realising that it produced not only images of water drops, but intriguing self-portraits, as the photographed drops reflected the surrounding space upside down. Thogersen therefore decided to alter the set-up and build a transparent case for it, so that people could move around the device and explore its reflective capacities.

The Exploratorium also conducts plenty of research on its development practices, as illustrated by the book *Fostering Active Prolonged Engagement* (Humphrey and Gutwill 2005), which consists of case studies on exhibits, each describing the idea behind designs and common responses among users, which were recorded with video and by tracking visitor paths. Such careful documentation demonstrates that developers are not experimenting for its own sake, but to ensure that the viability of designs is appropriately ascertained. This view is echoed in the writings of Dewey (1997a, 87), where experimentation involves "keeping track of ideas, activities, and observed consequences. Keeping track is a matter of reflective review and summarizing, in which there is both discrimination and record of the significant features of a developing experience".

However, although an exhibit might appear stable enough to deserve its place amidst other installations on the floor, this status is always temporary and open to reconsideration. Many exhibits, for instance, are deliberately left incomplete to foster user engagement, as developers believe that exhibits with a degree of unpredictability invite active experimentation and learning. A fitting example is *Marble Machines*, which has remained a prototype for the whole 10 years of its existence. Only seemingly simple, the exhibit is composed of a set of perforated hardboard walls onto which visitors can build paths for marble balls. The exhibit, which has proven a success among visitors of all ages, encourages the creative use of scrap materials by allowing people to design various types of contraptions to investigate complex chain-reactions. As Sebastian Martin, one of the developers of the wall, describes its inventive potential: "You do something that you think is going to work and then you test it out. Almost never is it going to work out the way you thought it would. And then you have to tinker with it."[5]

The practices of developers rest on the assumption that exhibits are sites of incessant tinkering, as there are always new ways to respond to an installation and hence scope for alteration and improvement. This thinking is supported by the fact that most developers seek to sustain, as King expresses it, "a connection to the floor" by closely following how exhibits are used and by taking part in maintenance work, which can range from cleaning up and changing light bulbs to carrying out extensive repairs. Such practices offer plenty of opportunities to revisit exhibits and explore how they might respond to changing understandings of phenomena and shifting user requirements. Developers therefore regard their exhibits as *perennial prototypes* by refusing to elevate them to the position of independent and separate objects. Instead, they view exhibits as enmeshed within a wider set of relations, which together shape what an exhibit is and could become.

Sustaining the perennial prototype asks for its own kind of design skills and sensibilities, which are illustrated by the *Tinkering Studio* that was founded in 2013 with the aim of elaborating on the pedagogical potential of the experiments unfold-

[5] http://tinkering.exploratorium.edu/marble-machines

Fig. 5 *Tinkering Studio* in the Exploratorium, image by author

ing in the workshop and on the museum floor (Fig. 5).[6] The studio is located in a separate space next to the workshop and offers classes for groups of children and adults, where

> people get to [...] make something that *they* want to make. [...] It extends the exhibit design experience in new ways by enabling people to explore their curiosity with the tools to make that happen. [...] tinkering enhances one's sense of design [...] It lets one create a feeling for materials, their affordances, and how to work with them (Semper 2013, 10; original emphasis).

The *Tinkering Studio* engages visitors in a variety of activities, which revolve around the composition of improvised things, such as drawing devices that use small motors to mobilise crayons. The classes are based on the idea that the outcome of making is not always as important as the learning it facilitates. One of the challenges of tinkering is to resist closing off the process prematurely and to remain in the experimental mode, where ideas and materials are allowed to shape each other in surprising ways and thinking happens alongside making. Tinkering invites patience, skill and a willingness to embrace uncertainty—qualities that Dewey (2005, 2015b) highlights in his accounts of design and making, both of which can immerse people in the "unified situational whole" (Johnson 2007, 72) of experience. In doing so, tinkering creates an environment for the perennial prototype to thrive and suggests that objects can only be held together through continuous effort.

[6] http://tinkering.exploratorium.edu/about

4 Conclusion

Arguing that Dewey advances a unique design philosophy that foregrounds form-making as an arena for experimenting with transformative experiences, this chapter has applied two of his concepts, *experience* and *experiment*, to describe how developers and visitors of the Exploratorium co-produce new affects, sensations and thoughts by composing and sustaining prototypical arrangements. In doing so, the chapter has indicated that the philosophy of Dewey and exhibit development reinforce each other in diverse ways, three of which appear particularly useful for contemporary design practice and theory.

First, as speculative approaches are gaining ground in design (Brassett and Marenko 2015; Dunne and Raby 2013), inviting explorations of alternative futures through temporary material interventions, Dewey could add to this work a critical methodological awareness by placing emphasis on "keeping track" (Dewey 1997a, 87) of design experiments through observation and recording to enable continuous learning. Exhibit developers also think through their prototypes with comparable attentiveness, ensuring that these are properly tested and monitored during each iteration, so that the experiences they facilitate "lead ever onward and outward" (Dewey 1997a, 88), instead of falling back on received ideas. Both the experimentalism of Dewey and the perennial prototypes of the Exploratorium therefore offer plenty of insight into alternative practices of speculative design.

Second, Dewey's thinking can contribute to recent work on "user experience" (Koskinen et al. 2011, 26–27), which seeks to understand the feelings of individuals as they engage with designed objects and spaces. Dewey, however, proposes a more radical notion of experience as an "impersonal" (Dewey 2005, 193) and ever-expanding field of relations that is shared with the broader environment. Exhibit developers advocate similar thinking when they engage with various kinds of materials, organisms and processes to produce experiences that provoke visitors to form new attachments with the earth. Thus, Dewey and exhibit developers promote design that does not take experience as given, but rather seeks to hold it open to new and unexpected human-nonhuman encounters.

Third, when designed objects are understood as relational entities, the primary "ethical challenge" raised by them is "to accompany their development in adequate ways" (Verbeek 2011, 165). Exhibit designers, for example, are committed to the maintenance of their artefacts, literally keeping them alive, as is the case with the *Glass Settling Plate* discussed above, which contains marine organisms from the surrounding bay. Allowing their pieces to remain perennial prototypes ensures that objects receive the care that they need, as this gives them space to evolve with changing institutional circumstances and encourages developers to enhance their potential to facilitate new relations. On this view, then, design is an ethical endeavour that recognises the incomplete character of all entities and supports their fragile transformation as they pass from one set of relationships to another (Bates et al. 2017).

References

Adams, C. (2013). Techne and logos at the edge of space. In L. Valentine (Ed.), *2013. Prototype. Design and Craft in the 21st Century* (pp. 99–113). London: Bloomsbury.
Allen, S., & Gutwill, J. (2004). Designing with multiple interactives: Five common pitfalls. *Curator: The Museum Journal, 2*(47), 199–212.
Bates, C., Imrie, R., & Kullman, K. (Eds.). (2017). *Design and care: Bodies, buildings, cities*. Oxford: Wiley-Blackwell.
Brassett, J., & Marenko, B. (2015). Introduction. In B. Marenko & J. Brassett (Eds.), *Deleuze and Design* (pp. 1–30). Edinburgh: Edinburgh University Press.
Cole, K. C. (2009). *Something incredibly wonderful happens. Frank Oppenheimer and his astonishing Exploratorium*. Chicago: The University of Chicago Press.
Dalsgaard, P. (2014). Pragmatism and design thinking. *International Journal of Design, 1*(8), 143–155.
Dewey, J. (1991). *The public and its problems*. Athens: Swallow Press.
Dewey, J. (1997a). *Experience and education*. New York: Touchstone.
Dewey, J. (1997b). *How we think*. New York: Dover.
Dewey, J. (2004a). *Essays in experimental logic*. New York: Dover.
Dewey, J. (2004b). *Democracy and education*. New York: Dover.
Dewey, J. (2005). *Art as experience*. New York: Perigee.
Dewey, J. (2015a). *Experience and nature*. New York: Dover.
Dewey, J. (2015b). *The school and society / the child and the curriculum*. New York: Dover.
Dunne, A., & Raby, F. (2013). *Speculative everything. Design, fiction, and social dreaming*. Cambridge, MA: The MIT Press.
Gunn, W., Otto, T., & Smith, R. C. (Eds.). (2013). *Design anthropology. Theory and practice*. London: Bloomsbury.
Gutwill, J. P. (2008). Challenging a common assumption of hands-on exhibits. How counterintuitive phenomena can undermine inquiry. *Journal of Museum Education, 2*(33), 187–198.
Hein, H. (1990). *The Exploratorium. The museum as laboratory*. Washington, DC: The Smithsonian Museum Press.
Hein, G. E. (2016). *Progressive museum practice: John Dewey and democracy*. London: Routledge.
Humphrey, T., & Gutwill, J. (Eds.). (2005). *Fostering active prolonged engagement. The art of creating APE exhibits*. San Francisco: The Exploratorium.
James, W. (2003). *Essays in radical empiricism*. New York: Dover.
Johnson, M. (2007). *The meaning of the body. Aesthetics of human understanding*. Chicago: The University of Chicago Press.
Koskinen, I., Zimmerman, J., Binder, T., Redström, J., & Wensween, S. (2011). *Design research through practice. From the lab, field and showroom*. Waltham: Morgan Kaufmann.
Kullman, K. (2013). Geographies of experiment / experimental geographies: A rough guide. *Geography Compass, 7*(12), 879–894.
Markopoulos, L., & McDougall, M. (Eds.). (2013). *Over the water: Fujiko Nakaya*. San Francisco: The Exploratorium.
Massey, D. (2005). *For space*. London: Sage.
Melles, G. (2008). An enlarged pragmatist inquiry paradigm for methodological pluralism in academic design research. *Art, 2*(1), 3–13.
Murdoch, J. (2005). *Poststructuralist geography. A guide to relational space*. London: Sage.
Rockwell, T. (2013). Over the water. In L. Markopoulos & M. McDougall (Eds.), *Over the Water: Fujiko Nakaya* (pp. 4–5). San Francisco: The Exploratorium.
Rylander, A. (2012). *Pragmatism and design research—An overview. The Swedish Faculty for Design Research and Research Education*. Royal Institute of Technology. http://www.designfakulteten.kth.se/sites/default/files/designfpragdesignrapport_18.4.pdf. Accessed 20 Dec 2015.
Semper, R. (2013) Preface. In Wilkinson, K. & Petrich, M. *The Art of Tinkering* (p. 10). San Francisco: Weldon Owen.

Shapin, S. (2010). Uncle of the bomb. *London Review of Books, 18*(32), 12–14.
Toon, R. (2005). Black box science in black box Centres. In S. MacLeod (Ed.), *Reshaping museum space. Architecture, design, exhibitions* (pp. 26–38). London: Routledge.
Verbeek, P.-P. (2011). *Moralizing technology. Understanding and designing the morality of things.* Chicago: The University of Chicago Press.
Yaneva, A. (2009). *Made by the Office for Metropolitan Architecture: An Ethnography of Design.* Rotterdam: 101 Publishers.

Sketch Representation and Design as Generative Transformation

James Andrew Self and Gabriela Goldschmidt

Abstract We discuss the role and importance of sketching during conceptual design ideation and position it as instrumental in understanding what it means to design. To do this we first define design as a generative, transformative act. We then situate sketching as an effective means through which the transformative requirement of design is achieved as reason for its prolific use in design. Following this, in order to ground our theoretical discussion, two examples of sketch work are presented and discussed. The two examples provide an illustration of how sketching is used to both resolve the indeterminacy of the conceptual design situation, and establish a means by which the designer may navigate a design solution. Finally, we reflect upon the potential of a focus upon sketch representation to contribute to developing an understanding of what it means to design and implications for efforts towards building a philosophy of design.

Keywords Sketching · Design representation · Conceptual design · Generative transformation

1 Introduction

Design is by nature generative, deriving its generative nature from a fundamental concern for, 'courses of action aimed at changing existing situations into preferred ones' (Simon 1996, p111). The Simon (1996) definition clearly articulates the transformative nature of design, which in turn requires the generation of and reflection upon potential solution candidates as possibilities for preferred situations. Although this clear-cut definition is contentious in its over simplification of a term that, in reality, encompasses a far wider scope of activities (Heskett 2002), a definition of

J. A. Self (✉)
School of Design and Human Engineering, UNIST, Ulsan, South Korea
e-mail: jaself@unist.ac.kr

G. Goldschmidt
Faculty of Architecture & Town Planning, Technion - Israel Institute of Technology, Haifa, Israel
e-mail: gabig@technion.ac.il

© Springer International Publishing AG 2018

P. E. Vermaas, S. Vial (eds.), *Advancements in the Philosophy of Design*, Design Research Foundations, https://doi.org/10.1007/978-3-319-73302-9_10

design as preferred transformation provides a useful conceptual anchor and point of departure for our current discussion of design sketching and its role in reasoning towards generative design ideation.

For, during an initial phase of conceptual design, we shall argue the visualisation of ideas, through sketching, is instrumental to an ability to engage in generative conceptual design, the evolution and establishment of a direction towards preferred transformation. Although much work has been done to understand how design representation as sketching provides opportunities for the designer to reflect-in-action upon potential conceptual solutions (see for example Schön 1983), little work exists to relate a necessity to sketch with a contemporary understanding of what it means to engage in design. Notable exceptions are the work by Ferguson (1992) and Herbert (1993). Ferguson stressed the role of sketching in engineering design. In his words: "… the designer uses sketches to try out new ideas, to compare alternatives, and (this is important) to capture fleeting ideas on paper." (p. 96). Herbert wrote about architectural sketches, analyzing the underlying forces that compel designers to produce strokes on paper.

Responding to this gap, we apply a contemporary understanding of reasoning during conceptual design ideation to a theoretical discussion of the ways in which sketching relates to and provides a supportive framework for the evolution of design directions through concurrent thinking between problem definition, solution ideation and development. Discussing two example sketches produced during conceptual design, we further illustrate how reasoning through sketching facilitates conceptual design ideation. In this our goal is to more clearly articulate the importance of sketching as driver for ideation through a theoretical discussion of its relationship to design reasoning. And, in laying the foundations for a theory of sketching based upon its relationship to reasoning during generative ideation, we contribute to the overarching aim of developing a philosophy of design. For, as we shall show, exploring the relationship between design sketching and design reasoning during ideation, provides opportunities to understand what it means to design and thus contribute to design theory and a philosophy of design.

2 Design as Generative Transformation

The activity of designing is often focused upon the provision of alternatives to the existing; its transformation into a state specified and articulated by the designer, but also that this change should bring about a preferred state of being. As Nelson and Stolterman (2012) suggest, in this transformative endeavour the designer's goal is not to identify a universal truth or single correct solution, but a most appropriate alternative to the existing situation.

As such, both the designer's approach to the execution of preferred transformation, and the nature and scope of the resulting change, may potentially be limitless. This is because design is an incremental activity of transformative change in which solution propositions are built step by step together with their rationales; assessment is made by way of reaction to the outcome of the latest step in the process. We suggest that no design, including the most radical and innovative, is conceived in the designer's head. Instead, an initial idea is subjected to a series of transformations before the outcome finally emerges as an appropriate response to the design problem. In this sense, every design is incremental. And, since the entity the designer reacts to has physical properties, for the most part complex sets of properties, it is not possible to react in the abstract. Therefore, in the assessment of potential solution candidates, the designer must resort to external representation such as sketches, for both perception and conception. With perception indicating how the designer sees the external representation (i.e. the sketch or illustration) and conception a reference to the ways in which seeing is interpreted as holding potential, given the designer's own criteria for evaluating preference in the appraisal of the sketch representation.

The particular transformative characteristic of design, and the role and use of sketch representation within it, is further clarified if we compare the aims of design with those of the sciences. For, while design is interested in appropriate transformation (the ought to be), science concerns itself with a comprehensive definition of how things are (Archer 1995; Simon 1996). On the other hand, due to the requirement of design to transform towards a not yet specified preferred, with little indication of a clear path to proceed or approach to be taken, the knowledge and information imported and applied may potentially come from any domain through any media to inform design at any point. If design is defined as focused upon preferred transformation, where potentially numerous approaches may be taken to arrive at any number of potential solutions, the design process can be described as *indeterministic.* This indeterminism often results in the designer's heuristic application of prior experiences as driver for a particular approach taken or the kinds of information both applied and extracted from the design situation. In this the use of sketching appears to provide opportunity for the designer to critically apply knowledge and experience to the current design situation.

These previous encountering with similar situations influence how and by what means both the current design problem is understood, and the approach taken in offering transformative solutions. For example, in a study aimed at understanding the relationship between design expertise and designerly ways of thinking and working, Gadamer (Gadamer 1986) defines a requirement to assess the appropriateness of design solutions as interpretation. Dorst (2006) further divides interpretation between the objective and subjective. The former describes the application of measurable specifications in the assessment of the appropriateness of solution attempts. Subjective interpretation, in contrast, is both a result of the indetermined nature of design, in that an element of interpretation is required to determine what elements of the design situation are most important, and the requirement for as yet unspecified change to drive preferred transformation. In terms of our current discussion of sketching, subjective interpretation appears most critical during the early phase of design where both opportunities for solution ideas and the direction of change they may potentiate are still very open.

As such, the production of sketch representations at a conceptual phase hinges on the designer's perception of the situation, as stated by the constructivist philosopher Nelson Goodman (1978, 138): "Briefly, then, truth of statement and rightness of descriptions, representations, exemplifications, expressions – of design, drawing, diction, rhythm – is primarily a matter of fit: fit to what is referred to in one way or another, or to other renderings, or to modes and manners of organization." Thus sketch representation provides a means through which the designer may assess a design solution as having the potential to result in a preferred state.

3 Sketch Representation and Design Ideation

Numerous tools and methods now exist to inform both objective (for example House of Quality, Ergonomic Design Criteria Index), and subjective interpretation (heuristic strategies) of design intent. However, all appear to have a common concern for the assessment of the propositional solution's potential to result in a preferred state. Transformation to the preferred and the generative expression of potential preferred states are therefore critically related in that the potential of a transformation to result in the preferred is assessed through the designer's interpretation of solution attempts. Due to both the complexity of the design problem and the potential for numerous solutions to emerge, it is through solution attempts that the designer is able to cope with the indeterminate nature of conceptual design.

These solution attempts may take many forms. However, the designer's ability to both interpret the situation and assess the potential for preferred change is often best achieved through sketch representation. This is because sketching provides opportunities to quickly approximate, at higher or lower levels of detail and fidelity, the form, characteristics and detail of potential solution ideas. Critically, these approximations then act as stimuli for further assessment. In this way the process of moving from the existing to the preferred is an emergent, generative activity, where solutions are incrementally developed and an understanding of the preferred state dynamically influenced by attempts made by the designer to acquire it. Here we position sketching as an effective instrument due to its ability to both provide the means through which the preferred is understood and potential improvements developed.

By way of example, Fig. 1 illustrates how sketching is used to stimulate and develop ideas towards a new truck brand and design aesthetic (Olofsson and Sjolen 2005). Using free-hand sketching and a perspective underlay the representations are achieved through a combination of pencil drawing accompanied by digital illustration software. The illustration (Fig. 1) provides an example of how sketching may be used to provide opportunities for feedback to facilitate transformation and a partial resolution of the indeterminacy of the design situation.

Figure 1 contains six complete sketches (each from the same perspective) representing design ideas towards a new truck's form language and identity. These are complimented by a further eight partial sketches of the same front-end detail. Two further perspectives of the cabin's interior, two side elevations of the cabin and various lighter line-work and annotations make up the complete page of sketching.

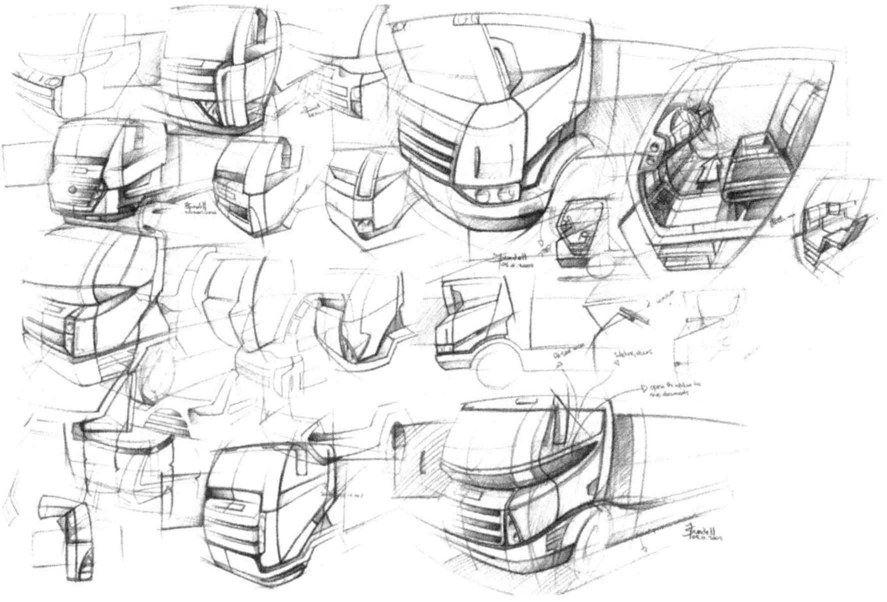

Fig. 1 Conceptual design ideation to establish identity and design direction (Olofsson and Sjolen 2005, 29)

Upon first inspection, the overall impression received is one of a rich interaction between thoughts and the expressive sketching of design intent. The designer appears to explore various forms and combinations of forms related to the truck's front-end design (i.e. headlight clusters, windscreen, front-grill and driver door detailing). These expressions may be thought of as vertical transformations of design intent (Goel 1995), where vertical transformation describes a process whereby the designer moves from one idea to a variation of the same idea. That is, the designer is focused upon reasoning towards the same general design features as they relate to a holistic, more established design direction (see Goel 1995). This is in contrast to lateral transformations, whereby the design moves from one idea to another, new idea. In engaging lateral transformations the designer is seen to take a more explorative approach to the problem space, often representing a number of different solution or sub-solution ideas in a relatively short space of time. An example of lateral transformations is presented in Sect. 4 below.

In an example of vertical transformation, or evolution of design ideas through sketching, while maintaining the overall proportions and perspective angle to each illustration, the designer's exploration of intent towards grill design appears to evolve through the sketch work running from top-left to bottom right (Fig. 2).

As indicated in Fig. 2, the designer attempts to resolve the indeterminacy of the design situation by exploring different solution directions. That is, a variety of combinations of grill, headlight cluster, door paneling and windshield designs are expressed, investigated and assessed. In this sense, it appears the act of sketching

Fig. 2 Conceptual design ideation, transformation of design ideas (detail of sketch 1)

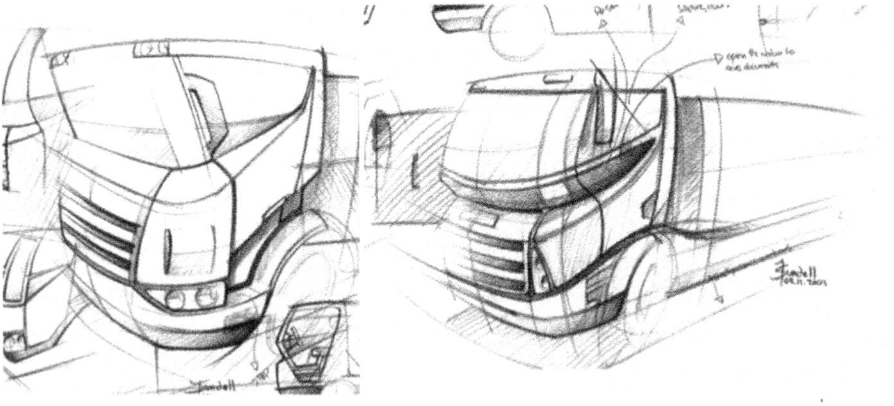

Fig. 3 Conceptual design ideation, detailing of ideas (detail of sketch 1)

has offered opportunities to reflect upon the potential of solution ideas. This has resulted in further transformations. In this way the indeterminacy of the design situation is constrained through the establishment of a design direction, facilitated by the proposition of potential solutions, and indicated in the ways in which each of the sketches illustrated in Fig. 2 appear to relate to one another.

The designer also produces two further expressions of intent towards cabin design at greater detail (i.e. increased shading and line work); a larger sketch representation (Fig. 3, left) and a representation detailing both cabin and trailer (right).

In Fig. 3 we find that, although the designer appears to settle upon the potential location for and combination of design features (front-grill design, lower skirt), others remain open to further exploration. For example, between the two larger sketches (Fig. 3), the front light cluster appears as a similar, triangular form but in two different orientations and locations; horizontal (left) compared to vertical (right). The position and form of the windscreen design also differs, from a convex design (Fig. 3, right), to concave (left). The door paneling also appears to alter in form and detail (i.e. location of handle, form of side panelling).

In this conceptual design work the designer's ability to control the indeterminacy of the design problem appears possible through the representation of potential solution ideas as sketches, with these sketch representations providing both opportunities to explore the potential of ideas, while at the same time controlling indeterminacy through the establishment of design directions. For example, as indicated in Fig. 3, the designer appears able to set a direction for moving towards the development of a solution (as seen in similarities between sketch representations), while, at the same time, leaving room for exploration of alternatives (i.e. alternative light cluster orientations, windscreen detail).

In this sense the illustrations shown in Figs. 1, 2 and 3 indicate sketch representation as instrumental in providing opportunities for the designer to establish a design direction, thereby coping with the indetermined nature of conceptual design ideation. And, at the same time, concurrently explore, assess and evolve the potential of the design direction through expression of alternatives. In this case, variations in the detail of design elements and their juxtaposition in relation to one another (Fig. 3). That is, it is through the external representation of intent as sketches that the designer is able to reason towards the potential of a complexity of design features (i.e. grill design, light-cluster design and location and the combination of these) in providing opportunity for a preferred design direction to emerge.

At some point in the session the designer switches attention from a focus upon exterior design to the cabin's interior (Fig. 4).

This shift in focus may have happened as a result of the larger illustration of the truck's front end (Fig. 3), indicated by the small arrows at the bottom left corner of Fig. 4 (boxed outline) and pointing from the exterior design to the interior.

As indicated in Fig. 4, reasoning may have shifted towards the exploration of ideas towards a different aspect of the design. Having been facilitated by the more detailed sketch of the truck's exterior, this shift in focus is an example of how sketching acts as catalyst for the designer's attention and interest. Rather than, or as well as, the designer deciding what aspect of the design to work upon next, it appears it is expression of intent that may have prompts a shift of focus from one aspect of the design solution to another. In other words, as thoughts are expressed through sketch representation, these expressions allow the designer to see further possibilities for development. In this the act of generative expression through sketch representation appears instrumental to the exploration of potential solution candidates. The sketch illustrations presented in Fig. 4 indicate how the expression of a potential solution provided the designer something to hold on to as anchor during design ideation. These same representations of design intent also appear to provide oppor-

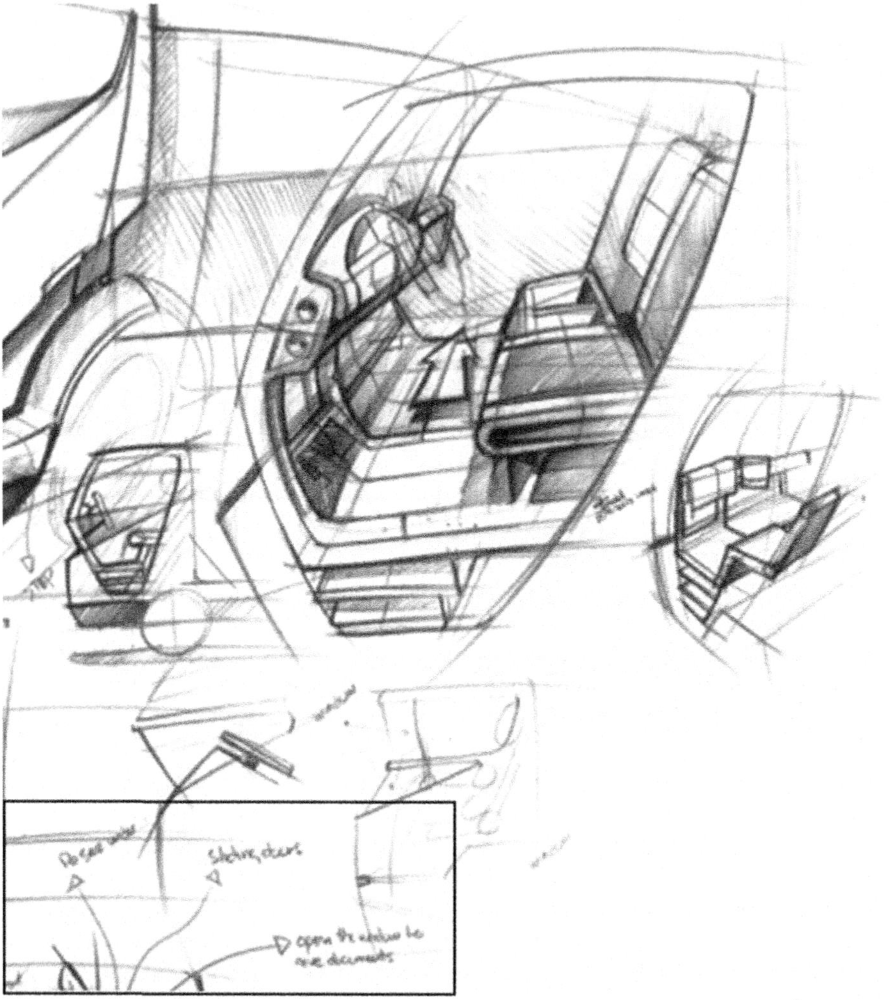

Fig. 4 Conceptual design ideation, focus of attention (detail of sketch 1)

tunities for reflection and further exploration along different dimensions of the design; as indicated through the designer's shift in focus from the external to internal cabin design.

Adopting a definition of design as a requirement to transform the existing to a preferred situation within the context of an indeterministic design process, the sketch representations illustrated in Figs. 1, 2, 3 and 4 can be positioned as a natural response to this requirement. Due to the indeterministic nature of design, it is through these types of expressive and underdetermined conceptual representations that the designer is able to resolve indeterminacy while, at the same time, readjust as required a design direction.

As illustrated in Figs. 1, 2, 3 and 4, our current thesis of design representation through sketching as instrumental to concept ideation focuses upon the iconic or figural (Goldschmidt 1997) expression of solution ideas so often seen as synonymous with design activity. That is, representations that attempt to approximate the attributes of potential transformative design propositions on one or more dimensions of interest (Ulrich and Eppinger 2012). The Ulrich and Eppinger's (ibid) notion of a dimension of interest is particularly suited to the current discussion of sketching in its reference to the ways in which sketch representation may attempt to approximate the physical, functional or use characteristics of solution ideas. In the example sketch work above, we see how different aspects and attributes of the design, in different combinations, are considered at different levels of detail. In this, the dimensions of interest, and the level of fidelity at which they are considered, are incrementally progressed during ideation. It is through these sketch expressions, on different dimensions of interest and levels of detail, that the designer appears able to assess the potential of solution candidates as drivers for preferred transformation.

Of course the media through which a sketch is made may influence the ways in which the representation expresses design ideas. For example, the use of digital graphics tablet compared to paper and pencil may influence the kinds of representations made and so the ways in which they express design intent (Self et al. Self et al. 2014). Although we admit the potential influence media of expression may have upon the designer's engagement with and assessment of solution ideas, we see a focus upon media as limiting. This is because we are then left with the media in isolation. For sketching, like any other tool or media, is only a tool in-so-far as it is used as such to achieve its purpose. In the case of conceptual sketching, the aim is expression of preferred transformation. A tool-focused approach (i.e. investigating the influence of media upon conceptual design ideation) results in circular arguments on the strengths and limitations of media of representation, with little account of the designer's critical role as driver for the expression of and reflection upon intent through design representation.

As such we do not attempt to conceptually split sketching as tool of design representation from the sketcher. Instead we use the term *sketch representation* as construct to indicate an interaction between designer and sketching as means of design expression. In so doing we focus upon an understanding of sketch representation as it relates to a particular type of generative reasoning during conceptual design.

Thus we define the scope of our definition of *sketch representation* as an act of representation that attempts to approximate the physical, functional or use characteristics of a solution candidate(s) for the purpose of assessing potential for preferred transformation. We do not deny the importance of other forms of design representation (written, oral, models, prototypes, etc.). However, here we focus upon sketch representation as the figural expression of dimensions of interest, offering potential for a greater understanding of sketching as a mode of expression synonymous with design activity.

4 Sketch Representation, Problem Definition and Solution Ideation

Central to design as defined by the transformation of existing situations into preferred states, is the notion of the ill-defined design problem. Ill-defined (or ill-structured) problems stand in opposition to well-defined (well-structured) problems. A well-defined problem is one in which the initial state is known, means to arrive at a solution (e.g., algorithms, sets of rules) are given, there is a clear goal state, which can be satisfied by a small and limited number of solutions, often only one. In contrast, for an ill-defined problem the initial state may be vague and lacking information, there is no given algorithm towards arrival at a solution, and the goal state is not known in advance but evolves in the course of solving the problem. The number of possible solutions may be very large (Holyoak 1990; Reitman 1964; Simon 1973). Dorst (2003, 2006) discusses the designer's interpretation of design problems, defining ill-definedness as the under-determination of design problems. Perhaps a useful way to understand this ill-definedness is to contrast design problems with the game of chess. As discussed by Dorst (2003), a chess problem can be described as well-structured in that the chess board, as well as the possible moves of all the pieces, is known a priori, with their own known affordances and constraints, within rules that are understood beforehand and applied during the course of the game. In contrast, none of the structure or rules for proceeding exists within a design problem. As Dorst goes on to discuss, "there is no fixed playing surface in design: we design in the real world, so outside influences can disturb our plans at any time" (Dorst 2006, p22). This ill-definedness also means that the perspective adopted in attempting to understand the ill-defined design problem will implicate how both the problem is seen and the nature and direction of potential solution ideas.

Through his work on the nature of design problems, Dorst (1996, 2003, 2006) attempts to express the underdetermined character of the design process in its central concern for exploration toward the identification and development of an appropriate solution. The notion of under-determination, then, is underpinned by the ambiguity of the initial design problem. As a result the knowledge and reasoning applied to the solution of design problems must necessarily provide opportunities to define and redefine the nature of the problem. This, Dorst (op cit) suggests, not only requires an analysis of the problem itself, but is also critically dependent upon the proposition, by the designer, of potential transformative solution candidates. That is, the nature of the design problem, as ill-defined, implicates directly the generative nature of designing (Cross 1990). Thus, the very indeterministic character of a design problem requires attempts made towards its resolution in order to frame the problem in a more manageable way. A similar view can be found in Goldschmidt (1997). In this we make the case for the generative act of transformation as a means through which ill-defined design problems may be addressed. Adopting this position, the importance of generative solution ideation in pursuit of preferred transformation becomes clear. For, if potentially numerous solution directions exist, the designer is left with little alternative than to engage in solution ideation as means to help define the initial design problem.

In this way, generative design solution ideation is not only a means to propose and develop appropriate solution ideas, but fundamental to understanding the initially ill-defined design problem (Cross 2011). This implies that the problem and its resolution are related in that they do not exist outside one another. Rather the nature of the design problem may only emerge from attempts made at its resolution.

For example Kruger and Cross (2006), in a study to examine solution versus problem-driven design strategies, observed that generative design was the most frequently engaged activity during conceptual ideation. However, their study also indicated the ways in which increased solution ideation provided opportunities for problem definition. Bjorklund's (Björklund 2013) investigation noted how more experienced designers tended to be better able to form a comprehensive mental representation of the design problem as a result of their solution attempts. This in turn appeared to provide opportunities for increasing proactive behaviour and a resulting increase in solution possibilities. Likewise, Darke (1979) indicates a relationship between the designer's initial framing of the problem as a *primary generator*. This Darke (ibid) describes as the departure point for transformative solution ideation, "When confronted with a new design situation, the designer imposes images of possible solutions to it…thus directing the actual development of the product form" (ibid, p38).

As a further example of how sketching provides the means through which a direction may be established in response to an ill-defined design problem, and how, the designer may apply knowledge to the design situation through heuristic strategies, Fig. 5 illustrates the exploration of potential solutions for a new product concept.

Taken from a case-study of the design and development of the Oxo Good Grips range of kitchen utensils (ibid), the Bottle Stopper/Opener sketches, illustrated in Fig. 5, express ideas towards a new product as part of a range of kitchen tools designed to address usability problems with existing designs in the same product category.

From top left (Fig. 5) it appears the designer begins to express potential solution ideas through sketches, illustrating the form of the stopper/opener's head as a flattened, bulbous and curvilinear form. A tapered stopper shaft is also included with three ridges expressed half way down its length. In these representations it appears the designer makes an initial attempt to express the form of a potential holistic solution. Three further sketches are included to the right of the larger sketch, illustrating a spherical head to the stopper and a concave lip between stopper head and shaft. The single annotation 'fingers' is added to the uppermost sketch. Here it appears the designer considers the problem of grip and leverage. Through these three further sketches the potential solution of a concave under-ridge is evolved and reflected upon.

In the second sketch from top right, a void space or hole has appeared in the head of the stopper. The design of the finger-grip also appears to evolve from a concave ridge to a short, thin line-like form. Looking down the page at the sketches below

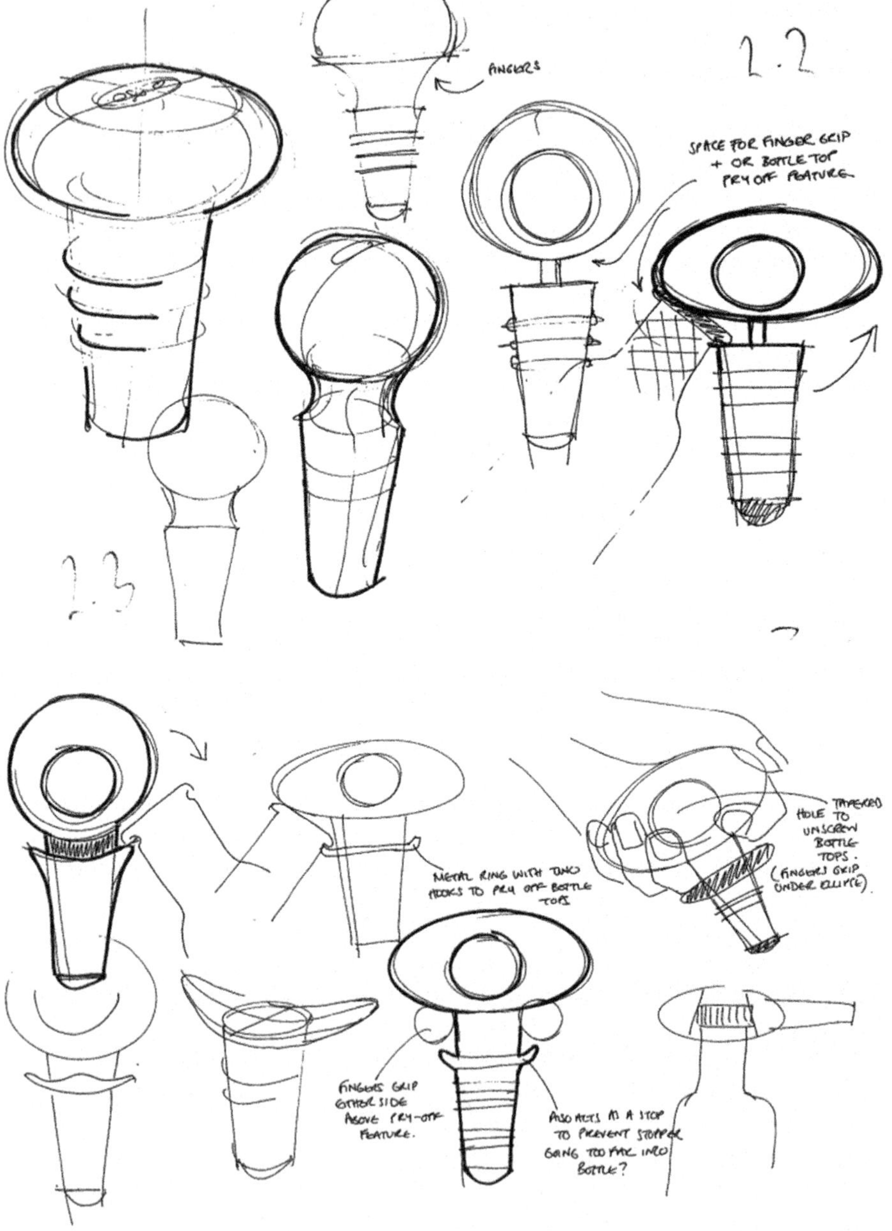

Fig. 5 Early sketches to explore potential solutions towards a new product concept (Industrial Design Society of America 2001, 16–19)

(Fig. 5) we see the conical hole is a feature maintained and developed. Further, a larger sketch (upper right), is produced together with an illustration of a bottle neck, top and three arrows; two running down the stopper head and a further arrow running up. The short annotation ‘space for finger grip + or bottle top pry off feature’ completes the sketch illustration. It appears the designer is considering the potential of the head and neck of the stopper to provide the added function of a bottle opener. In this way the designer appears to apply knowledge of both materials (Fig. 5, annotations indicating material choices) and heuristic strategies to address problems, and their associated solutions, related to interaction possibilities (expression of hand and grip alternatives). The sketch expressions of design intent are instrumental to this process in that they facilitate the application of knowledge of materials and interaction possibilities in the proposition of and reflection upon potential solution ideas. For example, the sketches in the lower set of illustrations (Fig. 5) appear to indicate a continuation of thoughts towards addressing problems around leverage and grip through the proposition of a ‘dual-use’ design.

In a further example of how the exploration of different solution ideas through sketch representation provides opportunities for resolving the indeterminacy of conceptual design ideation, Fig. 5 (lower set of sketches) indicates consideration of the opener function as the designer explores combining the finger-grip feature with the bottle opener. The head of the stopper/opener is also now expressed as a more bulbous, spherical form and the tapered shaft of the stopper shows a progression of detail in the depiction of a lifted lip or ledge, upon which an outline of a bottle is depicted. In this the designer appears to consider, in more detail, how the problem of the opener function may integrate with other design considerations (i.e. use and form). Here again, a relationship between the identification of a solution direction and exploration of potential ideas facilitated through sketching, provides the designer the means through which the uncertainty of the situation may be controlled while, at the same time, leaving room for further exploration, as seen through the designers’ lateral transformations of ideas towards various opener designs.

By way of a further example of how sketching has provided opportunities to apply knowledge to the design situation in this case a, “metal ring with two hooks to pry off bottle tops” is added (Fig. 5, lower series of sketches, annotation next to one of the sketches). Although comparatively ambiguous in their representation of detail, these sketches appear to provide space for the designer to continue to evolve an understanding of how the opener may be integrated within the design to address use and functional issues associated with the combination of the opener and the stopper functions. In this sketching appears to provide a catalyst for the designer to apply knowledge of materials and interaction possibilities to a developing design solution.

However, the designer’s sketch work does not attempt to propose resolutions to existing, well understood problems, but instead indicates an evolution of problem definition in parallel with solution proposition. For example, the problem of how to pry off a bottle top leads to the generation of a potential metal ring solution expressed as sitting under the head of the stopper. This problem is itself a product of the proposition of a potential design direction in the integration of a dual function. Likewise, the idea of the opener function appears to emerge from the sketching itself, rather than be pre-defined beforehand, or pre-existing in some way.

Two sketches in the lower set (Fig. 5) show the stopper design held in hand, and a further elevation sketch of the head of the stopper in use over a bottle-top. As before, in these sketches it appears the designer has continued to consider problems related to grip and leverage to explore the potential of using the hole in the stopper head as an opener (as an alternative to the aforementioned metal ring). In this it appears the expression of solutions through sketch representation have provided the designer opportunities to both explore the potential of solution ideas and identify sub-problems related to these solutions.

Throughout the sketches illustrated in Fig. 5, a progression of ideas is clearly evident through a continued identification of and attempted resolution for associated problems. That is, solution proposition and the identification of potential issues appear to evolve together, with solutions explored in terms of their associated problems related to function and use. Within this an evolution the sketch work indicates how both knowledge of materials and experience-based heuristic strategies are applied in an attempt to address problems around user interaction and their associated solution possibilities. The expression of ideas as sketches and illustrations appears to facilitate the application of knowledge in the establishment of the overall design direction for the Opener/Stopper concept.

5 Understanding the Benefits of Sketch Representation

As indicated by the ideation work illustrated in Figs. 1, 2, 3, 4 and 5, sketch representation continues to relate to solution ideation as means through which a designer is able to both initiate the search for potential solution candidates and reflect upon implications for the nature of the design problem. Representations thus also appear to be both internal and external. The former describes how representations may be generated internally in visual mental imagery in response to the design situation (Goldschmidt 1991). Visual imagery is an innate cognitive capability that enables us to create representations of things that are not present in our visual field, including things that do not exist (Kosslyn 1996). Therefore imagery has an important role in creative endeavors such as discovery, invention and design. The sketch, on the other hand, works as external expression. However, these two forms of representation do not exist outside one another. The designer does not externally represent ideas as sketches held in the mind. Instead the two modalities of representation work together, as external expression of design intent influence the internal representation of ideas and vice versa. This is seen in Fig. 5 as the designer appears to express design intent as external sketch representation before progressing initial ideas through alteration and revision of design features related to the use and function of the stopper/opener concept. Through the external representation of intentions as approximations of potential solutions, which in turn influence internal mental imagery, the designer moves across and between problem and solution spaces in pursuit of preferred transformation.

Schön's (1983) reflective paradigm for understanding design activity appears to add weight to our thesis in its focus upon the reflective and constructed relationship between design problem definition and solution ideation. That is, the designer must reflect-in-action upon the appropriateness of solution intentions in terms of their ability to potentiate preferred change. He refers to sketching and drawing as the designer's conversation, or dialogue, with the situation. Lawson (2006) also discusses the importance of a reflective ideation. From this perspective, the act of sketching is seen not as an externalisation of pre-existing ideas, but part of an interaction between the designer's internal representation of potential solutions and their external expression as sketches and illustrations. In this way, the designer is able to reflect-in-action upon sketch representations, where the representations themselves facilitate talk-backs, providing the designer opportunities to evaluate, revise and progress conceptual design ideas in light of issues emergent from and communicated through the sketches themselves. We propose to broaden the metaphor of conversation to include the relationship between designer and situated act of transformative design as it captures a requirement for transformation to proceed only through the proposition of and reflection upon externalised solution attempts. A conversation on the subject of appropriate transformation.

Dorst and Cross (2001) describe a co-evolution of the design problem and its resolution as the designer's ability to define core aspects of the design problem, leading to a crystallisation of a core solution, which in turn alters the designer's definition of the original problem. In this sense, co-evolution is a result of the ill-defined design problem. As the problem is not fixed, the designer is provided room to dynamically interpret both the problem itself and the kinds of solutions that may effectively address it. In the proposition of solution ideas the designer may realise that an interpretation of the problem may require revision, because the concept solutions do not appear to appropriately address the problem as defined. This then requires both a revision of how the problem is interpreted, together with the approach taken to its resolution. Thus, design problems and their associated solutions are related in that the nature of the problem is revised in response to attempts made towards its resolution.

Again, this appears to be the case in the design work illustrated in Fig. 5, and the designer's apparent identification of potential solutions together with their associated problems. This process of co-evolution thus provides opportunities for creative insight when a problem and potential solution align to create a frame within which subsequent, more refined, ideation may then proceed, as seen in the identification of a stopper/opener frame (Fig. 5). This then provides direction for all further ideation work.

Usefully, Dorst and Cross (ibid) provide the metaphor of a bridging exercise between problem and solution spaces, with the strength of the bridge providing opportunities for more relevant or appropriate solution ideas to emerge. In a study by Kruger and Cross (2006) a requirement to engage in solution driven activity appeared to provide greater opportunities to identify new problem related requirements during solution ideation. Likewise, Bjorklund (Björklund 2013) points to the ability to reason between a problem and its resolution as leading to more proactive activity, in turn allowing opportunities for increased options to emerge. Dorst (2006) appears to go further in suggesting that the design problem does not in fact exist as an objective reality. Rather it is seen as the amalgamation of different problems, with the kinds of reasoning required described as a breakdown in the normally fluent problem solving process resulting in opportunities for subjective interpretation and choice in the direction of transformation.

Thus sketching is able to bridge between problem and solution spaces in that it provides a means through which solution intent can be externally expressed. These expressions, or external representations, seen in the conceptual work illustrated in Figs. 1, 2, 3, 4, and 5, provide opportunities for greater definition of the design problem through the proposition of and reflection upon sketch representations. These representations then serve as means to evolve a potential design direction (as seen in the truck design work illustrated in Fig. 1 and as also illustrated in Fig. 5, with the problem of developing a new user-friendly stopper which doubles as an opener).

It is only because of the ill-definedness of the original problem that the designer is provided the space to reconstruct both the problem and solution in a new creative way, providing opportunities for better alignment. The ambiguous and incomplete character of the sketches illustrated in Fig. 1 also appear particularly apt in providing opportunities for the assessment of alignment, while at the same time allowing space for reinterpretation. In this the designer's ability to see problems as ill-defined is critical. This is because only through the building of alignments between a new interpretation of the problem and potential solution ideas, can creative (new) transformations begin to emerge.

As indicated in Figs. 1, 2, 3, 4 and 5, sketch representation as approximation of potential solution ideas appears effective because of the requirement to evaluate the potential of solutions against opportunities for preferred transformation, while at the same time offering a means to realign the ways in which the preferred is assessed. Sketch representation thus acts as facilitator for understanding the potential of solution ideas because the representations themselves provide the designer opportunities to read into or see information expressed within them, which they would not otherwise be able to see, and which may then point towards potential directions for further development.

Goldschmidt (1991) proposes that in design reasoning there are two modes of argumentation: *seeing as*, relating to the physical properties of the emerging design solution, and *seeing that*, which pertains to the rationale, often functional, for the choice of proposed physical properties. Both are revealed through sketching, which either expresses intention or serves as the basis for feedback. For example, when an architect says "this wall should be made of transparent glass, to enable visual continuity with the space outdoors," two arguments are put forth. The first is a *seeing as* argument – a glass wall. The second is a *seeing that* argument, which provides a rationale for the choice of glass – the ability to see through it, thus creating a desired visual continuity between indoors and outdoors.

Sketch representation supports reflection upon the suitability of propositional solutions, which may otherwise remain undiscovered, while concurrently refining the criteria through which suitability is assessed. Thus, it is a particularly effective means to readjust alignment in the co-evolution of problem and solution. This, we would suggest, is the central relationship between design sketching and design reasoning during concept design. With assessment continually readjusted in light of the sketch representation's ability to provide opportunities for (re)interpretation, ideation work through sketching provides the most opportune circumstances for bridge-building between design problems and their solutions.

We contend that sketch representation provides opportunities and departure points for understanding designerly reasoning and knowing, with the potential to provide insights into what it means to design and contribute to efforts aimed at building a coherent philosophy of design.

More work is now required to understand the role sketch representation plays during conceptual design ideation. We believe the study of sketch representation has much to offer in developing an understanding of, for example, how design is assessed, the nature of design expertise, designerly ways of knowing and the role generative ideation plays in creativity. Therefore, sketch representation's instrumental position as facilitator for conceptual design and, as we have shown, its relationship to the indeterminate requirements of conceptual ideation clearly warrant its recognition in any endeavour to establish the foundations for a philosophy of design.

References

Archer, L. B. (1995). The nature of research. *Co-design, 2*, 6–13.

Björklund, T. A. (2013). Initial mental representations of design problems: Differences between experts and novices. *Design Studies, 34*(2), 135–160.

Cross, N. (1990). The nature and nurture of design ability. *Design Studies, 11*(3), 127–140.

Cross, N. (2011). *Design expertise* (pp. 133–149). Berg: In Design thinking. Oxford.

Darke, J. (1979). The primary generator and the design process. *Design Studies, 1*(1), 36–44.

Dorst, K. (1996). The design problem and its structure. In N. Cross, H. Christiaans, & K. Dorst (Eds.), *Analysing design activity*. Chichester: Wiley.

Dorst, K. (2003). The problem of design problems. Paper presented at the *Expertise in Design: Design Thinking Research Symposium 6*. University of Technology, Sydney.

Dorst, K. (2006). Design problems and design paradoxes. *Design Issues, 22*(3), 4–17.

Dorst, K., & Cross, N. (2001). Creativity in the design process: Co-evolution of problem–solution. *Design Studies, 22*(5), 425–437.

Ferguson, E. S. (1992). *Engineering and the mind's eye*. Cambridge, MA: MIT Press.

Gadamer, H. G. (1986). *The rele vence of the beautiful and other essays*. Cambridge: Cambridge University Press.

Goel, V. (1995). *Sketches of thought*. Cambridge, MA: MIT Press.
Goldschmidt, G. (1991). The dialectics of sketching. *Creativity Research Journal, 4*(2), 123–143.
Goldschmidt, G. (1997). Capturing indeterminism: Representation in the design problem space. *Design Studies, 18*(4), 441–455.
Goodman, N. (1978). *Ways of worldmaking*. Indiana: Hackett Publishing Company Inc.
Herbert, D. M. (1993). *Architectural study drawings*. New York: Van Nostrand Reinhold.
Heskett, J. (2002). *Toothpicks and logos: Design in everyday life*. Oxford: Oxford University Press.
Holyoak, K. (1990). Problem solving. In D. N. Osherson & E. E. Smith (Eds.), *Thinking* (pp. 117–146). Cambridge: MIT Press.
Industrial Design Society of America. (2001). Oxo good grips bottle stopper/opener. In K. Goodrich (Ed.), *Design secrets: Products, 50 real-life projects uncovered* (pp. 16–19). MA: Rockport Publishes Inc.
Kosslyn, S. M. (1996). *Image and brain*. Cambridge: MIT Pess.
Kruger, C., & Cross, N. (2006). Solution driven versus problem driven design: Strategies and outcomes. *Design Studies, 27*(5), 527–548.
Lawson, B. (2006). *How designers think: The design. Process demystified* (4th ed.). Oxford: Oxford University Press.
Nelson, G., & Stolterman, E. (2012). *The design way: Intentional change in an unpredictable world*. London: MIT Press.
Olofsson, E., & Sjolen, K. (2005). *Design sketching*. Sweden: Keeos Design Books AB.
Reitman, W. R. (1964). Heuristic decision procedures, open constraints and the structure of ill-defined problems. In M. W. Shelley & G. L. Bryan (Eds.), *Human judjements and optimality* (pp. 282–315). New York: John Wiley.
Schön, D. (1983). *The reflective practitioner*. London: Ashgate.
Self, J., Evans, M., & Dalke, H. (2014). The influence of expertise upon the designer's approach to studio practice and tool use. *The Design Journal, 17*(2), 169–193.
Simon, H. A. (1973). The structure of ill-structured problems. *Artificial Intelligence, 4*, 181–201.
Simon, H. A. (1996). *The science of the artificial* (3rd ed.). London: MIT Press.
Ulrich, K., & Eppinger, E. (2012). *Product design and development* (5th ed.). New York: McGraw-Hill Education.

Models in Engineering Design: Generative and Epistemic Function of Product Models

Claudia Eckert and Rafaela Hillerbrand

Abstract Engineers interact with their products and processes largely through models, however rarely reflect about the nature of these models and how technical possibilities and actions are affected by the models' properties and characteristics. Models in engineering describe the product as well as its generating process, but at the same time also shape and create them. This clearly distinguishes them from scientific models that primarily aim to describe a certain target system. While over the last decade, there has been a growing body of literature on models in the sciences, much less research has been done on models in engineering design. In this chapter we aim to fill this gap by taking a closer look at models in engineering design from an epistemic point of view. In particular we suggest a classification of different types of models used in engineering design and compare them to models used in scientific research. Thereby we do not aim at an encompassing map of models in engineering practice, but we aim to identify key categories of models with regards to their relationship to their targets. We contend that the functions of models in engineering design cannot be fully captured when focusing on the representative aspects of models alone as done in contemporary philosophy of science.

Keywords Models · Models in science · Engineering design · Representation

1 Introduction

In engineering design, engineers interact with their products and processes largely through models, however rarely reflect about the nature of these models and how technical possibilities and actions are affected by the models' properties and

---

C. Eckert
Department of Engineering and Innovation, The Open University, Milton Keynes, UK
e-mail: c.m.eckert@open.ac.uk

R. Hillerbrand (✉)
Philosophy of Science, Engineering and Technology, Institute for Technology Assessment and Systems Analysis (ITAS) (ITAS) & Institute for Philosophy, Karlsruhe, Germany
e-mail: rafaela.hillerbrand@kit.edu

© Springer International Publishing AG 2018
P. E. Vermaas, S. Vial (eds.), *Advancements in the Philosophy of Design*, Design Research Foundations, https://doi.org/10.1007/978-3-319-73302-9_11

characteristics. Arguably, the design process can be seen as a process of creating, manipulating and using models during which the models themselves evolve and are repurposed multiple times. Both 'design' and 'model' are ambiguous terms in a very similar way (Poznic 2016). Both terms can be used to devise or plan a product (designed artefact/design or model) or a process (design*ing* or modell*ing*). This paper focusses on product models used in the design of physical engineering products, where often very complex physical products are created to meet physical requirements. By contrast software design is not concerned with a physical product unless it is embedded in one, so that some of the arguments discussed in the paper don't quite fit software engineering. While process models play an important role in design and raise interesting epistemic questions (see Eckert and Stacey 2010) they are beyond the scope of this paper. However unlike the science community,[1] the engineering design community has a strong interest in process models (see for example Browning and Ramasesh 2007; Wynn 2007).

Divergent interpretations of the context, content, purpose or role of models are a significant cause for (sometimes unnecessary) iterations in design processes; and therefore has a profound effect on both the product, and the effectiveness and efficiency of the process. Iteration plays an important part in exploring design alternatives and resolving resolve problems arising in design process (Wynn and Eckert 2017). For example designers create multiple models of the shape of a consumer product like a vacuum cleaner to discuss alternatives with each other. These may be in the form of sketches, technical drawing, computer aided design (CAD) models, or so-called blue foam models that are rapidly cut out three-dimensional shapes. Designers run focus groups with users to understand their responses to the product and find potential requirements they might have overlooked. However, iterations take time and resources and can jeopardize refining the design at the end of the process. Such iterations often arise from incomplete or ambiguous information contained in the models (Stacey and Eckert 2003), as well as a lack of understanding of the nature of models and the relationship a model has to reality (Eckert and Stacey 2010). Hence there is some hope that a closer look and better grasp of the modelling involved improves the whole design process.

Models also play a fundamental role in the sciences. They are central epistemic and pedagogical devices in the process of scientific discovery. For example, the standard model of particle physics provides the current standard explanation and understanding of the most fundamental physical processes at subatomic scale.[2] Bohr's model of the atom by contrast, which describes the motion of the electrons

---

[1] The philosophical literature does not dwell on the details of modelling the scientific practice. Scientific processes are discussed today rather in the context of the sociological study of science and technology. An example provides the actor-network theory that originally aims to describe processes of innovation and knowledge-creation in science and technology (e.g. Latour 1987). Psychological studies of scientific processes, e.g. Dunbar (1997) has been picked up in the artificial intelligence literature and in creativity research (e.g. Sawyer 2011; Holyoak and Thagard 1997).

[2] Notwithstanding its name, the standard *model* is often seen also as a *theory*. We will briefly turn to the intricate relationship between model and theories in the following section

around the nucleus in analogy to the motion of the planets around the Sun, is clearly outdated. However Bohr's model still proves useful in teaching science and thus even today fulfils a pedagogical function. It helps to understand, for example, as to how the subatomic movement is different from the classical planetary motion in that electrons do not move on fixed orbits.

While over the last decades, a growing body of literature on models in the sciences emerged, much less research has been done on models in engineering design. In this chapter we make a first step to fill this gap by looking at the epistemology of design from the model point of view. One of our guiding questions is whether and if so what the engineering design literature can learn from the models in science debate and conversely how philosophy of science can learn from design. In science as well as in engineering, various different objects function as models: From concrete material objects like scale models or animal models in the life sciences, where living special-breed animals are used to test for example drugs in place of the human organism, to abstract models that make use of mathematical equations or computer simulations. However there are also seeming differences between models in science and in engineering design. For example, engineering models typically model specific designs. Only rarely are engineering design models created with the intention of reusing them. Nonetheless, in practice the models will often be re-used in different circumstances from the original model. This may lead to problems and iterations in the design process as they were described above. At first glance, models in the sciences seem to be of a different sort. They often model repeatable and repeated phenomena, such as gravitational forces between two massive bodies. These models are usually on a rather high level of abstraction. However these are only one type of models used in the sciences. Also here models are sometimes constructed for a specific phenomena and a much lower level of abstraction. These models are often not intended to be reused. Examples can be found, for example, in the geosciences with models to explain specific rock formations or models for climate change. Just like in engineering practice, these models can get reused despite not being set up for reuse.

Philosophy of science has traditionally been very strongly influenced by physics as many of the philosophers of science themselves have a background in physics. However, in recent years the interest of philosophers in other sciences has grown. In particular climate science with its mixed methods and clear societal need has brought new issues into focus. The physics focus of philosophy of science suits reflection on the aspects of product models in engineering which are concerned with the physical properties and behaviours of products and their components. Modelling of physical properties of products have become a mature area and is now well supported with computer tools and modelling techniques. In recent years there has been a shift from thinking of engineering products only as physical products to thinking of them as socio-technical systems, which interact with humans and the environment. The behaviour of both humans and environment is increasingly modelled explicitly and often simulated in computer systems.

In studying models in design, the paper is organized as follows. In Sect. 2 we give a short overview over the philosophical discussion on models in science. Section 3 reviews the literature on models in design and tries to draw a first comparison to models in sciences. In Sect. 4 we develop a classification of models in design and zoom in on differences and commonalities between models in design and models in the sciences. Just as in the sciences, models in design represent a certain target system. However in design, models fulfil also other central purposes that cannot be reduced to their representative function alone. We refer to this as the generative function of models. Quite generally, however, most of the seeming differences between models in science and models in design prove on closer inspection to be differences in degree rather than differences in principle. Nonetheless awareness about these difference helps to understand better the design process as well as the sciences. The paper finishes with a conclusion in Sect. 5.

Before we begin our discussion, allow us a word of precaution. With this paper on models in design we want to reach to two rather disjoint communities: philosophers of science and engineering working on models as well as design scholar. The review on models in philosophy of science as well as on the use of models in design will thus at times remain cursory or introductory for one of the audiences.

2 Models in the Sciences

2.1 Models in Science and Models in Philosophy of Science

We can find a whole plethora of models in the sciences. Hesse (1963), for example, distinguishes between *material models* and *formal models*, where the latter can be *analytic* or *constructive* models. Examples of material models range from Watson and Crick's original tangible model of the DNA as a double helix, to animal models as used in the life sciences. Analytic models are mainly mathematical models. Examples comprise the standard model in particle physics, or the Black-Sholes model of the evolution of asset (e.g. stock) prices over time. Though of course here the mathematical equations are essential, most authors do not equate the model with the mathematics, i.e. the equations, alone (e.g. Frigg 2010). These analytic models are contrasted by constructive, i.e. computational models. Today, Hesse's two-fold distinction is sometimes replaced by a threefold one where computer simulations are explicitly distinguished from mathematical models (e.g. Weisberg 2013). A lot of the models that are implemented on a computer are first formulated as continuous mathematical equations. Take as an example the origins of lattice gauge theory. It originated in the differential equations of quantum chromodynamics. Computational models, however, often refer to those models that are first formulated in a discrete way, apt for direct numerical implementation. Paradigm example are the Schelling's Segregation Model or The Game of Life. Models that are first formulated in a discrete way are referred to as "discrete" models, while models consisting in differential equations that are then implemented numerically are referred to as "continuous models" (Hartmann 1996). Note that in Weisberg's or Hartman's sense, *simulation* models do not have to be numerically implemented or implementated on a computer. What is at the heart of a simulation model is that it mimics *dynamical* aspects. An example for a non-numerical simulation provides the FloWave Ocean Energy Research Facility at the University of Edinburgh, a large scale testing facility, which simulates waves and currents for marine energy devices.

While models have always been part and parcel of scientific reasoning, it took the Philosophy of Science a long time to recognize the importance of models in the scientific process. One reason for this may be philosophers' preoccupation with the context of justification. This detracted attention from actual scientific practice in which models are central both in theorizing as well as in experimenting. Moreover philosophers of science focused for a long time on those natural sciences where we find elaborate theories. For long (theoretical) physics seemed philosophers' favourite pet. Not surprisingly theories in physics and often natural laws were seen to be at the centre of scientific activities, while models were treated merely as orphans. In the few exceptions in which classical philosophy of sciences discusses models, mechanical as well as mathematical models were seen as inferior to theories and even as a disturbance of scientific reasoning (e.g. Duhem 1954).

Arguably this changed at least for abstract models with the so-called semantic view of scientific theories. While the alternative conception of theories, the so-called syntactic view of the logical empiricists, perceived theories as axiomatizations in first-order logic, the semantic view interprets a theory as the sum of all its models, where a model is an interpretation on which all the axioms of the theory are true. This notion of a model derives from mathematical logic, and in this sense a model represents a particular theory. Consider as an example Newton's laws and theory of gravitation. Then the model of the Earth moving around the Sun can be seen as one instantiation of this theory. Note that the semantic account of model coming from a mathematical understanding of models may seem far off the way scientist themselves use the term model. But in this way, the shift to a theory's models still allows for the axiomatization of a theory and for the syntactic view, while at the same time it solves certain problems that plagued the syntactic account. For example, the sentences that form the theory in the syntactic interpretation are uninterpreted and need to supplemented by so-called correspondence rules. These assign empirical meaning to the theory's theoretical terms. These correspondence rules, however, may not be uniquely defined nor can they be specified in every context. For the semantic view, on the contrary, the theory is a set of its models as the structure is already interpreted. One and the same theory may be formulated in different languages as long as its models are the same. There is hence no need for correspondence rules.

The semantic view was developed by Suppes (e.g. 1961, 1962) and later Suppe (1977) and has many followers. Though they all defend various different versions of the semantic view of theories and as such also different notions as to what a model is exactly, they all agree that models are the central unit of scientific theorizing. As a logic notion, the term model is defined in terms of truth here: A model of a theory is defined in such a way that all the theory's axioms are true. In these accounts models are seen to derive (more or less) straightforwardly from some overarching theory (e.g. Suppes 1962; Mayo 1996). For example, the model of the Earth's motion around the Sun is seen as the interpretation of an abstract calculus, in this case Newton's theory and Newton's law of gravitation. The model is then interpreted as a realization of the more general theory, and the set of all its models is the theory. All proponents of the semantic view agree that models are as a whole non-linguistic entities, but what they are exactly varies. For Suppes, whose approach we mainly followed in this short sketch of the semantic view, models are set theoretical structures, while for example van Fraassen (1980) views them as possibilities of how a possible state space evolves.

2.2 Models after the Practice Turn

The semantic concept of a model carries over directly from logic and seems to not be apt to accommodate all the various ways in which scientists use the term model. First of all, it seems impossible to extend it to concrete models like the mouse model, i.e. using standardised lab mice as a model for humans. But even for abstract models the semantic view seems sometimes far from the lab practice. For example, unlike in the semantic view often models do not derive in a straightforward way from theories. Further specifications are needed. This even holds true for the seemingly simple model of the Earth's movement around the Sun. In applying the theoretical knowledge to the planetary system, a lot of simplifying assumptions have to be made: Other gravitationally interacting bodies (e.g. planets, the Earth's moon, asteroids) are neglected; the bodies are assumed to be perfectly spherical; and the mass distribution inside Earth and Sun is assumed to be homogenous. All these assumptions do not follow in any straightforward way from the theory. They are additional modelling assumptions, independent of the theory. One example is the assumption that the mouse model, the actual mice used in a laboratory experiment, indeed mimics human bodies well enough so that they can be used to predict human responses to the tested drug. No theory supports this claim.

With the so-called practice turn, which began in philosophy of science around the 1980s, philosophy of science got more attentive to the actual practices with their fields of studies. Arguably this turn led philosophers to more caution about theorizing what scientists actually do. For sure, this turn led philosophers to engage more with the actual practice of scientists and reconstructions for actual work. These case studies gained more attention and with them various other areas of science, from the Life sciences to economics, come into focus. As a consequence, many contemporary philosophers of science turned away from the interpretation of models accroding to the semantic view. It got acknowledged that in many scientific contexts models are central epistemic tools that may not be subordinate to theories (e.g. Hughes 1999; Morrison 1999; Cartwright 1999; Suárez 1999; Bailer-Jones 2003). We follow these authors of the practice turn. Thus, unlike in the semantic interpretation, we see models (also) as non-linguistics entities. We will use the notion *model of theory* for the models of the semantic interpretation and now turn to what is known as *models of phenomena.* Such phenomena are "relatively stable and general features of the world that are interesting from a scientific point of view and for empiricists have to be observable" (Frigg and Hartmann 2012). A typical phenomenon is the motion of the Earth around the Sun, but as philosophy of science after the practice turn opened up to various areas of sciences, such phenomena do not have to be repeatable or naturally repeated events, but may also be singular phenomena like the world's climate, a specific rock formation or the extenction of the dinosaurs.

Though this last sentence may be not worth much more than a footnote for philosophy of sciences, it becomes crucial when comparing models in design to models in science: As already noted in the introductory section, at first glance there seems a difference between the universal models in science and the apt-for-purpose models in design. Indeed, design models are most often constructed with one particular design in mind, without the idea of re-using the model when constructing it. However, in practice they can often get reused later on. Models in science seem to be designed in an already universal way, for example the models of Newton's laws and his theory of gravitation. A rather high level of abstraction assures here a kind of universality, a readiness to be re-used. But closer inspection reveals that also the sciences know models that are much less universal than the standard example of Newton's gravitation and much less aimed at generalization and in this sense closer to models in design.

Next to models of phenomena, *models of data* are distinguished. Models of data transform the raw experimental data into interpreted data. Important steps are here the transformation of the so-called raw data by eliminating outliers due to disturbed or faulty observations etc. This step is referred to as 'data reduction' and essential in most experimental processes. The other essential step is `curve fitting.' For example when observing planetary motions, first points that seem to be based on flawed measurements as they lie outside the range of all other data points are eliminated, then a smooth curve, e.g. an ellipse, is fitted to the remaining observational points (cp. Frigg and Hartmann 2012). Often the data models are much more sophisticated and particularly for complex experimental setups also entail what Morrison refers to as the experimental model, containing at least a model of the measuring apparatus. In the end scientists do not compare a theory to an observation, but models with models (Morrison 2009, p. 49): The model of a theory is tested against a data model.

Morgan and Morrison (1999) developed the influential "models as mediators" account. It contends that models are autonomous agents in that "(1) [they] function in a way that is partially independent of theory and (2) in many cases they are constructed with a minimal reliance on high level theory" (Morgan and Morrison 1999, p. 43). Models here can mediate between theory and the real world by virtue of their partial independence from high-level theory. The model of the Earth moving round the Sun illustrates this lucidly. As already pointed out, many assumptions, simplifications and idealizations enter the model that do not derive from the theory (Newtonian mechanics and Newton's theory for gravitation) itself. Moreover, models are also partially independent from the raw data. Through this mediating role, models function as tools or instruments that enable us to learn more about both theories and the real world. Morrison (2009) argues that simulation models can function as "measurement instruments" that enable one to extract information from the apparatus under consideration in an experiment. In this sense, simulation models function as "mediators" between the theory and the material system.

As will become clear in the following sections, we hold that the more vague use of the term model by philosophers of science after the practice turn, as depicted above, is not only more adequate for models actually used in the sciences, but also for models in design. Nonetheless the semantic view in which models foremost represent theories has contemporary followers. For example Fraassen (1980) or Giere (1988) defend special versions of the semantic approach. As they are also philosophers who have made the practice turn, their views on models may be seen as a way to combine the more formal approach to models of the semantic view with the more recent discussion on models. That sometimes models of theory as well as models of phenomena or data are combined in a single model, is illustrated nicely by global climate models that project the Earth's climate over the course of the twenty-first century. Parts of these multi-modal models are derived in a more or less straightforward way from underlying theories, like the thermodynamics of the atmosphere. Other model part like modelling cloud formation and their interactions in the atmosphere are much more heuristic.

2.3 Models and Representation

With the above reconstruction of scientific practice as a model-based endeavour, it seems time to come back to the original mistrust that philosophers had of models. Some of their intuitions seem to be hard to deny. Models in some sense or another seem to be inferior to theories: They sometimes contradict accepted knowledge, they can be internally inconsistent, etc.

In order to capture how models can be more than merely heuristic devices despite all their shortcomings, the term representation was introduced into the model discussion. Models represent a phenomenon, some experimental data, or even a theory.

Models fulfil various different aims (Morgan and Morrison 1999), though arguably, the representation of some target system seems is the unifying feature of all concepts of models we encountered: The target system may be a theory, a set of data, or a phenomenon. Let us now focus on models of phenomena. These represent some phenomenon as their particular target system. Such models are not only of instrumental use, but they want to mimic some aspect of reality in a certain way. A model's target system most often is not the whole system under consideration, but only a certain aspect of it. Consider again the model of Earth around the Sun. The simple two-body model discussed above represents well the day–night cycle, but cannot account for most of the seasonal variation. Note that in representing a target systems, models of phenomena just as models of data make use of idealization and abstractions. In the Earth-Sun model we can explicitly formulate these assumptions and as to why these are valid, i.e. the forces exerted on the Earth by all other celestial bodies are smaller by order of magnitude than the force exerted by the Sun (cp. Hillerbrand 2015).

In contemporary philosophy, two accounts of how models represent can be distinguished: informational and pragmatic views. While informational views point to similarities between the model and the target system and put the relation between the model and target, understood as some form of structural similarity, at the centre, pragmatic account of representation holds that the (intended) use of the model by the scientists is prior to any established relation of representation (e.g. van Fraassen 1980). Note that informational and pragmatic views are not mutually exclusive but can be seen as complementing each other (cf. Chakravartty 2010). As to what exactly this structural similarity is, there is no consensus in the literature. Formal notions of isomorphism, partial isomorphism, homomorphism, or other mathematical mappings have been suggested just the same as less formal types of similarity such as analogies, similes, or resemblances (cp. Poznic, in press). Hesse (1963), for example, has developed an account of how analogies can prove useful for modelling. She distinguishes between positive, negative and neutral analogies between model and target systems. Consider the mouse model. Then the positive analogies are the known similarities between target and model, i.e. a similar hormonal cycle between mice and humans. The negative analogies are those where we know that the model does not represent well the target system, i.e. when the mice in the lab experiments do not develop AIDS like humans do. Following Hesse, the really interesting things for scientific progress are the neutral analogies. These comprise all properties or relations of the model of which it is not known yet whether they indeed have a correspondence to the target system.

2.4 Computer Simulations in the Sciences

Computers takes centre stage in many areas of modern life, so also in sciences and engineering. So let us turn to this specific type of modelling in more detail. To distinguish simulations from calculations or equations used in analytic models Weisberg, Hartmann and others use the term simulation to denote the imitation "of one process by another" (Hartmann 1996; Weisberg 2013; Parker 2009). Here, 'process' refers to some temporal sequence of states of a system, thereby stressing the dynamic aspects of (not only computer) simulations. By contrast, Humphreys (1991) adopts a broader notion of computer simulations: "A computer simulation is any computer-implemented method for exploring the properties of mathematical models where analytic methods are unavailable". But computer simulations may be of great value even where analytic solutions are known, for example via computer aided visualizations. However, in design, the term "simulation" is mainly reserved for mimicking dynamics processes and we thus follow Hartmann's definition given above.

Note that not all numerical investigations aim at "simulations" in the strict sense that refers to genuine dynamical aspects. But generally at least within the sciences, the most interesting computer experiments, however, are simulations in the sense that they mimic a dynamic sequence of states. Note moreover that most scientific investigations seem to be concerned with dynamic processes. Even when explaining such stationary phenomena rock formation, for example, one often falls back on dynamic explanations and thus simulations in the standard sense.

Computer simulations are used for various different purposes and we want to distinguish different epistemic aims that computer models commonly fulfil in the sciences. These different epistemic aims are not limited to computer simulations, but can be extended to abstract or concrete models as well, and parallel to some extent the distinction of structure, function, and behaviour we will distinguish for engineering design models in the following sections. Here we take computer models as used in the sciences to elucidate the different aims. Following Hillerbrand (2012), we distinguish three types of computer models as regards their epistemic content or aim.

Proof' by Simulation This type of simulation refers to simulations that aim at information about abstract, most often mathematical systems. Often the systems under investigation are differential equations that cannot or cannot yet be solved analytically. The search for finite-time singularities in the three-dimensional incompressible Euler equations provides an excellent example where the numerical investigation of an abstract system, i.e. the Euler differential equation, is of practical importance for research in the empirical sciences, in this case fluid dynamics (e.g. Grauer et al. 1998). In this first sense, simulations yield a (possibly preliminary) alternative for a lack of theoretical understanding.

Proxy-experiment Other simulations provide information on systems that cannot or cannot yet be accessed experimentally or are simply very hard to access in real laboratory or field experiments. Examples here are very diverse. (a) Physicists may use this type of simulation for analyzing turbulent flows on scales too small to access in laboratory experiments, but information on the behaviour on these small scales is very important for refining or testing existing theories. (b) In numerical experiments, certain aspects can be singled out that cannot be untangled in material experimenting. When analyzing inertial particles in any sort of flow, for example, a numerical simulation has the advantage that one may focus on the particles' inertia only while neglecting aspects like gravitational interaction or the particles' finite-size effects that cannot be decoupled in real experiments. Moreover (c) the analysis of some real, i.e. material experimental data may rely on simulations, usually on the form of Monte Carlo simulations. In this sense, simulations may be seen as a (possibly preliminary) replacement of experiments.

Prognosis This third type of simulations may be seen as a kind of instrument for prognoses or forecasting. Here, simulations are used for predicting the behaviour of real, usually complex systems for which (a) no accepted analytic description exists, or for which we are (b) certain that the theoretical description implemented numerically is correct (within the desired precision). Typical examples for the former arise in the engineering sciences and in weather and climate predictions, while the latter type is often studied in astrophysics (cp. Morrison 2009) or engineering sciences (see next section). Note that often, scientists use the very same simulation for various purposes. Practically in applied sciences the aims of "proof by simulation" and "proxy experiment" seem to mix fairly commonly.

3 Models in Design

Engineers in all fields, from nanotechnology to household appliances to nuclear technology use a huge variety of models in the process of defining and evaluating a new product. They create models of components, of sub-systems, and of entire products: from sketches to computer models of parts, from scale models in a wind tunnel to prototypes to the actual industrially manufactured product. Throughout the design processes the models become closer to the final product as the process progresses. Models are fundamental epistemic tools for designers and engineers without which modern engineering would not be possible. While engineers use models all the time, they rarely theorise about models or modelling as such; however some sub-communities with a more mathematical background have engaged more than others. In the modelling and simulation community Tolk and Turnitsa (2012) assert "Modelling is the purposeful process of abstracting and theorizing about a system, and capturing the resulting concepts and relations in a conceptual model" (p. 2), while Pidd (1999) defines a model in operations research as: "an external and explicit representation of a part of reality as seen by the people who wish to use that model to understand, to change, to manage, and to control that part of reality in some way or other" (p. 120). Both of these definitions assume that the target of the model either exists or is well developed and highlight both abstraction and selection of relevant features as vital aspects of representation.

Design is often described as the co-evolution of problem and solution (see Wynn and Clarkson (2005) for an overview of general design models). While many design processes particularly in engineering start with a clear statement of the problem or specific requirements, the co-evolution paradigm captures the fluidity of many design processes. Only by looking at a potential solution, it is possible to understand whether the problem has been posed in a suitable way, which makes design processes highly iterative. At the same time only by externalizing and reflecting over a potential solution it is possible to distance oneself sufficiently from an existing and just conceived idea to see another potential design in ambiguous or incomplete representations (Schön 1984). This notion of design puts representation at the heart of design as a cognitive activity. This not only includes the representations designers actively generate, such as sketches, but also other products or past designs that act as models for products that are being designed (Eckert et al. 2005). Designers frequently draw analogies to other designs or natural phenomena, that they are familiar with. In this case designers make a choice to use something as a model. As the target is often only emerging, we argue that rather than something being a model of its target, it is people who decide that something acts as a model.

In industry this fluidity has to be managed, and design processes are structured, for example, into stages, work packages or milestones often associated with particular deliverables, which prescribe activities and the models generated through them (e.g. French 1999; Pahl and Beitz 1996). These processes are adapted for individual companies and typically disseminated throughout an organisation through high level process models. These are typically complemented by Gantt charts of the specific process activities and a multitude of plans generated by individual designers and team leaders (see Eckert and Clarkson 2010). The overall logic of most engineering processes is one of fixing the overall design and concentrating on details as the design process progresses. However in practice both requirements and solutions change throughout the entire design process (Eckert et al. 2004). The options for designers to make changes become increasingly limited as key parameters or components are frozen to fix dependencies between different parts of the design or to accommodate long lead times (Eger et al. 2005). Thereby the epistemic status of the model changes through the decisions that people take.

In engineering design as in science, models thus take centre stage: Consider for example a scale model of a car in a wind tunnel. This model is not meant to represent all aspects of the car on the street: It aims to represent the air flow around the real car, while other features like its driving characteristics or the noise inside the driver's cabin are not targeted. Modelling involves a selection of the aspects of reality that are represented. This example hints that though models in engineering may have a more instrumental character, the representative function is essential in order to ensure, amongst others, the instrumental success.

Unlike science, engineering design research does not have a large body of literature reflecting on the nature of models in design. Visser (2006) sees design as the construction of internal and external representations, which is influenced by the structure of the representations and the actions they afford. This point is also argued by Galle (1999), who rejects the view of design representations as descriptions of possible or future things to avoid implying the existence of such non-existent things. Design representations frequently have a dual role as a means of communication and a vehicle of exploration. The interpretations of representations depends on the viewpoint and background of individuals (Bucciarelli 1994). Other authors, such as Schön (1984) and Ferguson (1994) highlight the role of representations in the idea generating process of individuals.

Models of the design product which are used during the design process, are classified in the literature in many different ways according to their purposes. The classification of design model we want to suggest follows a classical distinction of aspects of a design into *structure, function*, and *behaviour* (Gero and Kannengieser 2004; Umeda et al. 1996; Goel et al. 2009). Here structure is the physical elements of a product and the relations between them; function is the product's purpose; behaviour refers to the actual behaviour of the design product, irrespective of whether this is a desired or not. These terms are interpreted in different ways (see Vermaas 2013) that in itself can cause problems in using models (see Eckert 2013).

For example function can be seen as the purpose of a product, e.g. the purpose of a hairdryer is to dry hair or to bring in profit, but its main action is to heat up air. However this discussion is beyond the scope of this paper.

We use the distinction between function, structure and behaviour to distinguish different design models. Models of the *structure* of the product range from simple product sketches to CAD models of parts to physical and virtual prototypes of the entire product. Models that express the *function* of the product often try to avoid presuming the product's structure. For example a functional model for a car would include a function to "propel the car", without making assumptions about whether the car is an electric car or has an internal combustion engine.[3] Models of the *behaviour* of the product include simple performance equations, virtual analysis models, such as finite element models to carry out a stress analysis as well as detailed physical or virtual models to test the behaviour of the system. Many of the simulation models used in engineering model the structure of product in different circumstances to assess the behaviour of the product before it is built.

Note that the classification into one type of model is not necessarily unique. For example, during later stages of the product development process when the product is more or less defined, the same models can be used to describe and analyse structure and behaviour. Just as in the sciences we find concrete, mathematical, and computational models of designed artefacts. Since the introduction of computers, the growing number of increasingly more sophisticated simulation models tend to reduce the number of physical 3D models. 2D technical sketches, for example, have all but disappeared in favour of 3D Computer aided design models. Numerical models often reduce costs of building and testing when compared to physical prototypes; as well as the cost of rework to other parts which are affected by changes due to test results. One numerical model often represents both the structure and the behaviour of the product. Companies sometimes express this aspiration of not building multiple physical prototypes under the catchy, usually not quite accurate slogan of "right first time".

Let us turn to computer or simulation models in design. The academic research around the use of computer models in design currently concentrates on several different angles. Industrial need drives discussions about the extent simulation models can replace physical testing and failure mode analysis of products (e.g. Tolk 2012). Engineering simulations are most often a combination of proxy-experiments and prognosis as defined in Sect. 2.4, because they typically vary input conditions or use conditions to predict the performance of products or components over the life of the product. Rapid prototyping, which allows the creation of parts from three dimensional computer aided design data usually using 3D printing or additive layer manufacturing has begun to blur the boundaries between physical and virtual models. Over the last few years the technology has improved sufficiently that it can be used to create production parts, making the transition between virtual models, prototypes

[3] However this example also shows that in practice it is often unrealistic to assume that (at least the basic) structure remains unspecified, as no car company begins this process without knowing what type of car they will build

and final design seamless, so that product components become indistinguishable from their prototypes unless their role is known.

More academic research addresses the generation of structure. There is a debate around the role of computers in design synthesis, where computers are used to create new designs according to certain prescribed rules or heuristics or by inference from similar designs. In some areas such as the design of electronics chips, this is now nearly fully automated. Some computational approaches, such as shape grammars (Stiny 1980), which encapsulate rules of design based on analogous designs, aim at understanding as to how a designed object is influenced by the structure of the representation through which they are generated. Other research is interested in studying human creative processes by building computational models of them.

4 The Relationship Between Models and Their Targets in Design and in Science

The relationship between models and representations and their target systems is more problematic in design, as they bring the product into being (e.g. Galle 1999). This section unpacks this relationship between model and target system arguing that in design the target systems of models are frequently other models.

Regarding representations, a first difference between the design models and scientific models is the direction of fit (Poznic 2016): While in the sciences the fit is from the target to the model and hence in Searle's (1983) words a world-to-model fit, for design it is the other way round as the future designed artefact has to fit the model and not the other way round. Here the direction of fit therefore is from the model to the world. This can be illustrated through the relationship between an architectural drawing and the building. The building is built in such a way that the building represents the technical drawing. The technical drawing is derived from the earlier sketches. There will be many sketches that show alternative designs that have never been built. Rather than the models being fitted to world, as in science, in design the world is fitted to the model (cf Poznic 2016). The scientific phenomenon usually exists before it is being modelled, whereas an engineered artefact in brought in being through a sequence of models. Engineering models often need to fit ideas that engineering designers have in their mind, however the nature of that fit is very different as mental models tend to fluid and can change without the person being aware of it (Kosslyn 1994). At the same time phenomena can appear in models that engineers than become interested in.

Two different direction of fit hint at a central role models play in design. In design models are used to generate another model or the final artefact. The same models are used to analyse the properties of a target system, in a similar way to how models are deployed in science. Again the example of the model of a car in the wind tunnel highlights these two aspects of design models. The wind tunnel models serve the overall purpose of analyzing the properties of the new car. The results might lead to a modification of the shape of the car and further testing.

4.1 Classification of Models in Design

However, not all models in design have model-to-world type of fit. Some models are there to evaluate or test another model, a specific prototype or design. We thus propose to distinguish in design between *generative* and *evaluation* models.

Generative Models describe aspects of a potential product or process during the design process. The generative model can describe function, behaviour, or structure of a product. Throughout the design process the product is defined through a series of generative models. However not every generative model leads directly to a product or part of the product. For example an architect might make multiple drawing, cardboard or computer models of an office building or a school. Many alternative potential products, which might be represented by alternative models, are considered in the course of the product development. The same model can be used to represent different design ideas, for example different colours or scales; or models can be combined in different ways to form various products. This can be illustrated with the cardboard models for buildings, where parts of the building would be scaled, added or repeated. Generative models can be seen as suggestions for a potential product. Whether they become an eventual product is down to decisions taken by the designers in course of the design process.

Evaluation Models are used to analyse the behaviour of a design defined by a generative model. Their closest parallel in the sciences are the models of phenomena. The design literature distinguishes (see for example Eppinger and Ulrich 1995) between three different forms of evaluation: analysis, verification and validation. Analysis investigates the properties an emerging design has. Verification addresses whether the design meets the requirements and validation makes sure that the product addresses the underlying need or opportunity. For example wind tunnel models were traditionally used during the analysis phase to learn about aerodynamic properties. This role is now largely taken over by computer models. However a physical model in a wind tunnel is still used to validate the computer analysis. For design we can divide evaluation models into *analysis models*, which are typically mathematical and aim at analysing the behaviour of the product or component, e.g. stress analysis of a beam; *simulation models*, which are computational, often probabilistic and address different conditions and circumstances, e.g., simulating different load conditions and environmental conditions on a beam; and *physical prototypes*, which can be tested on a test rig or in the "the field", e.g. by loading a beam to breaking point.

The issue of validation and verification has come to prominence in the context of reliability of simulation (for a review see e.g. Sargent 2005) and have recently received some attention within the philosophical literature on simulations (see, e.g., Klein and Herskovitz 2005; Küppers and Lenhard 2005). Note that in the science, design or philosophical literature, the terms "validation" and "verification" are not used consistently. Within science or philosophy of science, "verification" often refers to the process of testing whether or not the simulation program of the real system under consideration is executed correctly on the computer. "Validation" refers to the process of substantiating whether the simulation program is an adequate representation of the real system. Here an adequate representation means that the simulation results mimic closely enough the features of the real system. In design, however, it is typically the final product that is being verified and validated. Here verification tests whether the final product indeed meets the requirements, while validation asks whether the product solves the problem the requirements tried to encapsulate. Verification can be done through simulation or physical testing. Evaluation activities occur through the design process; however only once a commitment has been made to the final design, it makes sense to evaluate whether this particular design needs the needs and requirements. For simulation, design engineers talk about fidelity.

Analysis and validation can give rise to changes in the product definition. Analysis models are typically numerical models set up to understand the behaviour of the design either through calculations or through simulation methods such as finite element analysis. Engineering companies often use the analysis models as starting points for simulation models, which aim to understand how a product or component would behave under different conditions of use. For example when a truck company is designing a new suspension, they start with calculations to understand the properties of a particular configuration. When they have a preliminary geometry, they carry out a finite element analysis to assess whether this design could meet the requirements. Later they use simulation to assess whether the design can operate safely under all load conditions and road conditions. Extreme use cases are bundled in so called worst case scenarios, which cover extreme points in the space of applications the product is sold for. In a simulation the engineers also check for misuse situations, for example if the load in a trucks shifts and or the truck is overloaded. Simulations enable designers to understand the point of failure, whereas in physical tests they make sure that the product meets the requirements.

4.2 Reuse of Models in Design

Design models usually have a primary purpose of being a generative model or an analysis model, however the distinction can be difficult to draw. Generative models allow designers to evaluate and reflect about their design as a form of informal analysis or evaluation. Many analysis models are based on generative models. For example the scale model of a car in a wind tunnel has been generated based on the geometry created in a CAD model to define the product. With some features added and subtracted for the wind tunnel analysis it is retested and will lead to modifications to the product.

A special case of design models worth considering in this context are past designs that designers often use as generative models. The overwhelming majority of designs are generated by modification from existing designs or based on other designs as inspirations. In fact it is extremely difficult to find an example of a design that is not based on a predecessor of some kind. Think for example of a car that was originally based on a horse drawn carriage which had existed for many years. Now cars are based on their predecessor models, the newest technology used in the company and on competitor products. Designers use references to past designs to communicate with each other, to generate ideas, but also to test designs or parts of designs (Eckert et al. 2005). For example to test the emissions of a diesel engine, engineers modify existing engines to run tests (Tahera et al. 2014). Thus the company's own existing designs are used as models. This also has the advantage that many of the models used in the design process of the predecessor design still exist and can be reused. Designers are therefore using models that appear finalised in early stages of processes. This is another example of the fluid repurposing of models by designers. Designers opportunistically reapply models. Our classification therefore refers to the intended purpose rather a property of the model.

This re-use of models in a different context may at first glance seem alien to the sciences. After all, here a model often aims to represent a phenomenon and thus often a naturally re-occurring natural event. However we also learned in Sect. 2 that in the data generation process, models are often needed to make sense of the experimental data. They may include models of the experimental setup. A case at hand are complex high energy physics experiments. Here a simulation simulates the real detector and determines what, following the underlying theory of particle physics, the detector is to see. These so-called event generators (and other simulations) are used to gauge the real detector and as such have a somewhat evaluative function as well. Moreover, for years, an event generator originally developed at CERN was reused for a wider range of high energy physics experiments all over the world.

4.3 *The Relationship Between Model and Target*

As already noted, in the design process, decisions are taken that finalize part of the design and thereby change the epistemic status of models. Only once such a decision is taken to keep an aspect of a design it makes sense to think about the mapping between the model and reality. However, there is a considerable amount of uncertainty and provisionality associated with design decisions. Some parts of the design are considered to be fixed from the beginning as components are reused, others are frozen at particular points in the design process. However, many decisions are undone in the course of a design process due to iteration and change propagation. While many aspects of a design are explicitly decided others are the results of other design decisions that are taken, for example when components are reused they strongly influence the overall architecture, the choice of materials and the manufacturing processes. Only once the design is completely finished can there be a product against which the model can be compared. Therefore the action lies in correspondences between models, as the design can change until it is handed over to manufacturing or the end customer. Before this point the relationship between the models and the target depends on the state of the design and the decisions the designers have taken. Still we can say that for generative models designers decide whether the whole model or part of it will be build.

Let us now turn to evaluation models. For analysis models the target of the model changes throughout the design process, as changes occurs and decisions are finalised in the design process. For example the shape of the car is repeatedly tested in a wind tunnel because engineering decisions can affect the aerodynamics of the car. While in that example the engineers are not allowed to change the external appearance of the car after the design freeze, the configuration of the parts behind the grill can still change affecting the aerodynamics. Initially analysis models analyze or simulate the behaviour of generative models. The results are compared both to the requirements and experience from the past.

For simulation models, designers are largely concerned with the fidelity of models as a measure for their validation. Tolk (2012) for example defines fidelity of a simulation as 'the accuracy of the representation when compared to the real world system represented. A simulation is said to have fidelity if it accurately corresponds to or represents the item or experience it was created to emulate: How realistic does the simulation react?' As monitoring real life systems such as traffic flow can require a lot of effort, often real live systems are used to calibrate the simulation model, where the simulation model is tweaked until the snapshot of the simulated processes meets the observed snapshot of reality. In engineering processes simulation models are calibrated against the results of physical tests of prototypes, unless companies are extremely confident about the quality and accuracy of their simulation models.

Companies still create some physical models of the product or parts of their products as soon as possible to compare the results of the analysis and simulation against test results from physical models to validate the design and calibrate the models, and to assess properties of the product that require physical interaction. At the beginning of the process, the physical models can be very crude or they are versions of the previous design, later they are prototypes of the later product, which can be parts of the final product or be produced with different manufacturing technology. This also applies to highly complex products, like oil platforms, which cannot be prototyped in their entirely. Here companies aim to reduce risk, by building physical models of components and systems and testing these to assure that key functional components are working. For example companies would not build prototypes of entire buildings or ships, but key components such as lifts or engines would be thoroughly tested.

To broaden the range, companies increasingly carry out so called hybrid tests where the product or part of it is tested physically, but also subjected to a computer simulation of different environmental or use conditions. The range of test conditions is usually tailored to the specific markets and customers that the company has. In highly regulated industries like the automotive industry the products are specifically tested against a set of explicitly described and standardised test conditions predefined by the regulator and which apply to all companies selling in the same market, which the product needs to meet.

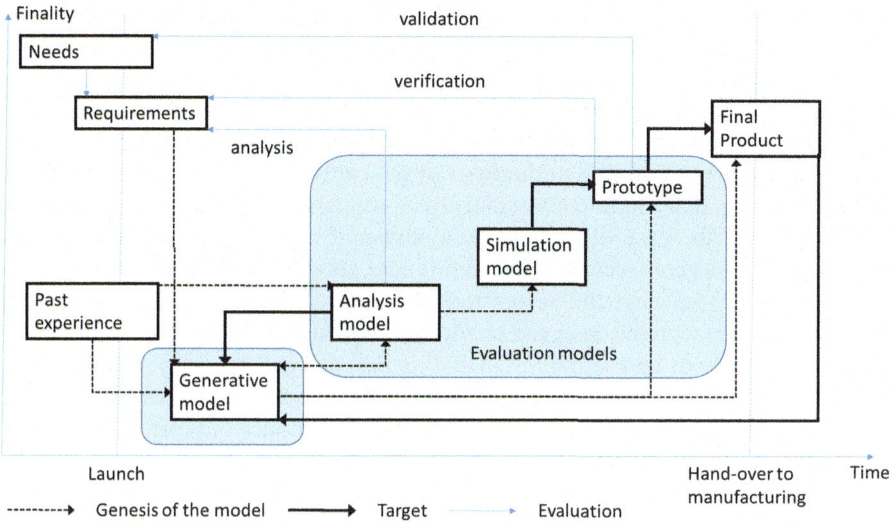

Fig. 1 Relationships of fit between different models through the design process (*black lines*), the genesis of models (*dashed line*) and the evaluation criteria (*thin lines*)

In design the artefact fits to the models, through which it is generated whereas in science the model is fitted to a pre-existing reality. Fig. 1 depicts the relationship between model and target in design. Here the bold black arrows show the directions of fit between the models and the target; the dashed arrows show generation of models based on each other. However on closer inspection these relations are even more complex than the figure shows. Typically before a design process officially starts, the organisation has identified the needs for a new product. These needs are then translated into requirements.[4] The generative models are derived based on requirements drawing on past experience. Analysis models are created to evaluate generative models or to identify inputs and improvements to generative models. Analysis models are developed into simulation models. The final design is created from the generative models. The thin lines show the evaluation processes going on. Analysis models are used to analyse a design against the requirement. The target of simulation models are physical prototypes. The target of the prototypes is the final product. Standing in place of the final product they are verified against the requirement. Towards the end of the processes, when the engineers are fairly confident of the product, it is validated against the original needs. The generative model is the target of first the analysis model and later the final design. However the target of the simulation model is the prototype that is generated based on the generative model.

Just like models of data and of phenomena in the science, models in design are partially independent from the target system they are meant to represent. As the

[4] Models of requirements can be thought of as part of the design models, however as they are often provided by customers or other teams outside the design organisation they are treated as external.

design process is highly iterative, the product or process models devised in the design process do not follow directly from a given design problem. Rather they elicit outside input and can lead to re-formulation or refinement of the original design problem. For example car designers like to maintain stylistic continuity across models, but might realise through wind tunnel experimentation or simulation, that they need to redesign the front of the car radically. Neither the available design methods, like a failure mode and effect analysis, nor the underlying scientific models, like in the case of the car thermodynamics for example, determine in a unique way how, given a certain design problem, its solution should look like.

Just as in the sciences the design models take a central role here in mediating between the artefact to be designed on the one hand and on the other hand the available design methods and resources, including past designs, relevant scientific theories and models. The models also mediate between the different agents involved in the design process, the organisations that design the product and often also the customer. Without models design would not be possible.

5 Conclusion

While models in the sciences have attracted the interest of philosophers for quite some time, models in design are rarely reflected upon. This paper aims at models in the design process and thereby tried to use the knowledge and insight gained from the philosophy of science debates on models. We have tried to compare models in design and models in science and, though differences remain, some of the seemingly obvious differences could not sustain a closer look at scientific reality. Models in science not only aim to represent repeatable phenomena like the interaction of gravitationally interacting bodies, but also aim at target systems that are much more concrete, such as specific rock formations or the evaluation of the Earth's climate system. These models are much less aimed at generalization, and in this sense are closer to models in design where models are tools for that particular design, only rarely created with the purpose of reusing them. Nonetheless, quite generally science often has more claim to generality than design.

Based on this, we aim at a classification of models as they are used in design. A first distinction separates models of the designed artefact from models of the design process. Overlapping with this distinction we can in both categories distinguish generative and evaluative models. Note that the term "generative" was chosen deliberately as the models are not quite prescriptive as they may seem at first glance. Indeed, the fit is from model to target, i.e. the designed product is to resemble in some sense the model (unlike in the sciences where the fit is the other way round: from the phenomena to be modelled to the model). The propositions following from the models are not prescriptions in a strict sense. Note furthermore that generative models are generative as regards the designed artefact, but also more generally regarding the ideas of the designer who creates and shapes the final product as the models stimulate or narrow down her imagination.

With its classifications this paper helps to identify hot spots for future research, particularly in those areas where design studies can profit from the model debates within philosophy of science and vice versa. Quite generally we hold that the view of "models as tools", such as epistemic tools, which is prominent in Philosophy of Science, may also be a useful perspective for design where commonly models are not seen as tools but as products of tools. Here for example, CAD systems are seen as tools that create models. This is undoubtedly a valid picture, but we think that viewing the model as a tool can help to better understand the limits and potential of the use of models in design.

For example, just as in the sciences, design product models act as mediators. This is a point that merits further investigations. In design, models fulfil various central purposes that cannot be reduced to their representative function alone as becomes clear in our distinction between generative and evaluative models. We hold that this is one of the major difference between modelling in the sciences and modelling in design. In engineering the same models are often used for multiple purposes over time due to the effort involved in modelling; however engineers are necessarily aware of the influence that the purpose of the model has on the details in the model. In the absence of reflection about the nature of models, engineers think of models as positive analogies of their target systems, when in reality many models have other models as target systems. Moreover the lively discussion on computer simulation within the philosophy of sciences may proof useful for the design community.

Turning to the philosophy of science community, the debates on models and simulations may learn from design studies as well. Design researchers and practioners often aim at a very detailed picture of complex design processes and have a clear awareness that the process and their understanding of the process can have a profound effect on the product that it generates. By contrast philosophy of science even after the practice turn, has a tendency to abstract away from a lot of possibly equally relevant features of the scientific process.

References

Bailer-Jones, D. M. (2003). When scientific models represent. *International Studies in the Philosophy of Science, 17*(1), 59–74.

Browning, T. R., & Ramasesh, R. V. (2007). A survey of activity network-based process models for managing product development projects. *Production and Operations Management, 16*, 217–240.

Bucciarelli, L. L. (1994). *Designing engineers*. MIT press.

Cartwright, N. (1999). *The dappled world: A study of the boundaries of science*. Cambridge University Press.

Chakravartty, A. (2010). Informational versus functional theories of scientific representation. *Synthese, 172*(2), 197–213.

Duhem, P. (1954). *The aim and structure of physical theory*. Princeton: Princeton University Press.

Dunbar, K. (1997). How scientists think: On-line creativity and conceptual change in science. *Creative thought: An investigation of conceptual structures and processes, 4*.

Eckert, C. (2013). That which is not form: The practical challenges in using functional concepts in design. *Artificial Intelligence for Engineering Design, Analysis and Manufacturing, 27*(03), 217–231.
Eckert, C., Clarkson, P. J., & Zanker, W. (2004). Change and customisation in complex engineering domains. *Research in Engineering Design, 15*(1), 1–21.
Eckert, C. M., & Clarkson, P. J. (2010). Planning development processes for complex products. *Research in Engineering Design, 21*(3), 153–171.
Eckert, C. M., Stacey, M., & Earl, C. (2005). References to past designs. *Studying designers, 5*(2005), 3–21.
Eckert, C. M., & Stacey, M. K. (2010). What is a process model? Reflections on the epistemology of design process models. In *Modelling and Management of Engineering Processes* (pp. 3–14). London: Springer.
Eger, T., Eckert, C., Clarkson, P. J. (2005). The role of design freeze in product development. In DS 35: Proceedings ICED 05, the 15th International conference on engineering design, Melbourne, Australia, 15.-18.08. 2005.
Eppinger, S. D., Ulrich, K. T. (1995). Product design and development. 1995.
van Fraassen, B. (1980). *The scientific image*. Oxford: Clarendon Press.
Ferguson, E. S. (1994). *Engineering and the Mind's eye*. MIT press.
French, M. J. (1999). *Conceptual design for engineers*. Springer.
Frigg, R. (2010). Fiction in science. In J. Woods (Ed.), *Fictions and models: New essays* (pp. 247–287). Munich: Philosophia.
Frigg, R. and Hartmann, S., 2012 "Models in science", The stanford encyclopedia of philosophy (Fall 2012 Edition), Edward N. Zalta (ed.), http://plato.stanford.edu/archives/fall2012/entries/models-science/
Galle, P. (1999). Design as intentional action: A conceptual analysis. *Design Studies, 20*(1), 57–81.
Gero, J. S., & Kannengiesser, U. (2004). The situated function-behaviour-structure framework. *Design Studies, 25*, 373–391.
Giere, R. (1988). *Explaining science: A cognitive approach*. Chicago: University of Chicago Press.
Grauer, R., Marliani, C., & Germaschewski, K. (1998). Adaptive mesh refinement for singular solutions of the incompressible Euler equations. *Physical Review Letters, 80*(19), 4177.
Goel, A. K., Rugaber, S., & Vattam, S. (2009). Structure, behavior, and function of complex systems: The structure, behavior, and function modeling language. *Artificial Intelligence for Engineering Design, Analysis and Manufacturing, 23*(01), 23–35.
Hartmann, S. (1996). The world as a process. In: R. Hegselmann u. a. (eds.): Modelling and simulation in the social sciences from the philosophy of science point of view (pp. 77–100), Dordrecht.
Hesse, M. (1963). *Models and analogies in science*. London: Sheed and Ward.
Hillerbrand, R. (2012). Order out of chaos? A case study in High Energy Physics. *Studia Philosophica Estonica, 5*(2), 61–78.
Hillerbrand, R. (2015). Explanation via micro-reduction: On the role of scale separation for quantitative modelling. In B. Falkenburg & M. Morrison (Eds.), *Why more is different – philosophical issues in condensed matter physics and complex systems* (pp. 69–87). Berlin Heidelberg: Springer.
Holyoak, K. J., & Thagard, P. (1997). The analogical mind. *American Psychologist, 52*(1), 35.
Hughes, R. I. G. (1999). The sing model, computer simulations and universal physics. In M. Morgan & M. Morrison (Eds.), *Models as mediators: Perspectives on natural and social science* (pp. 97–145). Cambridge: Cambridge University Press.
Humphreys, P. (1991). "Computer simulations", In: A. Fine, M. Forbes & L. Wessels (eds.), PSA 1990, Vorl. 2, 497–506, East Lansing.
Klein, E. E., & Herskovitz, P. J. (2005). Philosophical foundations of computer simulation validation. *Simulation & Gaming, 36*(3), 303–329.
Kosslyn, S. (1994). *Image and brain*. Cambridge, MA: MIT Press.

Küppers, G., & Lenhard, J. (2005). Validation of simulation: Patterns in the social and natural sciences. *Journal of artificial societies and social simulation, 8*(4).

Latour, B. (1987). *Science in action: How to follow scientists and engineers through society*. Harvard university press.

Mayo, D. G. (1996). *Error and the growth of experimental knowledge*. University of Chicago Press.

Morgan, M. S., & Morrison, M. (1999). *Models as mediators. Perspectives on natural and social science*. Cambridge: Cambridge University Press.

Morrison, M. (1999). Models as autonomous agents. In M. Morgan & M. Morrison (Eds.), *Models as mediators* (pp. 38–65). Cambridge: Cambridge University Press.

Morrison, M. (2009). Models, measurement and computer simulation: The changing face of experimentation? Philos. *Stud, 143*, 33–57.

Pahl, G., Beitz, W. (1996). Engineering design: A systematic approach (edited by Ken Wallace and translated by Ken Wallace, Lucienne Blessing, and Frank Bauert).

Parker, W. S. (2009). Does matter really matter? Computer simulations, experiments, and materiality. *Synthese, 169*(3), 483–496.

Pidd, M. (1999). Just modeling through: A rough guide to modeling. *Interfaces, 29*(2), 118–132.

Poznic, M. (2016). Modeling organs with organs on chips: Scientific representation and engineering design as modeling relations. *Philosophy and Technology, 29*(4), 357–371.

Sargent, R. G. (2005, December). Verification and validation of simulation models. In Proceedings of the 37th conference on Winter simulation .Winter simulation conference. pp. 130–143.

Sawyer, R. K. (2011). *Explaining creativity: The science of human innovation*. Oxford University Press.

Schön, D. A. (1984). *The reflective practitioner: How professionals think in action* (Vol. 5126). New York: Basic books.

Searle, J. (1983). *Intentionality. An essay in the philosophy of mind*. Cambridge: Cambridge University Press.

Stacey, M., & Eckert, C. (2003). Against ambiguity. *Computer Supported Cooperative Work (CSCW), 12*(2), 153–183.

Stiny, G. (1980). Introduction to shape and shape grammars. *Environment and planning B, 7*(3), 343–351.

Suárez, M. (1999). Theories, models, and representations. In *Model-based reasoning in scientific discovery springer US* (pp. 75–83).

Suppe, F. (1977). *The structure of scientific theories*. Chicago: University of Chicago Press.

Suppes, P. (1961). A comparison of the meaning and use of models in mathematical and empirical sciences. In H. Freudenthal (Ed.), *The concept ad role of the models in mathematics and natural and social sciences* (pp. 163–177). Dordrecht: Reidel.

Suppes, P. (1962). Models of data. In E Nagel, P. Suppes and A. Tarski (eds.): Logic, methodology and philosophy of science: Proceeding of the 1960 International Congress. Stanford, CA. Stanford University Press, pp 252–261.

Umeda, Y., Ishii, M., Yoshioka, M., Shimomura, Y., & Tomiyama, T. (1996). Supporting conceptual design based on the function-behavior-state modeller. *Artificial Intelligence for Engineering Design, Analysis and Manufacturing, 10*, 275–288.

Tahera, K., Earl, C., & Claudia, E. (2014). Integrating virtual and physical testing to accelerate the engineering product development process. *International Journal of Information Technology and Management 9, 13*(2–3), 154–175.

Tolk, A., Turnitsa, C. (2012, December). Conceptual modeling with processes. In Proceedings of the 2012 Winter Simulation Conference (WSC)(pp. 1–13). IEEE.

Tolk, A. (2012). Truth, trust, and Turing – Implications for Modeling and simulation. In A. Tolk (Ed.), *Ontology, epistemology, and teleology for Modeling and simulation* (pp. 1–26). Berlin: Springer Berlin Heidelberg.

Vermaas, P. E. (2013). The coexistence of engineering meanings of function: Four responses and their methodological implications. *Artificial Intelligence for Engineering Design, Analysis and Manufacturing, 27*(03), 191–202.

Visser, W. (2006). *The cognitive artifacts of designing* (p. 264). Lawrence Erlbaum Associates.

Weisberg, M. (2013). *Simulation and similarity. Using models to understand the world.* New York: University Press.

Wynn, D. C. (2007). Model-based approaches to support process improvement in complex product development (Doctoral dissertation, University of Cambridge).

Wynn, D., & Clarkson, J. (2005). Models of designing. In *Design process improvement* (pp. 34–59). London: Springer.

Wynn, D. C., & Eckert, C. M. (2017). Perspectives on iteration in design and development. *Research in Engineering Design, 28*(2), 153–184.

Part III
Design Aesthetics, Design Phenomenology

Notes for an Aesthetics of Social Innovation: A Reading Through the Lenses of Jacques Rancière's Philosophy

Virginia Tassinari

Abstract The DESIS Philosophy Talks are an initiative of DESIS Network aiming to nurture the dialogue between design and philosophy. The idea is to match practical issues and topics emerging from design practice in the field of social innovation around the world (for instance, the ones developed within the framework of DESIS Network) with insights from the philosophical tradition. Several notions, such as beauty, public vs. private, community, etc., which normally belong to the field of social sciences appear to emerge from a kind of "phenomenological" study of different cases of design for social innovation. The DESIS Philosophy Talks want to explore them from a philosophical, theoretical perspective and see how the result of these discussions can add meaningful value to the design practice (www.desis-philosophytalks.org) In the past years we have been raising the question whether we can speak of a "sustainable aesthetics" emerging from the "phenomenological" observation of both cases of grassroots social innovation and design for social innovation. The French philosopher Jacques Rancière speaks of the idea of aesthetics as shifting the political paradigm. Can we describe this emerging new paradigm of aesthetics starting from Rancière's notion of aesthetics? Can we consider "sustainable aesthetics" today as moving the political paradigm towards a more participative and open society? If this is the case, what is the role and responsibility of design in embracing this emerging "sustainable aesthetics"? This paper presents a hypothesis on how such an aesthetics could be considered as a driver of social and behavioural change.

Keywords Aesthetics · Politics · Design for social innovation · Design and philosophy

---

V. Tassinari (✉)
LUCA School of Arts, Genk, Belgium
e-mail: virginia.tassinari@luca-arts.be

© Springer International Publishing AG 2018
P. E. Vermaas, S. Vial (eds.), *Advancements in the Philosophy of Design*, Design Research Foundations, https://doi.org/10.1007/978-3-319-73302-9_12

1 A New Sensitivity

Across the globe we see an increasing stream of bottom-up activities in which citizens take on challenges and initiatives together, in which both socially and environmentally they innovate their own local contexts (Manzini and Coad 2015; Thorpe and Gamman in Fisher and Sparke 2016). People spontaneously decide to be part of this wave of new initiatives not because they are compelled to, or because they want to make a political statement, but because it feels to them as the right thing to do (E. Manzini and Tassinari V. in Crocker and Lehmann 2013). Moreover, these spontaneous groups of people - often also referred to as *creative communities* (Meroni 2007) - enjoy the fact of doing something together which they believe to be meaningful. They set up new initiatives in a collaborative and spontaneous way to tackle societal issues which have become too complex to deal with in a top down fashion only, issues which have come to challenge the limits of traditional models of policy making (Manzini and Staszowski 2013; Selloni 2017).

Within these creative communities, people experience a certain satisfaction in taking their own responsibilities within the common realm (Baerten and Tassinari 2010). The American sociologist Richard Sennett notes that many people currently show an increasing appreciation for making things together. It is a binding element. Quoting Kant, he writes:

> The hand is the window onto the mind. … (it) shows, in sum, a recipe for binding people tightly together. (Sennett 2012, p.103)

This renewed enjoyment of doing things together can be considered a shift in the *sensitivity* of our contemporary societies: from individualism to the pleasure of doing things together, from the passivity of consuming a product which has been produced by someone else to the pleasure of being an active player in one's own context (Manzini 2014; Penin et al. 2016).

2 Design for Social Innovation

In the past years the design community has paid extra attention to initiatives in which citizens collaborate in order to improve the quality of life within their own contexts. Designers and design researchers have started to map out these initiatives - where possible amplifying and/or scaling them - and to create the preconditions for them to take root (L. Penin in Crocker & Lehmann 2013).

Within the design discourse, this approach to design has been labelled in many different ways, among which *design for social innovation* is the one term probably most acknowledged (Manzini 2014). The projects stemming from the DESIS Network[1] - the International network of Design for Social Innovation and

[1] www.desis-network.org

Sustainability - are one example of initiatives developed in the field of design for social innovation, where design schools work with other citizens in their local contexts to create the preconditions for social innovation to be more probable (Manzini and Coad 2015). These schools position themselves as spaces for experimentation in which design students, teachers and researchers work together with citizens, local associations and different kinds of stakeholders (from both the public and the private sector) in order to create the preconditions for social innovation to develop (or evolve and develop further when already in place) in a given context (Manzini and Staszowski 2013).

From within the DESIS network community, the need was felt to create a space of exchange of experiences and know how, a shared space in which to work and reflect on concrete projects in a collaborative way amongst global members of the network. To accommodate for this, DESIS Network started to develop the so-called DESIS cluster projects, contexts in which labs would work collaboratively and focus on common themes and issues, relevant to their individual contexts (Manzini and Staszowski 2013; Lee and Moore 2015).

Some of these issues are for instance the co-creation of *collaborative services* (Manzini 2008) between citizens, public servants and policy makers (DESIS cluster Public and Collaborative, documented in Manzini and Staszowski 2013),[2] the co-creation with elderly of solutions employing their ingenuity (DESIS cluster Ageing and Ingenuity, documented in Lee and Moore 2015),[3] the study of the use of storytelling in social innovation (DESIS in the mirror, documented in Bertolotti et al. 2016),[4] the co-creation of sustainable initiatives around food production, distribution and consumption (DESIS cluster FOOD)[5], new (open) forms of production (DESIS cluster Distributed and Open Production)[6] and the identification of *potential ideas for a new generation of services that could be designed from the perspective of underserved communities'*[7] (DESIS cluster Informal, Formal and Collaborative). Some of these projects are presented in exhibitions,[8] others in dedicated DESIS publications,[9] symposiums and conferences (Jin and Yongqi 2014 and Manzini 2015).

What emerged from these collaborations – besides sharing experiences, knowledge, tools and insights - is the opening of a dialogue on values and meanings generated by the projects. This kickstarted a process of knowledge co-creation within the field of design for social innovation. Key learnings from this exchange include the realisation that most projects challenge the meaning of several key

[2] www.nyc.pubcollab.org

[3] www.hkdi.desislab.vtc.edu.hk/#!desis-ageing/c16ku

[4] www.desisinthemirror.imagishub.it

[5] www.desis-foodcluster.org

[6] www.desis-network.org/clusters

[7] www.desis-ifc.org

[8] For instance in the Triennial of Design RECIPROCITY Liège www.reciprocityliege.be/reciprocity-2015-4

[9] Such as Manzini and Staszowski 2013 and Lee Y. and P. Moore 2015

notions in society - such as those of politics, society and citizenship - and ask for these meanings to be reinterpreted and reframed (Manzini 2016). A similar critique of and quest for new meanings can also be noticed in other disciplines such as for instance philosophy, sociology and anthropology. A collaborative closer look from such a variety of perspectives would benefit the discourse around these notions and meanings.

Urged by this need, the author - together with Ezio Manzini and Victor Margolin - started an initiative entitled the DESIS Philosophy Talks[10]: i.e. small-scale encounters for a transdisciplinary discussion regarding some of the issues arising within concrete projects of design for social innovation. In each DESIS Philosophy Talk the issues at hand are 'read' through the lens shaped by the point of view of a given philosopher or philosophical tradition (Bertolotti et al. 2016).

One of the key topics which has been featured in the DESIS Philosophy Talks is whether we can speak of this enjoyment of doing things together described in the first paragraph - this change of *sensitivity* - in terms of an emerging, shared *aesthetics*.[11] (Manzini E. and V. Tassinari in Crocker and Lehmann 2013). Can we liberate the word *aesthetics* from the overload of connotations which we inherited from our recent philosophical past and use it in a broader sense, to denote a change in *sensitivity* in the context of our contemporary lives? In order to try and formulate an answer to this question, one first needs to ask oneself whether it is still possible to speak of *aesthetics* and eventually in which terms (Manzini and Tassinari 2012; Markenko and Brassett 2015).

Since Heidegger's condemnation of aesthetics (Heidegger et al. 2002) - seen as the key responsible for the death of art - many contemporary philosophers have been claiming that this concept belongs to our past and can no longer be used. Yet, there is one contemporary philosopher who is currently claiming otherwise, i.e. Jacques Rancière.[12]

In this paper we will take a closer look at his understanding of *aesthetics* and explore whether it can be mapped onto the shift of *sensitivity* registered in creative communities and how this could eventually help us to understand better the phenomena of grassroots social innovation emerging in our society and of design practices dealing with them.

---

[10] See also http://www.desis-philosophytalks.org

[11] The author raised this question with Ezio Manzini and Victor Margolin back in 2012. The first DESIS Philosophy Talk series, entitled *Emerging Qualities* featured a first event, entitled *Emerging Aesthetics,* which took place on March 2nd 2012 at The New School for Design (New York, U.S.A). For this issue, please also see Manzini and Tassinari 2013

[12] The use of the categories of Rancière's philosophy in order to interpret the idea of aesthetics/ politics emerging from practices of design for social innovation, has first been researched during the seminar *the (New) Public Goods: Design, Aesthetics and Politics* organised by The New School of Design in New York, in the spring 2013

3 The Art and the Art of Living

Rancière's definition of *aesthetics* dates back to Schiller's definition, according to which this is a *new sensorium, a specific sensory experience that holds the premises of both a new world of Art and a new life for individuals and the community* (Rancière and Corcoran 2010, p.113) also called *the art of living* (Rancière and Corcoran 2010, p.113). In his interpretation of Schiller's words, *aesthetics* is not only an upcoming common *sensitivity* manifested by Art with a capital "A" but also by the *art of living* experienced in everyday life. Rancière embraces this idea of *aesthetics.*[13]

This unity between art and life is what Rancière calls aesthetics: namely, a unity between the *sensorium of art* (Rancière and Corcoran 2010, p.128) - created by artists - and the *sensorium of everyday life* (Rancière and Corcoran 2010, p.128), the common sensitivity shared and experienced in everyday life. This *new sensorium* (Rancière and Corcoran 2010, p.118) is *a place of exchange between every day life and the realm of art* (Rancière and Corcoran 2010, p.127).

Rancière recognises a *common sensorium, a plot... recurring in practical attitudes, in modes of individual perception and in social institutions* (Rancière and Corcoran 2010, p.115) as well as in Art. He wants to understand it *a posteriori*, and not to theorise it *a priori*. He observes the rise of this *new sensorium* and wants to make sense of it, to act as a phenomenologist - observing a change in the common perception of everyday life - and to draw a parallel with the sensorium manifested and driven by contemporary artistic practices. This *new sensorium* emerging both in everyday practices and artistic ones is what Rancière calls *aesthetics.*

Yet, what allows us to imagine that grassroots social innovation can be interpreted as *art of living* and that the new *sensitivity* manifested by initiatives of creative communities can be imagined as belonging to this *new sensorium*?

4 The Idea of Aesthetics in Rancière's Philosophy

Rancière says that both Art and the art of living manifest and further drive this *new sensorium* where everything is made accessible to everybody. Its key characteristic lies in the fact that the *self-formation of a new sensorium leads to a new collective ethos*, (Rancière and Corcoran 2010, p.118), a new way of living (together).

[13] He believes that in ancient Greece, art and life used to be connected. Artworks such as the Juno Ludovisi used to embody *the vital spirit of a community* (Rancière and Corcoran 2010, p.122). This unity between art and life broke with the divisions of the arts, and is reappearing in modernity. The way in which this unity manifests itself in modernity is called *aesthetics*. According to Rancière, Schiller's words foresee a time in which art and life will once again be re-connected as it used to be in ancient Greece. This time predicted by Schiller is now.

The creation of an inclusive and equal society, where none are excluded from the idea of a collective creation of a new *ethos*, is to Rancière the main consequence of the rise of *aesthetics*.

This *new sensorium* causes a more equal society because things are made visible and tangible - *available* - for everyone to partake. Rancière sees it as *a new partition of the sensible* in which the latter is finally made available to everyone who is granted the possibility to partake of the sensible and have a stake in it. In this *new sensorium* the individual citizens can finally have a stake in the common realm. This points towards a more participatory, horizontal and equal society.

Both Art and *the art of living* aim at *making visible what was unseen; at making what was audible as mere noise heard as speech and at demonstrating that what appeared as a mere expression of pleasure and pain is a shared feeling of a good and a evil.* (Rancière and Corcoran 2010, p.38) As such, they manifest and further drive a *shared aesthetics* (Rancière and Corcoran 2010, p.38). In other words, in the *shared aesthetics* emerging in our contemporaneity the *facts* - experiences, spaces, times, and access thereto - are made visible and tangible for everybody, not only just for the happy few. This makes possible for these *facts* to be matched outside of the logic of distribution of *the part and shares* (Rancière and Corcoran 2010, p.36) in which few have the power to decide over many, which has been mostly characterising our Western modernity and contemporaneity.

By means of the *partition* - or *distribution* of the sensible - inequalities in wealth and political influence characterising contemporary societies finally take centre stage. The *facts* of contemporary social contexts show how deeply the general look and feel of things is connected to the manifestation of power; in particular, the power to participate in, and to propose alternatives to, a common realm:

> The distribution of the sensible reveals who can have a share in what is common to the community based on what they do and on the time and space in which this activity is performed [;] it defines what is visible or not in a common space, endowed with a common language, etc. It is a delimitation of spaces and times, of the visible and the invisible, of speech and noise, that simultaneously determines the place and the stakes of politics as a form of experience. (Rancière 2004, pp.12-3)

Aesthetics is a new sensory field, in which the terms of a shared social existence are re-configured, so that individuals can have a greater stake in society than is presently the case. This *new sensorium* is the field of action in which new modes of living can be configured, and a re-framing *for the common world* (Rancière 2004, p.13) can be reformulated. This consists of:

> (…) a multiplicity of folds and gaps in the fabric of common experience that change the cartography of the perceptible, the thinkable and the feasible. As such, it allows for new modes of political construction of common objects and new possibilities of collective enunciation. (Rancière 2004, p.13)

5 Grassroots Social Innovation and Design for Social Innovation as Aesthetics

In Rancière's view, both the Art and *the art of living* open up new forms of *partition of the sensible,* defining a *new sensorium* in which individuals can have an active stake in societal life. As such, they both promise new ways to experience the world and to act within it, suggesting new forms of social interaction and supporting the emergence of more equality in contemporary societies.

If one looks at the contemporary context on a global scale, this bears close resemblance to the route taken by grassroots social innovation. In these collaborative initiatives citizens are opening up the space of decision making, from something taking place behind closed doors by a minority taking decisions over the many, to a participatory, inclusive process in which citizens can be directly and actively involved. They are re-appropriating a more inclusive *partition of the sensible*, so that *facts* can finally be made available to everybody and become building blocks from which to shape a more inclusive and equal society.

For instance, cases of grassroots social innovation documented by the DESIS cluster *Formal, Informal and Collaborative*[14] show how citizens living in underserved communities - in which public services are poorly provided by the local authorities - re-think and directly co-produce new services for their local communities (Manzini and Staszowski 2013). Citizens take a stake in the common realm and act together. In Rio de Janeiro, for example, the local DESIS lab studies how new kinds of collaborative services - for instance, a new policy for energy bills that are reduced when residents exchange recyclable materials for energy credits - are emerging in informal settlements (Manzini et al. 2013). These new services can inspire other communities across the globe. Similar research is also carried out in Africa, South America, Pakistan and India.[15]

What happens in these case studies is that citizens co-design and co-produce new initiatives by means of which they can contribute to the creation of *a new collective ethos*. This means that grassroots social innovation can also be imagined as a form of *art of living* manifesting and further promoting the rise of a new *common sensorium.* Yet, how can we say the same of design initiatives enabling social innovation?

Art - also called *the aesthetic domain of art* - is to Rancière broader than the field of fine arts. In his interpretation, Art is the field of action in which diverse creations render visible and tangible for everybody what previously was not. This is the case for paintings, pictures, films, but also for instance for objects and products.

---

[14] www.desis-ifc.org

[15] Some of these cases have been presented to an international public of design schools around the world in 2014 in the Cumulus conference *Design with the other 90% of the world* www.cumulusjohannesburg.co.za

Therefore, design can also be considered as form of Art, as long as it manifests to everybody what was previously manifested only to the few:

> Both industrial production and artistic creation are committed to doing something on the top of what they do - to creating not only objects, but a sensorium, a new partition of the sensible. (Rancière and Corcoran 2010, p.122)

According to him, also design can be considered to manifest and drive a *new sensorium.* This is exactly what happens in the case of design for social innovation. As much as grassroots social innovation, projects of design for social innovation also aim to create a new *partition of the sensible* in which *facts* are made visible and tangible. When designers work to visualise and render tangible those *facts* which are normally not shared with citizens and empower citizens to take a stake in them in order to improve the society in which they live, they basically also shape *a new partition of the sensible.*

From this line of reasoning one can infer that both grassroots social innovation and design for social innovation manifest and contribute to the development of a new *common sensitivity* - in other words, a new *aesthetics* - in which *facts* are shared in a more inclusive and equal way.

Seen from this perspective, the role of the designer is that to open up new domains of visibility and tangibility, manifesting *a new partition of the sensible.* He does not do it from scratch, but rather by listening to the signals of change emerging in the emerging *common sensitivity* - or, in other words, *aesthetics* - of the context in which he happens to work. When confronted with the rising of the new *sensitivity,* a designer can catalyse the changes which he acknowledges there, amplifying them and enabling them further. He needs to be a *kind of symptomatologist, delving into the dark underside of the unconscious of a society to decipher the message engraved in the very message of ordinary things* (Rancière and Corcoran 2010, p.127). In order to do so, he needs to engage in deep listening, though ethnographic research, empathy and field research in the different contexts in which he is working. By mean of this deep listening, he can sense the rise of a *new sensorium,* capture *the message engraved* in it and identify there its meanings and values, finally bringing them to the surface:

> (…) the new poetics frames a new hermeneutics, taking upon itself the task of making society conscious of its own secrets… by delving to the depths of the societal, to disclose the enigmas and fantasies hidden in the intimate realities of everyday life. (Rancière and Corcoran 2010, p.127)

When looked through these lenses, design is not so much a *problem solver* but rather a *meaning maker*: it does not so much aim to solve current society's problems and *fix* them, but rather it aims to question the *status quo,* and open up the possibilities offered by its potential re-configurations, so that these problems can eventually be tackled directly by or in collaboration with citizens, with the support of designers. By questioning the *status quo,* and contributing to the development of this *new sensorium* - making visible and tangible all the elements that are already in place (facts such as places, skills, knowledge, time…) but of which the potentialities for society are not fully developed yet - design enables and empowers actions which can lead to a more equal and inclusive society. It envisions what society *could* look like if this *new sensorium* were fully in place. By doing so, the designer does not find the solution to the present and/or future societal issues him/herself, but rather works to enable, empower and facilitate a given community to develop initiatives aimed at tackling these issues. *Problem solving* as such is not the direct aim, but rather a possible, often unpredictable consequence of *meaning making* (Manzini 2016).

One could see in such an approach a more flexible and resilient way of tackling societal issues, as it accommodates for organic growth and adaptation to different circumstances, avoiding the pitfalls of one-size-fits-all, blueprint-like approaches. By allowing citizens to discover and enjoy this *new sensorium,* such an approach enables actions - even divergent from the ones co-envisioned with designers - to take place in a spontaneous way. In a sense, this approach accommodates for necessary serendipity. When design works to listen to, promote and further develop this *new sensorium*, unexpected things are allowed to happen which contribute to the creation of *a new collective ethos*. We can therefore extend to design what Rancière says about art:

> Art recruiting the hierarchical division as the perceptible and framing of the common sensorium; of art replacing politics as a configuration of the sensible world; or art even becoming, in its very isolation, the guardian of the promise of emancipation. (Rancière and Corcoran 2010, p.133)

In Rancière's understanding, the idea of Art - to which design also belongs - coincides with politics, as he believes the latter to be a configuration of the world which enables citizens to have a stake in the common realm: *Politics, before all else, is an intervention in the visible and the sayable.* (Rancière and Corcoran 2010, p.37) Aesthetics can ultimately be seen as politics, inasmuch as

> politics revolves around what is seen and what can be said about it, around who has the ability to see and the talent to speak, around the properties of spaces and the possibilities of time. (Rancière and Corcoran 2010, p.13)

Grassroots social innovation and design for social innovation can therefore be seen as forms of manifestation and construction of *politics*.[16]

6 Aesthetics as Politics

Rancière speaks of *politics* as a rupture of the idea that few have the *arkhe*, i.e. the possibility to rule, to commence new initiatives. They do this on behalf of the rest of people who are not given this opportunity. These few are given knowledge on *facts*,

[16] This union of aesthetics and politics was there with the Greeks. In modernity, however, art and life, aesthetics and politics became separated into distinct spheres of thought and action. Politics became the prerogative of an oligarchy whose members takes decisions on behalf of the collective. Active participation of citizens was no longer required. Aesthetics, meanwhile, became the prerogative of art alone, and not also of a generalised art of living as in previous centuries.

so that they can make use of them accordingly to their own interests. He calls *pure politics* or - borrowing the term from Foucault - *police* the hierarchical form of politics in which few have their say over the many. This is a *univocal partition of the sensible* (Rancière and Corcoran 2010, p.41), where the political good is handed *over to governmental hierarchies enlightened by their experts.* (Rancière and Corcoran 2010, p.28) The *ratio* according to which the sensible - the *facts* - is partaken, is pre-defined by rules legitimated by an oligarchy, such as for instance *the principle of birth* and *the principle of wealth* (Rancière and Corcoran 2010, p.38). The majority of people is instead confined to the role of a formless entity of those who cannot decide for themselves, hence have no voice in the common sphere.

In contrast with *police, politics* - i.e. aesthetics - gives the opportunity to citizens to step outside of this anonymity and to find their own voice, finally becoming *political subjects*: people who have the *capacity for staging scenes of dissensus* (Rancière and Corcoran 2010, p.69). According to the definition of Aristotle quoted by Rancière, the citizen is he who can partake, *avoir-part* (Rancière and Corcoran 2010, p.28). He can partake in the common realm and have a stake in what can be seen or not seen, heard or not heard, undertaken or not undertaken:

> (…) the citizen is given a name defined by partaking (metexis) both in the form of action (arxhein) and in the passibility corresponding to that action (arkhestai). (Rancière and Corcoran 2010, p.69)

The political subject can commence something and undertake actions in order to bring into action what is potentially out there in society. He can rule, without commanding. Everybody here can rule and be ruled at the same time. There is no form of domination implied.[17] According to Rancière this is the essence of democracy.[18] This is not another form of politics between others: it is the only real form of politics. Democracy is a *rupture in the logic of arkhe* (Rancière and Corcoran 2010, p.31) as *legitimate domination* (Rancière and Corcoran 2010, p.21) of the few on the many, defining who has a stake in the common realm:

> Democracy is not just another form of politics, but the very intuition of politics itself- of its subject and the form of its relationship. (Rancière and Corcoran 2010, p.32)

---

[17] In *police* an oligarchy of subjects decides which part to have of the common goods and how to use the resources which are supposed to belong to everybody. Yet, to Rancière this is a truthful way to be subjects. This idea of subject is rather a mask for domination. According to him, there are no pre-defined subjects which detect the power juxtaposed to a formless collectivity which does not have a stake in the common sphere and has *no right to speak* (Rancière and Corcoran 2010, p.32). To become subjects is a process. There is no other way to be subject besides the political one. A political subject - someone who takes his own role and responsibility in the common realm - is the only authentic way to be a subject. One is not a political subject a priori. One can only become a political subject, by excerpting forms of *dissensus* in the common sphere, which makes possible to question the *status quo* of *police* and re-distribute the sensible in an inclusive and equal way.

[18] *Democracy is the specific situation in which it is the absence of entitlement that entitles one to exercise the arkhe.* (Rancière and Corcoran 2010, p.31) As nobody is entitled to rule, everybody can do this. Democracy is a limit figure, where one can *partake in ruling and in being ruled.* (Rancière and Corcoran 2010, p.31)

Democracy takes place only when the indistinct collectivity who does not have a stake in the common sphere becomes a *demos,* being the sum of the political subjects, from which nobody is excluded. Democracy cuts through the logic of partition of the sensible according to

> the qualifications (of parts of the population) for partaking in the community and open the common share that they are due by virtue of these qualifications. (Rancière and Corcoran 2010, p.33)

When this happens, the supplementary subjects (the political subjects) can rise, *inscribed as a surplus in relation to every count of the part of society.* (Rancière and Corcoran 2010, p.33) Consequently there is a peculiar value which is pushed forward with the rise of aesthetics/politics: the principle of equality.

7 Design for Social Innovation as Dissensus

Politics is not a sphere, but a process. This is a trait it shares with design. Yet, reading design for social innovation through the eyes of Rancière's idea of aesthetics, allows us to see that design and *politics* have much more in common.

Design's manifestation of this *new sensorium* - making visible and tangible what is not visible and tangible yet to all citizens, so that it can be partaken in a more equal way - is also a political action, in the sense given by Rancière to the word *politics*. By contributing to the rise of a new aesthetics, design for social innovation also puts forward a new political paradigm: that of *politics.*

The new sensorium driven by design caters to the original, real meaning of politics: a field of action in which everything is made visible and tangible to everybody, and therefore new actions can be commenced by everybody. It is not just any kind of politics, but *democracy*: a distribution of the sensible that makes *facts* visible, that enables participation in the common realm which empowers citizens to have a stake there and co-construct together a more equal society. It is an act of *dissensus* towards the *consensus*, i.e. the logic of domination belonging to police: *governmental oligarchies enlightened by their experts.* (Rancière and Corcoran 2010, p.28) The logic of *consensus* is a logic of domination, as chosen groups of subjects have the ability to know, and then to decide upon others. It is *the power of those who know over those who do not know.* (Rancière and Corcoran 2010, p.31) While *police* is a univocal partition of the sensible, *politics* enable an inclusive partition of the sensible, where none are excluded. In *police* nothing is left to the ones which do not have the *arkhe*:

"The essence of the police is to be a partition of the sensible characterised by the absence of emptiness and supplementarity: society consists of groups devoted to specific modes of doing, of places where these occupations can be performed, of modes of being corresponding to these occupations and these places. In this adequation of functions, places and ways of being, there is no place for any void. It is this exclusion of the 'there isn't any' which is the policing principle at the very core of state-sanctioned practice. (Rancière and Corcoran 2010, p.36)."

In *politics* instead the void is made visible and tangible, so that political subjects can partake of it:

> The essence of politics consists in disturbing the arrangement by supplementing it with a part of those without part, identified with the whole of the community. (Rancière and Corcoran 2010, p.37)

In manifesting and further empowering this *new sensorium*, design for social innovation can be considered as a form of *dissensus*. It does not aim to solve social problems as such, but to create the preconditions for political subjects to tackle problems themselves in a collaborative way. This can happen because design challenges current meanings - such as those of politics, democracy, citizenship and equality - and calls for alternative ways of interpreting them, which differ from the ones employed by the unilateral and instrumental point of view of *consensus:*

> Consensus does not simply mean the agreement of the political parties or of social partners on the common interests of the community. It means a reconfiguration of the visibility of the common. It means that the givens of any collective situation are objectified in such a way that they can no longer lend themselves to a dispute, to the polemical framing of a controversial world within the given world. (Rancière 2004, p.48)

Design for social innovation can be considered to be disruptive towards the idea of a society in which citizens are mere passive spectators of the political discourse.[19] It challenges the *consensus*, encourages citizens to take on an active role in the common realm and empowers them to solve societal issues and contribute to a more inclusive and rightful society.

As a form of *dissensus,* design for social innovation can question the *consensus* and be disruptive, challenging the status quo and opening up new possibilities of re-configuration for the partition of the sensible. This happens when design makes visible and tangible what is normally not accessible, presenting what is potentially out there - but has not been acknowledged yet by citizens as their part - as a resource: services, physical resources, skills, knowledge and common goods of any kind. As long as design for social innovation makes those things tangible and visible, so that they can be acknowledged and appropriated by citizens who can finally partake them, can it be considered to manifest and further drive *politics*. This surplus is normally not visible in the perspective of *police*. It is a *gap* in the sensible:

> The essence of politics is dissensus. Dissensus is not a confrontation between interests and opinions. It is the demonstration (manifestation) of a gap in the sensible itself. (Rancière and Corcoran 2010, p.38)

This *gap* is what in the logic of *police* is not partaken of by all political subjects. In other words, the *gap* is the sum of the potentialities of everyday life which are not yet made available and manifested to all political subjects. By making this gap visible and tangible and these potentialities available to all political subjects, design as aesthetics/politics can empower actions in which the political subjects are part of the construction of the common sphere. By showing what is normally not made

[19] For more on this see also Meroni et al. 2013

visible in *consensus*, it can make it available to empower political subjects of *arkhe*: in other words, to initiate actions which can be disruptive towards the *police*.

This is what happened for instance also in various projects of the former Helsinki Design Lab,[20] in which citizens were invited to co-design and co-produce new public services.[21] That which was the prerogative of an oligarchy supported by experts - i.e. to design and produce services - has been opened up to the collectivity. In its being a *meet in the middle* between top down and bottom up approach to public service design and policy making, an initiative like the Helsinki Design Lab could at first glance being considered as not particularly disruptive towards the *status quo*. In the logic of Rancière's idea of aesthetics/politics this is the case however, as its initiatives can be seen as a form of *dissensus*, of disruption of the *consensus*. In the ideas co-designed and co-produced, there has been an operation of displacement there where what is public has been taken out of the hands of *police* and placed back into the hands of citizens. In the co-creation of public services - as it in happens for instance for the projects belonging to DESIS Public and Collaborative cluster[22] - the mechanism of the functioning of the state is made transparent to citizens so that they can match their needs with the already existing services and adapt them to their own needs and expectations (Manzini and Staszowski 2013).

This cluster project, developed by DESIS labs or schools connected to the DESIS network – such as for instance Politecnico di Milano, The New School of Design, Aalto University, Malmö University, University of the Arts London, Les Ateliers and LUCA School of Arts – illustrates how design researchers and students have been collaborating on the co-creation of services, together with citizens, local civil servants and policy makers. In some of these experiences – such as for instance in New York and in Saint Gilles (Liège, Belgium) – these collaborations led to formulate the need for physical places or settings allowing these experimentations to take place in a 'safe' environment: public innovation places, where citizens can meet civil servants and policy makers, and design researchers and student can facilitate the co-creation of new services and initiatives (Manzini and Staszowski 2013).

In the city of Liège, for instance, students and design researchers of LUCA DESIS lab and invited designers[23] have been working in a temporary public innovation place, where they imagined - together with citizens, policy makers, civil servants and stakeholders from the local associations - how the city could work in a more inclusive and collaborative way. For instance, they envisioned new solutions in which to use food as a way to connect people in the city, co-created a mobile puppet theatre to facilitate citizens' communications about their challenges and ideas for solutions/aspirations amongst each other and with the local authorities as

---

[20] www.helsinkidesignlab.org

[21] For a full explanation of the methodology that was used in the Helsinki Design Lab, please see the book Boyer et al. (2011)

[22] www.desis-network.org/publicandcollaborative

[23] Such as for instance Elisa Bertolotti, François Jégou and Pablo Calderón Salazar.

to eventually inspire future steps of a neighbourhood master plan, and developed a street magazine together in order to give voice to the ideas of the citizens (Bertolotti and Tassinari 2017 and Bertolotti et al. 2017).

In all these examples there is *a meet in the middle* between bottom up and top down approaches, as authorities understand they need to change their traditional hierarchical *modus operandi* and look for new ways to interact with citizens to match their real needs and expectations, also making use of the resources which can be provided by the same citizens, such as for instance time, knowledge and skills.

In many cases however, this *meet in the middle* does not happen from the start. It is often provoked by designers who initiate actions in order to spur a (co)productive dialogue with local authorities, displacing their traditional positions and roles as well as those of the citizens. Yet, this dialogue does not always necessarily come. Sometimes it is simply impossible, e.g. when the given political situation do not allow forms of *dissensus*. In these cases, the disruptive character of design for social innovation becomes even more accentuated, and can take the form of an open contrast. This happens both with initiatives of grassroots social innovation, but also with projects of designers working to stimulate social innovation in given highly consensual contexts around the world.

However, design for social innovation - on both ends of the spectrum, from collaboration to juxtaposition with authorities - is generally disruptive, in its being a form *of dissensus*, which creates new forms of visibility of what previously passed below the radar, it always questions the status quo and pushes for change. As *dissensus,* its aim is

> *to put into question the received distribution of the relations between the distinct and the indistinct, the pure and the mixed, the ordinary and the exceptional, the same and the other* (Rancière and Corcoran 2010, p.205).

8 Design and Melancholy

The goal of design manifesting and driving aesthetics/politics is that of *an effective re-configuration of the political field* (Rancière and Corcoran 2010, p.71). Rancière considers general Western history as a history of domination. *Police* has been the rule, and *politics* is its *state of exception* (Carl Schmitt quoted in Rancière and Corcoran 2010). History is always an interplay of the two, where *politics* is the state of exception in which alternatives to the *status quo* are imagined and prototyped.

By looking through the lenses of Rancière's philosophy, design for social innovation can be considered one of the forms of Art in which *dissensus* manifests itself in the contemporary world. It envisions alternatives to the *status quo* and helps to imagine what alternatives to *consensus* could look like, through prototyping. Yet, these states of exception are not guaranteed to ever replace the contemporary mainstream, the *consensus*. It is merely a possibility, yet unproven. This brings about a serious risk to be instrumentalised and become a tool in the hands of the *police*, of an oligarchy for its own benefit or its retraction from its societal (co) responsibilities. In order to avoid this risk, the fragile, non dominative, *exceptional* character of social innovation needs to be preserved. As social innovation is becoming increasingly 'trendy' and fashionable, this risk grows along its popularity. In order to avoid this, one needs to remind oneself of the *political, dissensual* character of social innovation, whose aim is not the interest of a few, but the co-creation of a new *ethos* enabling all to be involved.

Rancière says that the political aspirations are not yet fulfilled. This means that design for social innovation, though aspiring to co-create a new ethos, not necessarily succeeds in doing it:

> Aesthetic art promises a political accomplishment that it cannot satisfy...those who want it to fulfil its political promise are condemned to a certain melancholy (Rancière and Corcoran 2010, p.133).

The fallibility which Rancière attributes to Art is likely to help designers to become more realistic in managing the effects of their projects, and avoid triumphalism. The *melancholy* which he attributes to Art can help designers to deal with the high - and often unrealistic - expectations often encountered when collaborating with authorities, especially in a time in which design is often seen as a *panacea* which can solve all different kinds of problems. They clearly cannot. This *melancholy* may help keep in mind the necessity to cultivate humility and realism, and question both the role and responsibilities as designers. One cannot guarantee a transformation of society, but one can try to facilitate small scale transformations to take place, learn from these experiences, and try to scale them up, without having real evidence of a possible success. Anything which promotes *dissensus,* is also signed by this melancholy and frailty. This is also the case for design for social innovation. This frailty should yet not discourage. On the contrary, the awareness of being actors of *dissensus* should help to embrace increasingly one's political responsibilities as designers.

To promote *dissensus* means to create the preconditions for a *count of the uncounted or of the part of those without part* (Rancière and Corcoran 2010, P.35). In the peculiar historical period in which we are living, in which Western societies are increasingly confronted with a growing amount of those not having part - e.g. political refugees and minorities - the relevance of this task grows in importance. As individual designers we ought to be aware of this. As a design community we should engage more broadly and deeply, from this awareness of our (co)responsibility, and work to co-design this new *ethos* in which none are excluded.

References

Agamben, G. (2003). *Stato Di Eccezione: Homo Sacer, II, 1.* Torino: Bollati Boringhieri.

Baerten N. and V. Tassinari, (2010) *Design for Togetherness*, in Shanghai Cumulus Proceedings.

Bertolotti, E. and V. Tassinari (2017). *Experiments in the co-creation of "theatres" for social innovation.* A reading through the lens of Hannah Arendt, NORDES Proceedings.

Bertolotti, E., Daam, H., Piredda, F., & Tassinari, V. (2016). *The pearl diver, Designer as Storyteller*. Milano: DESIS Network Association.

Bertolotti, E., Piredda, F. & V. Tassinari (2017). *Storytelling in design for social innovation and politics: A reading through the lenses of Hannah Arendt* in EAD 12 Rome Porceedings.
Boyer, B., J.W. Cook & Steinberg, M. (2011). Recipes for systemic change.. Sitra.
Crocker, R., & Lehmann, S. (Eds.). (2013). *Motivating change: Sustainable design and behaviour in the built environment*. London: Routledge.
DiSalvo, C. (2015). *Adversarial design. Design thinking, design theory*. Cambridge, MA: MIT Press Limited.
Fuad-Luke, A. (2013). *Design activism: Beautiful strangeness for a sustainable world*. New York: Taylor & Francis.
Heidegger, M., Young, J., & Haynes, K. (2002). *Heidegger: Off the beaten track*. Cambridge: Cambridge University Press.
Hinderliter, B., Maimon, V., Mansoor, J., McCormick, S., Apter, E., & Rancière, J. (2009). *Communities of sense: Rethinking aesthetics and politics. E-Duke books scholarly collection*. Durham: Duke University Press.
Jin, M., & Yongqi, L. (2014). *Emerging practices in design. Professions, values and approaches*. Shanghai: Tongji University Press.
Keshavarz, M., & Mazé, R. (2013). Design and dissensus: Framing and staging participation in design research. *Design Philosophy Papers Journal, 11*(1), 7–29.
Lee, Y., & Moore, P. (2015). *Ageing, Ingenuity and Design*. Hong kong: DESIS Network.
Manzini, E. (2008). Collaborative organisations and enabling solutions. Social innovation and design for sustainability. In F. Jegou & E. Manzini (Eds.), *Collaborative services. Social innovation and design for sustainability* (pp. 29–41). Milan: Edizioni POLI.design.
Manzini, E. (2014). Making things happen: Social innovation and Design. *Design Issues Winter, 30*(1), 57–66.
Manzini, E. (2015). *Design, when everybody designs*. Cambridge, MA: MIT Press.
Manzini, E. (2016). Design culture and dialogic design. *Design Issues Winter, 32*(1), 52–59.
Manzini, E., & Coad, R. (2015). *Design, when everybody designs: An introduction to design for social innovation, Design Thinking, Design Theory Series*. Cambridge, MA: MIT Press Limited.
Manzini, E. & Staszowski E. (2013). Public and collaborative.. DESIS Network.
Manzini, E. & V. Tassinari (2012) *Discussion Paper: Is Sustainable Social Innovation generating a new Aesthetic Paradigm*? parsons DESIS lab. Desis philosophy talk no 1 (é March), Parsons the new School for design, New York.
Manzini, E., & Tassinari, V. (2013). Sustainable qualities: Powerful drivers of social change. In R. Crocker & S. Lehmann (Eds.), *Motivating change. Sustainable design and behavior in the built environment*. London, Routledge.
Manzini, E., Cipolla, C. & Melo, P. (2013). Collaborative services in informal settlements. A social innovation case in a pacified favela in Rio de Janeiro, *1st Social Frontiers Research Conference*, Glasgow Caledonian University in London, 14th and 15th November 2013.
Marenko, B., & Brassett, J. (2015). *Deleuze and design*. Edinburgh: Edinburgh University Press.
Meroni, A. (2007). *Creative communities. People inventing sustainable ways of living*. Milano: Edizioni POLI.design.
Meroni, A. et al. (2013). *Design for social innovation as a form of design activism: An action format*. NESTA Social Frontiers, The next edge of social innovation research.
Penin, L. (2013). Amplifying innovative sustainable urban behaviours. Defining a design-led approach to social innovation. In R. Crocker & S. Lehmann (Eds.), *Behaviour change, consumption and sustainable design, Series Sustainable Design*. Earthscan/Routledge.
Penin, L., Staszowski, E., Brown, S. (2016) A new generation of transdisciplinary thinkers and practitioners of design-led social innovation. Design and Culture Journal, 7, 3., 2015, Routledge.
Rancière, J. (2004). *The politics of aesthetics: The distribution of the sensible*. London and New York: Continuum.
Ranciere, J., & Corcoran, S. (2010). *Dissensus: On politics and aesthetics*. London: Bloomsbury Academic.

Schiller, F., & Elias, J. A. (1967). Naive and sentimental poetry, and on the sublime: Two essays. In *Milestones of thought in the history of ideas*. New York: F. Ungar Publishing Company.
Schmitt, C. (1996). *The concept of the political*. Chicago: University of Chicago Press.
Selloni, D. (2017). *CoDesign for public-interest services*. Springer.
Sennett, R. (2012). *Together: The rituals, Pleasures and Politics of Cooperation*. New Haven: Yale University Press.
Thorpe, A., & Gamman, L. (2016). *What is "Socially Responsive Design and Innovation"?* In F. Fisher & P. Sparke (Eds.), *Routledge companion to design studies*. Routledge.

Conceptualizing Aesthetics in Design: A Phenomenological Framework

Mads Nygaard Folkmann

Abstract The aim of this chapter is to introduce and discuss aesthetics as an approach to understand how design frames experience. In doing so, the chapter combines two philosophical interests in design, design phenomenology and design aesthetics, in order to promote a framework for discussing the impact of aesthetic meaning construction on experience. First, the chapter raises the phenomenological question of the relationship between design and experience, specifically, how design conditions experience. Second, in looking at aesthetics in terms of (a) the sensual appeal of design, (b) design objects as aesthetic media that frame modes of understanding, and (c) contextual factors, such as media, influencing what is regarded as aesthetic, it is the thesis of the chapter that a reflective concept of design aesthetics can be employed to differentiate between three different ways in which design frame our experience: We can look at sensual, conceptual, and contextual aesthetic dimensions of design and examine their contribution to the framing of experience, that is, how different dimensions of meaning articulation in design offer different framings of the experiences promoted by design objects and solutions. Further, the concept of aestheticization is introduced and discussed as the way in which objects are construed as 'aesthetic'. A central insight is that the contextual aspect of aestheticization can promote a cultural construction of new conditions and new categories for the way we experience relevant meaning properties of design objects.

Keywords Design phenomenology · Aesthetics · Aestheticization · Aesthetic categories

1 Introduction

As we live in a world of design, and with the design theorist Gert Selle's word, "swim in an ocean of design" (Selle 2007: 9), design frames and stages human experience. The first step is to acknowledge this as a condition for the way humans interact with the modern world; next, we can ask *how* this takes place and

M. N. Folkmann (✉)
Department of Design and Communication, University of Southern Denmark, Kolding, Denmark
e-mail: mnf@sdu.dk

© Springer International Publishing AG 2018

P. E. Vermaas, S. Vial (eds.), *Advancements in the Philosophy of Design*, Design Research Foundations, https://doi.org/10.1007/978-3-319-73302-9_13

investigate common traits of the conditions of human experience in our contemporary "age of design" – to use the phrase of the editorial programme of the journal *Design & Culture* (2009-).

In this chapter, I will raise the phenomenological question of the relationship between design and experience, specifically, how design conditions experience. I will do so by embedding the discussion in the evolving discussion of a specific design phenomenology, which can be characterized as a theoretic framework inspired by classic phenomenology but aimed at conceptualizing the conditioning of experience by designed objects, mostly material but also with an extension to more immaterially operating design solutions, such as service design or digital design, for example in interfaces. To specify the phenomenological framework, that is not only state *that* design frames experience, but also ask *how*, I will introduce concepts of aesthetics to describe different layers of experiencing through design. Design phenomenology and design aesthetics can be described as two different kinds of philosophical interests in design, but can productively be combined.

I start out with design phenomenology as I see it as the main entrance to understanding how design sets the scene for humans to have an interaction and engagement with the world and, as an aspect of this, also for the organization of aesthetic meaning in design (see also Folkmann 2013). After having introduced design phenomenology as a framework, I will specify the frame by looking at aesthetics on three levels: (a) On a sensual level, the focus is on sensual appeal, i.e. how the design object by its sensual and tactile effects, its outer shape and its use function creates an appeal to human experience. This is not a matter of style but of the object relating to and framing the conditions of experience; (b) on a conceptual level, the staging of meaning by design objects is in question which regards, on the one hand, the appeal to understanding by the human subject and, on the other hand, the ability of the object to reflect its own character as a site of meaning articulation, (c) on a contextual level, the focus is on the contextual factors of the aesthetics of the design, where design by different means is reflected, regarded and positioned as 'aesthetic'.

In so doing, it is my thesis that a concept of design aesthetics can be effective to differentiate between different ways in which design relates to and frames experience: We can look at sensual, conceptual and contextual aesthetic dimensions of design and examine their contribution to framing experience. In proposing aesthetics as an entry to understanding and conceptualizing meaning construction in design, we may ask how different dimensions of meaning articulation in design offer different framings of the experiences promoted by design objects and solutions.

In its combination of phenomenology and aesthetics in relation to design, the chapter offers a research contribution to two types of discussions: In relation to design phenomenology, the chapter introduces design aesthetics as an approach to differentiate the otherwise abstract framework. In relation to the expanding field of design aesthetics (e.g. Forsey 2013; Stockmarr 2014), the chapter points out the specific relation of aesthetics to experience. The chapter builds on several theoretical approaches to aesthetics (sensual aesthetics, art-related aesthetics and cultural aesthetics) and thereby challenge a trend in contemporary design aesthetics where experience is foremost understood in relation to the creation of emotional appeal by means of sensual elements of design objects. The chapter contains a broad view on aesthetics which encompasses philosophical issues of meaning articulation (e.g. the ontology of design objects as sites of meaning, the question of representation through design object, etc.) and cultural-contextual issues of mechanisms of aestheticization and will investigate how these affect modes and conditions of experience. In the end, the purpose of the chapter is to promote a framework for discussing the impact of aesthetic meaning construction on experience.

In the following, I introduce the concept of design phenomenology as an entry to the discussion of how design relates to experience, and then introduce the framework of aesthetics and discuss how it can produce insights into the way design frames experience. In continuation with the presentation of the contextual dimension of aesthetics, I discuss the cultural role of aestheticization, that is, the construction of something as 'aesthetic' by factors external to the object. In this discussion, we may ask which aesthetic categories are produced by the context of contemporary culture.

2 Design Phenomenology

As a discipline of philosophy, phenomenology deals with the conditions of human experience in relation to what is experienced. The term phenomenology was coined by the German philosopher Edmund Husserl based on the Old Greek etymology of the theory, *logos*, of that which shows itself, *phainomenon*. Essentially conceived in a unity of subject and object, that often is reflected as a dichotomy, phenomenology asks about the phenomena as they appear to the human subject; the way to the phenomena goes through our experience of them. For Husserl, his search for the being of the phenomena, the world of objects, continuously led him back to an investigation of the formation of the mental structures that perform the conditions for meaning to come into being. Through this, phenomenology rests on a paradox therein that it strives to get to the objects and the core of things, *zu den Sachen*, but refers back to the formation of the conditions of experience in consciousness.

When we proceed to *design phenomenology*, the perspective on meaning production changes to the actual objects. Even in the later phenomenological philosophy of Maurice Merleau-Ponty, which by virtue of its corporeal starting point gets one step closer to actual experience, the reflection often remains transcendental in the sense of being oriented toward the basic conditions of experience. Merleau-Ponty does, however, speak of access to the world of objects and of "getting back to the objects in themselves, that is, getting back to this world before consciousness becomes the constantly speaking consciousness" (Merleau-Ponty 1945: iii), even if he does not focus on the specific constitution of the world of objects.

Design phenomenology may designate an approach to design with the focus on how design, in its many types of appearance and its creation of the tactile and visual surfaces of the modern world, affects and structures experience. In relation to this, an interest in the role of objects has emerged in recent years, as Actor-Network Theory has claimed objects to possess active agency in networks with humans, for example in guiding behavior (cf. Latour 2005), and Material Culture Studies have pointed to the steering role of the "material environment" with regard to the "development of social forms" (Dant 1999: 12).

Further, in a reversal of the interest in the human subject in phenomenology, the philosopher Peter-Paul Verbeek employs the concept of postphenomenology (originally initiated by Don Ihde) as a way of pointing to and acknowledging the role of the object in shaping the conditions of experience: "Things – and in our current culture especially technological artifacts – mediate how human beings are present in their world and how the world is present to them; they shape both subjectivity and objectivity" (Verbeek 2005: 235). Or, to further underline the point: A postphenomenological reflection such as Verbeek's is, in part, an attempt at deconstructing the dichotomy of subject and object in experience and thus rejecting the notion of human subjectivity as the origin of the structure of experience.

In relation to design, though, I suggest the use of the term "design phenomenology". Thus, to employ the term "postphenomenology" is to emphasize the internal philosophical debate of phenomenology; to engage in the development of design phenomenology is to relate insights from the phenomenological outlook on the conditioning of human experience to design. Using the concept of design phenomenology, we can examine the impact of design on the conditions of experience: We can examine how we design things, and how we are, in turn, designed by the things we design. This dual perspective is suggested by Prasad Boradkar when he states that the title of his book *Designing Things* "refers to a reciprocity of agency and an ambiguity of design's locus of action. People and things configure each other" (Boradkar 2010: 4). Further, the philosopher Stéphane Vial has proposed that we focus on the *effects* of design in the context of experience and view design as more event than being, more impact than thing, more incidence than property (Vial 2014). The effects of design contribute to the creation of the space of experience, which is mediated and structured by the actual objects of design. In his book *L'être et l'écran*, 'being and the screen', Vial looks at the changes in our structures of perception brought about by new digital media that, for example, offer spaces of virtual perception (Vial 2013).

To apply a phenomenological approach to design is to focus at the dual question of how design, as a medium of meaning formation, both relates to and possibly changes the constituents of experience. Whereas Verbeek's approach aims to investigate the material effects of objects within a framework of "material aesthetics", that is, to look at the fundamental mediating role of objects with regard to the specific shaping of human "experience and existence" (Verbeek 2005: 211), my attempt in the following is to look at the framing of experience by means of a broader conceptualization of aesthetics.

3 Aesthetics

In this section, I will investigate aesthetics in design as a phenomenological issue and relate the discussion of the framing of experience to the concept of aesthetics in design. First, I discuss how the aesthetic does not have a site or an essence per se but can be seen a meaning-creating relationship between the objective and the subjective which is itself phenomenological; second, I introduce an object-focused theoretical framework of aesthetics in order to provide concepts for my thesis: that a concept of design aesthetics can offer tools for describing different ways in which design frames experience. In this kind of approach, aesthetics does not deal with beauty or art, as the focus is on parts of the tradition of aesthetic theory, but with the articulation, communication, and staging of meaning in various ways through and around the object.

3.1 The Aesthetic Relationship

In its philosophical tradition, reaching back to Kant and beyond, to Baumgarden and the English empiricists, aesthetics deals with human experience, judgment, and appreciation of specific sensually or cognitively appealing phenomena. Consequently, a dominant discussion in aesthetic theory has been about the *location* of the aesthetic, its site in the act of aesthetic appreciation. Kant's seminal *Kritik der Urtheilskraft* (1790) is symptomatic of this discussion: On the one hand, he speaks of value judgment and taste, that is, of matters of subjective concern. On the other hand, the purpose of his thorough philosophical investigation of the field of aesthetics is to search for trans-subjective criteria for aesthetic evaluation in a *sensus communis*. In his conception, the judgment of taste is bound to objective condition and not submitted to arbitrary subjective evaluation.

To take this reflection beyond Kant: On the one hand, we may ask what kind of special subjective experiences aesthetics calls for, e.g. coherence, harmony, and unity (Dewey 2005; Shusterman 2000) or a feeling of "pure presence" (Seel 2000) or "moments of intensity" (Gumbrecht 2003). In this way, it is a central question in the "first full treatment of design in the field of philosophical aesthetics", Jane Forsey's important book *The Aesthetics of Design*, what the nature of aesthetic experience is in relation to the use-oriented medium of design (Forsey 2013). On the other hand, determining which concrete elements in design objects are capable of evoking aesthetic experiences is an analytical question. In this sense, the literary scholar Gérard Genette speaks of objects with an "intentional aesthetic function" (Genette 1999: 2), aimed at being perceived and experienced aesthetically. In a general reflection, we can state that aesthetic meaning evolves and can be described as a *relationship* between a subject with an intention to see and perceive something as aesthetic, to have aesthetic experiences, and an object with certain aesthetically coded features.

Genette regards this relationship foremost as functional in the sense that meaning always only may be given in the interrelation of subject and object. The relationship can, however, also be seen as a feature of phenomenology and its interest in this interrelation as a basic premise for human experience. Also, phenomenology, at least in the version of Merleau-Ponty, stresses the fact, that experience always that takes place in a body which is embedded in the context of the world. Merleau-Ponty deconstructs the dichotomy (and relation) of subject and object and places them in a sort of continuity that he calls the flesh, *la chair*. He speaks of density of the flesh as a means of communication between the viewer and the object. Similarly, the body is located in a chiastic structure within the world: "The body participates in the order of things, and likewise the world is universal flesh" (Merleau-Ponty 1964: 179). Experience, in Merleau-Ponty's phenomenology, is an ongoing exchange between subject and object that takes place in the common material of *la chair*. Further, *la chair* is a means of relating experience to the world and thereby state that experience is never without context.

This latter point about the context is important in relation to aesthetics. Often philosophical speculations of aesthetics may only investigate the relationship of subjective appreciation and evaluation on the one hand and the aesthetically appealing object on the other hand. All experience is, though, embedded within a context which in many ways influences the horizon of the actual experience and how it is evaluated. My point is here that the relationship of subject and object is a basic tenet for the location of the aesthetic and that this structure of aesthetic experience must be seen within the contextual factors which conditions it, e.g. habits, cultural valorizations, conventions, etc., and that these factors also belong to the phenomenological framework as they define how experience can take place.

This said, and taking the relational as a defining point of aesthetics, I will in the following focus on the coded features of the design objects. On the one hand, all sorts of objects (chairs, refrigerators, tables, garden gnomes, smartphones, lemon squeezers, works of art), may serve as vehicles for aesthetic appreciation according to individual and idiosyncratic taste even if the context also may contravene this. There are, e.g., strong conventions to regard garden gnomes as bad taste and, hence, in a traditional sense as outside an aesthetic appreciation. Still, in principle, virtually anything can be regarded as aesthetic if we choose to view it with an aesthetic perception and have an aesthetic experience of it. On the other hand, there are differences in the *degree of aesthetic coding*. This raises the question of how some design objects may encourage aesthetic appreciation more than others. Since my question is how design frames experience, I want to take a closer look at how design objects do this, by various means. That is, my approach to aesthetics is to look at the coding strategies of the objects rather than at the subjective experiencing.

In the following, I present an interpretive framework for investigating the formation and articulation of aesthetic meaning in design. I present three levels of aesthetics that are also reflected in aesthetic theory, and which can be related to the framing of experience by design: a sensual, a conceptual, and a contextual dimension of aesthetics. Elsewhere, I have referred to these levels as a sensual-phenomenological level, a conceptual-hermeneutical level, and a discursive-contextual level (Folkmann

Dimension of aesthetics:	Sensual	Conceptual	Contextual
Impact on level of experience:	*Framing by means of sensual design aspects, both in terms of objects and in terms of use and material interaction*	*Framing of understanding through concrete objects and solutions*	*Framing of systems of meaning; ideology*
Impact of aestheticization	*Massive sensual effect – even to the point of becoming 'an-aesthetic'*	*Staging of reflective meaning as a strategy of design objects*	*Creation of new aesthetic categories*
Required discourse about how design objects relate to experience	*Description of sensual means and effects in design*	*Description of degrees and strategies of design objects for staging meaning*	*Description of the various aesthetic categories produced*

Fig. 1 Dimensions of aesthetics in relation to design

Fig. 2 Beolit 12 portable music device. B&O Play. Design: Cecilie Manz

2013). Figure 1 summarizes these levels and how they relate to the following discussion of the contextual dimension of aestheticization and point to a new emerging discourse on how design objects relate to experience.

As a running example, I relate my discussion to the portable music device Beolit 12 (Fig. 2) by the electronics manufacturer Bang & Olufsen, which praises the product as "a small but mighty music system that sports a powerful punch in a stylish package" (B&O 2015). The Beolit 12 is a combined amplifier and loudspeaker with a wireless connection to Apple's iPhones based on the AirPlay technology. It appears as a reinterpretation of the traditional transistor radio, as reflected in the name, which refers to a line of transistor radios reaching back, among others, to the (not portable) Beolit 39 (1939) and the compact Beolit 400 (1970). The Beolit 12 is a part of B&O Play which is a sub-brand of Bang & Olufsen aimed at consumers who prioritze mobility and a contemporary design expression. The object carries the signature of the well-renowned Danish designer Cecilie Manz, who plays a central role in the media representations of the object. As is typical of high-profile lifestyle-oriented types of design, the designer name is employed as an asset in the meaning production of the product (cf. Julier 2014).

3.2 Sensual Aesthetics in Design

The sensual level of aesthetics in design takes its starting point in an investigation of the sensual communication of the form and material-tactile dimension of the designed object. This interest is rooted in a trend in aesthetic theory that revisits Alexander Baumgarten's original idea of applying aesthetics to sensual matter in the work *Aesthetica* (1750–58; in Old Greek, *aisthetá* means 'that which can be sensed'). It is found in works by Martin Seel (2000, 2007), Gernot Böhme (2001, 2013), and, drawing on John Dewey's Pragmatist aesthetics (2005), Richard Shusterman (2000). Also, it relates to investigations of the role of the body in aesthetic experience (Bhatt 2013) and to a general interest in the nature of the aesthetic experience in its generic character of emotional and psychological responses independently of the object in question (Schaeffer 2015). In general, though, these inquiries deal with aspects of sensual appeal and the question of appearance, that is, on the one hand, how people respond to certain kinds and structures of appearance, and, on the other hand, how these are constituted in order to evoke response.

What is important, however, is to acknowledge differences in traditions and concepts of "experience" in this context. The Pragmatist, Anglo-American tradition has a tendency to be interested in experiences as special moments of sensation; Dewey points to the special character of having "an experience" which, for instance, can be facilitated by works of art (even though Dewey aims at general experience, he often points to works of art as examples of special catalysts). In reference to experience, works of art may "concentrate and enlarge an immediate experience" and present a "pure experience" freed "from factors that subordinate an experience as it is directly had to something beyond itself" (Dewey 2005: 285–6). In contrast, European Continental tradition has, in the aftermath of Kant, been interested in questions of perception and epistemology as a structure of experiencing. In English, the difference may be blurred, as *experience* is the single term to designate what is at stake in both traditions. In German, for instance, the difference can be detected in the difference between the words *Erlebnis* (being a special moment of experience) and *Erfahrung* (pointing to the structure of experiencing).

In relation to design, this overall interest in "experience" has several outlets, mostly related to the Anglo-American discourse on the special moment of experience. It has, among other trends, led to an increased understanding of the role of the *body* in design, as "an increased somatic awareness of the body and its surroundings can enrich and deepen everyday experiences" (Bhatt 2013: 4), which are seen as being shaped by design. Also, the interest in *pleasure* may be seen within this context (Jordan 2000). Pleasure and the pleasurable are at the core of the Dutch project UMA (Unified Model of Aesthetics) which with its model seek to predict "that the aesthetic pleasure we gain from designed artifacts arises out of a delicate balance between ranges of opposing forces" (UMA 2015), e.g. unity vs. variety and typicality vs. novelty. Also, the interest in the *emotional appeal* of design can also be seen in this context; as a trend, it marks a broad desire to understand the aesthetic qualities of the nonfunctional, emotionally appealing factors in design and how they affect the process of designing (Norman 2004; Hekkert 2006; Desmet and Hekkert 2007; Hekkert and Leder 2008; Desmet 2010). In these approaches the focus is on how aesthetic objects may create some kind of sensual appeal and effect, that is, a way where sensual elements of the design aim to create a positive and affirmative response in the recipient. Also, a recent thesis on design aesthetics concludes by suggesting that "a deepening fusion of ideas from the emotional theories of design and the philosophical notion of the aesthetic as sensual experience would be able to take the discussion of the aesthetic in design in new directions" (Stockmarr 2014, 202–3), even though it does not show specifically how this can be accomplished.

Below I will challenge this as being the only trend in design aesthetic. For now I will point the investigation in the direction of the role of the design objects in the sensual relationship, that is, how the framing of experience by sensual qualities of the design objects may be seen on more levels.

First, it deals with how the look, the feel of the texture, the application of materials, the execution of the detailing in the assembly and seamless fittings, and the overall physical presence appeal to people and conditions their approach to the world. This approach to looking at the aesthetics through the material and formal appearance of the object resembles the classical aesthetic appreciation of distance and disinterestedness, even though we also with the contemporary aesthetic philosophy of Martin Seel may ask how the object in question may appeal to enable a kind of "aesthetic perception", *ästhetische Wahrnehmung*, that not only invests itself in the immediate appearance of the object, but also "intensifies the appearance of the pure present that is otherwise inaccessible to ordinary perception" (Seel 2007: 13). To Seel, aesthetic perception is a matter of looking in a certain intent way that involves "attention to the play of appearances" (14) and this looking can be encouraged by the sensual qualities of the object.

Second, it is not enough only to investigate the material form of the object but also broadly its creation and framing of experience which in design also touches upon its *use* and the "material interaction" (cf. Dant 2008) the design object encourages to. Thus, the question arises how the form of the object is related to the creation of experiences when in use or interacted with.

The Beolit 12 is illustrative in this connection as it in its appearance presents a design object with a strong emphasis on formal and material qualities *and* employs this to stage new experiences when used. In its sensual appearance, the Beolit 12 reflects a strong emphasis on look, texture, high-end materials and quality workmanship as evident in the detailing of the assembly and seamless fittings. It is designed as an object with a closed extension and a formal expression as an enigmatic block whose function may be hard to detect at first glance. With regard to materials, the designer has used plastic in order to keep the product lightweight, mobile and affordable; at the same time, however, the more expensive material aluminum is used on the front, in accordance with the company's signature design idiom. Next, the creation of experiences goes beyond its immediate sensual appeal, as the function of the product is to provide mobile music experiences for the user. Thus, the kind of experience offered by the object is not obtained simply by an appreciation of its external formal and sensuous qualities (whether experienced remotely by looking or up-close by touching) but revolves around the possibilities enabled by the use of digital technology to enhance and create new experiences for the user. In case of the Beolit 12, this involves experiences of mobile music transmitted wirelessly by an iPhone to the Beolit 12 and perhaps to the iPhone via a streaming service. In the perspective of the creation of experiences, then, the aesthetics of the object should be judged not only on its external qualities but also on its specific capability of easily, innovatively or provocatively creating new experiences.

Establishing a discourse about the appeal of form, material and evocation of new experiences in design requires a differentiated vocabulary for describing sensual effects and their impact. In a project at the Design School Kolding, design researchers have examined how a group of users/consumers articulate aesthetic qualities and preferences in relation to the sensual, tactile qualities of textiles (Riisberg and Bang 2014). While wearing blindfolds, a group of young users/consumers were asked to verbalize their tactile sensation (Fig. 3). The study claims that more reflective users may help inform designers about how to design to achieve greater appreciation by the users, but it also shows that it may be a struggle to create a nuanced language for a kind of sensory appreciation that is often overwhelmed and dominated by visual impressions. In sum, it is pivotal for the sensual dimension of design aesthetics to encompass all dimensions of sensual appeal; the visual, tactile, auditory and olfactory (and maybe also the gustatory even it does not seem relevant for design).

3.3 Conceptual Aesthetics

The conceptual level of aesthetics involves analyzing artefacts as media for the construction of meaning and new patterns of understanding. This notion of aesthetics can be traced back to Kant's *Kritik der Urtheilskraft* (1790), where aesthetics was conceived as a basic aspect of epistemology bridging sensual appearance and conceptually formulated meaning. In Kant, this was transmitted through the experiences of the beautiful and the sublime, but it has subsequently developed into a more general question of meaning construction without any association with beauty in the art-oriented aesthetic theory influenced by Kant, especially in a German-language context (e.g. Adorno 1970; Bubner 1989; Menke 1991, 2013; Seel 2000). A key aspect of this line of aesthetic theory has been an immense interest in the extra- and trans-communicative effects of the aesthetic artefact beyond its normal

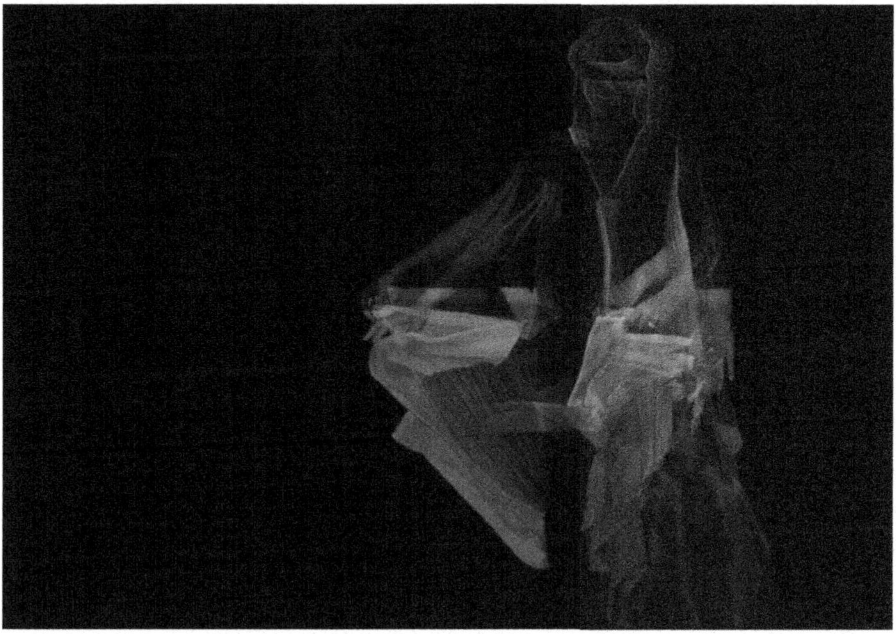

Fig. 3 Graphically reworked representation of the experiment with tactile sensing of textiles at the design school Kolding. Image design: Laura Locher and Alina Breuil Moat

communicative capacity. This interest thus goes beyond the artefact to explore the effect of communication and anchors the artefact as an integral element of the communication of the specific aesthetic medium. Consequently, the main focus is not on cognitive questions of understanding but on the role and capabilities of the specific aesthetic medium in question.

This approach to the aesthetics of design is not common in design theory where the focus is on the sensual qualities and the appeal to emotions and pleasure, as stated above. The ambition of this trend in design aesthetics has been to describe what is specific about an aesthetics of/for design and, in turn, what differentiates it from other kinds of aesthetics, especially the aesthetics of art. My claim is, however, that important insights about the specific aesthetics of design objects can be found also in the general, but art-related aesthetics and that this claim also has the advantage of challenging the close attachment of the concept "aesthetics" with beauty, which often is the case in design. Doing so, my ambition is the opposite of (re-) conceiving design as art, as that is a dead-end reflection which neglects the specifics of design, e.g. the role of use and functionality. Instead, my proposal is a framework for understanding aesthetics specifically in design, which explores and reconnects with the potential of a broader field of aesthetic theory, as much aesthetic theory – with inspiration from Kant – deals with general questions of the complexity and communicative capabilities of the aesthetic medium.

In relation to the framing of experience by design objects on a conceptual level, the key questions are how and by what means design objects enable and construct meaning and appeal to understanding, and what the nature of this meaning is, for example if it transcends any limitations, and what its implications are. Further, in the process of constructing meaning and appealing to understanding, design objects may prove to be reflective of themselves as sites of meaning construction.

Many design objects not only create a sensually operating framing of experience but also engage in framing and challenging patterns of understanding. An example of a product that operates in this manner is Philippe Starck's highly profiled, almost non-functional *Juicy Salif* (1990) (as it is not only a lemon squeezer but also challenges our understanding of it by reflecting the very idea of a lemon squeezer). Relevant in this context is Martin Seel's discussion about the capacity of aesthetic media to create new frames of understanding and serve as media for comprehending and meeting the world. Focusing on the function of human perception in the process of confronting something "other", he claims that the capacity of aesthetic media is to "bring forward otherwise unrepresentable circumstances" and that this capacity has to do with "ways of human commitment in the real or the unreal, in conditions of the world in the past, the present, or the future. *Ways of meeting the world* [*Weltbegegnung*] are put forward, whereby *ways of meeting the meeting of the world* [*Begegnung mit Weltbegegnung*] will be possible" (Seel 2000: 184, his italics).

Unlike art, design objects are often not obviously self-reflective of their own being as creators of meeting points between us and the world; nevertheless we may ask how design objects are conceived and operate as such. As objects of use, design objects have the effect of staging dynamic meetings with the world, that is, they produce a meeting with the world that *may* enable new kinds of meeting of the world, to rephrase Seel. This is an aspect that touches upon the element of interaction with the design, not just in terms of how to handle a specific physical object, but also in terms of how the object may alter its character in our interaction with it.

The Beolit 12 creates a meeting point between us and the world both through its static, material character and through its dynamic potentials of use. In its immediate appearance, it is an object that is designed for physical handling and to be carried around. As a distinct signature element, it is designed with a leather strap. The leather is a material that the designer has brought into the design of consumer electronics from her original field as a furniture designer. Also, the use of the leather strap refers to the design history of portable devices, e.g. the portable transistor radio/phonograph TP1 (1969) designed by Dieter Rams for Braun, which features a leather strap. From a sensuous perspective, the leather makes carrying the product more comfortable. Thus, the organic material of the leather strap softens our concrete, tactile meeting with an otherwise hard and geometrically distinct object. Next, the Beolit 12 also serves as a dynamic access point to experience by allowing music to be carried around and consumed in new and, from the perspective of the company, hopefully, innovative ways, as that would allow the company to project a self-image of being innovative. In its scope of operation, the Beolit 12 may encourage new kinds of mobile use. The potential of design objects to be used in different and maybe unforeseen ways is enhanced by the opportunities offered by digital technology, adding renewed relevance to the question of how aspects and situations of use can be designed, and how use can be seen as design (Bredies 2014). Thus, the real impact of creating new world meetings lies in the product's inner potentialities of being not merely a static, material object (which it also is) but also a dynamic object of interaction that through its use by different users is constituted as a personalized, changeable design object, capable of creating new patterns of engagement in its wireless interaction with, first, an iPhone and, second, surroundings that are increasingly pervaded by wireless information.

Seen in the perspective of conceptual aesthetics, a product such as the Beolit 12 may have an aesthetic effect in challenging the habits of use, the way we understand an object (what is it for?), and the way we meet the world through the object. This meeting builds upon sensuous effects but also on the more or less self-referential creation of the object as an interface between us and the world.

3.4 Contextual Aesthetics

While the two previous dimensions of aesthetics deal with meaning aspects of the given design object, the contextual dimension of aesthetics focuses on the wider implications of the circulation of objects on a cultural, social, and political level. This aspect also points to the role of *aestheticization* which will be introduced in the next section.

According to Jacques Rancière, who has influenced the contextual trend in aesthetic theory, aesthetics can be investigated as a political power issue in relation to the distribution of sensual material and the ability to determine "what presents itself to sensory experience": Thus, aesthetics can be seen as "a delimitation of spaces and times, of the visible and the invisible, of speech and noise, that simultaneously determines the place and the stakes of politics as a form of experience" (Rancière 2004: 13). Rancière seeks to discuss possibilities that are determined by aesthetic media; these media not only behave as transmitters of new possible meaning but also produce possibilities by defining and conditioning domains of the sensual experience.

In this conception, aesthetic media, such as design, have the capacity to radically reconfigure and transfigure the territories of "the visible, the conceivable, and the possible" (41) and propose possible models for accessing and experiencing the world in new ways. The contextual dimension of aesthetics deals with the *ideological* aspect of framing experience as it affects whole systems of meaning. Further, this is a matter of *cultural reflection* as the surrounding cultural context may circumscribe what is defined and regarded as aesthetic. Ways of effectuating this may be found in branding strategies and in the cultural production of "media-environments" (Lash and Lury 2007), where things turn into and are engulfed by media expressions in a manner where there is no outside, as seen from the perspective of these environments; everything is ordered in new ways and delimited as a space of experience.

Fig. 4 Product setting of the Beolit 12 by the company Bang and Olufsen (2012)

From a perspective of consumption (and not just use situations), media contexts are active in framing aesthetic experience.

To illustrate with the Beolit 12, we may ask how the aesthetic meaning of the product is culturally produced by means of actors and powers outside the design object, in the commercial context. As a start, Beolit 12 is positioned as aesthetic as a matter of *discursive*, performative action, as when Manz is quoted as saying that she thinks "great design should be 90% functionality and 90% aesthetics" (B&O 2015), thus pointing to aesthetics as a central concept for the product. Further, *visual* media may play a central role. I will point to two examples.

The first example is a static visual mediation of the Beolit 12 on the website of the sub-brand B&O Play (Fig. 4). Here, the product is displayed in a setting that reflects the company's view of the optimal value context of the product: co-branding with Apple products; coffee table books, perhaps study books or books on art or design, designating cultural capital; a coffee-to-go cup, signalling mobility; and posters and a bike in the background connoting a modern, urban, active lifestyle. The product is staged in a simplified and stylized set-up in bright, harmonized colours which at the same times signals urban mobility. This kind of visual mediation positions the object as aesthetic through a beautification of the object (in an association with values of youth, mobility and trendiness), which is not uncommon in contemporary design culture as a strategy of aiming to let the objects look their best.

The second example is a 2012 commercial for the Beolit 12 playing on the theme of Little Red Riding Hood getting lost before eventually finding her way again (B&O 2012). In a desolate, urban industrial setting, the Little Red Riding Hood character is carrying the Beolit 12 as her basket of goodies and seems to feel stalked by sudden flashing, rectangular lights. Eventually, the lights surround her, but at the end of the 1:18-minute film she is able to control the lights and make them go away by touching her iPhone, revealing that the flashing lights were iPhone screens. At the point of resolution, a female voice whispers, "Wireless sound system for your iPhone – by B&O Play". In this commercial, the object is not positioned as beautiful but as intriguing and interesting as the Beolit 12 acts as an object of resolution that integrates all the supposedly wireless streams of information indicated by the rectangular lights.

The two examples may seem contradictory but state how an actor as a company may employ different strategies at the same time in order to attach aesthetic meaning to the object. Other actors may be the designer or consumers communicating with each other in e.g. blog communities. Aesthetics is never de-attached interests or ideology but contextually dependent on factors surrounding the object.

4 Aestheticization

In the following, I will specify aestheticization as a *cultural meta-level of aesthetics*. Aestheticization can be employed as a concept to further describe the cultural process by which design is made or conceived of as aesthetic by factors that are often external to the concrete objects themselves. In relation to the framing of experience by aesthetic means, I point to aestheticization as a ubiquitous process of distributing sensual meaning with an emphasis on the overall impact of aesthetic media, such as design, on experience. The basic point is that the contextual aspect of aestheticization can promote a cultural construction of new conditions and new categories for the way we experience relevant meaning properties of design objects. The cultural aspect of aestheticization relates it to anthropology and studies in Material Culture as these deal with how the culture in its material manifestations frames human experience (see Dant 1999; Miller 2010). Often, however, the question of aesthetics is not in direct focus in these approaches even if aesthetics and aestheticization may be seen as an entry to the understanding of material culture.

4.1 Dimensions of Aestheticization

The concept of aestheticization brings additional dimensions of aesthetics into play. First, it describes the moment in cultural history when calculated sensual appeal begins to gain in importance as a means of communicating a new world of goods to contemporary consumers, and we see the boundaries of high culture break down, as devices and form repertoires of art enter the experiential domain of everyday life, and new, calculated surfaces of visual appeal and imaginary simulation appear (cf. Featherstone 1991). Thus, the "aestheticization of everyday life has provided changed conditions for the organization of material culture" (Lury 1996: 26) and, hence, the conditions for how to approach and experience the world. In effect, aestheticization can designate a high degree of aesthetic coding in our everyday surroundings and the transmission of this process through specific media. By this, the concept of aestheticization connects to concepts of *everyday aesthetics* (e.g. Oldemeyer 2008; Saito 2010).

Second, and as a consequence of the first meaning, aestheticization is the cultural construction of an everyday domain such as design as aesthetic, that is, as something that is attributed and imbued with a kind of meaning that can be labeled "aesthetic". Here, the emphasis is on a cultural analysis of design and its formative contexts (e.g. Hebdige 1994, du Gay 1997, Julier 2014, Folkmann 2016). Designers, manufacturers, retailers, design magazines, blogs, etc., all the actors in the cultural circuit of design, may employ the term "aesthetic" in relation to the design object in question and install a notion that the object may be regarded as aesthetic and thus be experienced as such in a special way – in the manner Cecilie Manz did in the example of Beolit 12.

Third, we may ask about the *specific strategies of coding* in aestheticization on the different levels of aesthetics and its relation to experience. In so doing, we may use aestheticization as a concept that describes a high degree of aesthetic coding in the everyday surroundings of human beings and the transmission of this process through specific media. In this way, to enter a discussion of aestheticization is to examine how the interface that we apply when we meet the world is changing as a result of strategies of making objects and surfaces more aesthetically coded and asking how this process affects the conditions of experience.

4.2 Aspects of Aestheticization

As a discussion of the role of aestheticization in connection with experience, I will now relate its cultural process to the sensual, conceptual, and contextual level of aesthetics and, consequently, the modes of experience addressed and conditioned by these levels.

On the level of *sensual appeal*, the logic of aestheticization is to seek sensual effects, even to the degree of overwhelming calculation where everything is designed to have maximum impact on the senses. This aspect of aestheticization can be seen in contemporary commercial design culture where more and more design objects and brands are staged as sensually appealing in different kinds of commercial settings and displays, e.g. flagship stores with a thorough calculated staging of the products of the brand in question. In a cultural-critical perspective, the reverse side of this kind of aesthetics has been described by Wolfgang Welsch as *an-aesthetic*, a state of being where the "elementary condition of the aesthetic, the ability to feel, has been negated" (Welsch 1990: 10). In this conception, the an-aesthetic ranges from the "zero phenomenon to the hyper phenomenon of the aesthetic" (11), that is when aesthetics runs the risk of becoming either too little or too much.

On the level of *conceptual meaning* in design, aestheticization can be seen in the tendency to stage design as a self-reflective medium that implicitly or explicitly debates its own conditions as a medium. This tendency is related to the historical development in aestheticization in the sense that design as, a medium of popular and everyday culture, employs strategies of debating and staging meaning through concrete objects that are normally found in art. In this way, many of the explicit endeavors to create reflective design, such as the 1960s Italian Radical Design movement or the more recent Critical Design movement (now dealing with "Speculative Everything", Dunne and Raby 2013), are affiliated with mechanisms and discourses of art. On this level of aesthetic meaning, aestheticization points to the cultural logic of not just creating forms for sensual appeal to experience but also reflectively positioning them as such.

Finally, and most important, on a *contextual level*, aestheticization implies the creation of new experiential patterns and new categories for seeing and regarding something as aesthetic and, hence, as important for the conditioning of experience. This approach has been promoted in recent years by the newer conceptualizations of aesthetics. In a philosophical approach to everyday aesthetics, the philosopher Terry Leddy, for example, speaks of the extraordinary in the ordinary and proposes a broad view on aesthetics including "major league aesthetic concepts such as harmony and balance, but also minor league ones such as neatness and messiness" (Leddy 2012: 259). Further, the cultural theorist Sianne Ngai productively associates cultural trends with aesthetics and states that "aesthetic experience has been transformed by the hypercommodified, information-saturated, performance-driven conditions of late capitalism" (Ngai 2012: 1). She argues that the change in conditions in society also changes the aesthetic categories, and that in our "hyperaestheticized world", aesthetic experience has changed from deriving its models from art and the beautiful/sublime to being based on "the stylistic triviality and verdictive equivocality of the zany, the cute, and the interesting" (19). In her perspective, the aesthetic categories are filters for experiencing the world in different ways. In the process of total culturalization and radical commodification, aesthetic experience is about the *ordinary* and not, as has traditionally been the case, about distance, play or disinterestedness. Further, Ngai points out that styles are culturally produced and codified as such, which has implications for "our perception of them as stylistic qualities" and, vice versa, that our perception affects "our language of aesthetic judgment" (29). Ultimately, style is not just a matter of the object in question but can be understood as a way of "perceiving an object" (29); that is, it can be seen as producing a specific perceptual setting.

With Ngai, we clearly see that the aesthetic categories of contemporary design culture have changed, even if her chosen categories may be debated. We experience the design through their aesthetic categories or, rather, *our experience of the world through its design is steered by aesthetic categories related or attached to the design in question.* This aspect can be seen in the visual mediations of Beolit 12 discussed above. By being intriguing, the latter example relates to the aesthetic category of the interesting while the first example more traditionally confirms beauty as an aesthetic category in design. In appealing to beauty, design may often be 'old-fashioned'

Fig. 5 fritz-kola ad, 2014. The slogan goes: "Doing cocaine is so 1980s". The ad can be seen as an example of how experience can be framed by branding

in its aesthetic appeal which may be in contradiction to the examples primarily from art which Ngai bases on. But an important point is that the aesthetic categories are not just given and final but are an object of interpretation and constantly can be culturally produced, e.g. by companies through their staging of objects in different kinds of mediations.

To further illustrate the cultural construction of aesthetic categories and their impact, the German cola company fritz-kola may serve as an example. By presenting the cola product on a black background with the slogan 'koksen ist achtziger' (doing cocaine is so 1980s) (Fig. 5), the brand and the campaign steer our perception and delimit our space and frame of experience (effects of pleasure are vital; everything is black). The company actively aims to create the aesthetic categories through which the product is to be evaluated. In the visual language of commercials and in its social communication, which actively defines cultural parameters, fritz-kola establishes ways of perceiving the product, that is, its style and our reflective judgment of it. Through its reflectivity, fritz-kola engages in – and produces – the activity and flow of the *zany* and the temporal-anticipatory of the *interesting*, to employ some of Ngai's categories. Thus, fritz-kola creates a structure of meaning around its product that produces the categories for how we can perceive the relationship between the brand as a vehicle for the consumers to engage actively with and the overall context of contemporary culture.

In general terms, aestheticization on the level of contextual aesthetics can mean a cultural production of new conditions for experiencing relevant meaning properties of design objects.

5 Conclusion

The purpose of the chapter has been to introduce to design phenomenology and design aesthetics as approaches to understanding design and to let them cast a reflective light on each other in the overall question of the chapter: *How does design frame experience*?

In the chapter, the phenomenological question of the relationship between design and experience has been raised, that is, how design on a general level conditions experience. In looking at aesthetics in terms of (a) the sensual appeal of design, (b) design objects as aesthetic media that frame modes of understanding, and (c) contextual factors, such as media, influencing what is regarded as aesthetic, design aesthetics has been employed to differentiate between three specific ways in which design frames our experience: We can look at sensual, conceptual, and contextual aesthetic dimensions of design and their contribution to the framing of experience, that is, how different dimensions of meaning articulation in design offer different framings of the experiences promoted by design objects and solutions. What is important on this point is to acknowledge the role of the cultural context in defining what is made available for our experience and how the material for experience is differentiated in different ways as aesthetic. Here, a central insight is that aestheticization promotes a view on the cultural construction of new conditions and new categories for the way we experience relevant meaning properties of design objects.

By looking at sensual, conceptual, and contextual aesthetic dimensions of design and exploring their contribution to the framing of experience, we may, as a next step, further develop discourses for describing the sensual means and effects in design, the degrees and strategies of design objects for reflectively staging their meaning and the resulting aesthetic categories. The latter point seems particularly important: Today, design is not a question of things being beautiful but of the production of meaning and new ways of framing experience.

References

Adorno, T. W. (1970). *Ästhetische Theorie*. Frankfurt a.M.: Suhrkamp.

B&O. (2012). Little Red Riding Hood - BeoLit 12 Commercial. https://www.youtube.com/watch?v=DURcvvn2PgI. Accessed 28 Nov 2015.

B&O (2015). Beolit .12. http://www.beoplay.com/products/beolit12. Accessed 28 Nov 2015.

Bhatt, R. (2013). *Rethinking aesthetics. The role of body in design*. London: Routledge.

Böhme, G. (2001). *Aisthetik. Vorlesungen über Ästhetik als allgemeine Wahrnehmungslehre*. München: Wilhelm Fink Verlag.

Böhme, G. (2013). *Atmosphäre. Essays zur neuen Ästhetik*. Frankfurt a.M.: Suhrkamp.
Boradkar, P. (2010). *Designing things: A critical introduction to the culture of objects*. Oxford: Berg.
Bredies, K. (2014). *Gebrauch als design*. Bielefeld: Transcript Verlag.
Bubner, R. (1989). *Ästhetische Erfahrung*. Frankfurt a.M.: Suhrkamp.
Dant, T. (1999). *Material Culture in the Social World*. Birmingham: OUP.
Dant, T. (2008). The 'pragmatics' of material interaction. *Journal of Consumer Culture, 8*(1), 11–33.
Desmet, P.M.A. (2010). Three levels of product emotion. *Proceedings of KEER 2010, International Conference on Kansei Engeneering and Emotion Research*, Paris, March 2–4.
Desmet, P. M. A., & Hekkert, P. (2007). Framework of product experience. *International Journal of Design, 1*(1), 57–66.
Dewey, J. (2005). *Art as experience*. New York: Penguin.
du Gay, P., et al. (1997). *Doing cultural studies: The story of the Sony Walkman*. London: Sage.
Dunne, A., & Raby, F. (2013). *Speculative everything. Design, fiction, and social dreaming*. Cambridge, MA: MIT Press.
Featherstone, M. (1991). *Consumer culture and postmodernism*. London: Sage.
Folkmann, M. N. (2013). *The aesthetics of imagination in design*. Cambridge, MA: MIT Press.
Folkmann, M. N. (2016). *Designkultur. Teoretiske perspektiver på design*. Frederiksberg: Samfundslitteratur.
Forsey, J. (2013). *The aesthetics of design*. Oxford: Oxford University Press.
Genette, G. (1999). *The Aesthetic Relation. Ithaca*. Cornell University Press.
Gumbrecht, H. U. (2003). Epiphanien. In J. Küpper & C. Menke (Eds.), *Dimensionen ästhetischer Erfahrung* (pp. 203–222). Frankfurt am Main: Suhrkamp.
Hebdige, D. (1994). Object as image: The Italian scooter cycle. In *Hiding in the light* (pp. 77–115). London: Routledge.
Hekkert, P. (2006). Design aesthetics: Principles of pleasure in design. *Psychology Science, 48*, 157–172.
Hekkert, P., & Leder, H. (2008). Product aesthetics. In H. N. J. Schifferstein & P. Hekkert (Eds.), *Product Experience* (pp. 259–285). Elsevier.
Jordan, P. (2000). *Designing pleasurable products*. London: Taylor & Francis.
Julier, G. (2014). *The culture of design*. London: Sage.
Lash, S., & Lury, C. (2007). *Global culture industry: The mediation of things*. Cambridge: Polity Press.
Latour, B. (2005). *Reassembling the social*. Oxford: Oxford University Press.
Leddy, T. (2012). *The extraordinary in the ordinary. The aesthetics of everyday life*. Peterborough: Broadview Press.
Lury, C. (1996). *Consumer Culture*. Cambridge: Polity Press.
Menke, C. (1991). *Die Souveränität der Kunst*. Frankfurt am Main: Suhrkamp.
Menke, C. (2013). *Die Kraft der Kunst*. Frankfurt am Main: Suhrkamp.
Merleau-Ponty, M. (1945). *Phénonénologie de la perception*. Paris: Gallimard.
Merleau-Ponty, M. (1964). *Le visible et l'invisible*. Paris: Gallimard.
Miller, D. (2010). *Stuff*. Cambridge: Polity Press.
Ngai, S. (2012). *Our aesthetic categories*. Cambridge, MA: Harvard UP.
Norman, D. A. (2004). *Emotional design*. New York: Basic Books.
Oldemeyer, E. (2008). *Alltagsästhetisierung: Vom Wandel ästhetischer Erfahrung*. Würzburg: Königshausen & Neumann.
Rancière, J. (2004). *The politics of aesthetics*. London: Continuum.
Riisberg, V., & Bang, A. L. (2014). Design practice and aesthetics. *Wassard Elea Rivista, 2*(2), 55–68.
Saito, Y. (2010). *Everyday aesthetics*. Oxford: Oxford University Press.
Schaeffer, J.-M. (2015). *L'expérience esthétique*. Paris: Gallimard.
Seel, M. (2000). *Ästhetik des Erscheinens. München*. Hanser.

Seel, M. (2007). *Die Macht der Erscheinung*. Frankfurt a.M.: Suhrkamp.
Selle, G. (2007). Design im Alltag. Vom Thonetstuhl zum Mikrochip. Frankfurt am Main: Campus.
Shusterman, R. (2000). *Pragmatist aesthetics: Living beauty, rethinking art*. Lanham: Rowman & Littlefield.
Stockmarr, P. (2014). *Det æstetiske i design*. [PhD thesis]. Copenhagen: The Royal Danish Academy of Fine Arts, Schools of Architecture, Design and Conservation.
UMA. (2015). http://www.project-uma.com. Accessed 28 Nov 2015.
Verbeek, P.-P. (2005). *What Things Do: Philosophical Reflections on Technology, Agency, and Design*. University Park: Pennsylvania State University Press.
Vial, S. (2013). *L'être et l'écran*. Paris: Presses Universitaires de France.
Vial, S. (2014). *Court traité du design*. Paris: Presses Universitaires de France.
Welsch, W. (1990). *Ästhetisches Denken*. Stuttgart: Reclam.

Phenomenology in Spatial Design Disciplines: Could it Offer a Bridge to Sustainability?

Emina Kristina Petrović, Bruno Marques, Natasha Perkins, and Guy Marriage

Abstract This chapter proposes that the philosophy of phenomenology is highly applicable, if not necessary, for a deeper and more integrated approach to spatial design disciplines in a world that aspire to be sustainable. The chapter develops upon the frameworks established by Heidegger, Jean-Paul Sartre, and Merleau-Ponty, and the works since which have attempted integration of such ideas in architecture and design, such as those by Aalto, Norberg-Schulz, Pallasmaa, Frampton, and Zumthor. It evaluates key historical and more recent phenomenological concepts for their importance in contemporary spatial design. Reflecting on these diverse views this chapter focuses on two strands and evaluates their usefulness in facilitating the uptake of sustainability in architecture and design. Although the two notions explored are different, jointly they show that there are significant constructive implications for phenomenology within architecture and design.

The chapter discusses the phenomenological concepts, their inherent relevance for spatial design disciplines and supports that discussion through a series of contemporary student projects in landscape architecture, furniture design, and architectural construction. Especially significant is a conceptual interpretation of pro-sustainable efforts as inherently and deeply reflective of many values already captured in phenomenology, and that through the active use of phenomenological concepts a more pro-sustainable design becomes possible.

Keywords Phenomenology · Architecture · Landscape · Design · Sustainability

1 Introduction

The application of phenomenology in architecture and design has a long tradition (Sharr 2007), and still attracts much attention as evident in the recent special issue of *Architectural Design* (2012) (Brislin 2012). Although there are strong overlaps and a good sense of collaboration amongst those using phenomenology in

E. K. Petrović (✉) · B. Marques · N. Perkins · G. Marriage
Victoria University of Wellington, Wellington, New Zealand
e-mail: Emina.Petrovic@vuw.ac.nz; Bruno.Marques@vuw.ac.nz;
Natasha.Perkins@vuw.ac.nz; Guy.Marriage@vuw.ac.nz

© Springer International Publishing AG 2018

P. E. Vermaas, S. Vial (eds.), *Advancements in the Philosophy of Design*, Design Research Foundations, https://doi.org/10.1007/978-3-319-73302-9_14

architecture and design, after more than half a century of such efforts much of the diversity has also been achieved. Phenomenology has been used to assert importance of human experience of architecture (Pallasmaa 2005), to justify importance of *genius loci* (Norberg-Schulz 1980), to call for an emphasis on tectonics and construction (Frampton 1990), and to engage with the poetic aspects of architecture (Zumthor 2006). The range of interpretations taken out of phenomenology has been increasing. This chapter proposes a new direction asserting that phenomenology can also play a critical role in facilitating uptake of sustainability.

The chapter opens with a general phenomenological emphasis on the importance of embodiment in perception and cognition of spatial design and thus effective, aesthetic and action-oriented experiences informed by environmental factors and bodily movement. The works of philosophers Jean-Paul Sartre and Maurice Merleau-Ponty and architects Christian Norberg-Schulz and Juhani Pallasmaa are examined to establish a field arguing for the importance of subjective experience as part of an embodied perception. Of specific emphasis is the the interactive physical exploration which introduces the body as the perceiver, the knower, the agent, by exploring the idea that the observer makes sense of the world through their bodily actions and participating with others in establishing meaning. This part of the chapter uses student work to evaluate the applicability of these to landscape and landscape architecture. The core proposition asserts that through bodily experiences of the landscape, the sense of separateness from the landscape decreases and ones desire to support natural environments increases.

The second part of the chapter develops from the ideas from Martin Heidegger and architects Alvar Aalto, Kenneth Frampton and Peter Zumthor by focusing on the materials in spatial design. It proposes that materials possess an inherent potential to play a significant phenomenological role in design. Rather than accepting an instrumental or scenographic use of materials, a more ontological approach to the materials and the process of creation is possible and needed. Materials can also be seen as supportive of enriching observers' experience of being. The applicability of such an approach to materials is examined in relation to student works in furniture design and architectural construction. The core proposition here is that through an ontological approach to design materials, they become ~~more~~ emancipated and the ~~consideration of the~~ totality of their impact on the natural environment becomes unavoidably apparent.

2 Importance of Phenomenological Experience in Landscape Architecture

The examination of phenomenology as the description and interpretation of human experience (Findlay 2009; Seamon 2000; van Manen 1990) through the comprehension of human situations, events, objects, meanings, and experiences begins this discussion. These understandings result from questioning the way we experience the world, how we relate to it and how it relates to us. This way of seeing the world we live provides spatial design disciplines with the conceptual and methodological means for the examination. By assessing the spatial, environmental, and architectural dimensions associated with human life a knowledge can be continuous and scaffolded.

The visual aspects of design disciplines and the excessive rationalisation of the design processes are currently leading professionals of architecture, interior architecture and landscape architecture. Pallasmaa argues that spatial design disciplines such as landscape architecture have become too dependent on the visual experience (Pallasmaa 1998). An excess of images both in quantity and pace afflict our society and that has been observed and named by philosophers as a rainfall of images, image addiction and the society of spectacle. This research argues that the adoption of phenomenological principles would serve as a link between the highly digitised images of student work with the sensual reality of the physical environment.

Designing a memorable experience is a unique task for a designer. This is acknowledged by phenomenology as sensory design, which can be translated into a manipulation of space, material, light and shadow to create memorable encounters between a subject and an object through the human senses. This creates an experience that is not only tangible, but also abstract, observed and perceived – an experience that relies on a sensorial approach embedded in an ontological view. It is by placing objects, ideologies, as central to both subjects and objects, where nothing has a special status but rather everything exists equally, that landscapes are perceived as things and exist independent of the human mind as defended in object-oriented ontology (Harman 2002; Bryant 2011; Bogost 2012).

2.1 Merleau-Ponty and Sartre: Subject, Body and Perception

Merleau-Ponty defends that phenomenology is existentialist in the sense that it deals with the existence of people in a pre-given world. It is a philosophy that is focused on the 'historical person' in as much as they engage with and live in the world (Martins 1992). Following Husserl's phenomenology but with a different perspective, Merleau-Ponty explores the return to the search for the essences of objects and their qualities. Merleau-Ponty sees these as part of the lived and experienced world, which is a world of things that have not been reflected upon and on which sciences are constructed (Merleau-Ponty 1945). Effectively, existentialism is dependent upon a common experience for the inner man.

However, for Merleau-Ponty, the truth does not inhabit the inner man, since there is no inner man (Merleau-Ponty 1945). Rather there are people in the world and it is by experiences within that world that people learn about themselves. The pre-reflexive, lived experiences of a human being in the world in the sense of being thrown into the world, with its intentionality, in an already existing world, ready but not thoroughly ready – are the founding concepts of Merleau-Ponty's phenomenology, which seeks to understand people and objects as being in a situation; never fully free but in a world never fully finished either. This supplants the notion of self-contained consciousness. It is not a matter of denying the inner world, as empiricists do, or denying the existence of the world outside, as idealists do (Merleau-Ponty 1942). The world has already been formed, although never completely.

Throughout his work (1942–1945), Merleau-Ponty deals with the body as a live entity through which one can be in the world and relate to other people and things. For him, the body is 'our anchor in the world' or 'our general means of holding on to a world' (Merleau-Ponty 1945, p.239). This foundation leads to a dialectic relationship between a person as a body and the world where it is located, which is neither exclusively objective nor exclusively subjective, and nor is it the sum of the two. The conditions of the world limit but do not determine a body; instead, people are in charge of determining themselves through their own choices. The body is therefore a vehicle for understanding the world. It is context dependent and crucial for the understanding of the user's experience, building an ontological wonder.

Similarly, Jean Paul Sartre defends the unity of body and consciousness, with the body as lived and experienced from within, from the 'first-person' perspective. This is neither pure consciousness nor a physical thing. The idea of a dialectic relationship between a being, provided by a body, and the world consisting of a feeling, means that a change in how one's body is experienced triggers a reflex-like reaction to a stimuli (Sartre 1989). As Moran describes, 'I do not know experientially that I have a brain or endocrine glands – that is something I learn from others. I have, as it were, a folk anatomy where I think my stomach is' (Moran 2010, p.136). The first person feels but needs to interpret what it knows.

Sartre believes the knowledge acquired from these encounters is real and exists for the development of mankind, which places the being as the most important entity and corroborates the importance of a lived body. This conception equally gives autonomy to the body and the possibility of projecting a consciousness embodied in perception and imagination. With this, Sartre introduces a new concept of perception – to perceive is not only seeing but also to be seen or 'touching-touched' – it is in touching the other (own body, other objects) that we encounter ourselves as located beings, it is to have the mind open to small meanings of reality. The world in which we are embodied is a world that has been humanized by us.

The idea of a dialectic relationship between the being as a lived body and 'the world' is better explained by Coelho (1991). He defends the body as simultaneously perceiving and being perceived. Using both this and Merleau-Ponty's viewpoint as guides, this chapter posits that the body casts itself into the world and knows the world through its movement, making the body a subject of perception (Merleau-Ponty 1945). This opposes to the positivist science view where people are seen as static beings composed of autonomous parts. This explains facts based on causality. Merleau-Ponty sees people as a set of possibilities that keep being realised through dialectical relationships with the world. He proposes a return to experience as basic data for building science, which creates a polarity between the person and the world – always in transformation, in a movement of endless search with each new phenomenon (Coelho 1991).

By relating with the world's objects, beings and things, a person is a being who perceives the world from different standpoints depending on the situation in time and space, who perceives particular perspectives that vary accordingly to the perceptual field – which is a horizon that is the place of perceptual experiences. We can perceive objects from different places, at different moments. These ideas are key concepts within this study such as perspective, field and horizon, which can be better explained with the example of perceiving a house that Merleau-Ponty uses in *Phenomenology of Perception* (1945, pp. 81–83):

> We perceive a neighbouring house we pass by it. When we come closer, firstly we see one side, then, as we walk by, we see the front of the house and next, the other side. If we went around the house, we would see its back and, if we could get in, we would see the inside from several angles according to where we were. As we have a different view from each angle and as we know that it is a house, we can conclude that the house exists by itself, independently from any perspective. At the same time any view we might have from any angle whatsoever would allow us to know that it is a house. Seeing the house is therefore seeing it from somewhere, at a certain moment, i.e. seeing it in a multi-perspective way, at a certain place, at a certain moment referred to as a horizon. Thus, seeing a house implies being able to see it from several perspectives, which are various possibilities.

Concepts related to the space-time structure refer to the phenomenological methodology based on Merleau-Ponty: when we ask several subjects for descriptions of a certain phenomenon being investigated, we understand that each subject will give this accordingly to their individual standpoint of how they perceive the phenomenon. Equally different people's perceptions, at different times and places, are given to us as several views from different perspectives of the same phenomenon, which cross over each other in inter-subjectivity and present common meanings that enable an understanding of the phenomenon's structure. The next stage, when as the researchers we make a phenomenological interpretation of these data, is when the phenomenon's structure is understood within our perspective. This is another perspective, another field, another horizon; that of scientific knowledge. This interpreted data allow us to reach a specific field of generalities, which we can say belongs to the general structure of the phenomenon.

The phenomenon, thus, depends on a person's perceiving perspectives. As something that alternatively shows and hides itself, it shows itself to whoever perceives it according to human perception, which means one's perception from different standpoints in time and space. One could say that a phenomenon is never seen in its totality, because this would be an abstraction; the convergence of several perspectives, however, leads us to perceive a phenomenon's structure.

As pedagogues, considering how design research is taught in architectural disciplines and how students live their experience of studying and designing an architectural object or a landscape, we try to understand what our students experience in a situation where they relate to an already given world, into which they are launched and which they will necessarily have to face. The student descriptions of their perceptions of landscape; of people and object at different scales, on meaning attributed by their consciousness in this experience and especially on the meaning of their relationships with landscape and design, form the focus of this analysis.

2.2 *Norberg-Schulz and Pallasmaa: Interactive Physical Exploration – Body as Perceiver*

Norberg-Schulz became best known for *Genius Loci: Towards a Phenomenology of Architecture* (1980), in which he defines architecture as an expression of the spirit of the place in which it is built – place-based architecture. His work follows closely the philosophy explored by Heidegger and criticises the lack of a genuine cultural and art historical insight as well as lack of sociological and psychological awareness regarding built environments and their influence on human beings (Van Nes 2014). Therefore, Norberg-Schulz breaks this silo and presents architecture as an expression of human experiences.

Where Norberg-Schulz differs is in situating the origin of those experiences in nature. According to his theory, the primordial structure of human experience is dependent on the landscape where it takes place. He equates those original experiences with the *genius loci*, such that in order to express the history of the place in a contemporary understanding and design, one should not necessarily look at historic layers and buildings, but instead should go back to the original source in the topography of the site or landscape – place (Crysler et al. 2012). Place is thus understood by Norberg-Schulz (1980, p.6) to refer to 'the concrete manifestation of man's dwelling' and is constituted by material substance, shape, texture, and colour, all of which gives 'character' or 'atmosphere.' It is through an understanding of all these factors – the total phenomenon of a place and the meanings which are gathered by a place – that an attentive reader of landscape or the architectural object is said to be able to arrive at an understanding of *genius loci* (Wilken 2013). As Norberg-Schulz (1971, p.19) states 'the places are goals where we experience the meaningful events of our existence, but they are also points of departure from which we orient ourselves and take possession of the environment.' Hence, these ideologies sought to promote concepts such as identity and place, which imply a mixture of memory, sensual experience, and interpretation to the design disciplines.

Nevertheless, Schulz advances the notion that visual diagramming is the key to the exegesis of the intellectual component of the landscape (Malpas 2009). However, other architects such as Pallasmaa have introduced the notion of multisensory architecture, demanding a less prominent role of vision in the experience of the architectural object or the place and criticising the fact that 'architecture has turned into the art form of an instant visual image' (Pallasmaa 1998, p.296). Buildings attempt to conquer the foreground instead of providing a supportive background for human activities and perceptions.

Pallasmaa explains that the 'experience of architecture is multisensory, being matters of qualities, space and scale are measured equally by the eye, ear, nose, skin, tongue, skeleton and muscle. Architecture strengthens one's sense of being in the world, essentially giving rise to a strengthened experience of self' (Pallasmaa 1996, p.28). He sees all the senses as extensions of the sense of touch with the body membranes of the other sensory organs all in some way being touched by stimuli, therefore presenting the senses as a generator of sensorial thinking as well as contributing to reason, deeply supported by the theory of phenomenology. Thus, the task of design disciplines is to place ourselves in the continuum of culture through embodied experience and memory of an object. Architects should think with their bodies in order to design.

Both Norberg-Schulz and Pallasmaa follow Merleau-Ponty's concept of body at the centre of perception and experience, reinforcing the importance of the collaboration of the senses in perception to create a fully embodied material and spiritual construct, and hence enhance future perceptions. The task of architectural disciplines is to provide resistance to cultural erosion and to design buildings and landscapes in an authentic existential and experiential manner.

2.3 *Experience, Idea, Design and Realisation: Design-led Research in Landscape Architecture*

Design-led research is seen as inclusive and exploratory, drawing from a number of research paradigms including historical, positivist, case study methods, action research and especially phenomenology and hermeneutics. Armstrong (1999) states that the reflective process through phenomenology and the richness of the interpretative realm using the hermeneutic circle is seen as action, reflection, interpretation, action, reflection, interpretation, and so on, which is supported by the works of Kvale (1983) and Corner (1991).

Phenomenology is interested in how people make meaning of their experiences (Crotty 1998), and the constructionist paradigm is concerned with the lived experiences of individuals. The work that will be presented seeks to understand the learning experiences of landscape architecture students in their design studios and the meaning that they attach to these experiences, with learning experiences being the phenomenon. It also explores how a lived experience can generate a perceptive model and how that feeds into the design process.

The design studio work of fourth year Landscape Architecture students at the School of Architecture of Victoria University of Wellington is used as an experimental endeavour in which future landscape architects explore their own inquiry into the phenomenological way of experiencing landscape and instigate a coherent design process. With a very broad brief allowing students to find their own way through phenomenology and design-led research, the selected township of Carterton in South Wairarapa, only 80 km northeast of Wellington, lies across a very flat floodplain of the Ruamahanga river as part of a wider catchment that connects with the third largest lake in the North Island of New Zealand – Lake Wairarapa. A three-folded process titled P.R.A. (Passive, Reflective and Active) has been used in this studio (El-Shafie 2011).

Within this vast landscape, students were able to make their own choice of the actual site that they felt more appropriate for the design they will undertake. With this, students began to explore the landscape, criticise their own attitudes towards the landscape and respond to the qualities of the site.

Fig. 1 Alison Baker, Perceptive model developed during the self-reflectiveness stage where site experience is summarised, 2015

The fact that a site can be observed from many different distances and viewpoints generates many different perceptions of it. Merleau-Ponty stated in *Phenomenology of Perception* (1962) that for each object, as for each picture in an art gallery, there is an optimal distance from which it requires to be seen, emphasising that an optimal distance must be sought to perceive the whole object and all its aspects. Therefore, the first part of the design studio was focused on representing student's response to the site. Experience was introduced to the students as one way to delve into the design process as a state of being emotionally and intellectually able to think and to express ideas creatively. This included collecting verbal and written information on the site and using their own body to measure and respond to the site as well as reading the site meaningfully.

The first stage in this design studio is self-passiveness in which students absorb, as much information as they can about the specific needs of the project, but without formulating ideas within a preconceived theoretical framework. Field trip experiences as well as relating to their peers and other people are invaluable inspirations for students at this stage.

By moving through the landscape, one experiences a sequence of different perceptions in time. Following Merleau-Ponty, at a certain distance we experience the optimal perception of the landscape, and moving from one position to another we perceive not only a fragment view of the whole, but also details and information not perceived before.

The second stage of the process is self-reflectiveness, where students reflect on what is identified as significant in the multi-sensorial information collected and recall their subjectivity to the task of relevant interpretation according to a collective background of knowledge and expertise. Students can perceive and relate themselves to different phenomena more readily when they reflect on their own lived experiences. They can also understand the experiences of other people, cultures, environments, and places. It is essential for them to learn how to bring forth the interaction with meaningful sketches and culminate with the creation of a perceptive model, which translates their experience into an object. Although experience of landscape is multi-dimensional and may be most effectively represented in a synthesis drawing, students were asked to summarise their site experience into a perceptive model (Fig. 1).

If the phenomenology of a site is to be found anywhere, it must be found in its ability to infuse with the landscape, and to be perceived and experienced in its multitude by the subject. The quality of a place is not a new phenomenon for design and results from the identity subjacent through connections and interactions between people and place (Fig. 2). The gradual perception joins the far, medium and short distance views, linking space, time, material and detail. Therefore, it is impossible to perceive either the significant of the selected site without its background of different unseen phenomena of local identity and cultural realities, or the meaning of these different unseen phenomena without their representation.

The last step of the process is self-activeness, which aims to reveal the unique phenomenal qualities of the place, by evoking students' creativity to produce a meaningful design. This requires them to translate their physical and emotional responses to the site into the design process (Fig. 3).

The process of self-passiveness followed by self-reflectiveness ending with self-activeness can lead to better design (Fig. 4). Self-passiveness involves the release of consciousness. Self-reflectiveness requires a distance. Self-activeness involves the understanding of a place and brings in the confidence to design. These three steps can assist the student to transition from the naïve amateur to a skilled and measured designer, a designer who is independent in their development process.

Phenomenology has been described as the interpretation of human experience (Findlay 2009; Seamon 2000; van Manen 1990) through the comprehension of human situations, events, objects, meanings, and experiences. In this part, student experience has been reflected upon to understand their experience as a catalyst for design. This studio assisted students in creating a new language for representing ideas and landscape. The ability to evoke emotional and intellectual meanings through design-led research is a necessary skill that a student of landscape architecture must be taught. The future generation of designers must be prepared to create places embedded in meaning to revive people's faith in their landscape and to learn from and with landscape. Designed landscapes should encourage people to discover and relate to a place as part of their identity, and that should be expressed through a design embedded on emotional and intellectual significance. Hence, phenomenology of landscape architecture can be defined as a systematic interpretation of human experience, where subjects and objects exist equally in a specific environment manifesting their nature and relations with one another as much as with ourselves.

This understanding results from questioning the way we experience the world, how we relate to it and how it relates to us. This way of seeing the world we live in allows spatial design disciplines to be provided with conceptual and methodological means for examining the spatial, environmental, and architectural dimensions of human life, permitting a continuously building up of knowledge in a process of development.

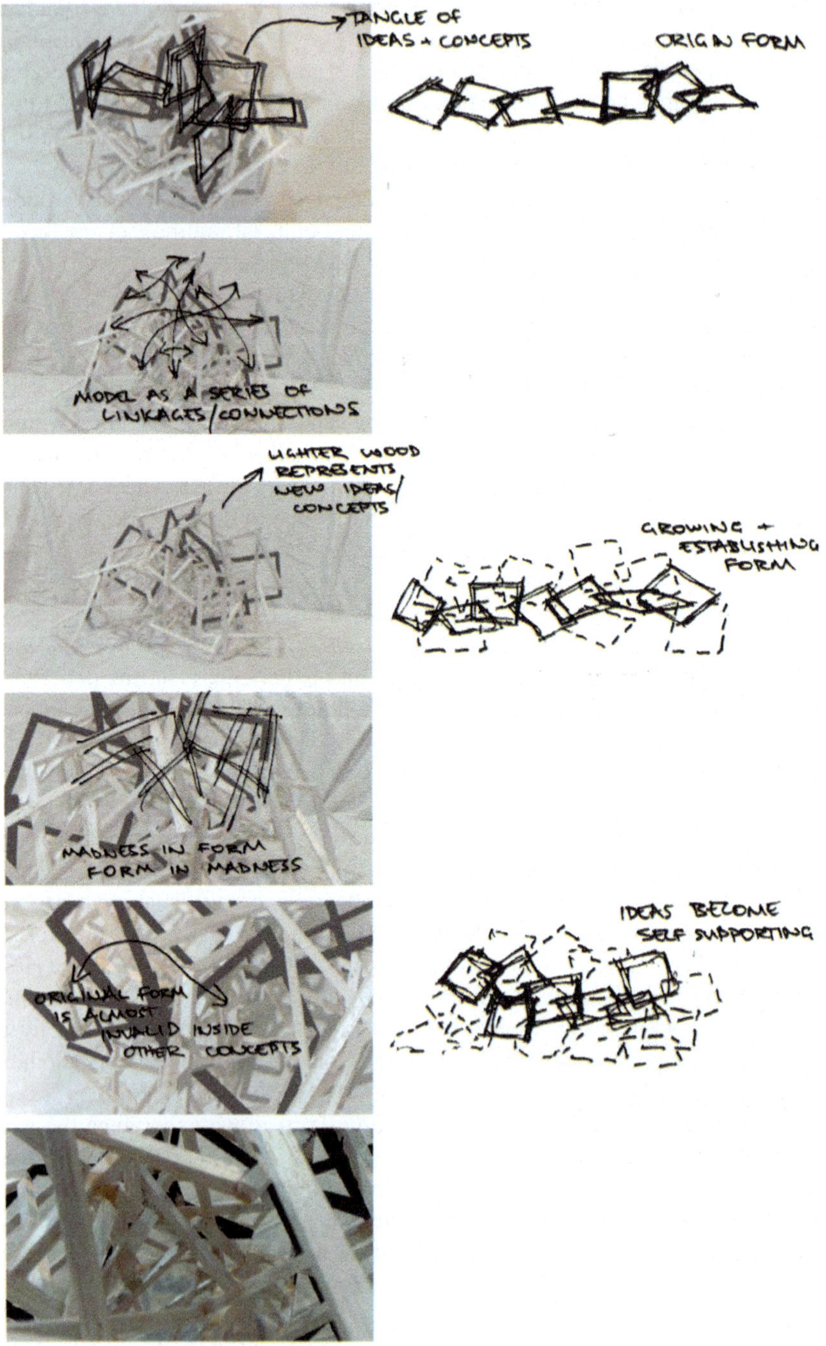

Fig. 2 Alison Baker, Exploration of connections and interactions of people and place through the perceptive model, 2015

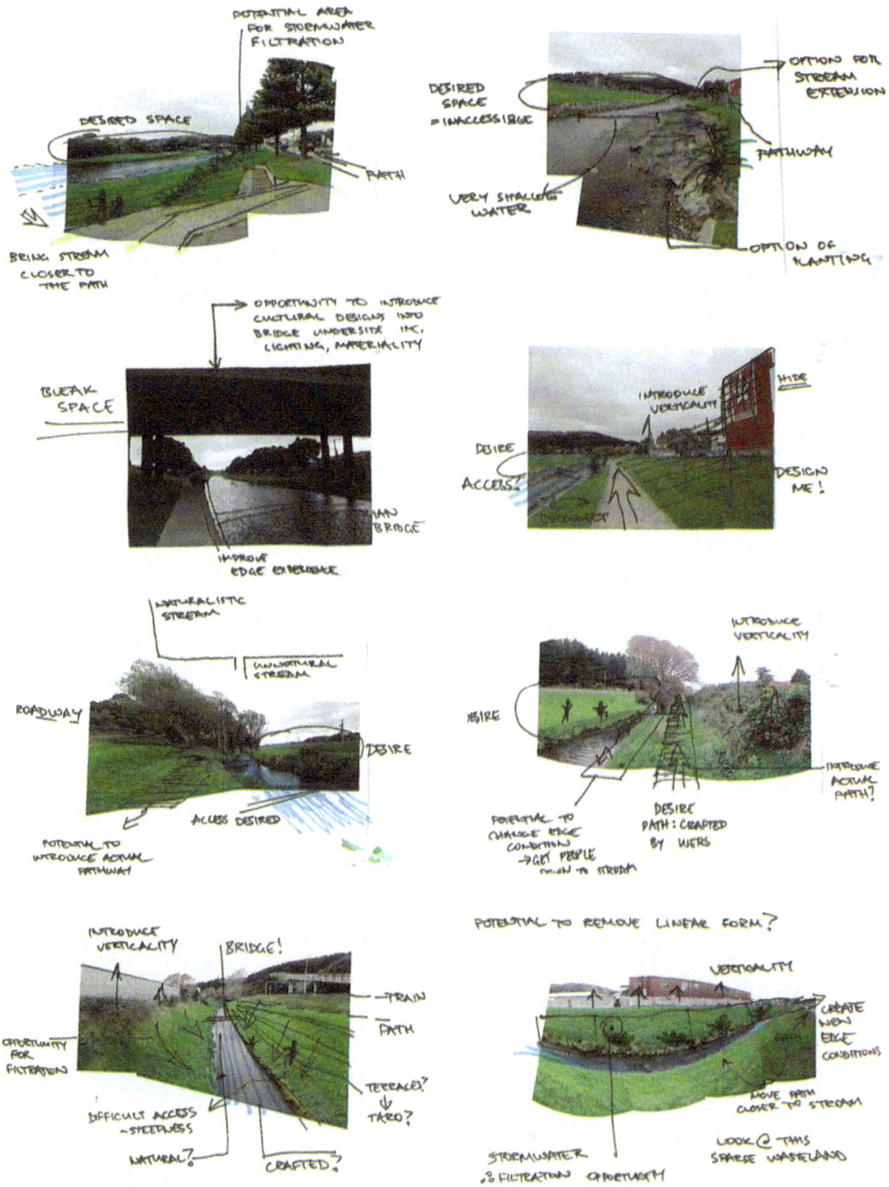

Fig. 3 Alison Baker, Application of self-activeness through the conversion of the physical attributes of the landscape combined with the emotional responses into a design strategy, 2015

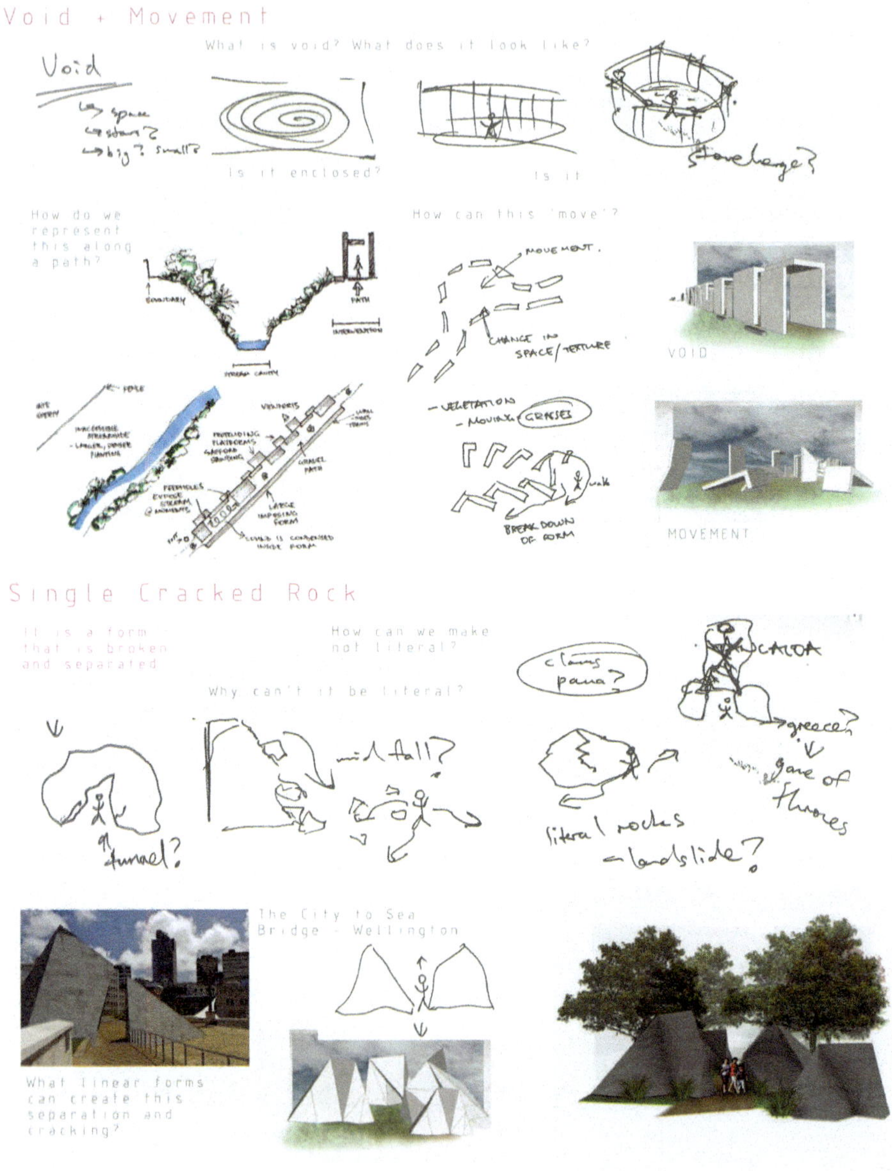

Fig. 4 Alison Baker, Final sequence of design strategies, 2015

3 Phenomenological Importance of Materials in Spatial Design

This part evaluates the significance of phenomenological concepts in relation to the use of materials in spatial design, proposing that materials contain within themselves a potential to play a significant phenomenological role in design, and that those could support pro-sustainable change. Such opportunities present themselves most clearly when evaluating Heidegger's work, which offers two significant interpretations. Firstly, materials play a more active rather than purely instrumental role during the process of creation, and secondly, materials present opportunities to enhance the phenomenological experience of being in the everyday.

This ability of materials to imbed phenomenological and ontological concepts is important from the position of design, because materials are necessary for achievement of any realized three-dimensional creations. Thus, materials are a given for design, and within them is equally given the inherent potential of phenomenological engagement with design. However, the role of materials in design has rarely received any significant theoretical consideration.

Although often aspired to, it is possible to question how well the design disciplines have succeeded to integrate a variety of essential design considerations. Rather, in some aspects it is possible to observe some compartmentalisation of focus and thus failure to provide holistic or integrated solutions (Petrović and Perkins 2016). This is particularly problematic when considering the importance of more complete implementation of sustainability in design, but is also concerning that design objects can negatively contribute to poor indoor air quality through their toxic off-gassing (Petrović 2014). However, deep engagement with the materials reflective of the phenomenological thinking offers potential to resolve a range of these issues, by asking for an active emphasis on the materials in all of their ontological and practical implications. Thus, a phenomenological approach to design materials can be actively used to help accommodate qualitative transitions facing contemporary society and design.

In the text that follows, the phenomenological importance of design materials is investigated. It opens by developing upon Heidegger's work and asserting the importance of emphasis on design materials sits well within the context of his work. Supporting this application, writing of other architectural thinkers drawing from Heidegger is also reviewed, with the emphasis on Frampton's influential writing in this area, and Zumthor's more recent design applications of similar ideas. Finally, applicability of such ideas is evaluated by reviewing a range of student design works which to various degree engaged with active or more passive considerations of materials in design. Because of the inherently unavoidable presence of the materials in all design work, this part proposes that even when the design assignments do not explicitly ask for direct investigations of materials, some students achieve a deep engagement close to the position of phenomenology. Taking these notions further, it is possible to assert that any three-dimensional design which does not actively aspire to a scenographic approach to materials, already has a deeply imbedded, even if unaware, drive to a phenomenological approach to the materials used.

3.1 *Heidegger's Ideas in Architecture: Aalto, Zumthor, Frampton*

For the type of analysis proposed here, the work of Kenneth Frampton can be seen as the critical mediator for inclusion of phenomenology in Heidegger's terms in architecture and design related disciplines (Pasnik 2003; Frampton 1990; Frampton 1995). In 1990, when starting his writing on tectonics in architecture, Frampton declared his criticism of surface-based scenographic, representational architecture of the decorated shed as his reason to assert relevance of the more tectonic, ontological approach, quoting Heidegger in support of this proposition (Frampton 1990).

Several years later, Frampton expressed similar sentiments which have been greatly expanded in his seminal work *Studies in Tectonic Culture: the poetics of construction in nineteenth and twentieth century architecture* (1995), providing a review of existing architectural works as expressive of concerns for tectonics through the use of visual, textural and tectonic properties of architecture. The key message of this book is foreshadowed by the subtitle in asserting the importance of poetics of construction and generally calling for 'a reconsideration of the constructional and structural models' of architecture, especially in relation to their expressive potential (p. 2). Here, Frampton revisited a number of genealogical myths generally accepted in architecture, greatly emphasising two mid nineteenth century German early modern thinkers: Karl Bötticher and Gottfried Semper (pp. 4–6). Expanding upon their dualism between the construction and form, Frampton develops new sets of dualistic pairs that interest him: ontological versus representational, and tectonic versus atectonic, clearly privileging in his discussion the construction, ontological and tectonic (pp. 16–21). This moves beyond interpretations already established in history and theory of architecture, proposing a new term of 'ontological construction,' and Frampton asserts such emphasis as fundamental for the modern architecture. The discussion that followed traced history of the twentieth century modern architecture from the perspective of 'ontological construction' and included a range of architects. Specifically emphasised are Alvar Aalto, Alvaro Siza, and Carlo Scarpa. Not much later, Frampton wrote about the success Peter Zumthor and Herzog & de Meuron as representatives of a recent Swiss German production with a similar ethos (Frampton 1997 in 2002).

Throughout these works Frampton regularly mentions Heidegger, but without providing any concentrated in-depth discussion of Heidegger's work itself, or explaining how and why Heidegger's concepts are applicable to architecture. Rather, Heidegger appears to be implied through the use of phrases such as the 'ontological construction,' 'phenomenological presence' (Frampton 1995), or '*thingness* of the constructed object' (Frampton 1990). Perhaps this absence of explicitly established discussion of

Heidegger's relevance in Frampton's work is not surprising, given that his main objective was to assert the importance of considering architecture in its tectonic, poetic expression of construction. Heidegger appears reasonably silent in this.

Clearer explanations of what is meant by some of these phenomenological terms can be found in the work of Gevork Hartoonian (1994), who developed under Frampton's influence. Hartoonian calls for an ontological approach to construction which integrates the scenographic with the actual construction (Hartoonian 1994). Unfortunately, Hartoonian retains Frampton's focus on architectural history and only mentions philosophy in passing, thus there is also no focused discussion of Heidegger's work, despite it being often mentioned. More valuable here is Hartoonian's emphasis that the separation between the design conception and actual execution occurred only since industrialisation (p.10). This shift in the production process corresponds to the shift from pre-modern to modern architecture and the same trends have only been increasing in importance with the proliferation of digital design and production processes.

Nevertheless, Frampton's work has had a lasting impact on architectural thought due to the attack on the scenographic and should be seen within its own historical context as rebellion against the post-modernist approaches which still dominated architectural production of the early 1990s. Since then, a body of work in architecture has developed emphasising the importance of tectonics, construction and through that, materials.

About the same time, Swiss architect Peter Zumthor was developing similar language through his built architecture in the Therme Vals, Switzerland (1996). In his subsequent writing, which started appearing from around 1998, Zumthor explicitly acknowledged inspiration from Heidegger's writing (Zumthor 2006; first published in 1998). Zumthor asserted that the sensual aspect of materiality found in his work is an expression of the engagement with being through dwelling and the 'thingly character' of architecture (Zumthor 2006, pp. 36–7). It is 'the reality of building materials, stone, cloth, steel, leather' which bring meaning and sensuousness to Zumthor's work (Zumthor 2006, p. 37). The connection with Heidegger is more explicit here, although often written in a poetic rather than philosophical style. Zumthor's Therme Vals have been referenced by many in architecture as a good example of this approach (Frampton 1997 in 2002; Sharr 2007).

However, the desire to use construction in an ontological and phenomenological way is not given. A distinctly different stream of examples which intensely use materials, details and construction come from product and furniture design focused on playful subversions of the conventional, with the 1990s work of the Droog Design group as a case in point (Droog Design 2013; Ramakers and Bakker 1998). Some of their projects explicitly focused on creative subversions of material properties, such as the 1997 'Droog Design for Rosenthal' collection (Ramakers and Bakker 1998). These works were created without any explicit desire to engage with phenomenology, yet the design outcomes can be seen to have properties desirable in terms of phenomenology, both through the process of creation and in everyday use. The reason for this could be inherent qualities associated with any emphasis on materials, details and construction.

3.2 On Phenomenology of Design Materials Based on Heidegger

When examined directly in Heidegger's work in relation to materials in design a richer and more relevant set of interpretations reveals itself. For Heidegger 'human beings are inextricably involved with things and people' (Dreyfus and Spinosa 1999, p. 51), and more specifically, any attempts to separate the mind from the world of things and other people would go against phenomenological understanding that it is precisely through those that human beings make sense of everyday things and of themselves. From this position, differentiation between instrumental entities without being in themselves and those entities that enhance the experience of being in the everyday underpins much of Heidegger's discussion. This is central for the consideration of materials in this paper.

Materials are considered in Heidegger's work in relation to three significant concepts: art, technology and dwelling. In his texts 'The Origin of the Work of Art' (1971a) and 'The Question Concerning Technology' (1977), Heidegger considers materials in relation to two dualistic pairs which he establishes between the equipment and the 'essence of technology' on one side, and the temple-work, art, and *techne* on the other side (Heidegger 1971a; Heidegger 1977). When seen as equipment, materials should perish as irrelevant, but when considered in relation to temple-work materials become more privileged, because they cause the work to come forth (Heidegger 1971a, p. 46). In this sense, the materiality of art work contributes to the essence of the message, or the truth of the work, thus being one of the essential aspects of the art work. Similarly, when considering technology, materials are seen as prone to be stock-piled in standing-reserves, and once something is in such a position Heidegger asserts that it stops its being as a 'thing' (Heidegger 1977, p. 17). In Heidegger's view while this perishing of ontological being for stock-piled materials is perfectly acceptable for situations when those materials are used as instruments, or equipment, it is also a missed opportunity to engage with the potential of materials to be much more than that, supportive of the very essence of creation. However, when materials are understood as part of *techne*, engagement with materials enables us to step out of the technological understanding of being, into the realm of art, which can be liberating (Heidegger 1977). This way of engaging with materiality requires quite a different, ontological approach to it and the process of creation, where the materials play a significant active role in bringing to the fore the art or temple-work aspects.

In practical terms, Heidegger differentiates between situations where materials are used as scenographic and irrelevant for the conceptual essence of the work, and those where the engagement with materials is an essential part of design, form, function and all other aspects of art or design. Due to his emphasis on phenomenological experience of the world, his commitment is clearly aligned with the production process which enhances experience in every step of the creation, through a deep engagement with all constituent parts of the process, and materials are seen as one of the most inseparable parts of this. In the contemporary architecture and design context, these ideas translate into a call for a design process that deeply engages with all constituent aspects of design, not allowing for a scenographic approach to any of these.

The second important message from Heidegger's philosophy focuses on materials as a vehicle that can enrich *our* experience of being, through the enhancement of phenomenological experience of being in the everyday. Heidegger's concept of dwelling is central here, as dwelling can be seen as an expression of our experience of things as they are, especially our experience of everyday things that we engage with on a regular basis, which are for Heidegger an essential part of being (Heidegger 1971a). Rather than being interested in a distant, objective observation on life's paraphernalia as an abstract system, as in an object, Heidegger attributed supreme value to describing life's paraphernalia as immersed in experience and use, as in everyday 'things' (Sharr 2007, p. 46; Heidegger 1971c). Heidegger often emphasized 'the thingly character of the thing' (Heidegger 1971a, pp. 19,32) and materiality as one of the fundamentally 'thingly' features of the thing, thus being essential to the being of the thing itself. Therefore, the everyday things bring us back to experience of the everyday, and all of the things which the everyday includes, as the foundation of an ontological experience of being. These connections are central for phenomenology, and materiality of the everyday 'things' plays an important role in these processes.

Therefore, while Heidegger's first notion was focused on the role of materials during the process of design, the second emphasizes the role of materials during use of finished everyday design artefacts. It asserts that materials can help support a psychologically deeper engagement with the everyday through dwelling. This can be seen as a proposition for a way of living which is supportive of a deeper, slower and potentially healthier engagement with the everyday. This provides a set of implications for design further than originally examined by Frampton, whose focus can be summed up as primarily focused on poetic and expressive notions imbedded in architectural materials and construction (Frampton 1995).

This shows that although poetic engagement with architectural materials and construction is very much possible, if adhering more closely to Heidegger's work a greater level of depth of engagement can be achieved. Unfortunately, such a deep engagement with Heidegger's work is not without its own issues. Firstly, it becomes necessary to acknowledge that the issues considered here are extremely close to the aspects of Heidegger's philosophical thought which are associated with Nazism, or as Tom Rockmore explains 'are interdependent and cannot be separated' (Rockmore 1997, p. 5). Secondly, the above discussion arrives at the point of similarity with the object-oriented ontology, as developed by Graham Harman (2002), because it is precisely an ontological understanding of materials that gives them potential to convey deep ideas. Similarly, as in object-oriented ontology, the interpretation proposed here accepts objects as important and independent from people's experiences of them. However, they are also seen as facilitators of a particular type of experiences and through that continues the relationship aspects which is criticised by the object-oriented ontology (Harman 2002).

3.3 Towards New Applications of Phenomenology in Design

While the work of Heidegger and Frampton provided crucial catalysts towards a more deliberate engagement with materials, and have stimulated some very important experimentation, these experiments maintained the tendency of considering only a smaller set of concerns related to materials at one time. Especially strong was the emphasis on the importance of the poetic, experiential and haptic aspects of architecture and design. This approach has already been criticized for its emphasis on haptic and visual aspects of materials, without sufficient acknowledgement of materials science and engineering (Fernandez 2006, pp. 10–11). Furthermore, it is relevant to develop and expand the range of ideas taken from phenomenology to support engagement with materials in design. It is especially relevant to give more emphasis to ontological understanding of materials, both in terms of ontology that they contain within themselves and which sensual experience of materials can trigger in those engaging with a piece of design. It is through such considerations that a stronger approach to sustainability reveals itself.

Sustainability in design and architecture is entering a new phase where the simplified gestural responses are inadequate and a much deeper and more complex range of considerations is needed. Currently, the most significant emphasis is placed on energy conservation through sustainable practices, which includes the energy needed when buildings are in use, as well as the embedded energy of the materials themselves. Waste, recyclability and reusability of materials are also frequently emphasised. While the energy and recyclability concerns are significant and should be retained, a broadening is needed to include a wider range of other issues currently only somewhat included in the contemporary sustainable practices, such as issues with toxic off-gassing of many of the design materials, or ecosystem implications both from the harvesting and the disposal practices.

It is essential to question expressive qualities achieved in design by using materials containing chemicals hazardous to humans or posing risk to the natural environment. What is the ontological value of such a project? It is precisely through the emphasis of an ontological engagement with the 'things' that surround us that that necessity of consideration of all conceivable issues becomes unavoidable. This is precisely the type of approach to design which is most needed in contemporary society. We have to consider all implications of design, and only through that comprehensive engagement with everyday objects can we have an ontological experience of them and ourselves. Of course without forgetting the functional, formal, and expressive aspects, but rather rejecting the supremacy of either form or function. Design materials have an extremely important role to play in development of more sustainable approaches to design, and phenomenological consideration of materials can greatly support a more comprehensive consideration of such views.

The question, however, remains whether the new generation of architects and designers is already operating from such a paradigm of thinking? The following text reviews a series of recent student projects in furniture design and architectural construction, evaluating the level of their engagement with a more comprehensive approach to design. Although none of the examples discussed here achieve the full phenomenological engagement with the materials which is conceptually proposed, these are useful illustrations of aspects which are already easily appearing. Even in their partial achievement these examples form a useful basis for expanding towards a fuller engagement.

3.4 Putting Concepts to Work: Materials in Furniture Design and Architectural Construction Education

This section evaluates the pedagogical opportunities for engagement with phenomenological understanding of materials. Two courses are evaluated: one in furniture design and one in architectural construction. Both courses are delivered at third year level at the School of Architecture, Victoria University of Wellington, New Zealand. The difference is that the architectural construction course is a core course for all third year students in architecture and building science, with a total enrolment of about 220 students, while furniture design is an elective which follows upon a second year elective, and tends to attract about 30 students enrolled in a variety of courses but mainly architecture, interior architecture, and industrial design. Neither of the courses specifically asked students to engage with phenomenology of materials, nor did they have such objectives integrated into the assessment. Thus, what is observed here is the result of students' own choice and initiative.

One of the most significant differences between the two courses is the scale on which they operate, and consequently how materials and construction are treated in relation to scale, and the works discussed here clearly indicate significant impact on the type of achieved engagement within the courses. The core pedagogical challenge is the same: teaching students to understand materials and their assembly into design artefacts (furniture or buildings). Similarly, the impact of digital visualisation of design has been clearly felt in both courses in recent years. From the position of this chapter, it is relevant to evaluate the direction of change, and inherent opportunities or challenges for a more phenomenological engagement with materials.

3.4.1 Furniture Design Course

The furniture course focuses on developing design towards resolution through material research, prototyping, production specification, critical review, communication and exhibition of the finished furniture pieces. Direct engagement with materials as one of the fundamental aspects of design process is explicitly encouraged by the course structure. The course provides an introduction to the common practices of industry and includes factory field trips which investigate processes and materials to support the independent design work, including presentations by manufacturers, sample material, and introductions to local contractors. In 2012 this emphasis was

Fig. 5 Yasmin Stewart, Lounger, laminated and bent Accoya, from advanced course in 2012. (Photo: Yasmin Stewart)

especially strong because the course was run with a focus on Accoya timber, a new acetylated wood product with sustainable properties (non-leaching, non-splinting, stable) after the Accoya company approached the school inviting the students to explore the design potential of this material. Rather than being supported in free exploration of various materials, students enrolled in the course during this year had to use Accoya timber, and were provided with additional information about it. Design results indicate that this was a successful approach, as the students produced design outcomes of reasonably high quality, while explicitly engaging with the characteristics of the material (Figs. 5, 6 and 7). These works are discussed here.

For many students this course is their first experience of more significant modelling, and thus there is much learning of the actual process of making. The course asks students to confirm their design with quick and dirty mock ups in full size (or as close as practical) to confirm a design object's real size and proportions. Materials are intended to play a significant role in this confirmation and refinement of design idea through a hands-on approach. However, due to lack of making skills, many students struggle to realize the actual properties of the materials that they are using. For example, it is hard for them to intuitively understand that light card at 1:5 scale is representative of sheet steel in terms of flex, compression, and stability. Similarly, they struggle to differentiate between various types of steel which have different properties. Thus, they need to increase their knowledge of basic engineering or structural principles, and their understanding of the properties of the materials. Finally, many lack an understanding that they are likely to need more than one sketch or mock up model before arriving at a developed concept design. This

Fig. 6 Larissa Meredith, Flower table, kerfed Accoya, from advanced course in 2012. (Photo: Larissa Meredith)

Fig. 7 Ella Bates-Hermans, Nest, reconstituted Accoya timber waste, from advanced course in 2012. (Photo: Ella Bates-Hermans)

tendency was always present, but it may now be exacerbated by the growing reliance on computers in design education; many students at our schools are accustomed to design only using computers.

However, once the students undertake some modelling and start learning about the material properties and actual making, their confidence in experimentation increases and learning accelerates. Potentially this indicates the importance of encouraging more modelling as part of general learning experience. Unfortunately, without such learning, some students are stymied in the iteration of the concept development, struggling to push the ideas past this stage. To compensate for these deficiencies in experience with making, the teaching framework encourages exploration of conceptually strong ideas, even if then many struggle to arrive at final design articulation within the allocated time period.

3.4.2 Student examples – furniture design.

In 2012 when the course specifically emphasized the use of Accoya timber, Yasmin Stewart's Lounger (Fig. 5) explored the properties of the material to reinterpret the lounger. Until this project, Accoya had only been used in architectural situations, so Stewart pushed the material's boundaries by bending and laminating thin strips of the timber together experimenting with thickness and bend radius. The achieved result successfully reinterprets the material and the traditionally recognizable form of the lounger. Larissa Meredith's Flower table (Fig. 6) also pushed the properties of the Accoya timber by using a traditional woodworking method of kerfing. This was an innovative exploration of Accoya timber, and with Meredith's strong evocative sensibility, it resulted in a poetic form. Throughout the design process, Meredith simultaneously explored the material, form, production process and evocative connotations, often reinterpreting the traditional craft and communication methods (she used drawing throughout) into a contemporary context using modern materials.

A poetic, but more constructive approach to material, is found in the Nest designed by Ella Bates-Hermans in 2012 (Fig. 7). For this project, the student used Accoya wood waste generated in the studio as the main material, by mixed resin with the wood chips to create a pulp. The idea of using woodchips was seen as linking the project to the materials that birds and wasps use to make their nests, which gives the design a more natural aesthetic, and also as helping to use waste well. As part of the process some sketch models were developed using papier-mâché to get a better sense of form, and a few experiments using PVA and woodchips to see what sort of textures from various densities of woodchip supported the development and refinement of the concept.

The student works described here reflect and perhaps even magnify the problems that can be observed on the theoretical level. The current trend of separating out of specializations on various issues associated with materials seems to be making it harder for the students to engage with totality of what is relevant for the materials. Their work often reflects the same struggles observed in literature with poor integration of information on what is sustainable and what is actually healthy for indoor environments. For example, while the Nest project by Bates-Hermans (Fig. 7) suggests an interest in sustainable practices by using wood waste, there is no clear indication that the resin component was fully understood for its impact both on human health and on the natural environment. Such evaluations are currently difficult, and probably too complex for third year students. The split between the formal and practical considerations is even greater in literature and likely to be more difficult for most students.

Similarly, although most of the design work discussed here can be seen as engaging with the tectonic properties of the materials, in many cases this is accidental rather than intentional. Even when somewhat intentional, the students seem unable to take such concerns a step further by considering the potential of phenomenological enhancement of the experience of everyday that these design pieces can offer. Although the phenomenological engagement with the materials is still largely absent, the emphasis on the process and craft seems to be providing some facets of such commitment. The lack of deep engagement in experimentation with materials could be at the core of this problem.

Some students in this course struggle with the transition to actual 1:1 making, and not surprisingly those who spend more time in the workshop get better results. Thus, for many students a deeper engagement with the materials is still difficult. However, experiments with teaching strategies indicate that by narrowing down the selection for students helps them to engage with the materials more effectively. In the 2012 Accoya timber projects, students not only had a material choice decided for them, they also received additional education from the Accoya supplier to gain more knowledge and understanding of the material and processes available to experiment with. The explorations that followed used the material and its properties, and students were able to focus on qualitative developments of those while also considering the formal, functional and poetic aspects. Much of that development was based on active consideration of craft and innovation and experimentation with the process of making. In many ways these examples could be seen as getting closer than many to a deeper and more complex approach to design.

These student works clearly indicate that increasing focus on materials, and aspiring to craft actual full-scale objects can lead to significant discoveries. Although it is hard to say that the student work discussed here epitomizes the theoretical approach proposed here, these examples show that some of the fragments are being engaged with, and that the process of crafting the material was the key for the most significant discoveries. However, this is currently challenged by the prolific use of computers in design, and generally accelerated requirements of design output in formats that are removed from the tangible, material considerations. The gap between the design outputs and tangible reality is widening, while what is needed is its narrowing. More experimentation is needed with a wider variety of strategies to encourage innovation and greater integration of materials in design.

3.4.3 Architectural Construction Course

The third year architectural construction examines the processes and building systems involved in medium to large scale buildings. The course looks at typical tall building systems emphasising efficient spatial planning and, where possible, sustainable practices in construction. The final outcome is a set of working drawings and a model of a segment of the façade. In preparation for the façade design students are encouraged to think about the structural expression of the building's support system, and the need to consider site and orientation to avoid excessive solar thermal input to the building interior. While the emphasis is to develop solutions that could be built, it is hoped that through development of an understanding of the importance of details, material selection, weatherproofing, appearance and sustainability, the students will produce a coherent, sustainable solution, and a clear visual aesthetic. It should be emphasised that although design aspects have a part to play this course, most of the focus is on students integrating various practical aspects as they conceptually 'put a building together.'

The issues observed in furniture design with students' limited hand-on experience of materials are also obvious here, but are exacerbated by the scale of architecture. There is no opportunity for full scale prototyping; in fact, every architectural project ends up being an expensive prototype. Furthermore, in architecture various contractors mediate between the design proposed by the architects and the realised, built building. Thus the architect is always one step removed from the actual construction, needing to use drawings and other visual instruments to communicate the design intention to a number of others who are to execute this on the site. Although historically much of architects' learning used to occur on real building sites, with the current health and safety rules it is increasingly hard to integrate any substantive site visits with students. The increase of class sizes, which is an international trend, adds to this problem.

Transition to digital visualisation and modelling of architecture, both helps and hinders the situation. The clearest advantage is that with the proliferation of digital communication accuracy in construction has increased to the nearest millimetre, which was never possible with hand drawings. Similarly, a digital model can in a reasonably short period of time lead to impressively resolved and visually appealing results. Visual accuracy of what is communicated can also be increased, as one does not have to deal with the representative use of materials in architectural models, such as cardboard standing for concrete. However, it is extremely easy for the digital models to be only suggestive of a good understanding the real materials and connections. In digital models there is no gravity, nor concern for weight or resistance to the size of fixings involved, and the inherent concern for buildability appears one step further removed. After all, with digital modelling, anything can be modelled and represented by vectors, with any materials mapped to the surfaces and crisp renders produced accordingly.

Small scale physical models, such as 1:25 models of segments of façades produced in this course, can help against such trends providing significant pedagogical relevance. Although in this scale glue replaces fixings such as screws, bolts, and nails, through the making at least some understanding of the nature of materials is achieved. This is especially the case in relation to gravity, necessity of reinforcing, and issues with suspending one material off another. The learning outcomes through the physical modelling outweigh those of purely digital fabrication, despite physical modelling being a poor substitute to real life construction and such models tending to be messier than the digital ones. That is why both approaches are used in this course, complementing and strengthening each other.

The assignment provided identical parameters such as floor to floor heights and common widths of structural grid. However, it also allowed for a great variety of

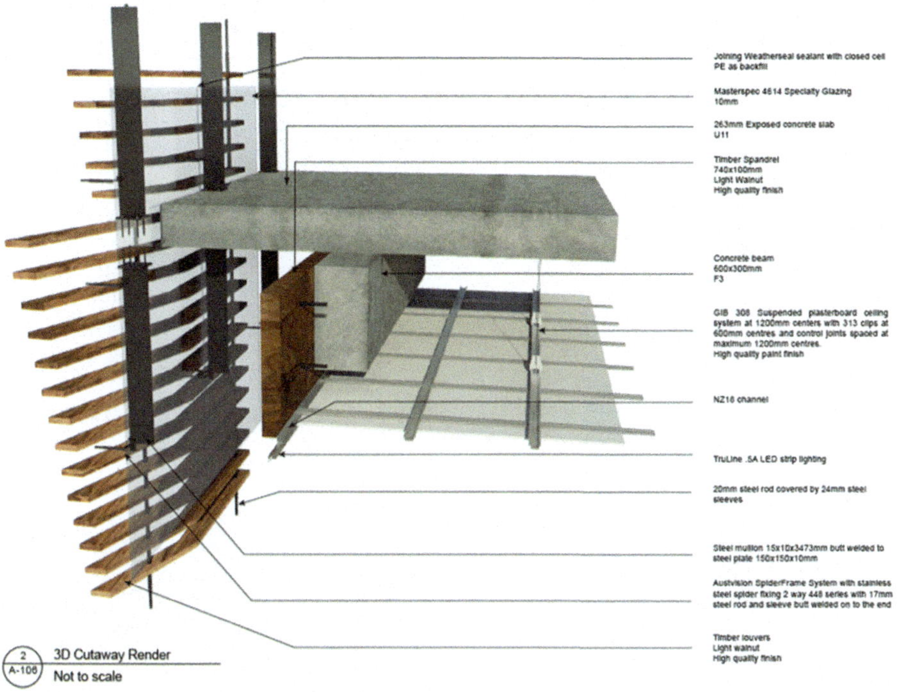

Fig. 8 Sophia Borissenko, Façade design, 2015. 3D cutaway detail. (Image: Sophia Borissenko)

elements to focus on and the students proposed a variety of architectural responses to the construction of façades. Many introduced external louvres to help reduce solar thermal load and consequently to decrease the need for mechanical cooling, aspiring to develop more sustainable design solutions. Design and orientation of louvres reflected the orientation of the chosen façade on the site. Some of the students quantified the impact of the louvres on the thermal load. Others committed to one preferred material as their main structural choice, and the façade details emerged reflecting this preference. The main point of the exercise enables the students to explore various aspects of construction and consider the appropriate level of detail and connections. Thus, apparently similar outline solutions allowed for significant differences when considered in detail.

3.4.4 Student Examples – Architectural Construction

Student works show that despite the fact that neither the materials nor poetic expression were central to the assignment, many works successfully engaged with those. For example, Sophia Borissenko (Figs. 8, 9 and 10) used a pallet of materials which allowed each to be emphasised and celebrated in their unique expressiveness. Concrete floor slab and beams are bare, finely executed and expressive in their

Fig. 9 Sophia Borissenko, Façade design, 2015. Photo of the model. (Photo: Sophia Borissenko)

Fig. 10 Sophia Borissenko, Façade design, 2015. Render of the façade. (Image: Sophia Borissenko)

materiality. Steel mullions are welded to steel plates and directly attached to the concrete floor slab. Louvres, spandrels and other timber elements are specified as light walnut. Each of the materials have only light finishes, allowing for their natural expressiveness to come through. The overall result provides subtle yet consistent reference to the materials and their expressive potential. The project reads as if this would not dominate the building, but be a reoccurring theme. Phenomenological experience of this project can easily be envisioned, and the use of well-known materials helps in this process.

Work by Bethan Davies-James (Figs. 11, 12 and 13) shows that more high-tech materials can also create a strong expression although different in nature. In this case louvres are made out of frosted tempered glass triangular panels on custom frames with motor activated hinges, allowing for mechanical control of the position of the panels. These are placed in front of double-glazed windows, providing three layers of glass, which require a number of supportive frames and accesses to function. The layering of glass, the use of frosted glass, and especially the mechanical

Fig. 11 Bethan Davies-James, Façade design, 2015. Photo of the model. (Photo: Paul Hillier)

Fig. 12 Bethan Davies-James, Façade design, 2015. Render of the façade. (Image: Bethan Davies-James)

Fig. 13 Bethan Davies-James, Façade design, 2015. Render of the detail. (Image: Bethan Davies-James)

control of the position of the louvres, all help give glass obvious material properties. While glass is often simply seen through, rather than looked at, here the glass itself gets an objectified material quality. Additionally, from both the inside and outside the glass is complemented with wood (floor and spandrels), allowing both materials to be highlighted. This brings a softer touch to this high-tech façade design, and additionally supports the clear opportunities for a phenomenological engagement with the glass.

Despite the emphasis of the course being on producing a set of working drawings of the façade, with modest opportunities to specifically contemplate materials or their expressive and poetic properties, the achieved works show that such engagements still occur. This supports the idea that as long as materials are required, and they are required in all architectural work, there is a latent potential for more a substantive and even phenomenological engagement with these. As seen here, this latent potential gets activated regardless of the specific pedagogical objectives at hand. This suggests that a more active engagement with phenomenology as part of design process is possible. Especially important for future pedagogical development is experimentation with ontological appreciation of materials which can strengthen the sustainable dimension.

Although louvres featured prominently in many of the works in the course and these can be seen as reflective of the sustainable concerns, there are relatively few other indications that sustainability has been greatly considered. For example, Davies-James does not discuss the environmental impact of layering of glass, as a material high in embodied energy, nor does Borissenko indicate consideration of the environmental impact of sourcing of hardwood walnut timber, which would have to be imported to New Zealand, where none is grown. Similar partial engagement with the sustainable issues is present in other works.

However, the issues with development and representation of materials in architectural construction are clear in these works. Many subtle details of finishes and connections are not easy to communicate, and consequently students tend to omit some of it. It is impossible to get a clear sense of the finishing treatment of walnut timber in Borissenko's work, although this is still more defined than precise detail or frosting of glass in Davies-James' work. While renders of the computer details read as more clear and pristine than the actual models, the quality of the materials and their poetic expression are stronger in the actual models.

While for furniture design it is feasible to propose more hands-on approach and 1:1 scale modelling, such propositions are impossible for the architectural construction. Rather, with the decrease of ease of construction site visits, and the accelerated development of new building materials (Schörpfer 2011, p. 19; Brownell 2006, p. 6), this gap is only widening. Potentially more of any kind of hand-on experience would be useful, but the trap there is that many smaller projects, such as furniture, operate on the scales where the details and the sense of scale of the materials are extremely removed from those of architectural construction. Still, unless there is an acceptance of with what Frampton termed 'abstracted atectonic' of architectural construction, and consequently all buildings, the real life experiences of materials, through sight, touch and smell, appears to be very much needed. It is through this more direct engagement with the materials that a more phenomenological engagement with architectural construction becomes possible.

This part evaluated the theoretical and philosophical frameworks for a more phenomenological approach to design materials and tested these ideas on the examples of student works. Although none of the student works achieved great level of resolution on the considered criteria, they show that a more intentional engagement with the materials can generate outcomes that are successful on multiple levels. Furthermore, the examples from the architectural construction course where the intentional engagement with the materials was more implied than explicitly required, clearly show that some such engagement is unavoidable in contemporary architectural production. Finally, this part shows great opportunities for these approaches to be developed further towards a more comprehensive consideration of design materials. An active engagement with the material choices in relation to phenomenology can have deep and qualitatively significant implications on all design creations.

4 Conclusions

The discussion of the importance of the direct experience and perception of the landscape through the bodily experience, which were discussed in the first part, reflect well established lines of application of phenomenological thought in architecture and design. These ideas were the first ones embraced by the architecture and design professionals, but still hold relevant (Brislin 2012). Such considerations of landscape also inherently support strong links with pro-sustainable efforts. Phenomenology strives for balance between person and world, designer and phenomenon, feeling and thinking, and experience and theory. Psychological identification with the landscape increases the desire to support all natural environments, through a decrease of desire to dominate the landscape (Milfont et al. 2013). The central desire is to explore and interpret mutual relationships through examining behaviour, experience, environment, materiality and meaning in a descriptive, interpretive manner as they happen in their everydayness. Further development of such concepts is needed to establish a coherent system on importance of bodily perception for pro-sustainable action.

Although the ideas associated with tectonics can be seen as another well-established application of the phenomenological principles in architecture and design. The second part of this chapter pointed that by going directly to Heidegger's work and taking such considerations beyond the rejection of the instrumental and scenographic use of design materials, it is possible to propose frameworks for a deeper, more ontological engagement with the materials. Such deeper considerations also support an active appreciation for the ontological presence of all everyday things, and from there can help facilitate a more sustainable engagement with the everyday world. Further development is needed in this area also.

The student examples provided some insights that engagement with the phenomenology in architecture and design is very much possible, even if often only partial.

For the landscape example this was primarily through the thought process and explicitly encouraged by the assignment framework. In examples from the furniture design and architectural construction such assignment requirements were not present and the level of engagement was less explicit. The student works indicate that some of the inherent aspects of design are closely related to phenomenology, such as bodily experience of the landscape, or materials for furniture and architectural construction, and thus provide a permanent easy link for phenomenology in architecture and design.

The importance of the depth of engagement was especially critical for the quality of phenomenological achievement by the students. It is precisely through this depth of engagement that more complex considerations, which use a wider range of otherwise compartmentalised issues, become possible. Furthermore, it is only through this comprehensive engagement with everyday objects that we can have an ontological experience of them and ourselves. That is the place where pro-sustainable efforts are the only option, as all human activity is seen as part of one great whole.

References

Armstrong, H. (1999). Design studios as research: An emerging paradigm for landscape architecture. *Landscape Review, 5*(2), 5–25.

Bogost, I. (2012). *Alien phenomenology or what it's like to be a thing*. Minneapolis: University of Minnesota Press.

Brislin, P. (2012). Introduction: Identity, place and human experience. In special issue: Human experience and place: Sustaining identity. *Architectural Design, 82*(6), 8–13.

Brownell, B. (Ed.). (2006). *Transmaterial: A catalog of materials that redefine our physical environment*. New York: Princeton Architectural Press.

Bryant, L. (2011). *The democracy of objects*. Michigan: Open Humanities Press.

Coelho, J. R. N. (1991). *Merleau Ponty: filosofia como corpo e existência*. São Paulo, Brasil: Escuta.

Corner, J. (1991). A discourse on theory II: Three tyrannies of contemporary theory and the alternative of hermeneutics. *Landscape Journal, 10*(2), 115–134.

Crotty, M. (1998). *The foundations of social research: Meaning and perspective in the research process*. Thousand Oaks: Sage.

Crysler, G., Cairns, S., & Heynen, H. (Eds.). (2012). *SAGE handbook of architectural theory*. London: Sage Publications.

Droog Design. (2013). Know more Droog 9: things you need to know about Droog. http://issuu.com/droog/docs/droog_brandbooklet2013_issuu. Accessed August 2013.

Dreyfus, H. L., & Spinosa, C. (1999). Coping with things-in-themselves: A practice-based phenomenological argument for realism. *Inquiry, 42*(1), 49–31.

El-Shafie, M. (2011). Phenomenology of site design as a fundamental concept in architectural education. *International Journal of Civil and Environmental Engineering, 11*(6), 64–70.

Fernandez, J. (2006). *Material architecture: Emergent materials for innovative buildings and ecological construction*. Boston: Architectural Press.

Findaly, L. (2009). The body's disclosure in phenomenological research. *Qualitative Research in Psychology, 3*, 19–30.

Frampton, K. (1990). Rappel a l'ordre: The case for tectonic. *Architectural Design, 60*(3/4), 19–25.

Frampton, K. (1995). *Studies in tectonic culture: the poetics of construction in nineteenth and twentieth century architecture*. Cambridge, MA: MIT Press.

Frampton, K. (2002). Minimal moralia: Reflections on recent Swiss German production. In K. Frampton, *Labour, Work and Architecture: collected essays on architecture and design*. London/New York: Phaidon Press. First published in *Scroope,* Cambridge Architecture Journal. No. 9. 1997.
Harman, G. (2002). *Tool-Bring: Heidegger and the Methaphysics of Objects*. Chicago: Open Court.
Hartoonian, G. (1994). *Ontology of construction: On nihilism of technology in theories of modern architecture*. Cambridge: Cambridge University Press.
Heidegger, M. (1971a). The origin of the work of art. First published in German in 1950. In M. Heidegger (Ed.), *Poetry, language, thought* (pp. 15–87). New York: Harper and Row Publishers.
Heidegger, M. (1971b). Building dwelling thinking. In M. Heidegger (Ed.), *Poetry, language, thought* (pp. 143–161). New York: Harper and Row Publishers.
Heidegger, M. (1971c). The thing. First published in German in 1951. In M. Heidegger (Ed.), *Poetry, language, thought* (pp. 163–186). New York: Harper and Row Publishers.
Heidegger, M. (1977). The question concerning technology. First published in German in 1954. In M. Heidegger (Ed.), *The question concerning technology and other essays* (pp. 3–35). New York: Harper Torchbooks.
Kvale, S. (1983). The qualitative research interview: A phenomenological and a hermeneutic mode of understanding. *Journal of Phenomenological Psychology, 14*(2), 171–196.
Malpas, J. (2009). Place and human being. *Environmental and Architectural Phenomenology*, 19–23.
Martins, J. (1992). *Um enfoque fenomenológico do currículo: a educação como poíesis*. São Paulol: Cortez.
Merleau-Ponty, M. (1942). *La structure du comportement*. Paris: Presses Universitaires de France.
Merleau-Ponty, M. (1945). *La phénoménologie de la perception*. Paris: Gallimard.
Milfont, T. L., Richter, I., Sibley, C. G., Wilson, M. S., & Fischer, R. (2013). Environmental consequence of the desire to dominate and be superior. *Personality and Social Psychology Bulletin, 39*(9), 1127–1138.
Moran, D. (2010). *Sartre on embodiment, touch, and the 'double sensation'. Philosophy Today, 54*, 135–141.
Norberg-Schulz, C. (1971). *Existence, space & architecture*. London: Studio Vista.
Norberg-Schulz, C. (1980). *Genius loci: Towards a phenomenology of architecture*. London: Academy Editions.
Pallasmaa, J. (1996). *The eyes of the skin: Architecture and the senses*. London: Academy Editions.
Pallasmaa, J. (1998). Logic of the image. *Journal of Architecture, 3*, 289–299.
Pallasmaa, J. (2005). In P. MacKeith (Ed.), *Encounters: Architectural essays*. Rakennustieto: Helsinki.
Pasnik, M. (2003). Introduction: The material autograph. In O. Riera Ojeda (Ed.), *Materials* (pp. 8–11). Gloucester/Rockport: Hi Marketing.
Petrović, E. (2014). Building materials and health: A study of perceptions of the healthiness of building and furnishing materials in homes. A thesis submitted to the Victoria University of Wellington in fulfilment of the requirements for the degree of Doctor of Philosophy in Architecture.
Petrović, E., & Perkins, N. (2016). Materials in furniture design: Towards a new conceptual framework. *ii Journal, 4*.
Ramakers, R., & Bakker, G. (1998). *Droog design: spirit of the nineties*. Rotterdam: 010 Publishers.
Rockmore, T. (1997). *On Heidegger's Nazism and philosophy* (2nd ed.). Berkeley: University of California Press.
Sartre, J. P. (1989). *Being and nothingness*. (H. Barnes, Trans. London: Routledge.
Schörpfer, T. (Ed.). (2011). *Material design: Informing architecture by materiality*. Basel: Birkhäuser GmbH.

Seamon, D. (2000). Phenomenology in environmental-behavior research. In S. Wapner (Ed.), *Theoretical perspectives in environmental-behavior research* (pp. 157–178). New York: Plenum.

Sharr, A. (2007). *Heidegger for architects*. London: Routledge.

van Manen, M. (1990). *Researching lived experience: Human science for an action-sensitive pedagogy*. London: Althouse.

Van Nes, A. (2014). The heaven, the earth and the optic array: Norberg Schulz's place phenomenology and its degree of operationability. *The Foot, 2*(2), 113.

Wilken, R. (2013). The critical reception of Christian Norberg-Schulz's writings on Heidegger and place. *Architectural Theory Review, 18*, 3.

Zumthor, P. (2006). *Thinking architecture*. Basel: Birkhauser. First published in 1998.

Part IV
Design Research, Design Epistemology

The Specificity of Design Research: How Practice-Based Design Knowledge Can Enter the Great Archive of Science

Paolo Volonté, Lucia Rampino, and Sara Colombo

Abstract In this chapter, we call into question the nature of academic design research. A reconstruction of the debate over the role of academic research in the field of design shows that its origins created the bias of attempting to model design research on the historically contingent *form* of scientific research rather than on its deeper *reason*. Indeed, design academics often imitate what scientific disciplines do when they do research (i.e. applying codified methods), yet the discussion about why such disciplines behave that way is still limited. According to science studies the answer to this why lies in scientists' habit of making the results of their research public, thus building what we refer to as the Great Archive of Science (GAS). By analyzing the dynamics of the GAS, we show that the rules, methods, and models typical of the research environment have as their main purpose to make the reliability of researchers' knowledge claims as durable as possible. Regarding design research in general, and research through design more specifically, we thus argue that what turns designers' work into academic research is not just the application of scientific methods but primarily the participation in the grand game of the GAS, whose dynamics enables a circumscribed corpus of knowledge to be held reliable by a community.

Keywords Design research · Research epistemology · Scientific method · Design knowledge

1 Introduction

Academic design research, as distinguished from customary research in professional design, arose when the main field of design had already been established for a long time. It has been developing for some decades now and its nature has been

P. Volonté (✉) · L. Rampino
Department of Design, Politecnico di Milano, via Durando 38/a, 20158 Milano, Italy
e-mail: paolo.volonte@polimi.it; lucia.rampino@polimi.it

S. Colombo
Massachusetts Institute of Technology, Cambridge, MA, United States
e-mail: scolombo@mit.edu

© Springer International Publishing AG 2018
P. E. Vermaas, S. Vial (eds.), *Advancements in the Philosophy of Design*, Design Research Foundations, https://doi.org/10.1007/978-3-319-73302-9_15

widely debated throughout the design field. The wideness of such debate proves that academic design research raises specific epistemological issues that need thorough examination. Notwithstanding the extension of the debate, it is our view that this issue still deserves further discussion.

Our reasoning takes its cue from a brief reconstruction of the origins of the debate over the role of "academic" or "scholarly" research in the field of design, striving to show how such origins created a bias that has become hard to overcome. The bias consists of attempting to model design research on the historically contingent form of scientific research rather than on its deeper reason. Indeed, apparently, design academia has predominantly tried to imitate what scientific disciplines do when they do research (i.e. applying codified methods) without questioning thoroughly why such disciplines behave that way.

According to several theories in the sociology of scientific knowledge (Bourdieu 2004; Latour 1987; Merton 1957; Ziman 2000), the answer to this why lies in scientists' habit of making the results of their research public, thus building what we refer to as the Great Archive of Science (GAS, cf. Volonté 2012). By analyzing the dynamics of the GAS, we want to show that the rules, methods, and models typical of the research environment have as their primary purpose to make the researchers' knowledge claims as durable as possible within their social setting. In other words, we argue that the original motive of the contingent forms of scientific research is the improvement of knowledge claims' reliability and durability within a social framework.

The claim we are making is thus as follows: if research activities modeled after the idea of scientific research are to find a place in the field of academic design research, such effort is worthwhile insofar as they seek the same objectives as research in consolidated scientific fields, i.e. making new knowledge claims reliable and durable. We thus seek to clarify how design can be an activity that has to do with knowledge production, the essential feature of research itself. Finally, relying on examples, we discuss some criteria that facilitate design-research results to enter the GAS, so as to become lasting claims for new knowledge.

To avoid any misunderstanding, we want to stress that in no way we assume that academic design research has to be somehow 'scientific'. Nevertheless, we noticed that in several design schools as well as in a large part of design-research literature (the part that is hereafter discussed), the idea and practice of design research is supported with research methods drawn from other academic fields (i.e. natural and social sciences). Our aim is to widen the discussion about the sense and benefit of modeling design research in such a way.

2 The Process of Design Academization

Several years' debate about design research has brought to light the multiplicity of activities lumped under this umbrella concept (Frayling 1993; Grand and Jonas 2012; Jonas 2004; Koskinen et al. 2011; Krippendorff 2007). This debate has burgeoned in the last two decades for a series of reasons, some intrinsic and others extrinsic to the field.

To start with the former, the traditional concept of research, with its claim to apply 'scientific' method, did not belong to the design world. There are two different classes of reason for this.

On the one hand, best practices in design have always implied research as premise and complement to the truly creative moment (Brown 2009, 69). However, such research has a different purpose than academic research. For all that it may be serious and exhaustive, this seeking aims not at producing new knowledge for a reference community but at acquiring knowledge useful to the designer for a specific design task. This is the reason why this type of research does not need to pay much attention to the careful application of a (scientific) method. It is no accident that the German term for such research, *Recherche*, is different from the noun for scientific and humanistic research, *Forschung*.

On the other hand, for a long time design has resisted topicalizing research as a methodologically structured activity dedicated to producing new knowledge for the reference community (*Forschung*). The reason is that design was dominated by the idea that design practice as such could be a science, and therefore should be based on universally recognized, codified proper methodologies (Simon 1969; for a reconstruction cf. Cross 2001, 50–52, and 2007, 119–122; Koskinen et al. 2011, 15–18). If the initial premise is that design is to be a science, the issue of defining design research as a special niche activity does not arise, whereas the question becomes how design methods are to be defined. As a result, design research starts becoming an object of attention only in the last decades of the twentieth century, when the positivist idea of 'design science' ceases to be the thematic core around which discussion of design is organized.

The main extrinsic factor raising the issue of design research and lending it the emphasis it still carries is the progressive academization of design. Beginning in the nineteen-seventies, design in its various forms gradually became part of academia, staking its claim as a field unto itself, opening bachelor's degrees, schools, and eventually PhD programs. Following this process, the issue of what avenues are open to academic design research within the practice of design became particularly topical (Buchanan 2001, 3–7; Schneider 2004, 5–8; Scrivener 2002; Vaughan 2017).[1]

Margolin (2010, 74) notes that the roots of this debate, and of how it was reckoned with, lie in the fact that academic design research (*Forschung*) arose from doctoral degree programs rather than, as would seem reasonable, the PhD in design originating from research. In other words, academic research in the field of design results from academizing design, not because design, during the course of its history, had raised questions whose answers demanded academic research. By the same

[1]Jonas (2012, 29) cites this reconstruction by John Langrish: "Late on the scene were the art colleges, absorbed into 'Polys' and then becoming universities and finding themselves with the research assessment exercise. Within present people's lifetimes, this sector had to work out what is an honors degree, then what is a Master's, and then cope with PhDs against the background of an educational activity which encouraged creativity and discouraged scholarship. So you still find people arguing about what is research and getting very confused about research and practice."

token, the world of academic design research (and doctoral design programs) is often divorced from the world of design professionals (Dorst 2008). In most of the cases, professional designers feel no need to be up to date on the plethora of methods for design practice developed within academia (Roedl and Stolterman 2013). Nor are they calling for research (*Forschung*). As a result, it is the models, procedures, and expectations typical of codified academic research, "for example university PhD regulations or the criteria of science funding organisations" (Rust 2007, 69), that dictate not only the need for a design research project, but also the form that it is to take on (Agnew 1993, 121–122; Bonsiepe 2007, 28; Schneider 2007, 213).

Succinctly put, the current debate over design research grew out of at least two different transformative processes in the field of design. On the one hand, a natural evolution in design itself led to abandon the idea of a 'design science,' thus opening the way to delving into the various methods for doing research within its own multi-faceted world. On the other hand, an extrinsic process of design academization fostered a need among academically trained designers to find something comparable to scientific research for their own discipline, and that something proved to be academic design research.

As sometimes occurs, the genesis of this phenomenon determined what form it took. Indeed, for the historical and cultural reasons noted above, design is often called upon to produce research modeled on the research that characterizes well-established academic disciplines, be they nature and life sciences or history and social science. And this "fetishistic approach" (Jonas 2015, 34) towards the sciences may represent a problem.

3 Categories of Design Research

The debate on design research, once it had thus arisen, focused first and foremost on analyzing the many ways in which we can speak of research in the context of design. Because this is the most well-known facet of the issue, we can concisely sum it up here (referring readers to Rampino and Colombo 2012 for a more thorough reconstruction).

At the outset of debate are contributions from Archer (1981 e 1995), Agnew (1993), and especially Frayling (1993), that recognize the heterogeneity of research carried out in the field of design. Myriad subsequent articles set out to criticize, extend or reformulate Frayling's categorization, but without substantially altering its significance. We recall here such categorization, relying on Jonas' (2004) reformulation, which we consider especially lucid, because it allows us to define closer the area of application of our epistemological issue. We do not claim that all research activities in design must be included into these categories, but we hold that they highlight three different ways of carrying out academic research in the field of design.

First, there is research that deals with design as an object of study. The scholars carrying out such research are not normally designers but typically those whose training was in social science (anthropology, economics or sociology) or the humanities (art history, semiotics). Their attitude is primarily cognitive, in the sense that, as Jonas (2004, 29) phrases it, they "try, whenever possible, not to change their object" but merely to become better acquainted with it. Taken as a whole, their research makes up the broad, varied field of design studies, which Frayling (1993) terms "research into design" and Archer (1995) calls "research about design."

Second, there is research at the service of the design profession (Buchanan 2001). This consists of all the research done to allow designers to design in innovative and accomplished fashion. In this case, the researchers may well coincide with the designers themselves. However, when the focus of the research waxes particularly complex or specialized, they typically come from other technical, scientific, or humanistic fields. In this case, again, a cognitive attitude prevails, since the aim of such research is to gather information and knowledge to be employed during design work proper. It includes market surveys, materials tests, ergonomics experiments, but also the mere "gathering of reference materials" (Frayling 1993, 5) and the simple "search for information" (Buchanan 2001, 17). In all such cases, the output is not 'research about design' but 'research for design.'

Finally, design activity itself may turn into a research tool. This means that the act of designing, using its own appointed methods and approaches, may take the qualifying role in a research process that yields knowledge precisely by planning out new products and services (Zimmerman et al. 2010, 310). Here, we may refer to "research through design," wherein design is neither the subject nor the objective of the research, but is its agent and tool.[2] Of course, the designers themselves usually coincide with the researchers in this case, while their output coincides, in ways that we will need to specify, with the output of research.

If we return now to the statement in the previous section, that academic design research becomes a topic of discussion starting with the academization of design and that, as a result, in the expression "design research" the term "research" refers to a work modeled on the research work typical of scientific disciplines, we will need to add the following considerations.

Research *into* or *about design* is not brought into question. Scholars for whom design is a research subject generally stand on the solid ground of some established, codified discipline. The legitimacy of their referring to an ideal of scientific research, especially that of historical-social science, is not disputed. Though the specific subject is design, design studies are epistemologically analogous to other research areas, such as innovation studies, fashion studies, and so forth. Nor was there a need to await the academization of design for the authors of this kind of research to become aware of its existence and rules. Research into design is thus a codified, legitimated field of endeavor that does not call for our attention.

Research *for design* presents a rather more complex case. Here, contingent, unstructured activities on the part of the designer, including googling for ideas and

[2] To be precise, it is the 'design milieu' (i.e. the designer, the design studio, etc.) who is the agent, while design as planning activity is the tool of research through design.

information as part of a newly commissioned design[3], coexist alongside scientific research activities in the traditional sense, including the laboratory experiment through which a cognitive psychologist hopes to discover the reaction of a certain kind of consumer to using a new prototype. But despite such variety, there are reasons to think that even the kind of work that falls into this category does not really amount to a problem for the issue of design research. What we have just stated about research into design holds equally true for all the scientific investigation carried out for design purposes, as Archer (1995, 11) has already noted: insofar as they are the product of consolidated disciplines, they are conducted according to established research canons. On the other hand, the investigations the designer undertakes in a haphazard or unstructured way in doing his/her job do not raise any epistemological issue, for they come no closer to scientific research than does, say, investigative journalism. There is, however, a broad middle ground where these two kinds of activity mix and mingle, where the collaboration (at times substitution) of designer with specialist becomes ingrained, and where the line between professionalism and dilettantism blurs. Brenda Laurel (2003) describes how designers, in doing their professional work, conduct research that may indirectly improve the design community's knowledge, although the primary goal remains increasing a product's chances for success on the market. The case of the Ideo Method Cards, a collection of design tools for creativity improvement developed and sold by the well-known design studio Ideo, may be mentioned in this regard. We are dealing here with a niche of great interest to the debate on design research, but one whose focus we can nevertheless say "is still on design as a practice and not as a research discipline that makes contribution of knowledge" (Zimmerman et al. 2007, 496). For this reason, we will not delve into it specifically.

We wish to concentrate rather on research *through design*. For it is only here that we find the research design *produces*. Indeed, this is the area where design, while continuing to be itself (which is to say planning work), may at the same time turn into knowledge production[4]. This kind of research requires an epistemological foundation of the knowledge production processes, which are not by nature the same of other (epistemologically founded) disciplines, but design-intrinsic processes. Indeed, this shift towards considering design as a research activity can be done only if we consider both the *process* and the *results* of the design practice (e.g. concepts or prototypes) as epistemic tools useful to produce and share new knowledge. Therefore, it is in research through design that the epistemological question of how academic research can get a space within the realm of design becomes urgent.[5]

[3] On the basis of a series of interviews with design professionals, Roedl and Stolterman (2013, 1954) found that "the most commonly mentioned sources of learning were coworkers, Twitter feeds, blogs, and practitioner-focused online magazines."

[4] For an overview of examples of what is meant by "research through design", see the proceedings of the RTD conference series: researchthroughdesign.org/2015proceedings/

[5] In the last decade, the epistemological implications of this practice-oriented research approach has attracted a great deal of attention within a broad scholar community, encompassing both the design and the HCI field. A number of new definitions have thus emerged, aimed at stressing different features of research through design. It is worth mentioning: *exemplary design research* by

4 Academic Design Research

Let us restate our guiding question. The concept of academic design research, when it is not limited to defining avenues of conventional scientific research, and thus especially in the case of research through design – which, from now on, will be our specific focus –, implies that it is possible "for designers to produce knowledge based on the skills and capacities of the design field itself" (Bang et al. 2012, 2). Yet, as a matter of fact the theory and practice of design research not infrequently have just attempted to hew to the models of codified academic research, adopting these models as their guiding lights. Often this happens when design researchers make use of research methods or strategies belonging to well-established disciplinary fields, such as ethnographic observation, case studies, focus groups, interviews, or trend analysis, to name a few.

Now, the question we want to answer is whether it is really by conforming to such models that design research can fully grow mature. In our view, such approach implies a shortcoming: design research can be exposed to the risk of adhering to the historical exterior aspect of scientific research instead of its epistemological reasons.

Historically, the exterior aspect of scientific research is its method, i.e. conforming research behavior to regulated, standardized models to be applied with the utmost rigor in order to achieve an understanding of real phenomena that is so precise as to allow the prediction of other phenomena that follow from them. Design-research theory often expressly strives to appropriate methods that have been developed and codified in consolidated, academic disciplines so as to apply them to design research work. Laurel (2003), for example, discusses a vast number of methods that can or have been employed in the design process, which are derived for the most part from the social sciences: ethnography and focus groups, quantitative and experimental methods, games and taxonomies, etc. Crouch and Pearce (2012) focus on four typically social-anthropological methods: observation, interviewing, case studies, and action research. Krippendorff (2006, 209–230) proposes a similar set of methods, but rightly recognizes that these essentially amount to methods suited to "science for design." Rodgers and Yee (2015, 2) maintain that "design should continue to borrow methods and approaches that fit from the physical sciences, the social sciences, and the arts and humanities".

In these cases and many others, codified methods from other disciplines are adopted by design as the tools for its own research: there seems to be an attempt in design to imitate what other academic fields do when they do research, to understand how they behave, without asking why they behave that way. Yet this can lead to a lack of foundation, as noticed by Hallnäs and Redström: "If our everyday

---

Binder and Redström, (2006); *experimental design research*, by Brandt and Binder (2007); *concept-driven interaction design research* by Stolterman and Wiberg (2010); *constructive design research* by Koskinen et al. (2011). The latter is often employed as a synonymous of research through design. With the term "constructive", Koskinen et al. intend to stress that this kind of research is aimed more at imagining new worlds and building them than at exploring the existent one.

understanding of a given method centres on the question of what to do and how, whereas the question of why is much more implicit as it belongs to, and is answered by, the general framework in which our practice exists, what happens when we borrow methods from other areas?" (Hallnäs and Redström 2006, 130–131).

In our view, taking advantage of existing research methods without addressing adequately the reasons of their original engendering is not an appropriate way for design research to free itself from the dominance of established academic disciplines. Seago and Dunne (1999) termed this dominance "methodological intimidation", recognizing in it the danger that "perplexed researchers in art and design will opt to 'play it safe' and, rather than risking the development and defense of really original hypotheses and methodologies [...] will choose academically acceptable and supervisable research topics with methodologies culled from established academic disciplines" (11–12). A similar issue is raised by Findeli et al. (2008) and Jonas (2015).

We believe that the way design sometimes appropriates other disciplines' methods bears the risk of taking what has historically been the exterior shape (or contingent form) of scientific research (namely, methods) instead of its grounds as the model. The issue is no small matter; for once the reasons underlying the ideals and practices of codified, scientific research have been disclosed, we will be free to decide in what way they can bear themselves out in the field of design – a way that may even be new, and better suited to the need, aims, and habits of design work.

5 Research in the Scientific Fields

It is now time to discuss the reasons underlying the ideals and practices of codified, scientific research. Of course, the topic is an extremely complex one, but we aim at focusing on just one aspect of it, apt to disclose the why behind the methodological attitude in established academic disciplines.

We will set off from science as a social activity, a set of people who act according to behavior models, thus starting from science-in-action (Latour 1987) rather than relying on established epistemological theories. This is an acknowledged approach in the sociology of knowledge and in science and technology studies (Barnes 1974; Knorr-Cetina 1981; Kuhn 1962; Latour 2005; Lynch 1993). We thus take a sociological attitude that tackles the factual social reasons rather than the ideal epistemological ones for scientists to behave the way they do. Nonetheless, we do not embrace a constructivist or relativistic stance. We rather share Pierre Bourdieu's attempt to overcome the juxtaposition of the positivistic and relativistic concepts of science. In his view, due to its specific features, the scientific field is "the historical site where trans-historical truths are produced" (Bourdieu 2004, 69). Accordingly, it is possible to explain the sciences' capability to produce trans-historical knowledge by inquiring the social dynamics of the scientific field, dynamics that are historically based. The enquiry of such social dynamics does not impede scientific knowledge from having trans-historical extent, because they strongly contribute to the binding character of long-term persuasion factors, such as consistency and evidence: "Logic itself, logical necessity, is the social norm of a particular category of social universes, scientific fields, and it is exerted through the constraints (especially the censorships) socially instituted in these universes" (Bourdieu 2004, 70).

Hence we will now discuss the social dynamics of science. This approach involves a shift of focus. When reflecting on a scientist's activity, we encourage concentrating not on his/her own knowledge, which is something very vague and intangible, impossible to detect, but rather on his/her knowledge claims: statements conveyed through oral (e.g. talks) or written (e.g. journal articles) communication (Ziman 1978, 6). Knowledge claims can be observed as well as other forms of human behavior. Furthermore, they are a typical feature of scientific research. A well-known characteristic of western scientific ethos, the "rejection of the ideal of secrecy" (Bernal 1939, 150), subjects researchers to an "imperative for communication of findings" (Merton 1957, 557). Claiming knowledge in public is no accidental aspect of scientific research.

Knowledge claims bear the quality of reliability. However, they can be more or less reliable. Yet on closer inspection, reliability is not a quality of the claim itself, but of the claim's audience. It is an effect of the audience's readiness to believe in that claim, or to act as if they believed in it.

In this paper we reserve 'reliability' to designate this kind of claims' character, whereas 'credibility' is the character of a subject laying reliable knowledge claims. Clearly, there is a circular dynamics of reliability and credibility: a credible scientist's knowledge claim will be considered more reliable than a less credible scientist's one, under pair conditions. And laying reliable knowledge claims improves the claimer's own credibility (Volonté 2006). It is a social dynamics of reputation that induces scientists to put the production of knowledge claims at the core of their activity (Bourdieu 2004, 55–56).

John Ziman (2000, 34) notes that the scientists' habit to claim knowledge, mainly in written form, has led to a 'notional archive' of scientific knowledge that is 'absolutely enormous'. Paolo Volonté (2012) calls this the Great Archive of Science (GAS). It consists of the entirety of scientific literature: books, journals, and data published online. But the larger an archive, the less accessible it is in practice. The Great Archive of Science therefore has an apparently contradictory feature: on the one hand, it is where knowledge claims are preserved from oblivion and handed down from one generation to the next, as well as the place where new claims are made public. On the other, owing to its immensity, it is also the 'graveyard' of scientific discoveries – or would be, were there not a meticulous system for classifying and retrieving the archive's knowledge claims.

There actually is such a system. It consists of library catalogs, book indexes, Internet search engines, and especially that portion of the GAS known as 'secondary literature', which makes up the bulk of the GAS. Secondary literature is a body of knowledge claims about other knowledge claims, where the latter are not only indexed, but also judged, evaluated, selected, combined, and relaunched in form of new claims soliciting attention by scientists and seeking for new reliability. For instance, today physicists assume that elementary particles are made up of quarks.

An American scientist, Murray Gell-Mann, claimed this idea for the first time in a series of articles published in the principal physics journals in 1964 (primary literature). However, it is through more recent claims, such as that advanced by Povh et al. (1993) in their textbook *Particles and Nuclei* (secondary literature), that most scientists learn such idea today. The GAS acts as if it consisted of two elements, a heavier GAS made up of knowledge claims about states of things[6] and a lighter GAS made up of knowledge claims about knowledge claims (second-order claims).

Secondary literature prevents the enormity of the GAS from turning into a graveyard for knowledge claims. It cannot, however, save all claims. The majority of the knowledge claimed since throughout science began now lies buried and forgotten in the bottom of the GAS. Only some of it has survived: the knowledge taken up by new knowledge claims. Although it is customary to think that scientific knowledge is known and shared because it is true, the reverse is more accurate: knowledge is considered true when it is shared. We believe $E = mc^2$ not because it is 'true' neither because Einstein said so, but because others have since argued persuasively that Einstein's equation is reliable. Second-order claims therefore bear the responsibility of selecting what part of the GAS will continue to be available – because considered reliable – over time.

This dynamics does not replace usual tools of stabilization of scientific knowledge, like consistency and experimental evidence. But it acts like a sort of filter. Scientists' knowledge claims do deal with states of things, but they do this *in front of* a peer community that has the power of sanctioning failure or success of the claimer (Bourdieu 2004, 45–55). The social relationship with the peer community is crucial here. The core value of a claim for the claimer is not its conceptual or factual content, but its reliability, because its reliability can enhance his/her position in what Latour and Woolgar (1986, 194-208) describe as the "credibility cycle". What comes first is not what the claimer says, but the fact that s/he is believed. Scientists strive to state claims that peers may consider reliable, not just in the immediate future, but also in the long term. The importance given to consistency in theories and to evidence in experimental research is an amazing effect of this dynamics, depending on the fact that consistent claims and claims based on evidence are generally more reliable in the long term. The social dynamics of reputation functions, in science, as the motor of the production of knowledge that turns out to be stable and useful.

We can now imagine the GAS behaving like in a huge imaginary tank. New claims enter from the top and lie on the surface, at first, but soon begin to sink inexorably to the bottom, where they are submerged by new claims, forming a thick layer of inert sediment. Every scientific knowledge claim is fated for oblivion in the dregs of the GAS unless new knowledge claims intervene by declaring it reliable, thus bringing it back to the surface. But these second-order new claims will start

[6] States of things encompass aspects of nature as well as culture: everything that can be enquired by natural, life or human sciences. This includes, for instance, pieces of literature that are considered not for their reliability, as it happens in secondary literature, but for other features like their textual structure (e.g. in semiotics) or aesthetic qualities (e.g. in the history of literature).

Fig. 1 The GAS tank

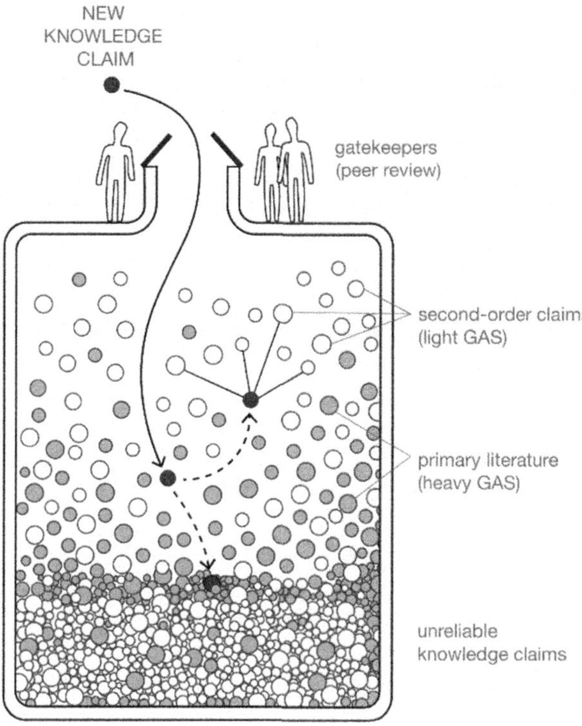

to sink, too, their fate hanging on yet more new claims. Thus, the GAS is animated by the knowledge claims scientists constantly advance, not just claims about reality but especially second-order claims. Absent the latter, the former would be destined to lie forever at the bottom of the GAS, which largely consists of claims that have simply been forgotten.

Two instruments, among others, are widely used in scientific practice to enhance a knowledge claim's chance to be admitted to and survive within the GAS: peer reviewing and referencing. Both of them justify the use of methods in scientific research (Fig. 1).

Peer review is a key feature of scientific research (Chubin and Hackett 1990; Cole and Cole 1981; Cole et al. 1978). To gain admittance to the GAS, a knowledge claim must first be scrutinized by 'well-informed colleagues.' The reason is that, to be claimed in public, most of times a piece of scientific knowledge must go through a formal communication medium. A speech at Speakers' Corner is not enough. Yet media have by their nature gatekeepers, people whose positions entitle them to decide what can be published, and how (White 1950). Science has institutionalized this general rule by making it part of required method: gatekeepers have to be scientists (the sole critics with the needed information) and must make decisions based only on scientific criteria (the claim's objectivity, its procedure, its originality vis-à-vis the GAS, etc.) while eschewing political, editorial, academic or other such criteria.

As a result, the reliability of any new knowledge claim admitted to the GAS has already been somehow vetted. It is in this process that research methods become important. Before submitting a knowledge claim to a gatekeeper, scientists meticulously refine it to forestall any possible criticism from referees. In the field of High Energy Physics, for instance, experiments are usually carried out by huge collaborations with hundreds or even thousands of scientists. Journal articles are signed by large numbers of authors. It is a standard, therefore, that articles' drafts are discussed in conference-like assemblies before they are submitted to the journals. The methodology that brought to the claimed results is scrutinized and discussed within the collaboration group to secure it from the reviewers' criticism. Hence, the smaller group in charge of writing the paper engages in producing clean data and sound claims of knowledge, to prevent criticism from the collaboration. Such clean data can be obtained only through the use of a robust methodology. Were there no peer-review process, scientists would probably give much less importance to research methods. The need to withstand informed criticism is a constraint on the upstream phases of research that forces them to rigorously comply with the rules of method (see also Ziman 1978, 6).

References form a web of knowledge claims that strengthen each other. Citations provide knowledge claims admitted to the GAS with specific locations, tying new claims' reliability to the reliability of their cited sources. When Stephen Olsen, at the beginning of his article "Observation of large *CP* violation in the *B*-meson system" (2002), writes: "Violations of *CP* symmetry were first observed in the decays of neutral *K* mesons in 1964" (Olsen 2002, 4), and references "Christenson et al., Phys Rev. Lett. 13, 138 (1964)" (Olsen 2002, 13), he ties his article, i.e. the knowledge claimed through it, to the so-called Standard Model in High Energy Physics, to which Christenson et al. contributed in 1964. Were the Standard Model eventually to lose favor in the community, it would drag all related knowledge claims toward obscurity at the bottom of the GAS. However, every new citation also reinforces the reliability of the claim being cited. A reference to Christenson et al. means their demonstration was deemed reliable once more, further reinforcing its reliability in the eyes of other scientists and bringing it closer to the surface of the GAS. The citation system is thus not only useful to the individual researcher but also a structural component of the scientific method. Its methodological importance has been institutionalized in recent decades by a number of citation indexes (e.g. ISI-Web of Knowledge, Scopus).

At this point, the reason scientists submit to rigorous, methodological research rules ought to be clear. Reliance on one method rather than another is justified not because that method is inherently 'scientific' nor because the principles of scientific research demand that it be used, nor even because it is functional to attain 'true' knowledge of 'reality.' The entire body of regulations that scientists live by, whether they actually submit to or secretly violate them (thus recognizing their legitimacy in the breach), can be explained by way of their sole truly vital purpose, to wit: allowing knowledge claims to survive by remaining afloat in the precipitating magma of the GAS. The purpose of all research's characteristic methods and behavior models is to solidify knowledge claims (despite the breakdown of the gas metaphor) to the greatest possible degree (Latour 1987, 22–29), i.e. to make them last as long as they can. However, the longevity of knowledge is bound fast to the human world that yields it up. More precisely, its ability to endure depends on the reliability of the corresponding claim acknowledged by the community. In the precipitating dynamics of the GAS, a knowledge claim can last only thanks to the 'collaboration' of players who see in its re-employment a chance to solidify their own claims, because they acknowledge its potential to last. Therefore, Nigel Cross (2007, 126) is right to affirm that "the whole point of doing research is to extract reliable knowledge from either the natural or artificial world, and to make that knowledge available to others in re-usable form." Every researcher aims to make the reliability of her/his own claims endure as long as possible; s/he seeks the willingness of others to place their own trust in what s/he has claimed, to believe in it, and to invest their scientific credibility in it. Scientific research, with all its quirks, obsessions, canons, and precepts, developed in response to the constraints imposed by the GAS as a collective tool for selecting knowledge that promises to endure in the long run.

We are aware that science as a social field is much more complex than the dynamics of the GAS. Leaving metaphors aside, the GAS has no definite limits. Nor we should imagine it like a field determined only by some formal characteristics (such as "the claim must be published in a scientific journal") or dominated by a kind of authority that decrees what is in and what is out. The limits of the GAS are set up by the GAS itself. What is in and what is out is established by those who take part in the game, i.e. claimers that bear some credibility in the eyes of their peers. And the means they use to establish what is in and what is out are second-order claims that appear reliable to the peers. This makes reliability the unifying force of the GAS. As said, reliability is a quality of a knowledge claim deriving from the attitude that people having access to the GAS have toward that claim. We need no assumptions here about the motives that drive people to consider a knowledge claim reliable or unreliable. However, the social dynamics of reputation implies that at least two main factors exert their force in producing the reliability of a knowledge claim:

a) the extent of social agreement (trust) about the claims' reliability, because this predicts the enrichment of credibility that the claim will permit;
b) the expectation about the duration of such trust, because this predicts the persistence of such credibility.

In conclusion, the prima facie validity of research findings is determined not by the use of equipment, strict logic, experiments or research methods, nor by negotiations and agreements among social subjects in a given institutional, economic and legal framework, but by the regulated consensus of a community of scientists. Methods are tools to achieve such consensus. Indeed, the materials deposited in the GAS are constantly filtered through those processes of intersubjective communication and evaluation that Robert Merton (1957, 560) called "organized scepticism".

6 Knowledge Production in Research Through Design

We have put forward the thesis that, by adhering to the historical exterior aspect of codified research represented by scientific methods, academic design research often risks overlooking the ultimate goal of those methods. We have then described such ultimate goal in terms of production and stabilization of the reliability of knowledge claims that aim at surviving in the GAS.

What turns a designer's work into research is thus not (just) her/his application of codified research methods but her/his participation in the grand game of the GAS, whose rules enable us to select – from among countless and sometimes contradictory knowledge claims – a relatively circumscribed corpus of knowledge that most researchers deem reliable and, through that very selection process, to ensure such knowledge will long endure. Any method may prove useful to that end, although it must be acknowledged that codified methods were codified precisely due to their particular utility.

Therefore, if academic design research is to exist not merely because it is mandated by the institutional context of academized design, but due to its own inherent *raison d'être*, enfranchised from the demands of other disciplines and from the risk of science's "colonization of design discourse" (Krippendorff 1995, 7), it needs to perform an essential task. It needs to produce an archive of knowledge available to the disciplinary community that is lent a certain reliability by the fact that many believe in it and presumably will do so in the future.

In brief, our point is the following: if the precipitating dynamics of the GAS can make sense of the gradual development of research methods and rules in the world of scientific research, we should resort to the same foundations to justify the idea of methods and rules for a design research. This means that the aim of academic design research should be to fit certain claims of knowledge into an interplay of claims analogous to the fluid dynamics of the GAS. Eventually, a great archive of design will have been created and it, too, will be part of the GAS – a very distinctive part where claims of knowledge production are shared not only through symbolic means, with words, numbers and graphics, but also through artifacts, i.e. models, prototypes, products, etc. (Schneider 2007, 216).

Yet, while it is a foregone statement that a designer creates artifacts (be they tangible – e.g. products –, or intangible – e.g. services), it is less obvious that a designer produces knowledge. Several scholars (e.g.: Bardzell et al. 2015; Bowers 2012; Buchanan 2001, 13–17; Friedman 2001; Friedman 2003; Hallnäs and Redström 2006; Höök and Löwgren 2012; Stolterman and Wiberg 2010) have already payed attention to the question of what does it mean, in practice, for the designer to "claim knowledge".

To begin with, it must be stressed that creating artifacts may also mean claiming knowledge, to some extent. As Agnew (1993, 121) noted, any human artifact "embodies extensive knowledge" because it is the result of a heritage of technical, aesthetic, and procedural knowledge without which it could not have taken shape, and that it makes visible and tangible, i.e. that it exemplifies. This is the reason why, by analyzing the archeological evidence of an object created centuries ago, we can understand traits of the population that created it. And not only the population's technical skills, but also its social and cultural features. Obviously, this holds not only for ancient objects, but also for more recent products, whose materials, shapes, colors, and manufacturing processes make us able to recognize, with a good approximation, the period in which it was produced. For instance, it would be easy – also for a non-expert person – to recognize if a car was manufactured in the '80s of last century or after 2010, just by looking at it.

Therefore, designing artifacts is an act that has to do with some forms of implicit knowledge, which are embodied and made evident by the object. In most cases, the professional designer does nothing but exploit this underlying body of knowledge to create a new design. But, in certain situations, the designer becomes a researcher and shifts interest from making a commercial product (or service) onto topicalizing this heritage to make artifacts that are, in turn, useful to the disciplinary community for designing new (more usable, more sustainable, more engaging) products. These artifacts indeed can be used as means to create new pieces of knowledge, as well as to challenge or complement an existing body of theoretical knowledge (Stolterman and Wiberg 2010) and to propose particular, intermediate-level ideas suitable to multiple situations (as suggested by Höök and Löwgren 2012 in their definition of 'strong concepts'). In such cases, the designer's work becomes outright *research*, specifically, *research through design*. The Audio System and the Digital Payment Terminal designed by Van Campenhout et al. (2013) illustrate in an exemplary way this strategy, which is also shared, among others, by the vast research activity of the idStudioLab at Delft University (Hekkert et al. 2000) or the Interaction Research Studio at Goldsmiths University of London (Jarvis et al. 2012). By designing and prototyping those artefacts, the researchers developed and tested a new design approach for digital products, which was then formalized and shared as new knowledge for the design community.

Design practice can therefore be used "as a way to perform experiments to develop theory" (Hallnäs and Redström 2006, 132). Furthermore, in research through design, concepts, models and prototypes help to envision unforeseen possibilities, suggesting future scenarios by making them embodied, tangible, and, therefore, also testable. Models and prototypes can be useful in explorative studies to foster dialogue between experts with different backgrounds and between users and researchers. Also in this way, artifacts are valuable tools to generate new knowledge. In some cases, they are left intentionally "unfinished" and "open-ended" by the designer, to avoid being seen as the ultimate solution. As an instance, one can look at the collection of "open-ended provocative design concepts" (hearing aids and insulin injectors), designed to present the results of a field study to two medical device manufacturers (Kelly and Wensveen 2014). The aim of such "unfinished" proposals is to stimulate comprehension and reflection, or, as Vallgårda puts it, "to explore new opportunities with the materials at hand, to develop new potentials, and to build examples that populate the new design space" (Vallgårda 2009, 11).

Moreover, a significant feature of design knowledge is that it often inhabits a nonverbal dimension (Mareis 2012, 63), one in the plane of what has, ever since

Polanyi (1958), usually been termed "tacit knowledge." That does not amount to an objection to the idea that research through design is possible. It is an achievement of the sociology of scientific knowledge that this is always inextricably made up in part of tacit knowledge, even in disciplines that are seemingly more abstract and formalized like physics (Collins 2001). At the same time, the pull toward verbalization and toward the formalization of the knowledge produced remains a characteristic feature of scientific research, because it contributes mightily to stabilize knowledge, to its reliability, and to its promised longevity (Volonté 2006). The crucial role of tacit knowledge in a research process matches with the need for the production of explicit knowledge within the GAS.

At this regard, even if a long way has already been covered, the design research community still needs to take some steps towards a fully shared definition not just of its proper modes of inquiry, but also of its proper modes of advancing knowledge claims. In general, there is a shared agreement on the need for research through design outcomes to be accompanied and wrapped up by written text, articles, dissertations, and the like (Bonsiepe 2007; Findeli 1998; Schneider 2007). However, for instance, there is still a debate on the hierarchy between images and text. Indeed, while visual supports are often considered a quite poor mode of capturing the details of artifacts, i.e. their dynamic and tactile qualities (Jarvis et al. 2012), and in the majority of design conferences formats regard written text as more important than images, a growing number of authors consider visual supports more important than verbal explanations (Bardzell et al. 2015; Bowers 2012; Durrant et al. 2015; Höök and Löwgren 2012).

What emerges clearly by this debate is that the research through design process is typically characterized by two different forms of output (Zimmerman et al. 2007, 497): "a concrete problem framing" and "a series of artifacts – models, prototypes, products" as well as interfaces, drawings, storyboards, videos etc. These two kinds of output are intrinsically correlated; for, together, they represent two complementary facets of the same knowledge outcome.

Zimmerman, Forlizzi, and Evenson's "problem framing" is the knowledge the researcher took pains to make explicit. Zimmerman et al. 2010, 313) clarify that such knowledge may become concrete in a "theory on design" or, more often, in a "theory for design" in the form of implications, of guidelines, of "design specifications for future products" (Frens 2007, 140) - which can be objects, interfaces, graphics, or services - or of new design processes and methods. But this research process's peculiarity lies in the fact that the designer-researcher's knowledge claims are made possible and validated by the existence of an artifact that embodies and exemplifies them (Mareis 2012, 67). Thus, the generated knowledge bases itself on producing "design exemplars" (Zimmerman and Forlizzi 2008) or "epistemic objects" (Mareis 2012), in the form of artifacts that "can be seen as the solid form of knowledge to be disseminated" (Bang et al. 2012, 7).

In brief, in research through design, artifacts can be seen as research tools useful for: (i) investigating future scenarios; (ii) developing new theories; (iii) embodying and exemplifying new pieces of knowledge.

7 The Dual Form of Design Knowledge

Design can produce explicit and implicit knowledge claims. It turns to academic design research when such claims join the dynamics of the GAS, thus when they are admitted to it because considered reliable enough.

Although the two products of research through design (the artifact and explicit knowledge) may even meet different fates once they have been admitted to such dynamics, they must be admitted together. For any explicit claim of knowledge that is not supported and demonstrated by the existence of a design outcome would not be a product of research through design. At the same time, to qualify as academic design research, the design output must be accompanied by the designer's reflection on his or her work (Bonsiepe 2007, 29) and a statement of the results of that reflection that can be re-used by someone else. Once such knowledge has been admitted to a kind of GAS, other researchers will have access to it not only by following Ariadne's clew of verbal claims and citations, but also through a 'knowledge bearer' in a class of its own: the artifact, whose existence is witnessed through images (in printed publications), by models or even directly (in conferences).

In paradigmatic terms, this multiplicity of research outputs is found, for example, in a research through design project carried out at ID-Studio Lab at Delft University, according to Keller's report (2007, 130): "The knowledge generated has been disseminated both through the regular process of scientific publication, and through the creation of working prototypes. The prototypes in this project serve the purpose of demonstrating the application of knowledge, but also as the generators of knowledge themselves." Furthermore, this visual (and possibly auditory and tactile) channel has the advantage of encouraging knowledge transfer (which is generally rather problematic) from academic design research to the design profession (Zimmerman and Forlizzi 2008, 4).

To take into consideration the typical dual form of research through design, already in 1997 the Sheffield Hallam University (UK) renewed its PhD regulations, allowing a more open definition of the thesis: "one in which artefacts are pre-eminent as the main evidence of investigation and outcomes and there is sufficient text to ensure that artefacts communicate appropriately" (Rust et al. 2000, 365).

Recent years have seen a growing interest in the design-research world towards the establishment of a designerly way to fill up the GAS, both with the rise of journals dedicated to the findings of research through design (the *International Journal of Design*, for example, cf. Chen 2007) and with changes in the structure of design conferences. One example that may stand for all is the establishment of the Research Through Design conference: the group of scholars chairing the two first editions (Wallace et al. 2013; RTD 2015) made a valuable attempt to create a balanced mix among constructive design critique, structured peer-review-process and design exhibition (Durrant et al. 2015). Another recent development that is worth mentioning is the introduction of a new Pictorials Track within the DIS (Designing Interactive Systems) conference, since 2014. The most interesting aspect here is that the submitted "pictorial" essays undergo a peer-review process modeled on the one already employed for "non-pictorial" papers. Therefore, even if in pictorial papers the hierarchy between written text and figures is reversed (Jarvis et al. 2012; Bardzell et al. 2015), still the authors are required to make an explicit claim of knowledge. As a matter of fact, without that claim, the gatekeepers would not have the tools required to judge whether said outcome, as the product of a research process, may actually be admitted to the GAS.

This ushers in a final issue we must deal with: By what criteria are the dual results of research through design to be deemed worthy or unworthy of admittance to the GAS?

8 Academic Design Research Assessment

Mattelmäki and Matthews' (2009, 3–4) state that "there is nothing like a research-through-design method (in the way that the scientific method has been idealised in popular culture). Rather, there are many ways that design methods, processes and products have been fundamental to (but not sufficient for) making a research contribution." We agree with this view and add that, for the reasons limned so far, what turns a design contribution into a "research contribution" is not reference to codified methods but inclusion in a GAS. From what, then, are we to recognize that this happens? What fosters inclusion in a GAS? What are the criteria and the parameters for orienting work that aspires to be academic design research?

Several authors have already taken up the issue of design research assessment. From our standpoint, one especially interesting contribution comes from Zimmerman et al. 2007, 499–500), who propose evaluating research through design contributions by referring to four criteria: process, invention, relevance, and extensibility. To conclude our reasoning and to make it more concrete, we will analyze these four criteria in detail, both vis-à-vis what we have argued up to now and vis-à-vis a number of research through design examples available in literature.[7] A similar examination could be extended to any further suggested criteria of design research assessment.

To begin with, Zimmerman, Forlizzi, and Evenson rightly believe it necessary for the designer-researcher to recount the process through which certain knowledge was generated, the reasons for choosing to employ certain methods, and the rigor they were applied with. Although we have observed that methods are not what

[7] Specifically, eight complete examples are considered (where by complete we mean that the research through design is described in an article or paper whose authors conducted the research): Lambourne et al. (1997), Petersen et al. (2004), Keller (2007), Frens (2007), Ross and Wensveen (2010), Andersen et al. (2011), Hobye and Löwgren (2011), Visser et al. (2011). Moreover we consider 21 examples mentioned in texts devoted more generally to the issue of research through design (one from Keyson and Bruns 2009; six from Bang et al. 2012; three from Binder and Redström 2006; three from Zimmerman et al. 2007; three from Zimmerman et al. 2010; five from Mattelmäki and Matthews 2009). In the latter group, the authors of the article often do not coincide with the research-project leaders.

primarily determines the soundness of research in general (and research through design in particular), describing the process through which certain knowledge was generated is a sine qua non for lending the claim reliability in the eyes of peers.[8] It must, however, be added that the two products of a given piece of research through design – the explicit claim of knowledge and the outcome of the designing – are produced by two different processes, which we believe need to be distinguished. The first is produced by the process of inquiry outright, which sets out from a research question for whose answer the researcher seeks to define an appropriate research strategy. To lend reliability to the knowledge claim, said process has to be documented and described in detail. The design outcome, be it a concept or a prototype, is brought forth, on the other hand, from a designing process that, as such, employs not the methods of inquiry but the methods of design-making. For the purposes of academic design research, this second process needs not necessarily to be described in detail. In this regard, Mattelmäki and Matthews (2009) refer to an example focused on designing interactive tiles for a playground and making observations of such devices while in use by children (Matthews et al. 2008). This observation enabled the researchers to contribute to the production of knowledge on issues such as how the designer can create new forms of interaction. In this example, indeed, "the process of the design of the tiles is entirely absent to the account (and inconsequential to the argument), since products-in-use are all that is required to make the points they argue." (Mattelmäki and Matthews 2009, 5).

The degree of detail used to describe the design process depends on the research question. In principle, if it is about analyzing specific product features and the user's interaction with the product, the description of the design process could be omitted. Thus, in most of the cases we analyzed, the design process was described generically, if at all. As an example, Hobye and Löwgren (2011) focus on users' behavior in reaction to a new kind of interactive product. Accordingly, though the prototype is described in detail so readers understand what is involved, the design process that went into it is not discussed. On the other hand, the process of investigating how the prototype was used in a real-world context is fully reported, as it should be. When the research question is about the designing process and the tools employed therein, however, describing that process becomes obviously indispensable. An example is given by the ViP (Vision in Product design) methodology developed by Paul Hekkert. The basic assumptions of this methodology were tested in an experiment described by Snoek and Hekkert (1999). In such description, a great deal of attention is devoted to illustrating the different steps of the design process (pp. 170–171). Jarvis et al. (2012) make an interesting point on this issue when they argue that the traditional format of academic design papers devotes little space to the documentation of the design process. According to them, this is a serious shortcoming, since the design process takes the majority of the time of the overall research process and generates myriads of insights and lessons. In order to fully share these lessons,

[8] It is worth noting that in two of the eight complete examples we analyzed, there is no systematic description of the research process. This shows how heterogeneous the research-through-design phenomenon is.

they argue the importance – in research through design – of paying attention to all the details of the design process, by documenting it through pictures in the format of a photo essay.

Zimmerman, Forlizzi, and Evenson's second criterion is that the research's output be significant invention. This is also a necessary feature for selection inside the GAS, because no one will rely on the uncertain durability of knowledge claims that do not seem to add anything to other claims that have already shown they can stay on the surface of the GAS. Therefore, also gatekeepers (e.g. reviewers) are likely to reject a knoweldge claim that merely say the same thing as already existing claims. This leaves moot the question of what means 'significant' in Zimmerman, Forlizzi, and Evenson's thesis. The way the GAS works shows there is no such thing as objective significance of knowledge claims. Ultimately, it is once again the fluid dynamics of the GAS that determines the significance of an innovation, though that, as we have seen, produces upstream constraints in how the claim is formulated by the author and received by the peer reviewers.

We consider this criterion especially useful when applied to an artifact produced by research through design, because it means that the researcher-designer has to make an argument why the artifact advances the current state of the art in the research community by specifying what design issue was dealt with and how it was resolved. This is an essential step because, as Stappers (2007, 87) states, artifacts "embody solutions, but the problems they solve may not be recognized." Therefore, it is important that the researcher explicitly declare the claimed advance both vis-à-vis solutions already on the market and vis-à-vis solutions that might be proposed by other research. This is prerequisite to allowing the gatekeepers of the GAS, first, and other researchers, hence, to gauge the relative significance of the invention and thus the foreseeable duration of the knowledge claim.

The above naturally holds true if the research through design focuses on the design outcome. When, on the other hand, it focuses on establishing new tools for designing (i.e. *how* rather than *what* the designer designs), the 'invention' will amount to a theoretical contribution or a proposal of new design methods. In this regard, we may refer to the newness of the knowledge contribution, just as in other disciplines. Mattelmäki and Matthews (2009, 4) cite a research project whose aim was to evaluate the various facets of design probes as a tool: "The role of designing in the cases appeared in the customising of the probes experiments, in designing the probes kits and 'communicational' artefacts as design objects. Design skills were also applied in seeking [...] design opportunities and elaborating alternative solutions in several cases (however these results were not well described in the research)." In such a case, the design outcome may actually not represent a significant invention vis-à-vis the state of the art but nevertheless prove useful for explaining how certain solutions emerged through specific design processes or through reliance on particular design tools and methods.

By the same token, generating design guidelines or design insights about, say, the user experience of a new prototype yielded by research through design might constitute the trait of invention. Such research thus focuses more on the knowledge generated than on the artifact that embodies it. In this case, too, the prototype might not be a significant invention, perhaps limiting itself to modifying some features of an existing product (e.g. its shape or color) to test its overall effect on users, for example with the aim of exploring how changes in product shape can affect perception of the product (Blijlevens et al. 2013) or understanding how different visual textures alter users' experience of an object (Fenko et al. 2011). As said, in all these cases the fluid dynamics of the GAS will determine what is significant.

The third criterion is relevance, i.e. the impact a given design outcome can be expected to have on the world. In the context of traditional scientific research, the foreseen impact of certain outcomes has increasingly become a key factor in distributing resources (research grants etc.) in recent decades. That has gone hand-in-hand with the evolution of scientific research toward more entrepreneurial forms in what Etzkowitz (1990) terms the "second academic revolution." Though this phenomenon is beyond the scope of this article, it bears noting here that the relevance of a knowledge claim is certainly a factor favoring selection in the GAS, if by "relevance" we mean its importance to the knowledge that interests a research community. No matter how clever or innovative a discovery, it is fated to lie dormant in the silt of the GAS if no one takes an interest in it.

Applying relevance to research through design means that the design process should preferably deal with a current problem by envisioning a preferred state of affairs. The design outcome and the explicit knowledge generated ought thus to meet a real need on the part of society, users, companies, designers or some such target. This is the reason why a lot of design research tackles issues universally considered "socially relevant," such as environmental sustainability or healthcare. Among the cases we analyzed, for instance, Andersen et al. (2011) present a new design concept within myRecord, an already existing web-based prototype of a patient-centric health record. The prototype of the new concept, was tested with three patients in cardiac healthcare, being at the same time "a solution to be evaluated and a research tool to generate new questions" (p. 8).

The fourth and last criterion applied by Zimmerman, Forlizzi, and Evenson is extensibility, which is to say the chance that others may use the knowledge outcome of the research. This lies at the core of the GAS. We may add that extensibility can take place only if the knowledge output has been properly conveyed as an explicit claim of knowledge. This is the prerequisite for research results, once they have been admitted to the fluid dynamics of the GAS, to be employed by others. For instance, the results of the "Static!" project (Backlund et al. 2007), consisting of a number of design concepts and prototypes aimed at giving feedback about energy consumption by a more sensory and less data-centered approach, have been taken as inspiration and reference by many other studies. Indeed, the knowledge they carry, i.e. the possibility to give eco-feedback in a different and more engaging way, has opened up new scenarios and new research directions. Just to cite an example, Strengers (2011) states how "inspiration can be taken from the Static! project for less data-oriented forms of feedback".

Several other examples could be referenced here. Stolterman and Wiberg (2010), for instance, mention three concept-driven design research projects that produced extensible knowledge for the design community: the DynaBook, the ActiveBadges and the Brick system. According to the authors, "researchers have regarded these three examples as seminal concept designs in the field, and they have been heavily cited. We see the success of these examples as a sign that these concept designs have had scholarly influence. According to Google Scholar, 528 papers have cited the DynaBook concept, 974 have cited the original paper describing ActiveBadges, and 297 cited the Bricks system. Even more important, researchers have also frequently cited follow-up projects built upon the same basic concept design" (Stolterman and Wiberg 2010, 108)[9]. We argued that remaining afloat on the surface of the GAS thanks to a vast number of second-order claims is the condition in order to stabilize knowledge claims. The extensibility of results, which partly depends on their being distributed through academic research channels, is a determining factor for research output to be deemed 'knowledge.'

9 Conclusions

We believe design research can be truly specific to the design discipline only when designing is its tool. For this is the platform on which design, while remaining a planning endeavor, may at the same time become an act of producing new knowledge. However, in order to establish its own field of research, design has often run the risk of taking up the research methods of other disciplines, without stopping to consider why such methods are considered 'scientific' there.

To better understand why academic research employs codified methods, we analyzed scientists' practice of publishing the results of their research. This has served to clarify that what determines whether research findings are valid is not the method in itself but the consensus of a community. Equally, what makes a designer's work research is not just having applied codified methods but also participating in the game of the Great Archive of Science (GAS), whose laws know how to select, from countless claims of knowledge, a relatively circumscribed corpus of durable knowledge.

In our view, knowledge claims based on research through design should normally consist of a design outcome (being it a product, a service, a game, an interface, etc.) equipped with an explicit declaration of the features that make it – or the process that brought to its creation, a new, valid, and trustworthy piece of knowledge.

Finally, in order to identify criteria for a design project to legitimately aspire to be deemed design research, we have discussed some suggestions drawn from literature. Criteria acquire relevance in so far as they facilitate the access to the GAS. For instance:

A description of the research process is indispensable to enable the knowledge generated by research through design to be evaluated as for what regards its reliability. A description of the design process, on the other hand, is required only if it is pertinent to the actual answer to the research question.

[9] In July 2017 the ActiveBadges system reached the sizable amount of 5118 citations.

The outputs of the research process should claim and advance of knowledge in order to foster the development of their own relative significance within the dynamics of the GAS. If the research focuses on the design outcome, it will need to bear the hallmarks of invention; if the research focuses on a more general knowledge outcome (for example, establishing new design tools), it will need to bear the hallmarks of newness.

The researcher's knowledge claims need to appear relevant to a community of peers. Naturally, that does not rule out that the research findings may prove interesting or relevant to another audience, as well, for example to other disciplines.

Respecting such criteria will make easier for a claim of knowledge yielded by research through design to be granted admission to the GAS. Its chances of remaining afloat in the fluid dynamics of the GAS will be directly proportional to the extensibility of the findings, i.e. to the likelihood that others take up as their own the knowledge claims it avers.

References

Agnew, K. (1993). The spitfire: Legend or history? An argument for a new research culture in design. *Journal of Design History, 6*(2), 121–130.

Andersen, T., Halse, J., & Moll, J. (2011). Design interventions as multiple becomings of healthcare. In I. Koskinen, T. Härkäsalmi, R. Mazé, B. Matthews, & J. Lee (Eds.), *Proceedings of the Nordes '11: The 4th Nordic design research conference* (pp. 11–20). Helsinki: School of Art and Design, Aalto University.

Archer, B. (1981). A view of the nature of design research. In R. Jacques & J. A. Powell (Eds.), *Design: Science: Method* (pp. 30–47). Guildford: Westbury House.

Archer, B. (1995). The nature of research. *Co-design: Interdisciplinary Journal of Design, January*, 6–13. (republished in Grand and Jonas 2012, pp. 109–122).

Backlund, S., Gyllenswärd, M., Gustafsson, A., Ilstedt Hjelm, S., Mazé, R., & Redström, J. (2007). Static! The aesthetics of energy in everyday things. In *Proceedings of design research society wonderground international conference 2006*. http://soda.swedish-ict.se/2608/1/WonderSTATIC.pdf

Bang, A. L., Krogh, P. G., Ludvigsen, M., & Markussen, T. (2012). The role of hypothesis in constructive design research. In *The art of research 2012: Making, reflecting and understanding*. Helsinki: Aalto University School of Arts, Design and Architecture. http://designresearch.aalto.fi/events/aor2012/download_content/selected_papers/anne_louise_bang.pdf. Accessed 30 Sept 2013.

Bardzell, J., Bardzell, S., & Koefoed Hansen, L. (2015). Immodest proposals: Research through design and knowledge. In *Proceedings of the 33rd annual ACM conference on human factors in computing systems* (pp. 2093–2102). New York: ACM.

Barnes, B. (1974). *Scientific knowledge and sociological theory*. London: Routledge and Kegan Paul.

Bernal, J. D. (1939). *The social function of science*. New York: Macmillan.

Binder, T., & Redström, J. (2006). *Exemplary design research*. Paper presented at the DRS Wonderground conference, Design Research Society, November 1–4.

Blijlevens, J., Mugge, R., Ye, P., & Schoormans, J. P. L. (2013). The influence of product exposure on trendiness and aesthetic appraisal. *International Journal of Design, 7*(1), 55–67.

Bonsiepe, G. (2007). The uneasy relationship between design and design research. In R. Michel (Ed.), *Design research now: Essays and selected projects* (pp. 25–39). Basel: Birkhäuser.

Bourdieu, P. (2004). *Science of science and reflexivity*. Chicago: University of Chicago Press.
Bowers, J. (2012). The logic of annotated portfolios: Communicating the value of 'research through design'. In *Proceedings of the ACM designing interactive systems conference 2012, DIS2012, 11–15* (pp. 68–77). New York: ACM.
Brandt, E., & Binder, T. (2007). Experimental design research: Genealogy, intervention, argument. In *Proceedings of the International Association of Societies of design research conference*, IaSDR07, Hong Kong, November 12–15.
Brown, T. (2009). *Change by design*. New York: HarperCollins.
Buchanan, R. (2001). Design research and the new learning. *Design Issues, 17*(4), 3–23.
Chen, L. (2007). International journal of design: A step forward. *International Journal of Design, 1*(1), 1–2.
Chubin, D. E., & Hackett, E. J. (1990). *Peerless science: Peer review and U.S. science policy*. Albany: State University of New York Press.
Cole, J., & Cole, S. (1981). *Peer review in the National Science Foundation: Phase two of a study*. Washington, DC: National Academy Press.
Cole, S., Rubin, L., & Cole, J. (1978). *Peer review in the National Science Foundation: Phase one of a study*. Washington, DC: National Academy Press.
Collins, H. (2001). What is tacit knowledge? In K. Knorr Cetina, T. R. Schatzki, & E. von Savigny (Eds.), *The practice turn in contemporary theory* (pp. 107–119). London: Routledge.
Cross, N. (2001). Designerly ways of knowing: Design discipline versus design science. *Design Issues, 17*(3), 49–55.
Cross, N. (2007). *Designerly ways of knowing*. Basel: Birkhäuser.
Crouch, C., & Pearce, J. (2012). *Doing research in design*. London: Berg.
Dorst, K. (2008). Design research: A revolution-waiting-to-happen. *Design Studies, 29*, 4–11.
Durrant, A., Vines, J., Wallace, J., & Yee, J. (2015). Developing a dialogical platform for disseminating research through design. *Constructivist Foundations, 11*(1), 8–21.
Etzkowitz, H. (1990). The second academic revolution: The role of the research university in economic development. In S. E. Cozzens, P. Healey, A. Rip, & J. Ziman (Eds.), *The research system in transition* (pp. 109–124). Dordrecht: Springer.
Fenko, A., Schifferstein, H. N. J., & Hekkert, P. (2011). Noisy products: Does appearance matter? *International Journal of Design, 5*(3), 77–87.
Findeli, A. (1998). A quest for credibility: Doctoral education and research in design at the university of Montreal. In R. Buchanan, D. L. J. Doorden, & V. Margolin (Eds.), *Doctoral education in design: Proceedings of the Ohio conference* (pp. 99–116). Pittsburgh: Carnegie Mellon.
Findeli, A., Brouillet, D., Martins, S., Moineau, C., & Tarrago, R. (2008). Research through design and transdisciplinarity: A tentative contribution to the methodology of design research. In *"Focused" – Current design research project and methods* (pp. 67–91.) http://5-10-20.ch/~sdn/SDN08_pdf_conference%20papers/04_Findeli.pdf. Accessed 23 May 2015.
Frayling, C. (1993). Research in art and design. *Royal College of Art Research Papers, 1*(1), 1–5.
Frens, J. (2007). Research through design: A camera case study. In R. Michel (Ed.), *Design research now: Essays and selected projects* (pp. 135–155). Basel: Birkhäuser.
Friedman, K. (2001). Creating design knowledge: From research into practice. In *Design and technology educational research and development: The emerging international research agenda* (pp. 31–69). Loughborough: Loughborough University Department of Design and Technology.
Friedman, K. (2003). Theory construction in design research: Criteria: Approaches, and methods. *Design Studies, 24*, 507–522.
Grand, S., & Jonas, W. (Eds.). (2012). *Mapping design research*. Basel: Birkhäuser.
Hallnäs, L., & Redström, J. (2006). *Interaction design foundations, experiments*. Borås: University College of Borås.
Hekkert, P. P. M., Keyson, D. V., Overbeeke, C. J., & Stappers, P. J. (2000). The Delft ID StudioLab. Research through and for design. In H. Achten, B. de Vries, & J. M. Hennessey (Eds.), *Design research in the Netherlands 2000* (pp. 95–103.) http://studiolab.ide.tudelft.nl/studiolab/wp-content/uploads/2014/05/2000+StudioLabHKOS.pdf. Accessed 8 Apr 2016.

Hobye, M., & Löwgren, J. (2011). Touching a stranger: Designing for engaging experience in embodied interaction. *International Journal of Design, 5*(3), 31–48.
Höök, K., & Löwgren, J. (2012). Strong concepts: Intermediate-level knowledge in interaction design research. *ACM Transactions on Computer-Human Interaction (TOCHI), 19*(3), 23.
Jarvis, N., Cameron, D., & Boucher, A. (2012). Attention to detail: Annotations of a design process. In *Proceedings of the 7th Nordic conference on human-computer interaction: Making sense through design* (pp. 11–20). New York: ACM.
Jonas, W. (2004). Forschung durch design. In Swiss Design Network (Ed.), *Erstes design forschungssymposium* (pp. 26–33). Basel: Steudler Press. (in German).
Jonas, W. (2012). Exploring the swampy ground. In S. Grand & W. Jonas (Eds.), *Mapping design research* (pp. 11–42). Basel: Birkhäuser.
Jonas, W. (2015). A cybernetic model of design research: Towards a trans-domain of knowing. In P. A. Rodgers & J. Yee (Eds.), *The Routledge companion to design research* (pp. 23–37). London/New York: Routledge.
Keller, I. (2007). For inspiration only. In R. Michel (Ed.), *Design research now: Essays and selected projects* (pp. 119–132). Basel: Birkhäuser.
Kelly, J., & Wensveen, S. A. G. (2014). Designing to bring the field to the showroom through open-ended provocation. *International Journal of Design, 8*(2), 71–85.
Keyson, D. V., & Bruns, M. (2009). Empirical research through design. In *Proceedings of IASDR '09* (pp. 4548–4557). Seoul: Design Research Society.
Knorr-Cetina, K. (1981). *The manufacture of knowledge. An essay on the constructivist and contextual nature of science.* Oxford: Pergamon Press.
Koskinen, I., Zimmerman, J., Binder, T., Redström, J., & Wensveen, S. (2011). *Design research through practice: From the lab, field and showroom.* Waltham: Morgan Kaufmann.
Krippendorff, K. (1995). Redesigning design: An invitation to a responsible future. In P. Tahkokallio & S. Vihma (Eds.), *Design: Pleasure or responsibility?* Helsinki: University of Art and Design. Reprinted by University of Pennsylvania Annenberg School for Communication Departmental Papers (ASC). http://repository.upenn.edu/ascpapers/46. Accessed 30 Sept 2013.
Krippendorff, K. (2006). *The semantic turn: A new foundation for design.* Boca Raton: Taylor & Francis.
Krippendorff, K. (2007). Design research, an oxymoron? In R. Michel (Ed.), *Design research now: Essays and selected projects* (pp. 67–80). Basel: Birkhäuser.
Kuhn, T. S. (1962). *The structure of scientific revolutions.* Chicago: University of Chicago Press.
Lambourne, R., Feiz, K., & Rigot, B. (1997). Social trends and product opportunities. In *Proceedings of CHI 97* (pp. 494–502). Eindhoven: Philips Corporate Design.
Latour, B. (1987). *Science in action: How to follow scientists and engineers through society.* Milton: Open University Press.
Latour, B. (2005). *Reassembling the social: An introduction to actor-network-theory.* Oxford: Oxford University Press.
Latour, B., & Woolgar, S. (1986). *Laboratory life: The construction of scientific facts.* Princeton: Princeton University Press.
Laurel, B. (2003). *Design research: Methods and perspectives.* Cambridge, MA: MIT Press.
Lynch, M. (1993). *Scientific practice and ordinary action: Ethnomethodology and social studies of science.* Cambridge: Cambridge University Press.
Mareis, C. (2012). The epistemology of the unspoken: On the concept of tacit knowledge in contemporary design research. *Design Issues, 28*(2), 61–71.
Matthews, B., Stienstra, M., & Djajadiningrat, T. (2008). Emergent interaction: Creating spaces for play. *Design Issues, 24*(3), 58–71.
Margolin, V. (2010). Doctoral education in design: Problems and prospects. *Design Issues, 26*(3), 70–78.
Mattelmäki, T. & Matthews, B. (2009). *Peeling apples: Prototyping design experiments as research.* Paper presented at the Nordic design research conference 2009 – Engaging artifacts, Oslo.

Merton, R. K. (1957). *Social theory and social structure: Revised and enlarged edition*. New York: Free Press. (first edition 1949).
Olsen, S. L. (2002). Observation of large *CP* violation in the *B*-meson system. In J. Lee-Franzini, P. Franzini, & F. Bossi (Eds.), *Lepton-photon 01* (pp. 4–13). Singapore: World Scientific.
Petersen, M. G., Iversen, O. S., Krogh, P. G., & Ludvigsen, M. (2004). Aesthetic interaction: A pragmatist's aesthetics of interactive systems. In *Proceedings of the 5th conference on designing interactive systems: Processes, practices, methods, and techniques (DIS'04)* (pp. 269–276). New York: ACM.
Polanyi, M. (1958). *Personal knowledge*. London: Routledge.
Povh, B., Rith, K., Scholz, C., & Zetsche, F. (1993). *Teilchen und Kerne. Eine Einführung in die physikalischen Konzepte*. Berlin/Heidelberg: Springer.
Rampino, L., & Colombo, S. (2012). Toward a taxonomy of design-research methods. In L. Rampino (Ed.), *Design research: Between scientific method and project praxis* (pp. 83–94). Milan: FrancoAngeli.
Rodgers, P. A., & Yee, J. (2015). Introduction. In P. A. Rodgers & J. Yee (Eds.), *The Routledge companion to design research* (pp. 1–6). London/New York: Routledge.
Roedl, D. J., & Stolterman, E. (2013). Design research at CHI and its applicability to design practice. In *Proceedings of the 2013 ACM annual conference on human factors in computing systems* (pp. 1951–1954). New York: ACM.
Ross, P. R., & Wensveen, S. A. G. (2010). Designing behavior in interaction: Using aesthetic experience as a mechanism for design. *International Journal of Design, 4*(2), 3–13.
RTD. (2015). 21st century makers and matrerialities. In *Proceedings of the 2nd biennal research through design conference*. http://www.researchthroughdesign.org/2015/proceedings.html. Accessed 15 Jan 2018.
Rust, C. (2007). How artistic inquiry can inform interdisciplinary research. *International Journal of Design, 1*(3), 69–76.
Rust, C., Roddis, J., & Chamberlain, P. (2000). A practice-centered approach to research in industrial design. In S. Pizzocaro, A. Arruda, & D. de Moraes (Eds.), *Design plus research: Proceedings of the Politecnico di Milano conference, May 18–20, 2000* (pp. 358–365).
Schneider, B. (2004). Design Forscht. In Swiss Design Network (Ed.), *Erstes design forschungssymposium* (pp. 4–13). Basel: Steudler Press. (in German).
Schneider, B. (2007). Design as practice, science and research. In R. Michel (Ed.), *Design research now: Essays and selected projects* (pp. 207–218). Basel: Birkhäuser.
Scrivener, S. (2002). The art object does not embody a form of knowledge. *Working papers in art and design 2*. http://sitem.herts.ac.uk/artdes_research/papers/wpades/vol2/scrivenerfull.html. ISSN 1466-4917. Accessed 7 July 2013.
Seago, A., & Dunne, A. (1999). New methodologies in art and design research: The object as discourse. *Design Issues, 15*(2), 11–17.
Simon, H. A. (1969). *The sciences of the artificial*. Cambridge, MA: MIT Press.
Snoek, H., & Hekkert, P. (1999). Directing designers towards innovative solutions. In B. Jerrard, M. Trueman, & R. Newport (Eds.), *Managing new product innovation* (pp. 167–180). London: Taylor & Francis.
Stappers, P. J. (2007). Doing design as a part of doing research. In R. Michel (Ed.), *Design research now: Essays and selected projects* (pp. 81–97). Basel: Birkhäuser.
Stolterman, E., & Wiberg, M. (2010). Concept-driven interaction design research. *Human Computer Interaction, 25*(2), 95–118.
Strengers, Y. A. (2011). Designing eco-feedback systems for everyday life. In *Proceedings of the SIGCHI conference on human factors in computing systems* (pp. 2135–2144). New York: ACM.
Vallgårda, A. (2009). *Computational composites. Understanding the materiality of computational technology*. Manuscript for Ph.D. dissertation submitted to the IT University of Copenhagen. http://citeseerx.ist.psu.edu/viewdoc/download?doi=10.1.1.369.8406&rep=rep1&type=pdf. Accessed 13 Nov 2015.

Van Campenhout, L. D. E., Frens, J. W., Overbeeke, C. J., Standaert, A., & Peremans, H. (2013). Physical interaction in a dematerialized world. *International Journal of Design, 7*(1), 1–18.

Vaughan, L. (2017). *Practice-based design research*. London: Bloomsbury.

Visser, T., Vastenburg, M. H., & Keyson, D. V. (2011). Designing to support social connectedness: The case of snowglobe. *International Journal of Design, 5*(3), 129–142.

Volonté, P. (2006). The social context of scientific knowledge production and the problem of demarcation. *Pragmatics and Cognition, 14*, 527–568.

Volonté, P. (2012). The GAS tank: Remarks on the real scientific method. In L. Rampino (Ed.), *Design research: Between scientific method and project praxis* (pp. 95–108). Milan: Franco Angeli.

Wallace, J., Yee, J., & Durrant, A. (2013). *Praxis + poetics. research through design 2013 conference proceedings*. Newcastle Upon Tyne: Northumbria University.

White, D. M. (1950). The 'gate-keeper': A case study in the selection of news. *Journalism Quarterly, 27*, 383–390.

Ziman, J. (1978). *Reliable knowledge: An exploration of the grounds for belief in science*. Cambridge: Cambridge University Press.

Ziman, J. (2000). *Real science: What it is and what it means*. Cambridge: Cambridge University Press.

Zimmerman, J., & Forlizzi, J. (2008). *The role of design artifacts in design theory construction*. Human computer interaction institute, Paper 37, http://repository.cmu.edu/hcii/37/. Accessed 31 July 2013.

Zimmerman, J., Forlizzi, J., & Evenson, S. (2007). Research through design as a method for interaction design research in HCI. In *CHI '07 Proceedings of the SIGCHI conference on human factors in computing systems* (pp. 493–502). https://doi.org/10.1145/1240624.1240704.

Zimmerman, J., Stolterman, E., & Forlizzi, J. (2010). An analysis and critique of research through design: Towards a formalization of a research approach. In *Proceedings of the 8th ACM conference on designing interactive systems* (pp. 310–319). https://doi.org/10.1145/1858171.1858228.

Design Research as a *Meta-discipline*

Anne Caplan

Abstract This essay is an attempt to enrich the discourse on an epistemological research through design based on Henri Lefebvre's *meta-philosophy* (Paris 1965). The *meta-philosophy* refers back to the human daily routine and the poetic exploration of practice as a premise for philosophical tradition. I use Lefebvre's thoughts in their relation of theory and empirical knowledge, to build a bridge to Design Sciences as a practice-oriented research. To underline the basic approach of a Design Science based on practice, the analysis is developed along a specific design project that operates with a "research through design" Frayling (1993/94). This allows to show, how concept- and theory development interlink with the designing practice of the researcher himself. The essay tries to develop a Design Science, which includes the subjective perspective of the designer himself as well as the phenomenological character of design processes.

Keywords Research through design · Participatory design · Artistic research · *Meta-philosophy*

1 Design- and Research Definition

In 1977 the artist Daniel Spoerri developed his exhibition-concept *Musée Sentimental*. His conception centers on every-day-objects and their stories. Not until the process of usage defining the mutual relationship between man and objects, an object becomes a memorable exhibition piece.

> "As the objects presented by the Musée differ from art stereotypes by corresponding to the 'used' and the 'common', a new democratic understanding of art evolves, turning every-day-life itself into the art-space." (Caplan 2015: 9)

The paper (including German quotes) was translated into English by Prof. Marion Digel and Simon Wolberg.

A. Caplan (✉)
Seminar für Volkskunde/Europäische Ethnologie, Westfälische Wilhelms-Universität, Münster, Germany
e-mail: anne.caplan@uni-muenster.de

© Springer International Publishing AG 2018

P. E. Vermaas, S. Vial (eds.), *Advancements in the Philosophy of Design*, Design Research Foundations, https://doi.org/10.1007/978-3-319-73302-9_16

Through the selection of the objects, the artist (Spoerri) succeeded in creating an urban typology. For instance, the *Musées Sentimentaux* reflected very specific urban spaces like Paris or Cologne. The conglomerate of sentimental connections to objects turned into an allegory of the individual, specific nature of the city and its living conditions.

By integrating fakes into his catalogue of objects with regard to their subjective, emotional approach, Spoerri stresses a process oriented and variable definition of 'Heimat' or home. In this respect, Spoerri's *Musée Sentimental* succeeds common exhibition conceptions. Moreover his conception can be understood as an artistic methodology bearing the potential to produce definitions of 'Heimat'.

As an artistic programmatic of how a design process can exemplify the reception of the urban 'Heimat', I adapted a design project in 2013/14 a so-called Folkwang *LAB,*[1] its results, forming the foundation for my analysis, will receive further explanation into this article. During the project, design students worked within interdisciplinary teams experiencing/testing the methods of Spoerri's *Musée Sentimental.* To do this, they questioned the residents of the Ruhr Area about their individual places of desire which allegorically formed their definition of 'Heimat'.[2] This selection of places with their every-day-stories becomes a characteristic through various urban visualizations of 'Heimat' in the Ruhr Area. This not only proves that the identity of the Ruhr Area is not limited to common clichés like coal, steel and soccer, but also that the narrative adapted by Spoerri makes it possible to reveal realities hidden behind collective clichés. Narration turns into a productive mode for presentation. For Spoerri the term 'place of desire' does not necessarily have to be understood in its spatial connotation, but it can also be expressed through an object or sensual stimulation. Essential is the distinctive perceived quality of a specific space. To make these qualities visible and tangible, the results of the design process were mostly Non-Products,[3] casually turning the attention to the parenthetical every-day incident, focusing on the emotional and sensual.

The *Musée Sentimental* enables a diverse understanding of the urban, applying to distinct reception of objects. The border between reality and fiction blends by integrating individual memories of the recipient into the process of understanding.[4]

---

[1] "The interdisciplinary Folkwang idea and the active, transdisciplinary exchange are fostered with particular emphasis in the Folkwang LABs. A Folkwang LAB is based on a general topic of social relevance, bringing together teachers and students of various faculties and disciplines. LABs are characterized by their experimental, research-oriented, artistic as well as practical components. LABs are integral components of the study courses." (see Folkwang University of the Arts 2016)

[2] Additionally the participating residents were given practical tasks by the designers, for example they had to photograph their sentimental everyday-locations using a one-shot-camera. Throughout the whole design-process, there was an exchange between them and the designers on interim results and their experiences.

[3] "Of course there are other forms of Non-Product-Design, which don't materialize in fabricated or otherwise produced objects, but which can be defined as a design of relationships. Social, urban, medial processes, are also designed, even if they don't result in materialized objects or products. That is a form of design, which produces Non-Products." (Borries and Fezer 2013a: 44 f.)

[4] "An innovative dimension of this subjective approach to research lies in its capacity to bring into view, particularized of lived experience that reflect alternative realities that are either marginalized or not yet recognized in established theory and practice." (Barrett 2007: 143)

The democratization of Art therefore takes place by directly integrating the implicit knowledge of the visitors of the exhibition.[5] In the *Musée Sentimental* each visitor develops his or her own vision of the city within the framework of an "experimental historiography" (Brock 2008: 197).

About the design project I would like to state, that

1. we adapted the artistic methods of the *Musée Sentimental* to negotiate theoretical questions about the emotional qualities of 'Heimat' within the design project.
2. we worked process-oriented without a specific design assignment.
3. as in the *Musée Sentimental*, we succeeded in a democratization of the design process through participatory integration of the residents.

For this discourse, we derive at an alternative extended definition of design, which has its

> "[…] root in the social-activist art of the 20th century. Actually Beuys's citation in which he stated that everyone was an artist, is wrong. It should have been: 'Everyone is a designer.' Because it really was all about the active design of our living conditions […]." (Borries and Fezer 2013b: 85)

Social-activist art, as for example Intervention Art, intervenes everyday-life, to ironize and agitate (Borries et al. 2012: 100).[6] Another goal of our design project was to perforate the urban surface by design intervention. Using artistic and design methods, the positions of the participating residents, distilled by the designers, became sensual and tangible esthetic-creative interventions. These were meant to sharpen the awareness and the perception for the city environment. In this context the designers became active initiators who identified their own missions. They intervened and linked people with the living space. For this process the democratization of the urban space is essential and integrates its residents in form of a bottom-up-process. The designer becomes a transmitter, who makes the residents input visible and tangible through subversive, temporary activities. Here design serves to create a connection between man and object and therefore it is not separable from the production process, moreover it is non-committed to an object (Jonas 2002). Beforehand I have described the role of the designer among others as translator, transmitter or as a link. Guy Bonsiepe's "ontological design-diagram" (Bonsiepe 1996: 19) is helpful to explain design understanding in this context. Gui Bonsiepe describes the activity-range of Design, by defining the design discipline as a gateway discipline, which joins tools, users and activities with one another. This leads to the expansion of the interactive design profession:

> "Everyone can become a Designer within his or her discipline. […] An entrepreneur or a manager, who organizes a company, is 'doing' design – without knowing it. A systems

[5] Later on there were also project approaches that involved the public in the selection process of the objects shown in the *Musée Sentimental*. See for that purpose *Museé Sentimental Krems/Stein* (2009) – Eine Stadt biografiert sich selbst!

[6] Friedrich von Borries refers here to Anne Pasternak.

> engineer, who conceives a process to reduce the misdirection of baggage during air-travel, is 'doing' design. A genetic engineer, who develops a new form of extremely resistant variation of wheat, is 'doing' design. Design topics are not limited to material products but they also comprise services. Design is a basic activity with capillary ramification into every form of human activity, so that no profession can claim a monopoly on design." (ibid: 25 f.)

According to Bonsiepes' "ontological design-diagram" (ibid: 19) design is / refers to an interface, which allows for action to take place and improve in any human communicative space. Design creates interaction based on practice. In our design project we developed interaction as a type of infrastructure, which made the articulation of individual conceptions about home possible. In order to do so, we focused on both the designers' and the participating population's practical knowledge, as the various emotional shades about home cannot be subsumed under a single normative concept. As mentioned before, the practical knowledge, or rather the tacit knowledge, is crucial for the *Musée Sentimental* to democratize the design process or to push a vigorous exchange. Approaches of authors, such as Hubert, Stuart Dreyfus or Donald Schön, are particularly suited to exemplify the meaning of the tacit knowledge:

> "The object of their studies are concepts and terms of understanding, such as 'expertise', 'intuition' or 'reflective practice', which are bound by experience and which seem to evade the descriptive competences of explicit rationalism and positivism." (Mareis 2011: 152)

Donald Schön and the Dreyfus brothers address the individual practical knowledge and thus put the relation between theory and practice as well as the relation between knowledge and ability to the test (ibid). The same applies to the *Musée Sentimental* by evolving an "experimental historiography" (Brock 2008: 197) through the mixture of fiction and reality as well as collectivism and subjectivism.

Against the backdrop of a social-activist form of art, the moment of participation is to be found in the relation between the perspective of experts and of users. Claudia Mareis states that an "asymmetrical perspective on design, which classifies designers as 'experts' and users as 'amateurs'" (Mareis 2013: 10), was induced with the rise of a professional design education at the beginning of the twentieth century.

From the late 1950s onwards, the lack of participation opportunities first became obvious in urban design (ibid: 10 f.). During the post-war period in Germany, participation models started to emerge along urban lines of conflict. Sebastian Haumann takes Cologne as an example to illustrate this development (Haumann 2011: 186). In the 1960s, the local association *Bund Deutscher Architekten (BDA)* began to promote citizen participation (ibid: 187). Haumann considered the support of an informed citizenship, or rather the promotion of a qualified formation of opinion, his field of action in order to initiate an active urban development policy and a broad sociopolitical discourse (ibid: 160). Within this process, the designer sees himself as a kind of translator who both structures negotiation processes and establishes an infrastructure between hitherto non-communicative levels, such as politics and citizens (Bonsiepe 1996: 19).

Design processes normally begin with the investigation of the recipients' behavior patterns. These patterns are shaped by the recipient's respective everyday knowledge and they are the starting point for interaction in this project. Elizabeth B.-N. Sanders refers to this design phase as "Fuzzy front end" (Sanders 2013: 68):

> "PD [Participatory Design, *A.C.*] takes place in the front end of the fuzzy front end of the design process where it is often not known whether the deliverable of the design process will be a product, a service, an interface, or something else. Here the goal of the pre-design phases is to find the problems to be solved and identify the opportunities to explore. The goal of the discovery phase is to determine what is to be, or should not be, designed and manufactured." (ibid)

Sanders divides the "Fuzzy front end" into the phases "Pre-Design" (ibid: 69) and "Discover" (ibid). While the development of opportunities takes place in the "Pre-Design", "Discover" distinguishes itself by research and translation of the tacit knowledge into design (ibid). Thus, in our project the phase of living took place during the "Pre-Design", in which the designers were the participating population's guests for one weekend. During this time the designers were not only able to experience everyday life with the participants, but they also found out about the participants' places of longing by interviews and participatory observation. During this phase the designers explored the detected destinations by different techniques, such as 2D mapping, a systematic cataloging of places of longing or by writing memory logs about the experience. In the phase of "Discover" the findings were consolidated by research, an evaluation of photos (which the hosts had taken with disposable cameras), the research of the historical framework and by the theoretical introduction to the methods of the *Musée Sentimental.* On that basis, the designers developed primal artistic translation devices. Within the framework of the "Fuzzy front end's" "Pre-Design" and "Discover" our main focus was to find out more about the emotional relationship between the participants and their destinations:

> "Emotional tools and techniques facilitate people's ability to remember and communicate their thoughts, feelings and aspirations for the future." (ibid: 77)

This became clear during the phase of living in our project, as some of the participants already had the idea in mind to show the designers the most popular sights in Essen long before they commenced the project. The secret places of longing did not come to light before an individual private interaction.

Relating to Sanders I base my theoretical foundations in the practice of design itself, especially at the beginning of a design process. By this, the theoretical reflection leads into the design process itself, to evolve into a thorough saturation of the topic. Theory in this case has a prospective function. This kind of approach not only impedes a superficial 'icing-on-the-cake-design', but it also makes design solutions more connective and inter-subjective. For "research through art and design", theory is part of the practical investigation. In the contrary case, in which theoretical reflection follows and reflects the design solution or product, it resembles an approach of "research for art and design"[7] or "research into art and design"[8] (Frayling 1993/94: 5), which stands above the practice of design. A similar tension between the projective and the reflective in the theory of design can also be perceived in art:

> "Research to the painter, he [Picasso, *A.C.*] said, equals visual intention. He's a maker not a researcher – and he doesn't even feel comfortable verbalizing his work." (ibid: 2)

Of course, at this point I object to this declaration that the designer or artist is a maker but not a researcher. Moreover, I declare that the "making" itself is research that directly enriches the methods of design practice.[9] Since Design articulates itself through design activities, it seems natural to focus on the practical design activity or the capability to design, in order to gather a scientific perception of the discipline. Therefore, in the orientation of "research through art and design" (Frayling 1993/94), my approach integrates design practice into design research (see Steffen 2011),[10] by referring theory and practice reciprocally to each other. If one understands research (in design) as a creative process, then this process is never led merely by the rational.[11] The designer as the operating subject cannot be omitted, since

---

[7] Here the designed product becomes an artifact, expressing the thoughts of the designer (see Frayling 1993/94: 5). This is comparable to that of a piece of art, which has incarnated knowledge that cannot be verbalized and it questions whether design or artistic action would even need a theory. Since I don't view design as art or as an artistic act, "research for art and design" is of no relevance to this paper. As soon as one were to investigate such an artifact theoretically, one would automatically get into a "research for art and design" discourse, which describes and analyzes the products of design similar to the discourse in art history.

[8] This means according to Frayling the historical, esthetic and perception research (Frayling 1993/94: 5).

[9] Considering the artistic research, see the study of the Internationales Forschungszentrum Kulturwissenschaften der Kunstuniversität Linz, e.g. "*Forschende Kunst statt bildende Kunst? Ein Workshop zum Stand einer Diskussion*", 19 March 2014.

[10] Steffen avoids the use of the expression "research through art and design" after Christopher Frayling in her dissertation. According to her, Frayling noted that "research through art and design" would rely upon practice-based research, which may enrich the methods of design-practice directly, however it would miss the reference to the appropriate research documentation. According to Steffen the practice-integrating research in design would have to be distinguished by a practice-referring scientific-theoretical dealing (see Steffen 2011: 107, 109).

[11] This differs from Siegfried Maser, who rationalizes design as science: "The work of the designer – to create products – is first of all a practical activity, consisting of action: The designer has to do something! This activity however – according to the thesis 'Design as Science' – happens on a scientific basis. Action becomes rational activity." (Maser 1976: 40). Maser bases his thesis of rationalizing designs on processes of "scientification" of design, which evolved in the twentieth century with the beginning of *Modernism:* "For example, in the early 1920s, the De Stijl protagonist Theo van Doesburg (1923) expressed this perception of a new spirit in art and design: 'Our epoch is hostile to every subjective speculation in art, science, technology, etc. The new spirit, which already governs almost all modern life, is opposed to animal spontaneity, to nature's domination, to artistic flummery. In order to construct a new object we need a method, that is to say, an objective system'. A little later, the architect Le Corbusier (1929) wrote about the house as an objectively-designed 'machine for living': 'The use of the house consists of a regular sequence of definite functions. The regular sequence of these functions is a traffic phenomenon. To render that

> "[...] the practice of the designer never consists completely of 'rational' action. Even when scientific knowledge permeates the design process and when design is comprehensibly grounded in theoretical research, it cannot be deduced from theories and a task-related framework. Rather design always relies on a synergy between non-scientific knowledge and subjectivity within the design process." (Steffen 2011: 75)

Therefore, in order to develop a science for the design discipline, it would be prerequisite to overcome the culturally conveyed dichotomy of reason and emotion.[12] So, the design in our project formed a synthesis of the fields 'art' and 'science'. This is especially relevant to devising a framework for the design discipline as a science, as it is often determined by "tacit knowing" (Polanyi 1974), which is usually the result of an emotional, personal and subjective motivation (see Barrett 2007: 143). Intuition is not only relevant to the reception of design, but also for its production. A true design science has to be able to include the subjective as well as the incremental of practice, and therefore scientific research in design always has to be of phenomenological nature as phenomenology does not define intuition as an individually conceived inner circumstance separating the subject from its surroundings. Rather, phenomenology makes "bodily sensitivity" and "affective dismay" a respectable part of the reasoning of philosophy. (Andermann and Eberlein 2011: 8).

2 Meta-philosophy

We find first indications of including the subjective in the 1970s in the work of pioneers of the *Design Methods Movement* like Christopher Alexander (1971) and J. Christopher Jones (1977), who criticized the behaviorist attitude of the representatives of a movement rooted solely in objectivity and rationality as well as in the logical regulation/conditioning of everyday phenomena, as a direct inheritance of modernism (Cross 2007: 120). Niklaus Schefer comments on this:

> "I describe this attitude as a rational functionalism. Classical modernism separates head from heart and reason from emotion. This division also marks logical positivism in science and technology, which goes back to Descartes' dualism. Analogous to this, an economy dedicated to rational principles of science omits the heart and assigns it to the social, cultural and artistic fields." (Schefer 2008: 128)

This kind of functionalistic critique is found in the social and political context of the late 1960s and early 1970s. These were marked by student protest movements and, associated with them, the rejection of conservative values. The art movements

traffic exact, economical and rapid is the key effort of modern architectural science.' In both of these comments, and throughout much of the Modern Movement, we see a desire to produce works of art and design based on objectivity and rationality, that is, on the values of science." (Cross 2007: 119)

[12] See Ulrich Heinen referring to the gap between theory and practice of Design. (Heinen 2008: 188)

during that time were represented by an avant-garde of the 'couleur' like the *Situationists*, who demanded the democratization of art[13] and the recapturing of urban space.[14] Central to this was the effort to overcome the alienation (Debord 1978) of culture in everyday life within *operative esthetics.*[15] The artist's role would be to bring back heart and emotion into everyday experiences. Artists' groups such as the *Nouveaux Réalistes* in 1960[16] elevated daily routine itself to a site of art, and Henri Lefebvre demanded that everyday life itself should be turned into a "philosophical place" (Lefebvre 1976: 1). In his 1975 *meta-philosophy* he criticized the segmentation of daily life into functional, organized and structured parts, such as work time, private time and recreation time, which was apparent in the development of housing and urban settlements (Schmid 2010: 118). According to Lefebvre, this development would lead to alienation in terms of passiveness, conditioning and non-participation in everyday life (Lefebvre 1975; 326 f.):

> "What is the aim of this critique? How should everyday routines be changed? Lefebvre received a first answer to his questions from surrealism, a possible antidote against alienation in everyday life: poetry." (Schmid 2010: 119)

The poetic quality in art, which was banned by *Modernism* according to the artistic avant-garde of the 1960s and 70s, should take place not on the canvas but in real life. Reality would be reunited with emotion. After Lefebvre, the conflicts between the rational and the irrational evolve in everyday culture along with the production of societal existence (ibid: 116). Therefore the research in practice, which is derived from everyday activity, is a prerogative to the development of a philosophical tradition for Lefebvre. Also the paradigm of alienation by Lefebvre is related to questions of everyday life as well as to philosophy[17]:

> "Diametrically opposed to the ideal of philosophy, everyday life presents itself as the real world. In the face of its practical banality and the triviality of daily life, philosophy aspires to the 'higher' and it detects its essence in the abstract, the absent, the distanced and the

[13] In 1979 Hilmar Hoffmann (1974) demanded a "culture for everyone" which would comprehend cultural practice as a democratic process.

[14] This can be witnessed among other things in the cuttings of Gordon Matta-Clark, with which the artist dissolved architecture as a monument and turned to the topics like demolition, erection, planning and improvisation. Matta-Clark sojourned repeatedly in Paris in the 1960s where he contacted the *Situationist* Guy Debord. (Ursprung 2009: 18)

[15] The *operative esthetics* stands for an activity-oriented approach. The recipient should be relieved from alienation of everyday life by self-dependent action and examination and by creative activity leading into a metamorphosis of society. See also Joseph Beuys: "Everybody is an artist". (see Beuys/Brügge 1984)

[16] The group of artists around the art reviewer Pierre Restany, consisting of Daniel Spoerri, Jean Tinguely, Yves Klein, Niki de Saint Phalle, Martial Raysse, Armand Fernandez, Jaques de la Villeglé and François Dufrêne (former *Ultra-Lettrist* around Guy Debord), published their first manifesto in April 1960.

[17] Lefebvre subsumes the thought of his whole philosophy under the expression *meta-philosophy*, which correlates to the expression of *metaphysics* of Aristotle on another level. This thought integrates theory and practice, system and totality, element and unity, alienation and abolition of alienation and so on. (Lefebvre 1972: 73, emphasis in the original)

indifferent – as a radical and at the same time futile critique of everyday commodities. Thus philosophy and everyday life contrast with each other, accumulating in alienation: philosophical alienation as truth without reality, everyday alienation as reality without truth." (Schmid 2010: 116)

Lefebvre deduces that philosophy can only be self-actualized, if it makes itself void. This happens when the "[...] separation[18] of the philosophical from the un-philosophical, the higher from the lower, the spiritual from the material and the theoretical from the practical [...] is not accepted anymore." (ibid). The shift from speculative philosophy to real practice is existential to the *meta-philosophy* of Lefebvre,[19] which he occasionally refers to the *meta-physics* of Aristotle. However there is a difference between *meta-philosophy* and *meta-physics*[20] in one crucial point: The first does not lay a claim on totality and finality and it radically rejects its fragmentation into separate sciences:

> "Thus the philosopher (or *meta-philosopher*) no longer claims to achieve finality, synthesis, totality. He rejects historical and societal philosophy just as much as the classical meta-physics and the classical ontology. It is his task to point out totality, as well as to remind us that it is impossible to accept segmentation and the division as final. Ruthlessly he criticizes finalism in general, but also singular finalisms, economism, sociologism and historicism. Philosophy turned into meta-philosophy no longer recognizes a 'finished' or derailed reality of 'people'." (Guelf 2010: 87; emphasis in the original)

The *meta-philosophy* of Lefebvre also breaks with categorized thinking in favor of interconnectedness. In contrast to *metaphysics* it is geared to more flexible structures of thinking (ibid: 222) and it perceives itself in this context not as a philosophy about philosophy, but as

> "[...] a 'profane project', self-actualizing in the world. It makes it possible to overcome the discrepancy of the philosophical and non-philosophical world." (ibid: 84)

As the process of 'becoming philosophical' is dialectically connected to the 'becoming-worldly' of philosophy, the realization of philosophy at the same time stands for its own loss:

> "Philosophy has to abolish itself: it realizes by abolishing and it abolishes by realizing. The 'becoming-philosophical' of the world creates space for the 'becoming-worldly' of

[18] The expression of separation was viewed by the *Situationists* against the foil of *Modernism*. They developed several art strategies in order to escape alienation with radical fantasy. Thus, based on the creation of situations and on the resistance against collective unifying systems they created the *urbanisme unitaire* (Engl. unitary urbanism). This criticized the separation of working, living, consuming and producing, as proclaimed by modern urban planning. The playful revival of the wholeness of all aspects of living in form of performative strategies in urban spaces was their goal, amongst other things.

[19] "This is how Lefebvre gains a sense of practice by examining early essays by Marx, who contrasts philosophy and speculation [...]." (Schmid 2010: 84)

[20] "Since Aristotele's first philosophy, it [metaphysics, *A.C.*] has always been the place for discourse where final substantiations were negotiated and reflections were driven to the non-plus-ultra. As the philosophical 'core-discipline' it [metaphysics, *A.C.*] contains all argumentation leading to final statements, distributing them into individual sciences and living-practice via basic and sector disciplines by work-sharing and specialization." (Geldsetzer 1974: 249)

philosophy, for the revolutionary realization and abolition of philosophy as such." (Lefebvre 1975: 25)

Applying this reasoning to the design-research context, the 'becoming-science' of design would be connected to a 'becoming-design' of science, which in consequence would make science obsolete in terms of its traditional meaning. This further supports my above thesis that a science of design can only develop from the design-practice itself, meaning in the form of "practice-integrating design-research" (Steffen 2011). If science becomes design, then the research-process would be comparative to the design-process.[21] Ranulph Glanville and Christopher Frayling take a radical constructivist stance by forming a hypothesis that scientific research, similar to the work in art and design, is led by intuition and subjectivity (Steffen 2011: 76). According to them, all rules by which science is applied, ultimately go back to self-developed social constructs (ibid: 77).

This can be exemplified by a simple example in graphic design. For the design of a pattern that mirrors a water surface, the designer can look for inspiration in Claude Monet's paintings of water-lilies, while listening to Bedřich Smetana's Vlatava and imagining the ripple of the fountains of Jean Tinguely and Niki de Saint Phalle in Paris or studying art-history essays about those works. She or he compares these sources with her or his own perceptions of water during the design process and with all of these develops the designs. She/he does not ask, if it would be permissible to mix *Impressionists* with the *New Realists* or if music and art may be integrated in equal proportions. She/he only chooses the sources that seem relevant for her/him in order to achieve the goal of the design task. In this way the design process is not only an interplay between theory and practice, but also phenomenological to the core. The esthetic activity of art or design is therefore permeated by subjective aspects of the designing personality itself.

In this notion, human beings with their individual experience are at the center. This can also be found in the *metaphysical* observations on human nature by David Hume, which form the foundations of his cognitive science. Hume, whom Lefebvre mentions along with his fellow philosopher Norbert Gutermann among others in "Essai sur l'Individu" (Elden 2004: 69), derives his conclusions anthropologically from human thinking, feeling and doing. However, against the background of the case presented here, it is Hume's approach of regarding man and her/his practical activities towards nature as the fundamental principles of his philosophical thoughts. Likewise Lefebvre found contemplations about the individual and her/his mindset central to his cause. This (Lefebvre's) is a central position to the definition of a design-science as a *meta-discipline*, if the subject and his or her "tacit knowing" (Polanyi 1974) are taken seriously in design processes as we did in the introduced project.

[21] This thought is framed (as others) by current considerations in science history. Müller-Wille and Rheinberger state: "[We want, *A.C.*] [...] to develop a picture of 'progress in science' as a form of creative productivity in which chance and variable design opportunities play a similar role as in the arts. This contradicts the self-conception of many scientists in various ways, at least being stated publicly." (Müller-Wille and Rheinberger 2009: 16)

The following parameters were established for the definition of a *meta-philosophy* so far: (1) the shift towards practice,[22] that is towards an everyday culture and thereby towards man as the acting entity, (2) process being the featured trait in consequence of the rejection of *finalism* as well as (3) a principal position that does not transcending the actual discipline but overcomes the difference between the philosophical and the non-philosophical world, respectively reason and emotion. 'Meta' deriving from Greek means 'in the middle of' or 'over' and thus describes not only the center of everyday context, but in the design-research context of this essay, as much as the generic structure of design in relation to other disciplines, short: the multi-disciplinary nature of design per se.[23] It formulates its position *in the interface* of various fields or disciplines:

> "If a designer is responsible for the knowledge-processes of others (externally), then it is a prerequisite that knowledge is distinguished by design (internally). This has been discussed extensively in recent years and formulated in relation to system, evolution and cognitive theories. It seems to be the preliminary consensus to position design research in the 'in-between' [...]." (Stephan 2010: 84)

The above mentioned overcoming of isolated disciplines and the position of the in-between could be mentioned as the last parameter (4) for the definition of a *meta-philosophy*. All of these parameters can also be applied to design-sciences. Regarding point number (4), the idiosyncrasy of singular sciences within the design science is being dissolved in favor of a border-crossing, combinative attitude.[24] This is also characteristic to the design practice. Cordula Meier states:

> "Design is tangential to every area of communicative and social practice. Similar to language, design is a basic mode of activity in the world and towards the world. It interferes ontologically with the microsphere of everyday stances and circumstances, macro-theoretically design is determining for methodological system requirements and theoretical frameworks, that is, design is orientation." (Meier 2003: 13)

Here, Meier positions a universal understanding of design which influences all activities in and towards the world. This also means that design is mode of everyday occurrence and a tangent to communicative and social practice.

---

[22] "For a thorough understanding of Lefebvre's sense of practice we have to first consider the linguistic meaning of these expressions: While 'activity' or 'action' are mostly neutral in German language, *activité* and especially *action* in French have a much broader horizon of meanings: Thus the expression *action* for Lefebvre is much more than mere 'activity', he also means the doing, the action, the resistance and with this introduces a liberating moment, he embraces the life's expressions of human existence in totality." (Schmid 2010: 81, emphasis in the original)

[23] Cordula Meier states, that design theory always has to be recognized as an interdisciplinary network of scientific systems. This is rooted in the nature of the design medium, its design processes being interdisciplinary per se. According to her the design relevant science system contains philosophy, linguistics, brain research, cultural science and German studies as well as phenomenological everyday culture, sociology (anthropology), communication theory, semiotics, design history, media history, art history, art theory, media theory and at times political science or religious science. (Meier 2003: 24)

[24] Also see Martin Zeiller (1986: 28) who states the same for the scientific attitude of the *Musée Sentimental*.

Point (1) of the parameters of *meta-philosophy* mentioned above describes the shift towards research through practice. So far the abolition of the separation between theory and practice, between physis and logos is just a marginal point in the contemporary debates on the parameters for a design-science.[25] This circumstance is not understandable against the background that almost all disciplines were in the position to establish their own science-philosophies and definitions. Dagmar Steffen gives a good insight in the current state of research of this; she observes the relation between theory and practice in different analyses of natural –, social –, care – and political science as well as in religious – and cultural studies and points out, that such investigations are missing in the design literature: she continues that in design theory and practice are often standing in an "adversarial relation to one-another". (Steffen 2011: 5).

The reason that there has been no valid research-interest in the relation between theory and practice in design up to now, may lie in the fact that design practice as well as design theory to some extent incorporate knowledge from other disciplines ("position of the in-between", Stephan 2010: 84). Each of these 'external' disciplines bring along their own philosophies of science on how to arrive at specific theoretical reflections. This is a crucial difference between design research and other sciences. It is also due to the consultation of these 'external' disciplines that design is not regarded as a genuine science. However, the question for me is, why doesn't design, like Lefebvre with the *meta-philosophy,* overcome a transcendental theoretical discourse by addressing its own practice, which especially incorporates the special, the explicit, instead of turning to what already can be found in other disciplines. That is to say, that the so-called "Designerly Ways of Knowing" (Cross 2007) cannot be found in any of the auxiliary sciences that are being consulted for theoretical reflection and they point out, that 'to design' is and has:

> "[…] something that distinguishes us from other animals, and (so far) from machines. The ability to design is a part of human intelligence, and that ability is natural and widespread amongst the human population. We human beings have a long history of design ability, as evidenced in the artefacts of previous civilizations and in the continuing traditions of vernacular design and traditional craftwork. The evidence from different cultures around the world, and from design created by children as well as by adults, suggests that everyone is capable of designing." (ibid: 49)

Compared to Lefebvre, who designates everyday life as a place for producing philosophy, Cross refers to everyday life and its human activities as a place for producing design. Even if we regard professional design as absolutely rational with the absence intuition, Cross's quote reminds us that the mere ability to design is not reserved to an exclusive circle of professionals. Against this background, participative projects involving non-professionals, seem fruitful for design:

[25] The insecurities concerning the theory-practice relationship in design research are also reflected in the contemporary means of production of designers. To this day it is hardly possible to do a practice-based doctorate in Germany. Procedure integrating a practice and a theory part are classified by special titles such as Dr. phil. in art and in this way they are clearly distinguished from classical doctorates. This leads to the conclusion that developing an advanced practice is not viewed as 'scientific' or as being worthy of a doctorate.

"A classic example are the 'Watts-Towers' – an environmental fantasy created by Simon Rodia in his Los Angeles backyard between the nineteen-twenties and -fifties. In architecture and planning, there have been moves to incorporate non-professionals into the design process – through design participation or community architecture. Although the experiments have not always been successful – in either process or product – there is at least a recognition that the professionals could, and should, collaborate with the non-professionals. Knowledge about design is certainly not exclusive to the professionals." (ibid: 39)

Participative projects like the "Watts-Towers" approach the concept of the "open artwork" according to Umberto Eco (1977). Open Art has no defined aim, since its evidence evolves from the creative process, whereby I arrive at point (2) of the parameters for a *meta-philosophy* stated above. The elements of improvisation or also disorganization are generally part of (participative) processes and apart from being conductive to their success, they can also be responsible for their failure. Generally the process is to be considered as an integrative if not intrinsic part of design. Thus the design concept develops and transforms several times during the working process and is hardly ever fixed within the first idea.

3 *LAB* Project

To illustrate point (3) of the above-mentioned parameters for the definition of a *meta-philosophy*, I would like to introduce the design practice project, called Folkwang *LAB*, that was explicitly intended to transcend the separation of reason from emotion and theory from practice. Since Lefebvre understands space as a contingent field of application for his *meta-philosophy*, I selected urban space as a suitable context for this practice-project:

"Urban rationality as contemporary form of thought stands for the (preliminary) end of an intellectual development and offers new prospects. This presumes the understanding of philosophy as a meta-philosophy, connected to the process of urbanization in standing and in action. The conception of space that fulfils both a social as well as an urban, economic and epistemological function can be adapted as a form of consensus between various fields of knowledge. Instead of centering on the mathematical or meta-physical interpretation of space, the focus should lie on the living space ('l'espace vécu') and social cohabitation." (Guelf 2010: 88)

Fernand Mathias Guelfs reflection of "Living space" as defined by Lefebvre, was the subject of an interdisciplinary *LAB* project I conducted at *Folkwang University of the Arts* in Essen in the winter semester of 2013/14 at its faculty of design. Students from the fields of communication design, industrial design, photography and urban planning investigated how living space constitutes itself as a beloved place or as 'Heimat' (home) and how one can actively produce these values using design processes. For this it was essential to open the design process to the perceived qualities of 'Heimat' (home). Corresponding to the transcendence of individual disciplines in *meta-philosophy* point (4), the students investigated within interdisciplinary teams, collaborating with experts from sociology, architecture and history. Places for investigation were communities in the north and in the south of the city of Essen, divided by the 'Autobahn A40' the so-called 'social equator'. The north, in this context, is the socially and economically weaker region with a relatively high immigrant population, whereas the southern communities in Essen are generally characterized by higher income levels and a mostly homogenous structure.

At the beginning of the project it was rooted in the design methodology and the concept of the *Musée Sentimental* (Paris 1977) by Daniel Spoerri. With his museum concept, the *Nouveaux Réalisme* artist revolutionized everyday culture by putting trivial everyday objects on a pedestal. It was especially decisive for our work, that Spoerri was able to communicate a specific urban character with the synopsis of these everyday objects. The artist illustrates how such objects become advocates of their own biography and how they can become memorials, charged with meaning in the process of their use.

Crucial to the object of investigation was the participative integration of the people of Essen, who reproduce their 'Heimat' locality within their everyday activities. Addressing the shift to practice, point (1), for one weekend during the project the students moved in with hosts from the Essen community, in order to participate in their daily lives and to learn more about their individual references of 'Heimat' and to hear their individual stories behind their conceptions of 'Heimat'.

The actual 'designing' began for the students after the living-in stage and after having finished the evaluation of their new findings during the stay with their hosts. The subsequent design process was characterized by constant feedback from their hosts. The students were now given the task of staging the found stories and places of memory by design, in order to visualize their individual subjective, emotional quality to an outsider. Since sentimental quality can derive from the senses as well as from memory, from a spatial context or from an object, the students were free to choose whichever medium of expression they felt appropriate. This can be seen in the results of this project, which contain classical products as well as installations, documented or staged photo line-ups alongside interactive applications for online media. In this context it is worth mentioning, that some of the students did not stick to their own discipline with its typical media but instead ventured into other design formats.

The design research in this project was determined by the dialectical connection of theoretical and practical thought. In the theoretical sense, the *Musée Sentimental* with its artistic methods, its contemporary context and connecting avant-garde-movements, was analyzed primarily at the beginning of the project. This aimed at making Spoerri's methods of sentimentality used in the Musée fruitful for design in the context of an urban cultural history. In contrast, practical 'thinking' was revealed in the form various techniques for creativity applied in the project. Some of these techniques were, for instance, the distribution of disposable cameras for intuitive shots of a memorable place by the hosts, the making of mental maps, the integration of the olfactory senses for memory-based reception and the hearing of memorable places as well as the systematic development of a catalogue for memorable places. The following physical activities in form of developing design concepts were due to interpret the data mentioned before. Theory and practice were related to each other reciprocally in the project.

Corresponding to Lefebvre's concept of practice, we centered our investigation in the 'sentimental urbanism' project around everyday life as the field of production for intuitive as well as professional design and as the form of practical energy needed for a social system to produce a sense of 'Heimat'. Lefebvre states:

> "The abolition of philosophy takes on a programmatic character as it becomes tangible. It expands philosophy by radically changing its position. It incorporates analysis of practice as well as the display of practice in totality (the actual and the potential, closings and openings, layers and de-alignment), simultaneously seeking practical energy as the collective forces capable of intervention. This program aims at taking the unhinged elements of philosophy and inserting them into (revolutionary) practice." (Lefebvre 1975: 329)

To Lefebvre practice and production are connected. If one understands design research as a laboratory, as a creative/creating (Greek poiein) or producing science, it reveals itself through the 'poiein', the *metaphysics* of Aristotle's Genesis[26] and the *meta-philosophy* of Lefebvre. Here poetry, art, design and all the applied expressions of creativity of the human spirit are recognized as producers of knowledge. The findings of our *LAB project* are therefore of 'poietic' nature. In 2000 Luz María Jiménez Narváez publicized an analysis about "Design's Own Knowledge". This text seems helpful to illustrate some of the following questions and findings of the *LAB project*.

Based on the 'poietic' nature of the project, Jiménez Narváez describes design knowledge as a common, practical knowledge with the potential for emancipation:

> "It [design, *A.C.*] also involves the theoretical elements to transform and act upon reality; ideally to improve it. Its sense is thus transformational, and its interest is emancipating. These sciences have the ability to distinguish how the foundations of socially dependent relationships have been ideologically established, yet also deriving the means to change these relationships." (Jiménez Narváez 2000: 44)

Since art and design reflect cultural and social reality, they are recognized as a suitable medium to guide their recipients away from alienation. In the context of our project we used Spoerri's artistic strategies for sentimentalization and relocated them out of the museum and art space, back into practice. Consequently Lefebvre's *meta-philosophy,* poetry and practice were united. In our *LAB project*, the knowledge produced by this connection, was mostly evident in the habitual. How does a sense of belonging to one's living surroundings develop in everyday culture? And how does it appear in its own material culture/production? As a basis of these questions the following conclusion of Jiménez Narváez could be used for the analysis:

---

[26] Luz María Jiménez Narváez refers here to José Ferrater (1971): "*Praxis* is, by definition, opposed to theory. Greeks determined the *praxical* character of all actions, transactions, and human affairs. Nevertheless, Aristotle distinguished three classes of knowledge: theoretical knowledge, *praxical* knowledge, and poietic knowledge. The object of the first class is knowledge, the object of the second one is wisdom with respect to moral action (politics), and the third has productive action as its object." (Jiménez Narváez 2000: 39, emphasis in the original)

> "[…] the presence of material culture relies on nonmaterial culture, and vice versa. It is, therefore, necessary to locate design and the studies it may originate within the space-time framework of 'material culture', i.e. the physical world and environment created by human beings and their social relationships associated with." (Jiménez Narváez 2000: 38)

Apart from the relationship between recipient and object/sentimental place, in this *LAB project* relations between designer and recipient, between designer and object/sentimental place and between object/sentimental place, city and mankind were also of interest. In the end all of these relations reflected the design products of the *LAB project.* At this point it should mentioned that the greatest impact on the designs derived from the personal and direct relationships between designers and the recipients. This can be explained for one thing by the adoption of suggestions from the 'intuitive' designer (recipient) by the professional designer, along with the fact that there were inhibitions because of the emotional connections with the hosts. For instance, there were stories connected to the sentimental places in question, which were in part political or religious, but were not openly discussed by the designers since they did not want to offend their hosts. As an example I would like to introduce the work by the photography student Anna-Marie Knüppel. Knüppel was hosted in the *Beginenhof* a women's spiritual community in the south of Essen. These women named the "Hof" as their sentimental place, where they experienced a sense of community and social cohesion. Since this collective is based solely on the women living there, to outsiders the *Beginenhof* comes across not only as a factor of inclusion but also of exclusion. Without judging the collective, Knüppel created a very sensitive production between performance and installation. The so-called 'collective-sweater', was a chain of several long-sleeved shirts, which were sewn together at the hand-openings. This sweater became a connective form, which brought joy, warmth and closeness to its wearers, but also claustrophobia, tightness and constriction.

Against the background of this work, I would not suggest that our project lacked critical capacity, far more I would point out positively that the direct relationship between the recipient of the design trained the sensibility of the designer and led her to an examination of her responsibility towards the user:

> "Design thinking is a holistic, synergetic, and continuous whole shaped according to the designers personality and social influence which also relies directly on the sensible, expressive, or communicative abilities required to accomplish an idea. The material structure of this act is the design project." (Jiménez Narváez 2000: 41)

Here Jiménez Narváez highlights not only the social influence of designers and their communicative abilities as transmitters within design projects, but also their personalities. At this point, it is once again evident that the autobiographical contexts of designers are incorporated into their products. Within the *LAB project*, this factor was especially stimulated by the individual knowledge of 'Heimat' on the part of the designers. Their own living or spatial experiences meant that each of the participating students was already familiar with the city of investigation. Each of them already had their own stories to tell from their own experience of everyday life. Independently of this, everyone, designer or not, has an individual and distinctive image of 'Heimat' or its materialization in sentimental places. Therefore each participant in the project was already an insider in the field of 'Heimat' perception. In the end the knowledge gained within the project was displayed with the design solutions developed.

> "The object is the tangible materialization of the *poietic* act of design because it reflects the emotional, volitional, and cognitive interests of the designer. Furthermore, the material object is considered an object of knowledge for the perceiver, because it is cognizable, it is real; it has a sensitive and communicative existence that enables it to represent the constant dialectics between the ideological condition and value of the designer and the user." (Jiménez Narváez 2000: 47; emphasis in the original)

Empirical research enabled us to visualize a sense of 'Heimat' and make it tangible by design itself. The products reveal the changeableness and the process-related character of 'Heimat' and through their materialization show that design-thinking about culture implies a proposition of what this culture might look like. I would like to exemplify this with the work by Torben Körschkes. The student of industrial design investigated *Katernberg's*[27] market square during the *LAB project.* For his host this was a place of diverse sentimental qualities. For her the market is linked to a feeling of community, relating to her experience of having participated in a demonstration against right-wing radicalism. She is actively engaged in several political or church groupings. To her *Katernberg* market as a place is a symbol for her own activities in the past. She often lingers at the *Katernberg's* market as well as in the so-called 'pit-man's cathedral' next to it. The market represents a memorial of the social systems in which Körschkes' host is engaged. Körschkes chose the collectivity, the liveliness of the market as a place for the freedom of speech as a motif for the dialogic development of his design. During his research he found out, that many market spaces remained unused during the weekly market-days. These 'market gaps' then became the entry point into his conceptions. Körschkes wanted to revive the market as a collective place and as a place for active exchange using a medium easily accessible to many people. He developed a radio in form of a do-it-yourself modular system which would be distributed by a circular to the people of *Katernberg*. This radio can receive only one channel, namely the "market-broadcast", which is a mobile radio station, broadcasting a neighborhood program live from the market. Branded with the *Katernberg's* tomcat, the neighborhood's mascot, in the form of a sculpture in *Katernberg* market, this do-it-yourself radio was also intended create a direct identification with the object for the user. Torben Körschkes was inspired by the childhood memory using cans with a string for a telephone and developed the design of the radio accordingly. The radio is aimed at people who are disabled and cannot come to the market square during market days themselves. Torben Körschkes comments his idea: "In the best case, my intervention will revive the *Katernberg's* market as the lively center of *Katernberg*, which it once was, where the residents of the neighborhood can get to know each other, exchange, shop and be informed."

[27] *Katernberg* is a neighborhood in the north of Essen.

Both examples from Anna-Marie Knüppel and Torben Körschkes show, that the distilled qualities of 'Heimat' as well as its designed materialization are snapshots whose meaning is further transformed by the recipients' use of them. They emphasize the course of the everyday experience and knowledge processing as in point (2) of the *meta-philosophy*, which turns against finalism but demands the procedural. In this context, the work of Torben Körschkes is explicitly aimed at producing new knowledge.

In summary one can state that on the one hand, the kind of practice-based design research practiced in the *LAB project*, united rational understanding with emotion by integrating the designer with his subjective perspective as well as an acting researcher.[28] On the other hand, the *LAB* was a poietical concept or art as a foundation for the interest in gaining knowledge. Apart from the strategies of functionality and objectivity of *modernism*, this form of phenomenological and subject-based research promotes

> "[...] a trend away from modernism and its assumptions about legitimate knowledge production toward a broadening of what counts as research." (Bullough Jr. and Pinnegar 2001: 13)

The subject-based approach, however, was not only fruitful for the investigation of the position of the designer, but it was just essential for achieving access to the normative 'Heimat' conceptions of the hosts. Since knowledge and ideas in design (poiesis) are generated by emotional discourses and not by objective reflection (see Ings 2013: 679), I propose design research as a *meta-discipline* that does not ignore its core practice but rather values it as a source of insight and a place for production of research that demands that designers analyze their own actions. As in the *meta-philosophy* the separation of episteme and doxa is dissolved here:

> "The ancient Greeks divided into two classes: one, the result of reflection, *episteme*; the other one, a result of daily living, *doxa*. Today, the limits between *doxa* and *episteme* have become more confusing and intricate, and less obvious. Pure sciences have developed into applied sciences, and knowledge has become habitual, usable, and practical." (Jiménez Narváez 2000: 36; emphasis in the original)

Knowledge is generated in design as a consequence and solution of everyday life. As a *meta-discipline*, design science recognizes everyday life as its place of production. It is not a theory transcending design, but an activity-oriented science which generates itself from practice. A design science as a *meta-discipline* concentrates on methods and specifications of the design process, it refines this by reflection and most of all enables the integration of the designing researchers intuitive/subjective knowledge into the process.

[28] "Autobiographic inquiries have an established although contested history. Such approaches to research place the graphic designer at the centre of the problem to be solved. They elevate his or her ability to utilise informed subjectivity, self-search and intuition as tools for discovering solutions to complex and often protean problems." (Ings 2013: 678).

4 Conclusion

In this paper a concept of design has been developed evolving from the social-activist art of the twentieth century. In order to do so, artistic methods within the framework of an *LAB project* have been transferred to this design. By including Spoerris' behavior patterns of the *Musée Sentimental* both the designer as an artistic author and his subjective perspective on the development and production of the design goal were accentuated.

Within the *LAB* we conducted a research by means of a design, whose epistemological level is based on Henri Lefebvre's *meta-philosophy*. The research through design draws on Lefebvre's demand to turn to practice as well as on the production of philosophy in everyday culture. For Lefebvre, turning to practice means overcoming the philosophical and non-philosophical world as well as overcoming ratio and sensation. Applying this to a research through design, the research must be able to include any subjective and incremental elements of any practice.

In order to develop this aspect, I avail myself of Spoerris' subject-oriented approach of the *Musée Sentimental*. I formulated an understanding of design, which distinguishes itself by integrating "tacit knowing" (Polanyi 1974) and emotion. This goes back to the "experimental historiography", which Spoerri pushed in the *Musée Sentimental* (Brock 2008:197) and which derives from individual emotional states and the overlapping of everyday knowledge and historical traditions. Such a hypothetical access to city and history involves process orientation, which reflects the variability of these terms as well as the notion 'home'. This process as a method is not only to be found in Spoerri's definition/concept/theory, but it constitutes a crucial element in Lefebvres' *meta-philosophy*, which clearly rejects finalism. Within the framework of the *LAB project*, the focus of the process was applied to the analysis of the topic 'home' in Germany's Ruhr area. The *LAB* showed that the designer assumes the position as mediator between people and city/object. The design's interface function was proved by means of Gui Bonsiepe's ontological design diagram (Bonsiepe 1996: 19). In part, this function can also be found in the *meta-philosophy*, which not only describes the context of everyday culture, but also the design between miscellaneous subject areas.

In conclusion, Lefebvres meta-philosophy proves very useful in order to develop a research design for a research through design and to substantiate it by strong arguments.

References

Alexander, C. (1971). The state of the art in design methods. *DMG Newsletter, 5*(3).

Andermann, K., & Eberlein, U. (2011). Einleitung: Gefühle als Atmosphären? Die Provokation der Neuen Phänomenologie. In K. Andermann & U. Eberlein (Eds.), *Gefühle als Atmosphären: Neue Phänomenologie und philosophische Emotionstheorie*. Berlin: Akademie.

Barrett, E. (2007). Foucault's what is an author: Towards a critical discourse of practice as research. In E. Barrett & B. Bolt (Eds.), *Practice as Research. Approaches to Creative Arts Enquiery*. New York: I.B. Tauris.
Bonsiepe, G. (1996). *Interface. Design neu begreifen*. Mannheim: Bollmann.
Borries, F. von & Fezer J. (2013a). Produkt und Non-Produkt. In ibid (Eds.): *Weil Design die Welt verändert... Texte zur Gestaltung* (pp. 42–46). Berlin: Gestalten.
Borries, F. von & Fezer J. (2013b). soziales & asoziales. In ibid (Eds.). *Weil Design die Welt verändert... Texte zur Gestaltung* (pp. 84–88). Berlin: Gestalten.
von Borries, F., Hiller, C., Kerber, D., Wegner, F., & Wenzel, A.-L. (2012). *Glossar der Interventionen*. Berlin: Merve.
Brock, B. (2008). *Lustmarsch durchs Theoriegelände. Musealisiert euch!* Köln: DuMont.
Bullough, R., Jr., & Pinnegar, S. (2001). Guidelines for quality in autobiographical forms of self-study. *Educational Researcher, 30*(3), 13–21.
Caplan, A. (2015). *Sentimentale Urbanität*. Die gestalterische Produktion von Heimat. Bielefeld: transcript.
Cross, N. (2007). *Designerly Ways of Knowing*. Basel: Birkhauser.
Debord, G. (1978). Die vollendete Trennung. In ibid. In *Die Gesellschaft des Spektakels* (pp. 3–34). Hamburg: Lutz Schulenburg.
Eco, U. (1977). *Das offene Kunstwerk*. Berlin: Suhrkamp.
Elden, S. (2004). *Understanding Henri Lefebvre*. London/New York: Continuum.
Frayling, C. (1993/1994). *Research in art and design*. London: Royal College of Art.
Geldsetzer, L. (1974). Metaphilosophie als Metaphysik: Zur Hermeneutik der Bestimmung der Philosophie. *Zeitschrift für allgemeine Wissenschaftstheorie, 5*(2), 247–255.
Guelf, F. (2010). *Die urbane Revolution: Henri Lefèbvres Philosophie der globalen Verstädterung*. Bielefeld: transcript.
Haumann, S. (2011). Schade, daß Beton nicht brennt ...Planung, Partizipation und Protest in Philadelphia und Köln 1940–1990. Stuttgart: Franz Steiner.
Heinen, U. (2008). Bildrhetorik der Frühen Neuzeit – Gestaltungstheorie der Antike. Paradigmen zur Vermittlung von Theorie und Praxis im Design. In G. Joost & A. Scheuermann (Eds.), *Design als Rhetorik. Grundlagen, Positionen, Fallstudien* (pp. 143–191). Basel/Boston/Berlin: Birkhäuser.
Hoffmann, H. (Ed.). (1974). *Perspektiven der kommunalen Kulturpolitik. Beschreibungen und Entwürfe*. Frankfurt a. M: Suhrkamp.
Ings, W. (2013). Narcissus and the muse: Supervisory impliocations of autobiographical, practice-led PhD design theses. *Qualitative Research, 14*(6), 675–693.
Jiménez Narváez, L. M. (2000). Design's own knowledge. *Design Issues, 16*(1), Cambridge, 36–51.
Jones, J. C. (1977). How my thoughts about design methods have changed during the Years. *Design Methods and Theories, 11*(1), 48–62.
Lefebvre, H. (1972). *Die Revolution der Städte*. Munich: List.
Lefebvre, H. (1975). *Metaphilosophie: Prolegomena*. Frankfurt a. M.: Suhrkamp.
Lefebvre, H. (1976). *Kritik des Alltagslebens* (2nd ed.). Munich: Hanser.
Mareis, C. (2011). *Design als Wissenskultur. Interferenzen zwischen Design- und Wissensdiskursen seit 1960*. Bielefeld: transcript.
Mareis, C. (2013). Wer gestaltet die Gestaltung? Zur ambivalenten Verfassung von partizipatorischem Design. In C. Mareis, M. Held, & G. Joost (Eds.), *Wer gestaltet die Gestaltung? Praxis, Theorie und Geschichte des partizipatorischen Designs* (pp. 9–23). Bielefeld: transcript.
Maser, S. (1976). "Theorie ohne Praxis ist leer, Praxis ohne Theorie ist blind!" Grundsätzliches über die Notwendigkeit einer Design-Theorie. *form. Zeitschrift für Gestaltung, 73*, 40–42.
Meier, C. (Ed.). (2003). *Design Theorie. Beiträge zu einer Disziplin* (2nd ed.). Frankfurt a. M.: Anabas.
Müller-Wille, S., & Rheinberger, H.-J. (2009). *Das Gen im Zeitalter der Postgenomik: Eine wissenschaftshistorische Bestandsaufnahme*. Frankfurt a. M: Suhrkamp.

Polanyi, M. (1974). *Personal knowledge. Towards a post-critical philosophy*. Chicago/London: The University of Chicago Press Books.

Sanders, E. B.-N. (2013). Perspectives on participation in design. In C. Mareis, M. Held, & G. Joost (Eds.), *Wer gestaltet die Gestaltung? Praxis, Theorie und Geschichte des partizipatorischen Designs* (pp. 65–79). Bielefeld: transcript.

Schefer, N. (2008). *Philosophie des Automobils. Ästhetik der Bewegung und Kritik des automobilen Designs*. Munich: Wilhelm Fink.

Schmid, C. (2010). *Stadt, Raum und Gesellschaft. Henri Lefebvre und die Theorie der Produktion des Raumes* (2nd ed.). Stuttgart: Franz Steiner.

Steffen, D. (2011). *Praxisintegrierende Designforschung und Theoriebildung: Analysen und Fallstudien zur produktiven Vermittlung zwischen Theorie und Praxis*. Inaugural-Dissertation am Fachbereich F: Design und Kunst der Bergischen Universität Wuppertal.

Stephan, P. F. (2010). Wissen und Nicht-Wissen im Entwurf. In C. Mareis, G. Joost, & K. Kimpel (Eds.), *entwerfen wissen produzieren* (pp. 81–101). Bielefeld: transcript.

Ursprung, P. (2009). Living archeology: Gordon Matta-Clark und das New York der 1970er-Jahre. In S. Folie (Ed.), *Die Moderne als Ruine: Eine Archäologie der Gegenwart* (pp. 18–25). Nuremberg: Verlag für moderne Kunst.

Zeiller, M. (1986). *Musée sentimental – Oberfläche, Organisation, Obszönität*. Magisterarbeit am Ludwig-Uhland-Institut für Empirische Kulturwissenschaft der Eberhard Karls Universität Tübingen.

Online Resources

Beuys, J., & Brügge, P. (1984). Die Mysterien finden im Hauptbahnhof statt. Spiegel-Gespräch mit Joseph Beuys über Anthroposophie und die Zukunft der Menschheit. Der Spiegel, 4.6.1984. http://www.spiegel.de/spiegel/print/d-13508033.html. Accessed 16 Mar 2016.

Folkwang University of the Arts. (2016). LABs, projects and graduation projects. http://www.folkwang-uni.de/en/home/hochschule/projects-labs/. Accessed 16 Mar 2016.

Internationales Forschungszentrum Kulturwissenschaften der Kunstuniversität Linz. (2014). Forschende Kunst statt bildende Kunst? Ein Workshop zum Stand einer Diskussion, 19 March 2014. http://www.ifk.ac.at/index.php/events-detail/events/forschende-kunst-statt-bildendekunst-workshop-zum-stand-einer-diskussion.html. Accessed 16 Mar 2016.

Jonas, W. (2002). Die Spezialisten des Dazwischen: Überlegungen zum Design als Interface-Disziplin. Tagung Medium Design, Bauhaus-Universität Weimar, 13. und 14. Dezember 2001. http://home.snafu.de/jonasw/JONAS4-58.htmlMuseé Sentimental. Accessed 16 Mar 2016.

Krems/Stein. (2009). *Eine Stadt biografiert sich selbst!* http://www.kunsthalle.at/de/kunsthalle-krems/veranstaltungen/09/eine-stadt-biografiertsich-selbst. Accessed 16 Mar 2016.

On Testing Engineering Design Methods: Explanation, Reverse Engineering, and Constitutive Relevance

Dingmar van Eck

Abstract In this chapter I draw on philosophical literature on (scientific) explanation to assess the goodness of engineering design methods. I focus this analysis on the engineering design practice of reverse engineering and redesign, and elaborate a constraint drawn from the mechanistic explanation literature to assess the goodness of reverse engineering practices and the content of design representations resulting from those practices. This constraint concerns the distinction between *causal* and *constitutive* relevance in mechanisms. I spell out two ways in which constitutive relevance assessments give traction to designing: reverse engineering explanation, and design optimization. I end by showing how this analysis fits within and extends recent philosophical work on the interplay between engineering design and explanation, indicating the (broader) relevance and promise of connecting philosophy of explanation and philosophy of design.

Keywords Mechanistic explanation · Constitutive relevance · Mutual manipulability · Reverse engineering · Testing design methods

1 Introduction

Two recent and related topics of attention in the philosophy of design concern the (disputed) distinction between science and design (Farrell and Hooker 2012, 2015; Galle and Kroes 2014, 2015), and the testing of design methods (van Eck 2014; Vermaas 2014). Vermaas (2014, p. 47) observed that concern about the scientific status of design by design researchers might be due to the concern that "design research does not live up to the standards of science", since "design research does not yet have the means to test and refute design theories and models". In this chapter I take up the second issue of the testing of design methods.

D. van Eck (✉)
Department of Philosophy and Moral Science, Centre for Logic and Philosophy of science, Ghent University, Ghent, Belgium
e-mail: Dingmar.vanEck@Ugent.be

© Springer International Publishing AG 2018

P. E. Vermaas, S. Vial (eds.), *Advancements in the Philosophy of Design*, Design Research Foundations, https://doi.org/10.1007/978-3-319-73302-9_17

There is recent philosophical interest in the connection between engineering design and explanation, both with respect to engineering itself (van Eck 2014, 2015a, b, 2017; Levy 2014; Calcott 2014) and with respect to the interface between engineering and branches of biology (Calcott 2014; Calcott et al. 2015.; Braillard 2015; Levy 2014; van Eck 2017). These issues are discussed in philosophy of science, particularly those branches dealing with explanation, yet have by and large not been picked up in the philosophy of design literature. I here discuss and further extend both aspects of this work on explanation. In so doing I offer a means to test the engineering design practice of reverse engineering and redesign, as well as the content of design representations resulting from that practice, and discuss a neglected conceptual connection between biology and electro-mechanical engineering design, viz. "evolvability" (cf. Calcott 2014).

Elsewhere I took up the related project of elaborating the structure and role of design representations in terms of insights from the philosophical literature on (causal-mechanical) explanation (cf. Van Eck 2014, 2015a, b). Here I elaborate a constraint drawn from the mechanistic explanation literature to assess the goodness of reverse engineering practices and the content of design representations as used in reverse engineering and redesign contexts, viz. the distinction between *causal* and *constitutive* relevance in mechanisms (Craver 2007).

This analysis is in line with Vermaas' (2014) work on the testing of design methods in the sense of invoking methods or approaches from philosophy of science to address this issue. Vermaas (2014) argued that work from the philosophy of the natural sciences, specifically Lakatos' (1978) approach towards falsification and research programs, provides a means to secure a scientific signature for design research and enables the testing of design methods. Vermaas' proposal is (still) programmatic however since such testing along Lakatosian lines is currently not being carried out in design research. Ultimately, Vermaas offers general guidelines that design researchers may pick up to start the project of comparative testing of design theories, models, and programs. This is an enormous task and long-term endeavor, for it would require fleshing out in plausible fashion, in the context of design research, all the key concepts of Lakatos' machinery, like theories' hard core, protective belt, associated positive and negative heuristics, empirical content, empirical success, as well as clear comparative measures between competing theories and models.

I rather choose to focus here on recent work from the (mechanistic) explanation literature to elaborate what we might call a 'positive heuristic' that designers are advised to follow in reverse engineering the workings of complex technical systems and in describing the mechanisms by which such systems (are taken to) work, viz. clearly distinguishing constituent parts of technical mechanisms from causal influences on them.

I start with briefly discussing the core tenets of mechanistic explanation in Sect. 2. I subsequently elaborate in Sect. 3 a key aspect in the construction of mechanistic explanations and assessment of the goodness of such explanations: *constitutive explanatory relevance* in mechanisms. I discuss this constraint against the backdrop of the *mutual manipulability account* of constitutive relevance in mechanisms

(Craver 2007) and show in Sect. 4 how this account, when properly extended, can be brought to bear on assessing the goodness of reverse engineering practices and resultant design representations. I spell out two ways in which constitutive relevance assessments give traction to designing: reverse engineering explanation, and design optimization. I then show in Sect. 5 how this analysis fits within recent philosophical work on the interplay between engineering design and explanation, indicating the (broader) relevance and promise of connecting philosophy of explanation and philosophy of design. One result is that the notion of "evolvability" or modifiability (Calcott 2014), in addition to software engineering, also marks a common core between biology and electro-mechanical design. I end this section with conclusions.

2 Mechanistic Explanation: Explanation by Decomposition

2.1 Mechanistic Explanation

By now, several accounts of mechanistic explanation are on offer in the literature. Although they come in different flavors, there is broad consensus on a number of key features: "All mechanistic explanations begin with (a) the identification of a phenomenon or some phenomena to be explained, (b) proceed by decomposition into the entities and activities relevant to the phenomenon, and (c) give the organization of entities and activities by which they produce the phenomenon." (Illari and Williamson 2012, p. 123). Mechanistic explanations thus explain how mechanisms, i.e., organized collections of entities and activities, produce phenomena (Machamer et al. 2000; Glennan 2005; Bechtel and Abrahamsen 2005; Craver 2007).[1] In the literature on explanation in the life sciences, it is now uncontested that mechanisms play a central role in explaining capacities such as digestion, pattern recognition, or the maintenance of circadian rhythms. The idea is that to explain such capacities, one provides a model, or more generally a description/representation, of the mechanism responsible for that capacity.

It is clear that mechanism discovery (a, b, and c) is key to the construction of mechanistic explanations (Machamer et al. 2000; Bechtel and Richardson 1993/2010; Craver 2001, 2007; Illari and Williamson 2010). Functional and structural 'decomposition' and subsequent 'localization' of operations/activities on components (Bechtel and Richardson 1993/2010) is probably the most extensively discussed discovery strategy or heuristic (cf. Machamer et al. 2000; Glennan 2005; Craver 2002, 2007). Structural decomposition concerns the process of decomposing a mechanism into its constituent working parts/entities, and functional decomposition gives a model of a mechanisms' constituent operations/activities. Mechanistic explanations are built by aligning these decompositions in terms of localizing mechanisms' operations onto working parts, i.e., by ascribing causal roles to the

[1] The precise lingo differs; some speak about 'entities' and 'activities', others 'working parts' and 'operations', yet others 'capacities'. These differences need not concern us here.

operations of working parts. These decomposition-localization heuristics are core explanatory business in life sciences like neuroscience, cognitive neuroscience, and parts of biology, where the workings of mechanisms are investigated in terms of a variety of intervention techniques and experiments, such as brain area stimulation studies and neuroimaging.[2]

Localization is crucial in all this. If done correctly (a non-trivial matter, if anything), one gains knowledge of which parts belong and contribute to the functioning of a mechanisms and how they do so, i.e., which causal or biological role(s) they fulfill in a mechanism. However, neither the conceptual machinery and the experimental practice of decomposition and localization give an unambiguous handle on the issue which component parts and processes are genuine constituents of a mechanism, and which ones are merely causal background conditions or irrelevant parts (Craver 2007). For instance, it is intuitively very clear that the windscreen wipers do not make a (constitutive) difference to the operation of a car engine, whereas the carburetor does. With respect to the mechanism of the car engine, windscreen wipers are simply irrelevant parts. But how to spell out relevance vs. irrelevance and constitution vs. causation in a clear-cut fashion? For most philosophers, constitutive relevance is a *non-causal* notion (Craver 2007; Craver and Bechtel 2007; Couch 2011; Baumgartner and Gebharter 2016; Baumgartner and Casini 2017). Constitutive relevance relationships are in the mutual manipulability account (see below) always bidirectional dependence relations – a change in either overall behavior or a component activity (due to an intervention) always is accompanied by a change in component activity or overall behavior, respectively. With causal relationships this is often not the case (exempting cases of feedback). In addition, the relata in constitutive relationships are not mereologically independent: the tokening of an overall behavior implies the tokening of component activity, and vice versa. Causes and effects in contrast are taken to be mereologically independent. Finally, constitutive relationships are synchronic: component activities or overall behaviors taking on a particular value are not temporally prior to one another, but happen concurrently. Causes however precede their effects. Craver's (2007) *mutual manipulability account of constitutive relevance* is an epistemic instrument devised to handle this problem of constitution vs. causation in empirical/experimental practice, and to assess when entities' activities are constitutively relevant, i.e., genuine components, of mechanisms rather than causal background conditions or simply irrelevant parts.

To be sure, constituency is crucial to mechanistic explanation. Explanation in terms of mechanisms requires clarity on the 'make-up' of mechanisms and causal

[2]Other techniques used in experimental practice and discussed in the literature, concern 'schema instantiation' in which abstract mechanism schemata are made less abstract and applied to particular cases, 'forward-backward chaining' in which gaps in the stages of a mechanism's operation are filled in terms of knowledge of a mechanism's operation in preceding and succeeding stages, respectively (Darden 2002; Darden and Craver 2002), and 'modular subassembly' in which known types of mechanistic modules are assembled to form a hypothetical mechanistic model (Darden 2002). These procedures depend, of course, on mechanistic knowledge procured by earlier functional and structural decompositions and localizations.

influences on their functioning. Without clarity on what comprises a mechanism in a given explanatory context, that is, what makes up the explanans, explanation becomes vacuous.

3 Mutual Manipulability and the Causal-Constitutive Relevance Distinction

3.1 Mutual Manipulability

Constitutively relevant factors are individuated by Craver (2007) in terms of mutual manipulability relationships. On Craver's (2007) account, an entity's activity is considered constitutively relevant to the behavior of a mechanism as a whole if that entity's activity is shown to be a spatiotemporal part of the mechanism, and shown to contribute to the behavior of the mechanism as a whole. Evidence for constitutive relevance is taken to be procured if one can change the overall behavior of the mechanism by intervening to change the entity's activity, and if one can change the activity of the entity by intervening to change the overall behavior of the mechanism. Somewhat more formally, a factor is considered to be constitutively relevant if two conditionals are met (Craver 2007, CR1, p. 155, and CR2, p. 159):

(CR1) When ϕ [an entity's activity] is set to the value of $\phi 1$ in an ideal intervention, then ψ [an overall behavior] takes on the value f($\phi 1$)
(CR2) When ψ [an overall behavior] is set to the value of $\psi 1$ in an ideal intervention, then ϕ [an entity's activity] takes on the value f($\psi 1$)

Craver (2007) defines an ideal intervention I on ϕ with respect to ψ as "a change in the value of ϕ that changes ψ, if at all, only via the change in ϕ" (p. 154, italics in original). The reverse holds for an ideal intervention on ψ with respect to a specific ϕ (Craver 2007, pp. 154–160). These conditionals cover both scenarios in which interventions change the manner in which ψ or ϕ occur, i.e., their value, as well as ones that lead to the occurrence or elimination of ψ or ϕ (cf. Craver 2007, p. 149). In the latter case, ψ or ϕ would take on the value '1' or '0', respectively. So mutual manipulability relations offer evidence for two sorts of constitutive relevance, i.e., difference making, relations, i.e., with respect to the *occurrence* of explananda phenomena, and with respect to the *precise manner in which explananda phenomena occur* or obtain (cf. van Eck 2015a). Note that although the mutual manipulability account is inspired by Woodward's (2003) account of causal explanation, constitutive relevance is, as said, a *non-causal* notion (Craver 2007; Couch 2011). In case of constitution, according to mutual manipulability, interventions on components directly alter overall mechanism behavior and vice versa, rather than standing in a temporal cause-effect relationship. Since interventions on components or overall behaviors alone fail to tease causal and constitutive relationships apart, the bidirectional intervention/mutual manipulability constraint is imposed on constitutive relevance assessments (Craver 2007). Mutual manipulability is devised as a general demarcation yardstick for mechanism individuation across sciences dealing with mechanisms.[3]

Craver's (2007) mutual manipulability account provided a major impetus for thinking about constitutive relevance in mechanisms. However, as recent analyses convincingly show (Baumgartner and Gebharter 2016; Baumgartner and Casini 2017), the account needs to be extended in a significant way in order to indeed be useful in individuating constitutively relevant parts of mechanisms. The key to this extension concerns dropping the 'ideal intervention' requirement in favor of the notion that interventions, in the case of constitution, are necessarily 'fat-handed'. I elaborate this 'fat-handedness' extension in the next section and subsequently illustrate in section four how mutual manipulability, appropriately extended, can be used for the testing of engineering design methods.

3.2 Fat-handedness and Mutual Manipulability Combined

After Craver's (2007) initial formulation, various extensions and criticisms of his mutual manipulability account have been formulated (e.g., Couch 2011; Leuridan 2012; van Eck 2015c; Baumgartner and Gebharter 2016; Baumgartner and Casini 2017;). The most recent extension to the account concerns doing away with the 'ideal intervention' requirement and, rather, elaborating constitutive relevance in mechanisms in terms of the idea of fat-handed or *common cause-interventions*. (Baumgartner and Gebharter 2016; Baumgartner and Casini 2017).

I endorse this recent alternative: mutual manipulability in itself is not sufficient to establish conclusive evidence for constitutive relationships. In order to have sufficient (abductive) evidence that an entity's activity ϕ is constitutively relevant for a mechanism's overall behavior ψ, mutual manipulability relations are needed but of a different sort than envisaged by Craver (2007). What is needed is a re-characterization of mutual manipulability in terms of the idea that both a mechanism's overall behavior ψ and its entity's activity ϕ are manipulable through a fat-handed or common cause-intervention that causes changes in both, *combined* with a demonstration that (i) there are *only* such *common cause-interventions* of the mechanism's overall behavior ψ and the entity's activity ϕ, and that (ii) there are *no surgical cause-interventions* that solely change the entity's activity ϕ or the mechanism's overall behavior ψ (Baumgartner and Gebharter 2016; Baumgartner and Casini 2017; cf. van Eck and Looren de Jong 2016). Let me explain these important technicalities.

Given the (assumed) non-causal, constitutive relationship between a phenomenon ψ and a mechanistic constituent ϕ, an intervention on either the phenomenon or a constituent will ipso facto alter the value of *both* the phenomenon ψ and the con-

[3] Of course, the interactions between component parts and operations in a mechanism are causal; the relationship between these components parts and processes and a mechanism's overall behavior (the explanandum phenomenon) is constitutive, i.e., non-causal.

stituent ϕ (since they occupy the same region of spatial-temporal space and are not related in terms of cause and – temporally later – effect). Such interventions are hence a *common cause* of the changes in both ψ and ϕ. i.e., ψ and ϕ are only mutually manipulable via *common cause-interventions* (cf. Woodward 2003, 2008; Baumgartner and Gebharter 2016). Thus, ideal interventions that change the value in a phenomenon ψ through an associated change in the value of a constituent ϕ, and vice versa, are not possible. The change rather is effected in both ψ and ϕ through a common cause-intervention. So ψ and ϕ are still mutually manipulable but only via *common cause-interventions*. Furthermore, assuming constitution, it is not possible to change solely the value of a phenomenon without altering the value of a constituent. If that is the case this implies that the relationship, if any, between ϕ and ψ is not one of constituency. Such *surgical cause-interventions* – interventions which only cause a change in a ψ *but not in a* ϕ – thus should not be possible. For instance, when intervening on memory formation (ψ) by engaging a subject in an experimental task would not lead to changes in the formation of long-term-potentiation (LTP) – synaptic changes in Hippocampal neurons associated with learning and memory processes –, such an intervention would count as surgical: it only causes changes in ψ, not in Hippocampal neurons (ϕ's). On the other hand, when such an intervention causes changes in the value of both memory formation (ψ) and Hippocampal neurons (ϕ's) – a much more plausible scenario – it counts as a common cause intervention.

However, the problem now becomes that it need not be the case that the observed correlations in changes in a phenomenon and some putative constituent are due to constitutive relationships between them; correlations might simply result from the fact that both ψ and ϕ are intervened on via the same (common cause) intervention. For example, an intervention that effects a change in both LTP characteristics of Hippocampal neurons (ϕ's) and some aspect of memory formation (ψ) might suffice to explain the correlated changes in ϕ's and ψ due to the 'common cause' nature of the intervention. It seems that there is no further empirical evidence on offer to conclude that constitution grounds the observed correlation:

> mutual manipulability via common cause interventions provides no empirical evidence in favor of the existence of constitutive dependencies. Thus, (MM) [mutual manipulability] is not sufficient to account for constitution on evidence-based grounds. (Baumgartner and Gebharter 2016, p. 20).

However, when one combines mutual manipulability through common cause-interventions with demonstrating that there are *only* such common cause interventions on a mechanism' overall behavior ψ and some constituent ϕ, and *no* surgical cause interventions (that would only alter ψ), this does provide sufficient (abductive) evidence for constitutively relevance. This constraint is labeled "the criterion of fat-handedness" (Baumgartner and Gebharter 2016, p 21). If only common cause interventions are observed, and no surgical cause interventions, i.e., if fat-handedness is satisfied, the best explanation for this feature is that the relationship between a mechanism' overall behavior and some putative mechanistic component is one of constituency. That is:

constitution provides the best available explanation for systems satisfying both mutual manipulability and fat-handedness (Baumgartner and Gebharter 2016, p. 2)

When systems-mechanisms satisfy both mutual manipulability and fat-handedness, the assumption of constitution is a superior explanation for these features than the idea of a common cause intervention, since *constitution also explains the absence of surgical cause-interventions* (cf. Baumgartner and Gebharter 2016). So when it is the case that the dependencies between a phenomenon ψ and some constituent ϕ cannot be screened of by surgical interventions, constitution offers the best explanation for the observed correlation. For instance, when it is the case that every intervention carried out on some aspect of memory formation changes that aspect of memory formation (ψ) as well as induces changes in LTP characteristics of Hippocampal neurons (ϕ's), and there are no interventions that change memory formation but leave LTP characteristics of Hippocampal neurons (ϕ's) unaffected, constituency explains these features of fat handedness and mutual manipulability between ψ and ϕ.

Note that this is an example of abductive reasoning, contingent on the current state of play in the relevant sciences. If for a given case only common cause-interventions are known and no surgical cause-interventions are available, one has fallible (abductive) evidence for constitution since it explains the absence of surgical causes best. Yet, this does not rule out that at some point in the future surgical causes might be found. This of course is a feature of all analyses that wage their plausibility on evidential, empirical grounds.

With mutual manipulability plus fat handedness, we have solid tools, or so I argue, to test the goodness of aspects of the engineering design practice of reverse engineering and redesign as well as the content of design representations resulting from that practice. This of course concerns the distinction between causal and constitutive relevance.

4 Testing (Reverse) Engineering Design Methods: Applying Mutual Manipulability

4.1 Mechanistic Reverse Engineering Explanation

In engineering, reverse engineering and engineering design go hand in glove (e.g. Otto and Wood 1998, 2001; Stone and Wood 2000). Otto and Wood's (1998, 2001) method for reverse engineering and redesign gives a clear illustration of this interplay. In their method, a reverse engineering phase in which reverse engineering explanations are developed for existing artifacts, precedes and drives a subsequent redesign phase of those artifacts. The goal of the reverse engineering phase is to explain how existing artifacts produce their overall functions in terms of underlying mechanisms, i.e., organized components and sub functions (behaviors) by which overall (behavior) functions are produced. These explanations are subsequently

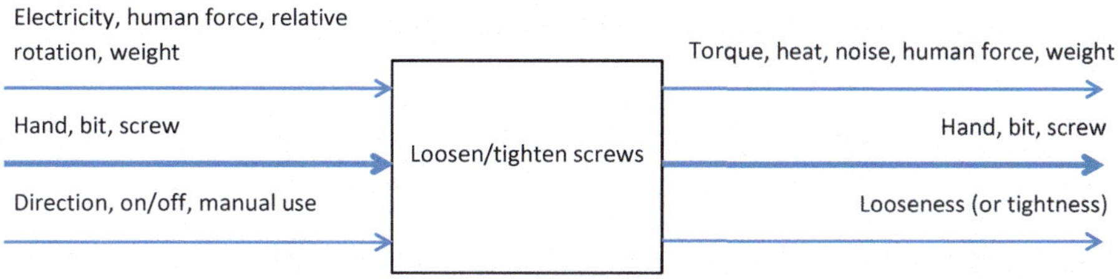

Fig. 1 Overall function of an electric power screwdriver. Thin arrows represent energy flows; thick arrows represent material flows, dashed arrows represent signal flows (Adapted from Stone and Wood 2000, p. 363, Fig. 2)

used in the redesign phase to identify components that function sub optimally and to either improve them or replace them by better functioning ones. Otto and Wood (1998, p. 226) relate explanation and redesign as follows: "the intent of this [reverse engineering] process step is to fully understand and represent the current instantiation of a product. Based on the resulting representation and understanding, a product may be evolved [redesigned], either at the subsystem, configuration, component or parametric level".

In the reverse engineering phase, an artifact is first broken down component-by-component, and hypotheses are formulated concerning the functions of those components. In this method, functions are represented by conversions of flows of materials, energy, and signals. After this analysis, a different reverse engineering analysis commences in which components are removed, one at a time, and the effects are assessed of removing single components on the overall functioning of the artifact. Such single component removals are used to detail the functions of the (removed) components further. The idea behind this latter analysis is to compare the results from the first and second reverse engineering analysis in order to gain potentially more nuanced understanding of the functions of the components of the (reverse engineered) artifact. Using these two reverse engineering analyses, a functional decomposition of the artifact is then constructed in which the functions of the components are specified and interconnected by their input and output flows of materials, energy, and signals (Otto and Wood 2001). Such models represent parts of the mechanisms by which technical systems operate, to wit: causally connected behaviors of components.[4] They are the end results of the reverse engineering phase and are subsequently used to identify sub-optimally functioning components and so drive succeeding redesign phases. Examples of an overall behavior function and behavior functional decomposition of a reverse engineered electric screwdriver are given in Figs. 1 and 2, respectively.

In the model in Fig. 2, temporally organized and interconnected behaviors are described. Components of artifacts are described in Otto and Wood's method in

[4]To be sure, as mentioned, most have it that the interactions between component parts and processes in mechanisms are causal; the relationships between component parts and processes and overall behaviors of mechanisms are non-causal, constitutive relationships (but see Leuridan 2012 for an alternative construal).

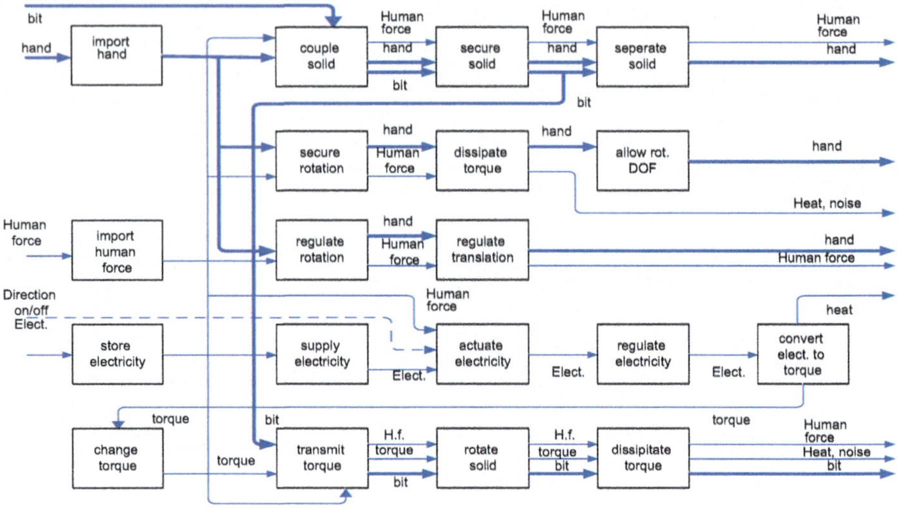

Fig. 2 Functional decomposition of an electric power screwdriver. Thin arrows represent energy flows; thick arrows represent material flows, dashed arrows represent signal flows (adapted from Stone and Wood 2000, p. 364, Fig. 4)

tables, what in engineering are called 'bills of materials', together with a model, called 'exploded view', of the components composing the artifacts. Taken together, these component and behavior functional decomposition models provide representations of mechanisms of artifacts.

After the reverse engineering of a technical artifact, aimed at providing detailed understanding of the mechanism(s) by which it operates, the redesign phase starts by identifying components that *function sub-optimally*, and, thereby, cause artifacts to manifest their overall functions in sub-optimal fashion. Redesign efforts are subsequently directed towards designs with improved functionality of these components (Otto and Wood 1998, 2000). Otto and Wood (1998) discuss an example of redesigning an electric wok. The (reverse engineered) artifact's desired behavior to "deliver a uniform temperature distribution across the bowl" failed to be achieved due to the fact that the electric heating elements of the wok, such as a bimetallic temperature controller, were housed in too narrow a circular channel (Otto and Wood 1998, p. 235). Redesign efforts were subsequently directed towards a design with improved functionality of the heating elements, inter alia resulting in a design with a thicker bowl and different shape than in the reverse engineered electric wok.[5] In sum, a reverse engineering – mechanistic – explanation of the operation of an existing electric wok was used to identify sub optimal functioning components – in this case, electric heating elements – which resulted in modifications to these components.

[5] This redesign step involves a lot of mathematical modeling, use of physical and technological principles, and/or prototype building (Otto and Wood 1998, 2000). These details need not concern us here.

4.2 Testing Case

The model in Fig. 2 of a reverse engineered electric screwdriver also gives a clear illustration were things can go wrong in reverse engineering explanation (and mechanism individuation and mechanistic explanation in general): not every component operation represented in Fig. 2 is a constituent part of the mechanism by which the electric screwdriver operates. This reverse engineered model is described in terms of a functional modeling language, called Functional Basis, that is taken to only represent device functions, i.e., operations-on-flows carried out by technical artifacts (Stone and Wood 2000; Hirtz et al. 2002; van Eck 2010). With respect to this model, Stone et al. (1998) state that the top chain of functions represents the insertion and removal of the screw bit, that the second represents the fastening of the screw bit, that the third represents the positioning of the screwdriver, and that the fourth and fifth represent the actuation of the device.

However, despite the model and the Functional Basis in general being advertised as describing solely device functions, not every operation-on-flow described in the model in fact represents a device function; quite a few represent operations-on-flows *carried out by users* (van Eck 2010). All the functions of the top function chain and the leftmost function of the second function chain of the power screwdriver exemplify the characterization of user functions given by Hirtz et al. (2002), i.e., operations-on-flows carried out by users. As can be seen in Fig. 2, the first function chain is represented in terms of four functions that transform the flows "hand", "bit", and "human force" from input to output. By representing the insertion and removal of the screw bit in terms of a sequence of functions that transform a material "bit" flow, a "human force" flow, and a "hand" flow, the (de)coupling of the screw bit is represented as a sequence of user functions. More specifically, the (de)coupling of the screw bit is represented as realized through human force applied through the hand, i.e., operations-on-flows carried out by a user. This analysis applies as well to the leftmost function "secure rotation" of the second function chain, which represents the manual fastening of the screw bit. In this function chain, the function "secure rotation" transforms a "human force" flow and a "hand" flow, describing that the securing operation is realized by human force applied through the hand.

Now, erroneously interpreting these functions as device functions leads to incorrect understanding of the functioning of the mechanism in question, which in turn is detrimental to redesign and optimization efforts, as well as design knowledge sharing. Mutual manipulability (properly understood) gives a handle on this issue. Although one can envisage (fat-handed) interventions that affect user actions and the overall functioning of the power screwdriver, say, applying too much or too little manual force when driving in screws, the reverse does not (necessarily) hold. For instance, intervening to change the overall functioning of the screwdriver by changing the materials or resistance of the materials in which screws are driven or removed may affect the speed with which screws are driven/removed, but need not have an effect on the hand grip of the user operating the device. This intervention certainly will not have an effect on the action of fastening or loosening the screw bit by a user. In other words, there here exist surgical cause-interventions that would enable effecting changes solely in a phenomenon – the driving of screws – without affecting putative components – user actions. User actions are hence not constitutive parts of the mechanisms of technical systems, here a power screwdriver (but of course they are relevant causal influences on the workings of such systems).[6]

Not only can the conflation of user actions and device functions be ruled out with mutual manipulability. It also can be put to work in teasing apart genuine device functions from (physical) inputs or causal influences on technical systems. In the Functional Basis method for designing, operations-on-flows that represent how input (materials, energy, and signals) enters a technical system also count as device functions (Stone and Wood 2000; Hirtz et al. 2002; cf. Ookubo et al. 2007). We saw above that such functional descriptions may refer to user actions rather than device functions. In other cases such descriptions may refer to input to or causal influences on a technical system, rather than specify device functions. Consider again the model of a reverse engineered electric screwdriver in Fig. 2. Human force is being modeled as being imported into the screwdriver. This of course is quite sensible, but do such operations-on-flows count as genuine device functions of the screwdriver? On Functional Basis terms they do, but applying mutual manipulability tells a different story. Without the input of human force the screw bit of the screwdriver cannot be fastened/decoupled ("regulate rotation") and the screwdriver hence will not perform its overall function of driving screws. However, intervening on this overall function, again say, by changing the materials or resistance of the materials in which screws are driven or removed will not have an effect on the human force recruited for fastening or loosening the screw bit by a user.

Not only are some operations-on-flows at the system boundary ruled out as genuine constituents of technical systems. Also some operations-on-flows at the 'center' of the mechanism description fail to conform to mutual manipulability via common cause interventions. Consider the two descriptions 'dissipate torque' in the second and fifth function chain. Interventions that affect the overall function likely also have an effect on the amount of torque recruited: for instance, increasing the resistance of the materials in which screws are driven or removed in all likelyhood impacts the amount of torque recruited by the screwdriver. So, torque likely is

[6] I my view, user actions are never constituents of technical mechanisms (whereas there are cases in which, by my lights, bodily movements are constituents in cognitive systems-mechanisms (Van Eck and Looren de Jong 2016). I suspect that each case, when regimented in term of mutual manipulability and fat handedness, will support this claim. Although giving a thorough defense for this claim is beyond the scope of this chapter, consider a brief additional example in support of this view. A good-old fashioned hand-operated screwdriver – without batteries or electrical wiring – that solely works by applying human force and hand-directed movements. By definition, such artifacts do not drive/remove screws when they are not manually used: user actions are vital to its operation. Yet, one can intervene on 'its driving/removing screws' function by, say, making the tip of the screwdriver sharper or blunter. This intervention does not elicit immediate, synchronous changes in user actions concerning the manual operation of the artifact (this intervention can, of course, cause temporally later hand-directed operation of the artifact to be more or less smooth). So, surgical cause- interventions here (again) exist which block constituency claims with respect to user actions.

constitutively relevant for screwdriver functioning, yet its spreading out is not! Interventions that affect the overall function to drive/remove screws, say changing the resistance of the materials, have no effect on the rate of torque dissipation (only on the amount of torque recruited). The dissipation of torque in power screwdrivers contributes to the detachment of screwdrivers from screwed connections, but not to the driving of screws. That is, torque dissipation occurs *after* screw bits are fastened (in the second chain) and *after* screws are driven or removed (in the fifth chain). Given the possibility of such surgical cause interventions, it becomes clear that the operation-on-flow descriptions "dissipate torque" do not refer to constituent operations of the mechanism by which power screwdrivers drive/remove screws.

Again, clarity on which features comprise a technical system's mechanism and which features are causal inputs to such a mechanism or comprise its "mode of deployment" (Chandrasekaran and Josephson 2000), are crucial for understanding its functioning. And, hence, crucial for redesign purposes and knowledge sharing.

4.3 The Goodness of Design Representations

The point of course is that good reverse engineering practices and resultant design models or representations highlight bona fide constitutively relevant components and distinguish these from (relevant) causal input, user actions, and irrelevant parts.[7] As alluded to in the case above, the value of making these distinctions lies in their ability to offer sound understanding of the workings of technical systems. We can make this idea precise in terms of a reverse engineering model or design representation's ability of offering adequate *counterfactual understanding*. The model in Fig. 2 is a design representation of the operation of a technical system, in casu a power screwdriver. It displays part of the mechanism by which the screwdriver works, i.e., some of its temporally ordered behaviors.[8] Such a partial description of a mechanism thus partially explains how the screwdriver works and realizes its product function. I've argued elsewhere that an important role of design representations is their ability to offer counterfactual understanding in terms of offering answers to *what-if-things-had-been-different questions* (van Eck 2015b; cf. Woodward 2003). For instance, returning to our screwdriver example, what would happen when say, some specifics of the conversion of electricity into torque were to be changed, say, when the function 'regulate electricity', or perhaps more precisely

[7] I use the term design representation in a broad sense, which may include models qua diagrams, physical models, drawings, cardboard models, etc. In addition, design representations may refer to extant artifacts as well as designs of to-be-built artifacts. Only the former have truth conditions; the latter cannot be assessed in terms of alethic criteria (see note 9).

[8] The concept of 'function' is used with different meanings in engineering design, notably 'purpose', 'effect of behavior', and 'intended behavior'. Product and basic functions in the Functional Basis method refer to 'intended behaviors' (Vermaas 2009; van Eck 2011).

'voltage regulation', were to be fulfilled by a 'voltage regulator' rather than a 'capacitor' (cf. Fig. 2).[9]

Models that include descriptions of spurious components of mechanisms – be it spuriously identified user actions, causal influences, or irrelevant parts as genuine components – partially fail with respect to this role. Spurious aspects procure incorrect understanding or none at all. For instance, asking how changes in the rate of torque dissipation affect the overall function of the screwdriver to drive/remove screws is an ill-posed question. Torque dissipation is irrelevant for understanding the screw driving mechanism of the artifact, hence no explanatory traction is gained by an inquiry into interventions on its rate with respect to the driving of screws.

Consequences of asking the wrong what-if questions with respect to the effects that interventions on user actions and causal inputs have are far more serious. Interventions that change the values of these parameters, of course, often have an effect on overall mechanism function, but is it crucial to know the nature of that effect. Changes in overall device function that result from changes in user actions or causal inputs but are incorrectly taken to result from changes to device functions, gives incorrect understanding of the workings of mechanisms. Misreading changes to an artifacts' mode of deployment as changes to its mechanism is nothing short of a category mistake. Redesign/optimization efforts, inter alia, are compromised if these different issues are lumped together, since:

> Giving good explanations is tightly coupled with our ability to manipulate and control the world [...] The better we understand the results of various manipulations on some system, the better we can explain how it works. And the better we understand how to control a system by manipulating its parts, the better we can design and build a mechanism with the precise capacities we desire (Calcott (2014, p. 296).

If, however, interventions on component device functions are collapsed with interventions on modes of deployment we have poor explanation and understanding, and designing and manufacturing are then the worst for it.

There is another way in which mutual manipulability can be of use in design optimization: robustness testing.

---

[9]Although the truth makers of answers to these questions are facts about artifacts that in a design phase still have to be build (and interventions on them, such as the replacement of components), answers can still be given to these questions in the design phase, the plausibility of which derives from sound knowledge of past designs, artifacts that have been built in terms of these designs, and scientific and technological principles governing them. Design models or representations thus assist in counterfactual understanding, and the understanding they procure in design phases can be assessed in terms of their plausibility. Alethic norms do not govern such assessments in cases were the artifact has not yet been built/produced (nevertheless such counterfactual understanding may lead to improved designs when *plausible* answers to what-if questions result in the selection of other, better components in the design phase than the ones originally conceived of) (van Eck 2015b).

4.4 *Design Optimization: Mutual Manipulability and Testing for Robustness*

Both in (systems) biology and engineering design the notion of 'robustness' looms large. Robustness is a key property both in biological and engineering contexts for it is "a property that allows a system to maintain its functions despite external and internal perturbations." (Kitano 2004, p. 826). Almost always, robust systems – ones resilient to perturbations to parts of the mechanism or the environment in which it functions – require complex sub systems dedicated to counteract perturbations (Kitano 2004). This holds both for complex biological systems as well as (most) engineered systems. Think for instance of all the sub systems of an airplane dedicated to counteract changes in order to make it fly in the appropriate manner.

Assessing or testing for robustness also plays a key role in design optimization (in redesign contexts). Otto and Wood's (1998, 2000) method for reverse engineering and redesign captures this point vividly. As we saw, in this method, reverse engineering – mechanistic – explanation drives the subsequent development of a *comparative analysis* between an extant technical system and a to-be-redesigned one with respect to how well a capacity of a component contributes to a specific high level capacity. For instance, the comparison of how well the capacity of a specific extant wok bowl to conduct heat contributes to the system level capacity of the wok to provide uniform heat distribution vis-à-vis how well a novel redesigned wok bowl with different properties is expected to contribute to heat distribution (Otto and Wood 1998). A relevant component in such comparative analyses are robustness tests. For instance, testing whether the heating elements of an extant electric wok and/or novel prototype are able to stand prolonged high temperatures and do not break down (Otto and Wood 1998). We can make this testing for robustness precise (and suggest testing procedures) in terms of mutual manipulability.

Mutual manipulability is, in addition to assessing constitutive relevance relations, also a means to test and compare whether engineered systems – extant ones and novel prototypes – are *robust* with respect to specific capacities, and to what extent.[10] That is, mutual manipulability can be invoked to assess whether specific (constitutive) relationships between component capacities and system level capacities are invariant or stable across a range of interventions (cf. Woodward 2003) or, rather, are fragile and easily break down after interventions on either component or system-level capacities. Components figuring in unstable relations are fragile ones that require improving either through parametric modifications or component replacements (cf. Otto and Wood 1998, 2000). So mutual manipulability thus is a means to test and

[10] Since mutual manipulability concerns interventions on extant systems, I restrict the application of mutual manipulability to robustness testing to extant technical systems and novel physical prototypes of technical systems. A lot of comparative work in Otto and Wood's method (1998, 2000), in addition to comparing extant systems with novel physical prototypes, concerns the counterfactual comparison between extant systems and conceptual redesigns with respect to the functional performance of extant and hypothesized components of these systems and redesigns. One cannot factually intervene on conceptual (re)designs of course.

compare different (extant or prototypical) engineered systems and spell out which ones function better, i.e., are more robust, with respect to specific desired functionalities, say, a wok's capacity to provide a uniform temperature distribution across the wok bowl (Otto an Wood 1998). Needless to say, such information is vital for design optimization.

For instance, if cooking a number of times with a specific wok (a range of interventions on the wok) quickly results in deformations on the wok surface due to which it then conducts heat in sup optimal fashion, and in turn uniform temperature distribution fails to be fully achieved, a change of materials, or properties of these materials of which the wok bowl is made, seems called for. Vice versa, one may track the effects of systematically changing, i.e., carrying out a range of interventions on, the thickness of specific wok bowls with respect to uniform temperature distribution; as it happens, thicker bowls contribute positively to this high level capacity since the capacity to conduct heat is improved when having thick rather than thin bowls, the former contributing better to uniform heat distribution across the bowl (cf. Otto and Wood 1998). Mutual manipulability is especially relevant for this testing since it gives, when extended with the fat handedness criterion, a handle on constitutive vs. causal relevance and on the robustness of constitutive relations.

5 Outlook and Conclusions

We have seen that import of concepts from the philosophical literature on explanation – here, mutual manipulability – has relevance for the testing of (engineering) design methods (cf. van Eck 2014). This connection also has relevance for the philosophy of explanation. One recent project at the interface of biology and engineering concerns elucidating, and re-characterizing, the nature of the relationship(s) between these domains (Calcott 2014; Levy 2014; Calcott et al. 2015). Historically, processes of designing have been likened to biological evolutionary processes (Calcott 2014). Such 'adaptionist' thinking has recently been criticized for providing misleading characterizations of (engineering) designing and, in effect, obscuring import commonalities between biology and (engineering) design (Calcott 2014). One important commonality that has been overlooked concerns the notion of "evolvability" or modifiability that is common to the development of both biological and engineered systems. As Calcott asserts:

> Complex integrated systems, whether evolved or engineered, share structural properties that affect how easily they can be modified to change what they do (Calcott 2014, p. 294).

Evolvable properties refer to features that affect how capacities of systems, engineered and evolved, change over time. Interestingly, although philosophy is only recently picking up on this theme, biologists and engineers alike have been stressing such joint principles governing change for more than a decade (e.g., Csete and Doyle 2002; Kitano 2004; Tomlin and Axelrod 2005). Modularity and robustness are two features that have gotten substantial attention in this context. Calcott (2014) analyzed this common core in the context of biology and software engineering.

The analysis given in this chapter extends this connection between biology and electro-mechanical engineering design. As in biology, evolvability also plays an important role in the context of the reverse engineering and redesign of electro-mechanical systems. Good reverse engineering explanations provide insight into the structure of extant technical systems, making it possible to modify or adapt parts such that optimization of system functionality ensues (cf. Sect. 4). Modularity here looms large of course, for this system feature makes it possible to optimize or change parts without affecting other functionalities of the system (in negative fashion). Ease of evolvability or modifiability is thus a desirable feature of technical systems, and good reverse engineering explanations, by highlighting the modular architecture of the functionalities of (genuine) constitutive parts, make it possible to evolve or optimize such systems.

This extension of the connection between biology and engineering, under the rubric of evolvability, is based on this chapter's main objective of elucidating the fruitful interplay between philosophy of (scientific) explanation and engineering design, specifically with regard to the testing of engineering design methods. As we saw, the mechanistic concept of constitutive relevance and its assessment in terms of the mechanistic mutual manipulability account, gives means to test the goodness of reverse engineering and redesign practices and the content of design representations resulting from them.

I suspect or at least hope that this is only the beginning. The philosophy of scientific explanation offers a rich source of diverse models of explanation that might prove relevant in the further elucidation and testing of design methods.

Acknowledgments I thank Pieter Vermaas for useful comments on previous versions of this chapter.

References

Baumgartner, M., & Gebharter. (2016). Constitutive relevance, mutual manipulability, and fat-handedness. *British Journal for Philosophy of Science, 67*(3), 731–756.

Baumgartner, M., & Casini, L. (2017). An abductive theory of constitution. *Philosophy of Science, 84*(2), 214–233.

Bechtel, W., & Abrahamson, A. (2005). Explanation: A mechanist alternative. *Studies in History and Philosophy of Biological and Biomedical Sciences, 36*, 421–441.

Bechtel, W., & Richardson, R. C. (1993/2010). *Discovering complexity: Decomposition and localization a strategies in scientific research*. MIT Press.

Braillard, P. A. (2015). Prospects and limits of explaining biological systems in engineering terms. In P. A. Braillard & C. Malaterre (Eds.), *Explanation in biology* (pp. 319–344). Springer.

Calcott, B. (2014). Engineering and evolvability. *Biology and Philosophy, 29*, 293–313.

Calcott, B., Levy, A., Siegal, M. L., Soyer, O. S., & Wagner, A. (2015). Engineering and biology: Counsel for a continued relationship. *Biological Theory, 10*, 50–59.

Chandrasekaran, B., & Josephson, J. R. (2000). Function in device representation. *Engineering with Computers, 16*, 162–177.

Craver, C. F. (2002). Interlevel experiments and multilevel mechanisms in the neuroscience of memory. *Philosophy of Science, 69*, S83–S97.
Craver, C. F. (2007). *Explaining the brain: Mechanisms and the mosaic unity of neuroscience*. New York: Oxford University Press.
Couch, M. (2011). Mechanisms and constitutive relevance. *Synthese, 183*, 375–388.
Csete, M. E., & Doyle, J. C. (2002). Reverse engineering of biological complexity. *Science, 295*, 1664–1669.
Darden, L. (2002). Strategies for discovering mechanisms: Schema instantiation, modular subassembly, forward/backward chaining. *Philosophy of Science, 69*, S354–S365.
Darden, L., & Craver, C. F. (2002). Strategies in the interfield discovery of the mechanism of protein synthesis. *Studies in the History and Philosophy of the Biological and Biomedical Sciences, 33*, 1–28.
Farrell, R., & Hooker, C. (2012). The Simon-Kroes model of technical artifacts and the distinction between science and design. *Design Studies, 33*, 480–495.
Farrell, R., & Hooker, C. (2015). Designing and sciencing: Response to Galle and Kroes. *Design Studies, 37*, 1–11.
Galle, P., & Kroes, P. (2014). Science and design: Identical twins? *Design Studies, 35*, 201–231.
Galle, P., & Kroes, P. (2015). Science and design revisited. *Design Studies, 37*, 67–72.
Glennan, S. (2005). Modeling mechanisms. *Studies in the History and Philosophy of the Biological and Biomedical Sciences, 36*(2), 375–388.
Hirtz, J., Stone, R. B., McAdams, D. A., Szykman, S., & Wood, K. L. (2002). A functional basis for engineering design: Reconciling and evolving previous efforts. *Research in Engineering Design, 13*, 65–82.
Illari, P., & Williamson, J. (2010). Function and organization: Comparing the mechanisms of protein synthesis and natural selection. In *Studies in History and Philosophy of Biological and Biomedical Sciences* (Vol. 41, pp. 279–291).
Illari, P., & Williamson, J. (2012). What is a mechanism? Thinking about mechanisms across the sciences. *European Journal for Philosophy of Science, 2*, 119–135.
Kitano, H. (2004). Biological robustness. *Nature, 5*, 826–837.
Lakatos, I. (1978). Falsification and the methodology of scientific research programmes. In I. Lakatos, J. Worrall, & G. Currie (Eds.), *The methodology of scientific research programmes* (pp. 8–110). Cambridge University Press.
Leuridan, B. (2012). Three problems for the mutual manipulability account of constitutive relevance in mechanisms. *The British Journal for the Philosophy of Science, 63*(2), 399–427.
Levy, A. (2014). Machine-likeness and explanation by decomposition. *Philosopher's imprint, 6*, 1–15.
Machamer, P. K., Darden, L., & Craver, C. F. (2000). Thinking about mechanisms. *Philosophy of Science, 57*, 1–25.
Ookubo, M., Koji, Y., Sasajima, M., Kitamura, Y., & Mizoguchi, R. (2007). Towards interoperability between functional taxonomies using an ontology-based mapping. In Proceedings of the International Conference on Engineering Design (ICED 07), August 28–31, 2007, Paris, France.
Otto, K. N., & Wood, K. L. (1998). Product evolution: A reverse engineering and redesign methodology. *Research in Engineering Design, 10*, 226–243.
Otto, K. N., & Wood, K. L. (2001). *Product design: Techniques in reverse engineering and new product development*. Upper Saddle River: Prentice Hall.
Tomlin, C. J., & Axelrod, J. D. (2005). Understanding biology by reverse engineering the control. *PNAS, 102*(12), 4219–4220.
Stone, R. B., & Wood, K. L. (2000). Development of a functional basis for design. *Journal of Mechanical Design, 122*, 359–370.
Stone R.B., Wood K.L., Crawford, R.H. (1998). A heuristic method to identify modules from a functional description of a product. Proceedings of 1998 ASME design engineering technical conferences, September 13–16, 1998, Atlanta, Georgia.

Van Eck, D. (2010). On the conversion of functional models: Bridging differences between functional taxonomies in the modeling of user actions. *Research in Engineering Design, 21*(2), 99–111.

Van Eck, D. (2011). Supporting design knowledge exchange by converting models of functional decomposition. *Journal of Engineering Design, 22*(11–12), 839–858.

Van Eck, D. (2014). Validating function-based design methods: An explanationist perspective. *Philosophy and Technology*. https://doi.org/10.1007/s13347-014-0168-5.

Van Eck, D. (2015a). Mechanistic explanation in engineering science. *European Journal for Philosophy of Science, 5*, 349–375.

Van Eck, D. (2015b). Dissolving the 'problem of the absent artifact': Design representations as means for counterfactual understanding and knowledge generalization. *Design Studies, 39*, 1–18.

Van Eck, D. (2015c). Reconciling ontic and epistemic constraints on mechanistic explanation, epistemically. *Axiomathes, 25*(1), 5–22. https://doi.org/10.1007/s10516-014-9243-x.

Van Eck, D. (2017). Mechanisms and engineering science. In P. Illari & S. Glennan (Eds.), *Routledge Handbook of Philosophy and Mechanisms* (pp. 447–461). New york: Routledge.

Van Eck, D., & Looren de Jong, H. (2016). Mechanistic explanation, cognitive systems demarcation, and extended cognition. *Studies in History and Philosophy of Science, 59*, 11–21.

Vermaas, P. E. (2009). The flexible meaning of function in engineering. Proceedings of the 17th International Conference on Engineering Design (ICED 09):2.113–124.

Vermaas, P. E. (2014). Design theories, models, and their testing: On the scientific status of design research. In A. Chakrabarti & L. T. M. Blessing (Eds.), *An anthology of theories and models of design*. Springer.

Woodward, J. (2003). *Making Things Happen*. Oxford: Oxford University Press.

Woodward, J. (2008) Invariance, modularity, and all that: Cartwright on causation. In S. Hartmann, C. Hoefer and L. Bovens (eds.) *Nancy Cartwright's Philosophy of Science* (pp. 198–237). New York: Routledge.

Research in Interior Architecture: Interdisciplinary Viewpoints and Research Approaches

Ann Petermans, Jan Vanrie, and Kris Pint

Abstract Until relatively recently, reflecting on interior environments was not regarded as a subject in its own right, but rather as an adjunct to architecture or an extension of decoration. During the last decades however, activities relating to interior architecture have become more visible, and have also become relevant topics for academic research. As the practice of designing interiors requires input from diverse areas of interest, ranging from humanities, social sciences to applied sciences, research in interior architecture and the construction of its body of theory should reflect this interdisciplinary character. However, the epistemological foundations of these various components tend to differ quite strongly and so do various research approaches within the discipline itself. As a consequence hereof, in this chapter we first discuss the 'identity' of the discipline of interior architecture whereby an explicit focus on exploring the human perspective is proposed. Phenomenology is discussed as a very valuable approach to the analysis and understanding of interior environments.

Next, we elaborate about two contrasting but complementary approaches for doing research in interior architecture: Design for Human Flourishing, and an arts-oriented approach towards the study of interiors. Both approaches differ in various aspects, but share the same core: the centrality of human experiences.

By comparing both approaches in terms of the underlying philosophical assumptions and methodological implementations, we illustrate the similarities and differences but also the added value that they can have to the further development of a more unified and proper body of theory for interior architecture.

Keywords Interior architecture · Phenomenology · Human experiences · discipline's identity · Diversity of research approaches

---

A. Petermans (✉) · J. Vanrie · K. Pint
Faculty of Architecture and Arts, Hasselt University (Belgium), Hasselt, Belgium
e-mail: ann.petermans@uhasselt.be; jan.vanrie@uhasselt.be; kris.pint@uhasselt.be

© Springer International Publishing AG 2018

P. E. Vermaas, S. Vial (eds.), *Advancements in the Philosophy of Design*, Design Research Foundations, https://doi.org/10.1007/978-3-319-73302-9_18

1 Introduction

The Bologna process, initiated in 1999, propelled an ongoing reform of European higher education. One of many practices that the Bologna process has instigated, is the integration of design schools – formerly belonging to a polytechnic or beaux-arts tradition – into academia. This presents different kinds of challenges to universities and design schools, but has also prompted different design researchers to reflect about the identity of their disciplines and fields of knowledge. Interior architecture is a relatively young discipline (cfr. Dickinson et al. 2009), which still lacks a specific body of knowledge, especially in relationship to architecture (Abercrombie 1990; Clemons and Eckman 2008, 2011; Edwards 2011). The process of integrating into a university context therefore provided an almost existential framework.

Starting from these premises, it is not surprising that research in interior architecture typically and mainly has relied on theoretical and methodological knowledge of relevant adjacent disciplines. By using information from these disciplines, researchers in interior architecture strive to add to design knowledge (Petermans and Van Cleempoel 2010). Indeed, until relatively recently, reflecting on interior environments was not regarded as a subject in its own right, but rather as an adjunct to architecture or an extension of decoration. During the last decades however, activities relating to 'interior architecture' or 'interior design' have become more accessible and visible. With interior architects worldwide, both in academia and the professional field, applying and producing domain-specific knowledge and developing their body of theory, the discipline is emancipating and growing more independent. In academic research in interior architecture, this is reflected in a very recent growth in publications contributing to the discipline's body of theory. However, up to date, there are no overarching theories or guidelines as to how interior architecture as a domain should be approached in order to capture its distinct nature. Different approaches seem possible – and often, also desirable.

Interior architecture as such requires input from diverse areas of interest: humanities, social sciences and applied sciences all affect the practice of designing interiors (Clemons and Eckman 2008, 2011). Research in interior architecture and the construction of its body of theory should therefore also express this interdisciplinary character. However, the epistemological foundations of these various components tend to differ quite strongly and so do various research approaches within these disciplines themselves. In this chapter, we firstly reflect on the issue of *identity*, that is, interior architecture as a distinct discipline. Next, we elaborate on *diversity*, i.e., different approaches employed within the discipline for performing research. We contrast two 'lenses' for doing research in interior architecture that differ in various aspects, but do share the same core: the centrality of the human perspective. Indeed, in line with other researchers before us (e.g., Abercrombie 1990; Bollnow 2011), we put forward a more human-centered approach in interior architecture; an approach inspired by phenomenology, which takes human experiences as the philosophical starting point of research in interior architecture. As will be demonstrated later on, Foucault's critical revaluation of phenomenology was particularly inspiring in this respect.

The core idea proposed in this chapter, is that the human dimension of design, that is, the particular attention for the interaction between user, designer and space is a crucial and distinguishing component in the design and understanding of an interior environment.

2 The Identity of the Discipline of Interior Architecture

During the last decades, theory of interior architecture has received increasing attention, although it was argued that it lacked a specific body of knowledge especially in relationship to architecture. At the same time however, academics in architecture noticed that existing academic architectural knowledge hardly filtered down to practicing architects (Neuckermans 2004). Aiming to assure that academic knowledge can be transmitted to interior architectural practice, while in the meantime aiming to support the development of the discipline's body of knowledge, Edwards (2011) pleas for combining the advancement of the discipline's theoretical knowledge base with insights into actual design practices. We agree with Edwards' point-of-view, but we also want to add an element to the discussion: taken into account the wide spectrum of interior architectural design practice, it can be argued that theory of interior architecture needs to reflect this diversity, both thematically and methodologically, without neglecting the human dimension of design. Therefore in this section, we first discuss what interior architecture entails, and what research in interior architecture focuses on so as to come to discussing the current 'identity' of the discipline. Next, we argue how, in our view, research in interior architecture can be strengthened in an effort to also reinforce this identity.

2.1 What Is Interior Architecture?

To date, there seems to be a semantic discussion regarding the question of which label, 'interior architecture' or 'interior design', best characterizes the activities which are key to the discipline.

Brooker and Stone (2007: 126) suggest that *interior design* is *'an interdisciplinary practice that is concerned with the creation of a range of interior environments that articulate identity and atmosphere, through the manipulation of spatial volume, placement of specific elements and furniture and treatment of surfaces'*. According to Edwards (2011), this definition emphasizes the link between the space, the materials and objects used in the process of designing the space, and the user. Interior design in the strict sense usually entails projects that require little or no structural changes to the existing building, although there are exceptions (Brooker and Stone 2007, 2010).

Regarding *interior architecture*, Brooker and Stone (2007: 126) state that it '*is concerned with the remodelling of existing buildings and attitudes towards existing spaces and structures, building reuse and organisational principles. It bridges the practices of interior design and architecture, often dealing with complex structural, environmental and servicing problems*'. This definition entails that interior architecture encompasses designing interiors for domestic, recreational and business usage, whereby architectural processes and principles can be applied (Edwards 2011).

As the definitions reveal, interior design and interior architecture can be considered as 'specialisms' within the practice of designing interiors. There are, however, regional differences. Edwards (2011: 2), for example, states that '*in mainland Europe, the interior architect generally undertakes the type of work that in North America would be recognized as interior design*'. As such, the differences that Brooker and Stone (2007) identify between what they label as 'interior design' and 'interior architecture' seem to be rather small. During the last decade, also in North America, the topic of how to describe most appropriately the activities undertaken by people who work with interiors continues to be a point of attention, with continuously more people seeming to prefer the label of 'interior architecture' (Carll White 2009).

2.2 Research in Interior Architecture

Taking into account the early stages of theory building and research in interior architecture, we first consider the broader epistemological framework of research by discussing a concise scheme put forward by Groat and Wang (2013).

Groat and Wang (2013) emphasized that every act of research is unavoidably framed by a particular system of inquiry, because all researchers make assumptions about the nature of the world and how knowledge about aspects in the world around them can be gathered. They introduced a figure (see Fig. 1) that illustrates how a choice for a particular 'system of inquiry' (sometimes this is also labelled a 'paradigm' or 'worldview', e.g., post-positivism) has repercussions for the 'school of thought' that inspires researchers. In this framework, phenomenology for instance is a 'school of thought' that has profoundly influenced research in various scientific disciplines, among which interior architecture. A particular 'school of thought' can in turn influence researchers' choice for concrete research 'strategies' and 'tactics' to set up research in interior architecture. 'Strategies' here concern the overall research plan or structure of the study, whereas 'tactics' relate to the actual research method(s) to be employed in order to answer the proposed research question (Groat and Wang 2013).

Studying how users experience a particular interior environment can be done from multiple angles. Generally, as Fig. 2 demonstrates, three research paradigms have been distinguished: qualitative research, quantitative research and mixed methods research (Creswell 2003; Johnson 2007).[1] Comparable to Groat and

[1] Mixed methods have reached maturity in the last decade (Creswell 2003).

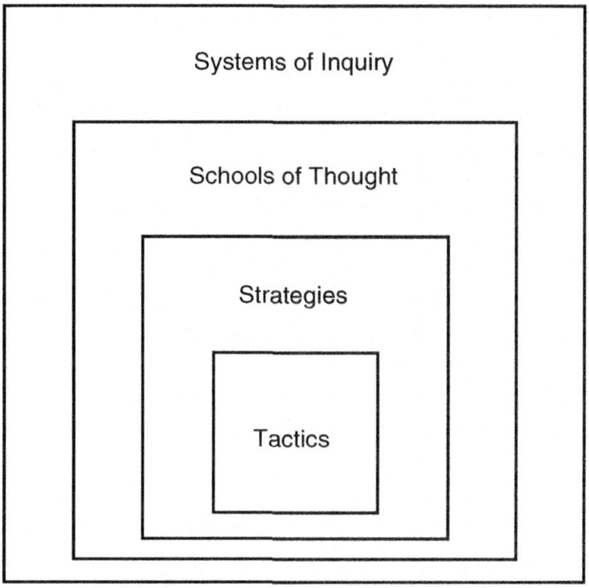

Fig. 1 The methodological practices of strategies and tactics are framed by broader systems of inquiry and schools of thought (source: Groat and Wang 2013, p. 10)

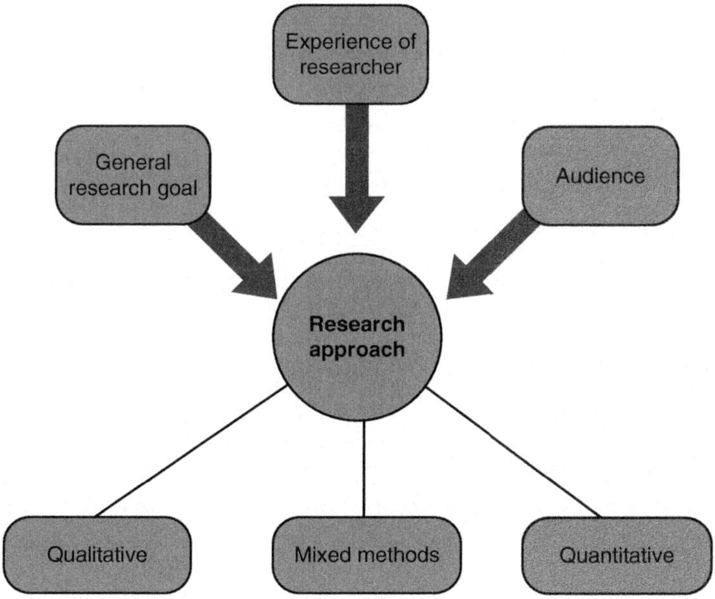

Fig. 2 Overview of different research paradigms

Wang's characterization, each of these major paradigms includes assumptions, principles and values concerning methods of doing research (e.g., experiments, ethnography), methods of data collection (e.g., interviews, questionnaires) and related philosophical issues (Morgan 2007). In the past, researchers often were considered as being representatives of the qualitative or quantitative research paradigm. Today, the situation has become less clear-cut, as *'the best that can be said is that studies tend to be more quantitative or qualitative in nature'* (Creswell 2003: 4).

According to Creswell (2003), three factors influence the ultimate choice for a particular research approach. First, the research question and the general research goal of the study. Certain types of research problems require specific research approaches (Creswell 2003). For instance, to find out which factors are the best predictors of particular outcomes, a quantitative research approach seems the most appropriate. On the other hand, to investigate a subject which has not been explored in its full detail yet, a qualitative approach seems to be more suitable (Ritchie and Lewis 2003). Mixed methods research, then, combines aspects of qualitative and quantitative research. This seems to be the appropriate research approach when a researcher for instance not only wants to develop a detailed understanding of the research problem, but also wants to generalize the findings to a population. Second, the researcher's personal training and experience usually also play a role in the choice for a particular research paradigm. Third, when deciding on the choice for a particular research paradigm, researchers also take into account the audience to whom they report their research findings. Audiences can range from professors over colleagues in design practice to journal editors. Together with the earlier discussed factors, the experiences of these audiences will help to guide the involved researcher to choosing a particular research paradigm.

Qualitative research has a long history in the humanities. As Fig. 3 illustrates, in the course of the twentieth century, different 'schools' of qualitative research have emerged (Ritchie and Lewis 2003; Snape and Spencer 2003).

Researchers, influenced by symbolic interactionism, are convinced that people act according to the meanings things and other people have for them. These meanings come forth from social interaction and are modified via interpretation. In their research projects, symbolic interactionists thus focus on exploring behaviour and social interaction, in order to find out how people react to their environment. Constructivist researchers highly value the thought that knowledge and reality is constructed by human thinking. As a consequence, researchers working in this tradition concentrate on revealing these constructed realities through the shared investigation (researchers and participants) of meanings and explanations. Researchers, working in critical theory, examine and criticize society and culture. In general terms, their research projects often aim to identify ways in which material conditions (e.g., economic, political, gender, ethnic) influence beliefs, behaviour and experiences (Snape and Spencer 2003).

Taking into account that we aim to do research on experience in a way that appeals to our holistically inspired design peers, we are particularly inspired by the phenomenological and ethnographic school of thought. From all disciplines where

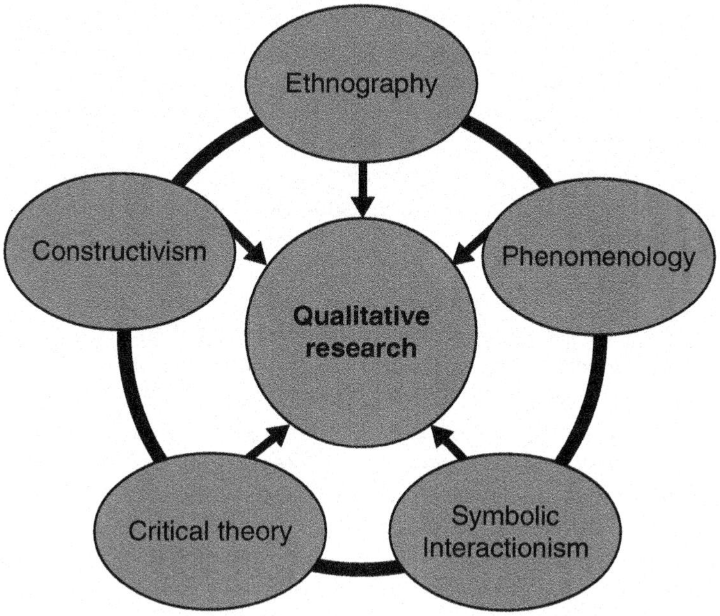

Fig. 3 Different schools of qualitative research (source: Snape and Spencer 2003)

holistic thinking has penetrated throughout time, phenomenology and Gestalt thinking offer holistic perspectives which help to understand the dynamic relations between designers, their creations, and those who use them.

2.3 *Strengthening the discipline's Identity: An Explicit Focus on Exploring the Human Perspective*

Designing interiors implies taking into account issues and reflections originating in humanities (e.g., philosophy, history), social sciences (e.g., sociology, psychology), the arts, architecture, and various applied sciences (e.g., environmental studies, technology and ICT). As these branches impact on interior design practice, it is evident that researchers focusing on the development and further growth of the body of theory of interior architecture seek and also find inspiration in these disciplines. However, we argue that not only an outline of the unique identity of the discipline is slowly emerging with at its core the centrality of the human perspective, but also that this identity, this unique human perspective on interior spaces, can be further strengthened by stimulating research approaches which also embrace this particular perspective.

Designing interior spaces always entails an interaction between a user, designer and a space, whereby the spatial design is (or, ideally should be) fostered via designer's dialogues with the concerned users, in an effort to take their experiences, wishes and longings into account in the design process. In this chapter, we focus on studying human experiences in domestic spaces, as these concern the most intimate spaces with which interior architects work. This brings about numerous challenges, both in terms of design research and the practice of interior architecture.

The literature that will be used to develop our arguments and illustrate our point-of-view is studied from an interior architectural perspective, which, however, does not imply that some findings could not be valuable for the field of architecture as well.

In design practice, everyday interior spaces are constructed by focusing on human experiences. This perspective is evident in Bachelard's work (1964/1994), who found a metaphor of 'humanness' in the house. He wrote (1964/1994: 4) '... *our house is our corner of the world... it is our first universe, a real cosmos in every sense of the word. ... If we look at it intimately, the humblest dwelling has beauty'*. In his view, houses are not only a geometrical object that originates from the combination of different angles; on the contrary. Bachelard's inquiry focuses on the house and different aspects related to the intimacy of domestic space. He elaborates for instance on the idea of a person's room, not only as a physical space, but also as a space for reflection and recollection. In his view, a space is a symbol of one's realness and certainty of being, and so, in his view, the poetics of a home place *define* us. Pallasmaa (2011) agrees in different publications (Pallasmaa 1995; 2011). In *The embodied image* (2011), he indicates that architecture today is being threatened by two processes: instrumentalisation and aestheticisation. In his view, on the one hand, our materialist, secular culture is continuously considering buildings as mere instrumental structures without meaning, for the purposes of utility and economy. On the other hand, in order to appeal to people, architecture is increasingly turning to the fabrication of aestheticised images that have no roots in our existential experience and miss references to people's authentic longings. Therefore he pleas that architecture needs to focus not only on providing shelter to people. Architects also need to reflect how buildings can become *'externalisations of our imagination, memory and conceptual capacities'* (2011: 119), as they help to structure our experiences and allow us to give meaning to them. Borch (2014) concurs by emphasizing the (re)turn towards the importance of atmosphere and the way people perceive or *experience* this in a spatial context. Although these concerns stem from an architectural background, in our view, they are even more pertinent for interior architecture. This also resonates strongly in recent handbooks attempting to capture and define this new domain of interior architecture, such as Edwards (2011), who indicates that this trend truly relates interior architecture with attention for human aspects, issues and perspectives, or, in the words of Brooker and Stone (2010: 60), *'The interior can be regarded as the narrative or backdrop to life. It is the manifestation of the occupier; ... it expresses ... the individuality of <u>those who inhabit</u> the space '*.

2.4 Phenomenology and Attention for Experience

To understand how interior architecture works, one has to take human experiences as a starting point, as the active participation of individuals gives shape, meaning and purpose to interior spaces (Edwards 2011). This brings us to phenomenology, as a very valuable approach to the analysis and understanding of interior environments. In this respect, the existential phenomenology of Maurice Merleau-Ponty is inspiring, as he focused on the experiences of concrete human beings in the concrete world. While phenomenology focuses on studying and describing reality as it emerges in concrete experiences (Seamon 2000), Merleau-Ponty made the human body the center of the experiential world and focused on studying people's sensory experiences of the world (Pallasmaa 2005a, b). In his view, a human being can never experience things independent of his experience as a bodily engaged being in the world.

In his writings, Merleau-Ponty (1964: 48) also pointed to the simultaneity of experience and sensory interaction: *'My perception is not a sum of visual, tactile, and audible givens: I perceive in a <u>total</u> way with my whole being: I grasp a unique structure of the thing, a unique way of being, which speaks to <u>all</u> my senses <u>at once</u>'*. Without labelling it as such, Merleau-Ponty here refers to a 'Gestalt' way of thinking, which is a key issue for theorists and practitioners in interior architecture. Considering environments as a 'totality' or a whole where various elements interact and together emanate a 'Gestalt' environment, was introduced in the field by Norberg-Schulz (1988). By introducing the concept of 'genius loci', he brought phenomenological thinking to design theory overall, and into the architectural field in particular (Wang 2015). 'Genius loci' entails that every place has its own unique qualities, whereby these qualities altogether determine an environment's character, and shape the place's essence. In a 'Gestalt' environment, users experience the whole as more than the sum of its constituent parts (Norberg-Schulz 1988). Approaching architectural design via a phenomenological lens thus allows to truly take the interdependent relationship between Bachelard's concept of 'humanness' and the world into account. These reflections demonstrate in what way phenomenological writings have had a major impact on architectural thinking and architectural theory.

In design practice, the 'Gestalt' way of thinking is also evident: interior architects look at interiors as Gestalt environments, which continuously interact with their users (Cupchik and Hilscher 2008). The importance of this concept demonstrates that spaces which need to be designed or studied empirically, need to be approached as holistic totalities, since the experience of the actual, 'total' environment will determine how a person feels and behaves in a space.

In addition, also the work of Michel Foucault is highly valuable to analyzing and understanding interior environments. However, the work of Foucault radically challenges the human-centered approach that phenomenology takes as its starting point, with the individual consciousness as the autonomous origin of experience. We want to argue that this critique is very relevant, but does not invalidate a phenomenological approach based on subjective human experiences, rather than on strictly quantifiable, objective parameters.

For Foucault, the 'human perspective' of phenomenology is an ideological construction that belongs to the modern *episteme*, the specific field of knowledge that structures our contemporary subjective experience. Foucault argues that individual experiences are always embedded in a discursive formation that renders them intelligible. In modernity, Foucault argues, the central conceptual figure in this framework is 'man', an epistemological invention of the human sciences, defined and determined by labour, language and the body. These sciences turned the human being into an object of research, whose 'true nature' was to be studied and discovered by economics, philology, and biology. 'Man' became the central object of research, and the research results were used, explicitly and implicitly, to define 'normal' human behaviour (see Foucault 2002). This implies that if we want to focus on the human dimension of design, we should also take into account how we as 'human beings' have been 'invented', 'designed' by the discourse of the human sciences. Foucault's *caveat* is particularly relevant for studies in interior design, precisely because the emergence of the modern interior coincides with the conceptual 'invention' of man at the end of the eighteenth century. Human behaviour, including its practices of living and dwelling, became the focus of scientific research, and the results of that research were used to legitimize the design of human dwelling, both public and private. Human sciences like psychology, biology, sociology and economics provided the framework for a view on dwelling and working based on economic efficiency, hygiene, and psychological and societal needs. They became the epistemological legitimation of a process of normalisation and standardisation in western societies, in which any serious deviance from the norm had to be corrected and adjusted. For Foucault, it is therefore not only the interaction between user and space that is crucial to understand interior environments; one should also take into account the role of discourse in this interaction. The subjectivity of the user is a secondary effect of both discourse and architectural design, as the spatial expression of such a discourse. In his *Surveiller et punir: Naissance de la prison* (1975), Foucault gives the example of the panopticon, a kind of prison typology designed by Jeremy Bentham. In the middle of the prison there is a watchtower, encircled by the individual cells of the prisoners. This allows the guard in the centre to observe each individual cell, leaving no possibility for the prisoners to hide from a controlling gaze that they themselves do not see. '*They are like so many cages, so many small theatres, in which each actor is alone, perfectly individualized and constantly visible. The panoptic mechanism arranges spatial unities that make it possible to see constantly and to recognize immediately. (...) Visibility is a trap.*' (Foucault 1977: 200). Obviously, the supervisors cannot observe each prisoner all the time, but the simple fact that one is never sure if one is looked at or not, suffices to control the behaviour of the prisoners. The panopticon is the extreme architectural realisation of the desire of the human sciences to make everything visible, accountable, and controllable. Not only prisons, but also schools, factories and to some extent also private interiors became the focus of a permanent surveillance, based on assessments, surveys, evaluation, monitoring. In our current society the panopticon has become the default typology. Whereas in the disciplinary society the controlling gaze was limited to specific places, like prisons, schools, mental asylums or working spaces, in what Deleuze calls the contemporary 'societies of control' the panoptical gaze is everywhere, due to technological innovations (see Deleuze 1992).

The panopticon serves a specific exercise of power which Foucault called 'bio-power', and which has replaced in Europe the older pastoral power of the Church as a way to control both the individual and the population: '*It was no longer a question of leading people to their salvation in the next world, but rather ensuring it in this world. And in this context, the word salvation takes on different meanings: health, wellbeing (that is, sufficient wealth, standard of living), security, protection against accidents. A series of 'worldly' aims took the place of the religious aims of the traditional pastorate*' (Foucault 1983: 215). In such a society, the figure of the priest has been replaced by the manager, the consultant, the technocrat and the expert: all working to optimize the organisation of labour, of recreation, of dwelling, based on what the human sciences implicitly have defined as normal human behaviour. As a Nietzschean, Foucault wants to go 'beyond good and evil', so his analysis of bio-power should not be interpreted as a simplistic rejection of it. But what Foucault does point out is the totalising nature of this bio-power, by a system of constant surveillance, its effort to make everything transparent, visible and measurable. Bio-power is virtually everywhere, and this makes it so difficult to notice its impact on the formation of subjective, individual experiences. Of course, a phenomenological, human-centered approach that focuses on wellbeing can critically address the negative symptoms of this culture of 'transparency' and 'efficiency', and its impact on the perceived quality of interior environments, the increase in social anxieties, uncanny feelings, fatigue, stress, burn-outs etc. A good example of this is the work of Pallasmaa mentioned earlier: his texts provide us with a very relevant critique of the dominance of a visual and instrumental approach to architecture (Pallasmaa 2005a) and the lack of imagination, of sensuality in modern-day architecture (Pallasmaa 2005b, 2011). But precisely because it takes the subjective experience as its starting point, it cannot address what precedes this subjectivity, its structural 'formation' as a modern subject in the society of control, which is already a form of disciplinary normalisation, and an implicit definition of what, e.g., 'happiness' should mean in a specific system of values. It overlooks the discursive framework that has created this individualised, meritocratic performing-subject that suffers from the social and physical design of the interior environments in which he dwells (see, e.g., Verhaeghe 2014; Han 2015). Following Foucault, we should not forget that the 'inner perspective' of phenomenology is precisely part of a broader discursive field that wants to explore the nature of 'man'. In a way, phenomenology too wants to discover a 'hidden dimension' of human existence, and is thus itself a form of *panopticism.* The desire to make everything visible, even such ephemeral things like personal imagination, experiences, moods and affects, runs the risk of mobilizing these subjective experiences to control individuals, to 'normalise' their behaviour.

Foucault's work urges us to keep in mind the historical and discursive nature of 'subjective experience'. It makes us aware of the implicit normative aspects of research in interior architecture and the danger of naturalising a specific contemporary discourse on dwelling, well-being and subjectivity. The active participation of (future) users in the research process can be one possible way to avoid this. By allowing them to express, verbally or visually, their outlook on the built environment in which they live or want to live, a specific normative discourse can be challenged. But even such an approach has to take into account the dangers of a panoptic desire to know, and the possible risk of turning this knowledge into a form of biopower, even with the best of intentions from the part of the researcher. Another way to integrate Foucault's critique into a phenomenology of the interior, is to explore the possibility of exploring and integrating the kind of experiences that go beyond 'normal' subjective experience. As we will try to argue, artistic research offers the possibility of such an experimental exploration of other modes of existence.

It is clear that phenomenology has an added value for research in interior architecture because it provides not only a valuable conceptual frame of reference, but it also offers levers towards research methodology which can assist researchers in interior architecture to examine spatial, experiential, environmental or architectural dimensions of life. In addition, phenomenological research in interior architecture can focus on revealing aspects with regards to experiences and our relationships with physical and spatial environments which typically would remain unnoticed if not studied via this particular frame of reference. Such information can consequently be of high relevance for designers in practice (Seamon 2015), all the more because it provides a particular lens to discuss issues which designers encounter regularly in their daily design processes (Wang 2015).

3 Diversity of Research Approaches in Interior Architecture

Figures 2 and 3 which were discussed in Sect. 2 already concisely pointed to the diversity of research approaches possible for doing research in interior architecture. Taken into account our focus on the human perspective, and our inspiration in phenomenology, Sect. 3 discusses methodological repercussions that phenomenology can have for research in interior architecture. We propose two approaches for doing research in interior architecture, both framed within this school of thought of phenomenology, and differing in various aspects, but that do share the same core: the centrality of the human perspective. First, we describe 'Design for Human Flourishing' as a research and design approach that focuses on the contributions of interior architecture to help people to be or become happy or happier in the environments wherein they reside. Next, we propose an arts-oriented approach towards the study of interiors as a way of thinking in and through images and spaces. This approach also takes 'human experience' as a starting point, but tries to integrate the poststructuralist critique of phenomenology by keeping in mind the discursive, historical dimension of human experiences. Artistic research can offer a way to explore what remains 'outside' the ideological framework that constitutes our contemporary subjectivities, and as such, it offers a valuable supplement to a more classical phenomenological approach to interior environments.

3.1 Studying Interiors Inspired by Phenomenology

In 2009, Poldma and Thompson indicated that interior spaces should be studied as the '*dynamic backdrop of human activity*' (468) whereby researchers should pay attention to conversations, perceptions and experiences of the stakeholders who are involved in the design process, which inevitable have subjective and tacit characteristics (Nelson and Stolterman 2003). Both in a literal and non-literal sense, the voices of these stakeholders are key to gain insight in experiences that people encounter in physical and social environments. Research that aims to elaborate stakeholders' voices is highly valuable for interior architecture, because answering the question of how people use and experience a space is essential for the relative success of a particular interior architecture project (Poldma 2015).

Trying to understand the complex nature of experiences in designed environments, it is necessary to look for research approaches which rely on entering people's natural life worlds while aiming to interpret the complexity of human experiences.

In the last few decades, phenomenology has often been connected with the conceptual and methodological possibilities of qualitative research (see Seamon 2000 for an elaborate overview). As discussed in Sect. 2, qualitative research has a long tradition in human sciences, where different researchers (e.g., Ritchie and Lewis 2003; Snape and Spencer 2003) have thought about clustering qualitative research approaches into separate groups. One of these groups concerns *ethnography*.

Researchers applying ethnographic research approaches are convinced that what individuals believe, understand and act upon, cannot be detached from its context (Riemer 2008). As ethnography takes place in the real world, it seems a valuable research approach to offer well nuanced insights into design practices. In that way, ethnography's task is '*not only to watch, but also to decode human experience*' (Mariampolski 1999: 18). Classically, ethnographic research is inspired by phenomenology, whereby ethnographic approaches aim to understand the insiders' perspective of reality (Fetterman 2008). As a consequence, it seems particularly well matched with viewpoints of researchers and practitioners in interior architecture, as discussed above.

Evidently, there is a wide scope of ethnographic research approaches, i.e., strategies and tactics, possible for doing research in interior architecture. In discussing 'Design for Human Flourishing' in the next section, we elaborate about the use of photo-elicitation, that is, a visual ethnographic research approach.

3.2 *Design for Human Flourishing*

3.2.1 Experience, Wellbeing, Happiness

As discussed earlier in this chapter, a current key question in different design disciplines among which interior architecture, is how people experience designed environments (see for instance, Petermans et al. 2013; Hassenzahl et al. 2013; Petermans and Pohlmeyer 2014). If this question is taken one step further, one can ask: can design empower people to flourish in the environments which they occupy, and in which they live, work and/or play? Can design stimulate them to undertake activities that contribute to their subjective wellbeing?

Such questions still go beyond most current common beliefs in design disciplines such as interior architecture, although the issue starts to surface. For instance, with regards to reflecting on interior architecture and the contribution that this discipline might have for wellbeing and happiness, Smith et al. (2012: 2) stated *'historically, wellbeing and interior design have been linked through the creation of hospitals, doctors' surgeries and other related facilities for health care and treatment of illness. Recent trends in interior architecture have broadened the scope beyond the medical model of treating the sick, aged and disabled to the* <u>*wellbeing*</u> *of all people in their everyday environments'* (Smith 2011).

In discussions concerning people's wellbeing, however, up to date, interior architecture still often goes unnoticed. The issue itself, however, is becoming more and more important. 'Happiness' and 'wellbeing' after all are major, if not the ultimate goals, for every human being.

To date, researchers from diverse disciplines have tried to point to the essence of wellbeing and happiness. Different philosophers, theorists and researchers from psychology, economics and neurosciences are interested in happiness, and also the last few years, different researchers from various design disciplines have begun to investigate whether their discipline can contribute to the happiness of people – and if so, what this contribution can look like or how it can be set up or produced. Although there is no consensus to date relating to the conceptualization of wellbeing and happiness (Lee et al. 2011; Desmet and Pohlmeyer 2013; Petermans and Pohlmeyer 2014), what different researchers seem to agree about is firstly, that happiness is determined for a large part by genetics (50%), life circumstances (10%) and intentional activities (40%) (Lyubomirsky et al. 2005; Lyubomirsky 2008). The fact that people can influence their happiness by focusing on the set-up of intentional *activities*, creates tremendous opportunities for design in general, and for interior architecture in particular.

A second point of agreement among various researchers is that happiness and wellbeing have an objective and a subjective component (Veenhoven et al. 2014; Petermans and Pohlmeyer 2014; Petermans and Nuyts 2016) (that is, objective and subjective wellbeing). Objective wellbeing can be understood as the degree to which external conditions – that can be objectively assessed – for having a high quality of life are met (Constanza et al. 2007). Subjective wellbeing can be generally understood as people's self-reported evaluations of their lives as a whole. Subjective wellbeing can lead to *human flourishing* if several wellbeing components, such as positive emotions, engaging activities, positive relationships, and meaning are present in combination (Seligman 2011; Huppert and So 2013; Desmet and Pohlmeyer 2013).

When thinking about wellbeing from an interior architectural perspective, one can thus consider the 'objective conditions' of a designed space, as well as people's subjective experiences herein. Without neglecting the contribution of objective conditions of wellbeing, it is valuable to study how a designed space can support people to engage in or to relate to activities that add meaning and pleasure to their life. Such activities can in turn lead to sustainable increases in happiness (Lyubomirsky et al. 2005; Sheldon and Lyubomirsky 2006).

In the field of 'Design for Human Flourishing', various research approaches or 'tactics' (see Fig. 1) can be employed in order to formulate answers to particular research questions, ranging from historical research, literature review, quantitative research to qualitative research. In what follows, we chose to elaborate one particular qualitative and ethnographic research approach that seems highly valuable to gain insight in this topic of inquiry: photo-elicitation.

3.2.2 A Visual Research Approach to Study Design for Human Flourishing

In today's society, the *experiences* that people have, or should have, at particular places has become a very important aspect of the design brief and consequently, the design process, of many architects and interior architects. However, experience and subjective wellbeing as such are very abstract concepts to work with; they are full of subject matter, but it is very difficult to grasp the exact contents of a person's experience and subjective wellbeing in a designed environment (Petermans 2012; Petermans and Pohlmeyer 2014). To overcome this difficulty with which various designers are struggling today, in this section we elaborate on the use of a visual research approach called 'photo-elicitation', as we believe that this particular research tactic can be truly inspiring for interior architects and support them in their design projects.

3.2.2.1 Visual Research

According to various authors (Harper 2002; Prosser 2007; Schroeder 1998, 2002), visual research can contribute to understanding both the symbolic and physical meanings of the built environment. Since Wagner's (1979) work on photography as a research method, different researchers have demonstrated an interest "*in handing the camera to those whose lives we wish to explore…because photography offers opportunities for research participants to express their subjectivities as – quite literally – their view of the world*" (Warren 2005: 865). From the first use of photography as a research method, visual research methods succeeded in disciplines with traditional ethnographic histories such as anthropology and sociology (Berg 2008). Relatively recently, also researchers in design disciplines have come to acknowledge the value that visual research might have for their field of knowledge.

Within the domain of visual research and its methodology, researchers (Harper 2002; Warren 2005) have identified different visual research approaches, such as using images to help participants to express their feelings, beliefs, perceptions and experiences, either as an aid to verbal narrative, or in place of it. One of these research approaches is "photo-elicitation".

3.2.2.2 Photo-Elicitation

Photo-elicitation refers to a visual research approach in which a researcher asks participants to take photographs in a particular environment in an effort to help them to capture some or multiple aspects of their experiences in the concerned environment (Warren 2005). Afterwards, the photographs are inserted into a research interview, and as such act as a communication bridge between the interviewer and the interviewee as they help the interviewees to explicate their thoughts, feelings and experiences about the research topic in question in the course of the interview.

Collier (1957) initially proposed its use in social science research, but the steady growth of interest towards photo-elicitation as a research methodology has been stimulated by various developments. First, there is everyone's growing and everyday exposure to and immersion in multiple visual signs, images and photographs. This makes photo-elicitation a research approach that is close to people's life-world and their everyday practice of life. Second, the interest in photo-elicitation can be explained by the influence of postmodernist thinking in social science, which provides methodological underpinning for visual research, as this research approach helps to overcome problems that research participants might experience with regards to articulation and discourse in other, more traditional research approaches (Pink 2015). Third, Warren (2005) indicates that for various researchers, issues such as reflexive practice, subjectivity, and immersion in the worlds which they research are important. The subjective nature of photography lends itself particularly well to take such issues into account, as "*the photograph almost literally acts as a lens through which we see what others 'see' and importantly, deem important enough to 'capture' with a camera*" (Warren 2005: 866).

Photo-elicitation can help to make the use, experiences and wellbeing of people in a space tangible for designers and design researchers. Recently, Doyle et al. (2015) for instance used photo-elicitation in a research project wherein she aimed to gain insight in aspects of home life of ordinary people, their home experiences, wellbeing, and daily routines in their respective home environments. Research participants for instance came up with images concerning positive issues of their home life, which often related to children, pets, hobbies or relaxing activities. Images relating to negative issues concerned household chores or disorderly spaces at home. Starting from these concrete visual stimuli prompted participants to elaborate in an interview on personal frustrations or other emotions, and allowed them to reflect to what extent such concrete, tangible issues contributed to their home experiences and wellbeing. Doyle et al. (2015) thus learned that offering the participants this creative task, and engaging them in the research, made them emotionally connected and enabled them to reflect about otherwise probably unconsidered issues of domestic life at home.

3.2.2.3 Added Value of Photo-Elicitation for Strengthening the Knowledge Base of Interior Architectural Theory and Practice

As Doyle's research illustrates, gaining insights in these issues can be highly inspirational for architects and interior architects firstly, as the method allows them to proactively engage with future users which can help them to steer their design process in a particular direction. Secondly, photo-elicitation can also be highly inspirational for designers in a post-occupancy trajectory, as the method then can allow them to learn from users as they offer 'sight and insight' in their perceptions, and experiences and wellbeing in the designed environment which the concerned designers delivered.

Thirdly, photo-elicitation also seems an appropriate research approach for researching uses and experiences of people in different sorts of environments, ranging from home places to museums and cultural centers to care-related environments.

Next to an added value for the further theory development of the discipline of interior architecture, photo-elicitation can also offer architects and interior architects the possibility to engage with users and learn about their experiences of the concerned designed environment. Using photo-elicitation can for instance help designers to get insight in user experiences with regards to the consequences of particular choices that need to be made in the design process. Via photo-elicitation, users can explicate their experiences concerning particular materials, lighting, furniture, organization of spaces, etcetera. Such information can distinctively extend knowledge of the meanings which users apply to the experience of the environments wherein they live, work and function. This can be relevant and valuable input that can help steer future design projects and add to the body of theory of interior architecture.

3.3 Artistic Research and the Exploration of Limit-Experiences

After discussing Design for Human Flourishing as an approach to doing research in interior architecture, in what follows, we elaborate about an arts-oriented approach towards the study of interiors. In so doing, we reconnect to Foucault's critical revaluation of phenomenology.

Despite his fierce rejection of phenomenology, Foucault's analyses share the same focus on the subjective experiences of our actual existence (see May 2005). But in his analysis of these experiences, Foucault wants to focus precisely on those experiences that are normally left out of phenomenological analyses, those experiences that are abnormal, eccentric, that emerge at the borders of our subjectivity. Foucault called them 'limit-experiences', paradoxical events where the subject experiences something that overturns it, threatens to destroy it. It were the kind of experiences he found described not in the works of Sartre or Merleau-Ponty, but in the transgressive literature of Bataille and Blanchot, inspired by the work of Nietzsche (see Foucault 2001: 241). The kind of radical phenomenology Foucault wants to practice, does not contend itself to express the inner self, but to radically change and transform it, to explore other modes of existence (Foucault 1997b). 'Limit-experiences' are important, because they reveal the cracks in the discursive formation of subjectivity, they hint at what remains unsayable, invisible in a given discursive formation. They provide a line of flight, a possibility for change. Such moments of resistance are potentially liberating, because they offer the chance to formulate an alternative. They are an invitation to cross the border and to experiment within a specific field of knowledge, to find out where change to the existing framework becomes both 'possible and desirable'. Artistic practice, and more recently, artistic research, can be a form of resistance to the totalitarian effect of the technocratic discourse of panopticism and bio-politics.

But of course, the question here is for *whom* this resistance is relevant. For whom should society change, if one, as Foucault did, utterly rejects the idea of an 'authentic' core that should be safeguarded, something essentially 'human' that is supressed and should be allowed to express itself, to 'flourish'? In existential phenomenology (Kierkegaard, Heidegger, Binswanger, Sartre …) the experience of the individual, the absolute freedom of his inner self, provides the ground from which to challenge the technocratic reduction of human experience in modern society. But what if this 'inner self' is merely a discursive construction? How then to defend the need and value of resistance? At the end of his life, Foucault began to understand that such a radical experimentation with ways of living, with modes of existence was only possible in relation to the individual's own existence. There has to be a kind of subject left that can experience its own limits, in order to transform itself and others. It requires a subject that wants to live 'truthful', whereby the truth of his existence is not legitimized by an external field of knowledge. Here the 'truth' of one's experiences is grounded in a life-practice that is in accordance to one's own discourse. Foucault was inspired by the existential praxis that the ancient Greeks called the 'care for the self': a set of techniques, exercises and rules an individual has to his disposal to transform himself (see Foucault 1997a: 223–252). Transposed to modernity, this 'care of the self' became the Nietzschean idea of an artistic 'self-creation', an aesthetics of existence. This notion of the 'care of the self' has also implications for the kind of research such a subject performs. Such research should be clearly engaged with one's own subjectivity, but only in order to challenge it, test it, change it. Artistic or designerly research wants to be such a borderline exploration, a method to express what is left out in our current and common-sense view on 'human experience'. Artistic research is a way to use art to think, and to perceive differently. Taking such research seriously does not only imply that artists, interior designers and architects have the right to approach their work with their own set of tools. It can also be seen as a valid and valuable way to understand reality. The challenge, and one could even say, the necessity here is to develop, through sketching, drawing, painting, writing, building, performing, a discourse that allows to trace those limit-experiences, those thoughts and experiences that could not be experienced or thought in the framework of the discourse of the human sciences. Artistic research in this sense by definition requires a 'human-centered approach', because it takes the individual life as a starting point to explore different modes of being, or rather: becoming, human. It can indeed be seen as a form of auto-ethnography, but of an experimental kind, in which the self changes in the process. Such an approach runs of course the serious risk of recycling the old romantic idea of the artist-as-genius, whose heightened powers of perception and imagination turn him or her into a kind of secular version of a prophet or guru. In a Foucauldian perspective, however, the artist is nobody special, and surely not someone who has access to a higher truth of being, which the others are too conformist to grasp and understand. The figure of the artist is also an effect of a specific discourse. Because artists and designers are more visually and spatially oriented, they are positioned at the margins of a knowledge-system still dominated by verbal, linear, logic communication. This 'marginal' position, combined with the aura of 'artistic freedom' has allowed artists of the past centuries to be transgressive, eccentric and radical, without losing their right to speak and be heard (unlike, e.g., the discourse of the madman, the criminal…). It still makes it possible for them to claim such a self-experimentation as a valid form of expression of other modes of existence.

3.3.1 An Exploration of Residual Spaces and Subjectivities

As a case study of such artistic research, we will take the work of Remco Roes (2016), whose research in the field of interiors is at the crossroads of scenography and adaptive re-use. The design problem with which Roes' research initially started, was how to deal with interior environments that do not fit within the discourse of efficiency and rationality, of production and consumption that dominate current architectural space. Either because their typology has become obsolete in a secular, post-industrial Europa (cloisters, factories), or because their program confronts us with life-experiences that are marginalised, that do not fit the image of an active and productive subject (i.e., a crematorium, the uncanny, painful experience of death and mourning). The initial design question was how to safeguard and potentially enhance the sensual and existential qualities of such interior environments. It soon became clear that any design solution would be a contradiction, because it would be a form of 'normalising' spaces to the very demands of rationality and efficiency that these places, these programs resist (e.g., to generate more touristic revenue, or to facilitate the process of mourning). Roes' research soon refocused on all kinds of 'residual spaces' that confront us with these borders of our subjectivity, and resist a smooth recuperation into the architectural texture.

Roes' artistic research more or less simultaneously followed two paths: one was the creation of a set of photographs and installations that explore these 'residual spaces', all part of our daily environments: storage rooms, offices, abandoned factories, empty stores, places that have escaped the constant effort to create agreeable, functional and meaningful spaces, with as few disturbance and friction as possible. There is nothing special to be found in these spaces – Roes' 'performance' consists merely of the very careful reshuffling of elements, in order to create an aesthetic spatial experience. He 'sorts out', and by doing this, he literally makes room for the kind of experiences that such residual spaces are able to generate: a sense of wonder, of subtle sensations that manifest itself in the useless, the banal, the vacant. These are intimate spaces where nothing is expected from us except being there – fragile spaces, destined to be destroyed, tidied up, re-used, re-designed. Roes' installations imply a series of characters, subjectivities that resonate with these 'residual spaces'. These subjectivities are not spatially present (albeit sometimes represented by pictures, film footage, figurines). These characters can be interpreted as contemporary incarnations of the Zen-monk, the Benjaminian *flaneur*, the concierge – at the same time idlers, wanderers and caretakers. They provide alternative subjectivities to the 'normal' characters that pass or dwell in those spaces and who actively perform a specific set of actions and undergo a specific set of experiences: the resident, employer, the consumer, the museum-visitor….

The implied subjectivities in Roes' work are clearly not the kind of transgressive subjectivities that haunt Foucault's philosophical imagination, like the madman, the sexual pervert, the drug-user, the criminal. Their limit-experiences are easily recuperated: their radical provocation of normality only confirms, by a clearly demarcated opposition, the prevalence of what is supposed to be 'normal'. The subjectivities staged in Roes' scenography are less extreme, yet arguably more transgressive, precisely because they challenge the *status quo* in a more subtle, non-oppositional way. Their limit-experiences are the result of a combination of free time spent in a free space, freed from the demand to perform (including the performance of provocation). They confront us with another mode of existence, not that far from our ordinary, daily existence, but fundamentally more careful to ourselves, and more attentive to what makes a space essential. And at the same time Roes does not provide us with a comforting discourse that legitimizes the apparent senselessness of these reordering acts (turning them explicitly in a form of Zen, of dandyism etc.).

These subjectivities also imply the personal involvement of the artist himself. The title Roes has given to a series of installations, 'exercises of the man', can be applied to his oeuvre as a whole. In order to be able to be exemplary, to be a modern-day 'man without qualities', the artist has to restrain himself and only strive for minimal interventions, enough to create a meaningful and aesthetic change in the quality of a space, but without 'framing' this experience into a specific artistic statement.

This requires a form of *ascesis*. Roes' interventions are the result of a lot of work, of deliberation, of concentration and self-critique. One can call the resulting works cynical, in the philosophical sense of the word. In his last lecture course at the Collège de France, *Le courage de la vérité* (1983–4), Foucault considers modern art as an heir to the cynical philosophical tradition: '*art itself, whether it is literature, painting, or music, must establish a relation to reality which is no longer one of ornamentation, or imitation, but one of laying bare, exposure, stripping, excavation, and violent reduction of existence to its basics*' (Foucault 2011: 188). That is indeed what Roes' work does – it forces us to revaluate our desire to intervene, and by that intervention to destroy the existential 'ground experience' these spaces have to offer: 'I exist and will die'. His work is a revaluation of what really counts in the light of our eventual disappearance into nothingness.

The second part of Roes' research consists of a written text, in which he discusses three spatial typologies: the garden, the monument, the home. What these typologies have in common is that they all imply a care of the self, and the activities that 'happen' there (gardening, commemorating and dwelling) are always in close relation with our own subjectivity. These typologies are explored by a set of four fictive characters: the scholar, the teacher, the artist and the 'living being'. Throughout the text, these characters are in a continuous dialogue with each other to explore the experiential field of thoughts and affects that is opened up by these typologies. Each character consists of a polyphonic texture of different voices, which allows the language of literature (Beckett, Musil…), philosophy (Zen, Tao…) and artistic practice (Cage, Reis…) to intervene in the discourse of interior studies. These voices allow other perspectives on interior environments, other ways of dwelling. Rather than to ask how existing spaces can be adapted, re-designed to better fit the needs and desires of a twenty-first century subject, Roes' work tries to formulate the existential question these places pose to our default subjectivity. What are the limit-experiences they can evoke, which different forms of subjectivity do they make possible in their actual form? How can they transform us? This experimental nature of artistic research thus provides the necessary critical supplement to a more classical phenomenological approach. It allows to explore, by the creating of images, of spaces, a different frame with which to understand and express the subjective experiences that are created by the interior environments that surround us.

4 Conclusion

In this chapter, we reflected about research in interior architecture, its identity as a discipline and the diversity of research approaches possible to study the interaction between a user, designer and an interior space. 'Design for Human Flourishing' and 'Artistic research' were proposed and discussed in more detail as valuable and diverse research approaches to add to the body of theory in interior architecture, both inspired by phenomenology, which allow to gain insight in human experiences. By proposing these research approaches and illustrating their value via concrete strategies and tactics, similarities and differences were illustrated. Design for Human Flourishing starts from a truly empathic stance, and takes the user group(s) for which one designs and their experiences as a starting point for the design process. Artistic research tries to look for the human experiences that occur on the edges, as a way to explore other possible modes of existence in interior environments. In a way, this approach 'uses' the sensibility and expressiveness of artists to expand the frame of 'normal' human experiences.

In the authors' view, both approaches can highly contribute to the further development of a proper body of theory for the domain of interior architecture.

As Fig. 1 demonstrated, research and research approaches are always framed by researchers' affinity for a particular system of inquiry which in turn influences the 'school of thought' which inspires researchers in a particular domain. In this chapter, it is clear that the authors are influenced by phenomenology, which influences their choices for particular research strategies and tactics. In the Design for Human Flourishing approach, research currently is often performed via ethnographic methodologies in an effort to generate rich data on diverse aspects of user experiences. Using the *tactic* of photo-elicitation in this respect can be highly insightful for researchers. At the same time, such methodology empowers research participants, as it allows them to verbalize their experiences in their proper vocabulary, while looking back and being inspired by their proper visuals which they made themselves in an effort to document their experiences. In such a research approach, research participants have to do more than rely on their verbal skills to take researchers along their experiences. However, the process of involving users is not only valuable for research in interior architecture; design practice also could be inspired by such an approach. In a recent publication, Poldma (2015: 466) pointed hereto as well, when she wrote: "*When designing interior spaces, visual languages often dominate the process. However, such languages alone cannot always sufficiently address how people experience the interior spaces that designers create. These experiences are best understood by listening to the voices of the people who live, play, an work in an interior environment. In other words, only by incorporating into the design process an understanding of how people's experiences frame their world-view can the design of an interior space be truly meaningful*".

The artistic research approach, another *tactic* for doing research in interior architecture, illustrated how the bias of normality is replaced by another 'bias': the specific subjectivity of the artist, or more accurately put: his or her singular confrontation with the borders of a given existential framework. When the experiment 'succeeds', a new way of thinking, feeling, perceiving or dwelling becomes visible that, however slightly and subtle, challenges the status quo. This is how modern art, and architecture, is able to revolutionize and reinvent itself. For example, Le Corbusier considered his modernist houses as the expression of a new mode of existence, supplanting anachronistic forms of dwelling that were for him out of touch with the demands and aesthetics of the 'machine age' (see Le Corbusier 2007). His work meant a radical, influential rupture with former views on architecture, but also provoked a lot of resistance with the actual residents of his houses, who tried to adapt and change his original design, in order to improve their personal experiences with these buildings (see Boudon 1969).

Evidently, as Fig. 2 demonstrated, it is clear that various other research approaches to do research in interior architecture are possible than the ones which we proposed in this chapter. However, we are convinced that a phenomenological approach defends what one could call the singular, subjective experience of dwelling in the age of 'big data'. It gives a voice to specific residents, to users, but also to artists, and these voices can be used for empowerment. They allow to resist a specific dominant view on what ideal dwelling should be, formulated by experts and decision makers.

The latter argument can, in fact, also be made regarding our initial starting point: the integration of the discipline of interior architecture into academia. Indeed, the decision to try and standardize the European higher education system in the Bologna-process not only gave interior architecture the opportunity to define its own specific identity, but has also confronted the human sciences with their epistemological borderlines. Artists and architects brought to their new academic 'territory' a set of perspectives, methodologies and practices that were unfamiliar for the traditional academic way of thinking. Where these new members of academia were immediately subjected to a process of normalisation, registration and academic overview, they were urged to understand and translate their experiences as architects, artists, and designers, in the established frameworks of protocols, methods and scientific jargon. However, this integration has also had the complementary effect of challenging the scientific discourse on interiors. The new context can help to legitimize designerly and artistic methods and add them to the set of methods with which we can create valid knowledge within the field of interior architecture.

Specifically with regards to artistic research in interior architecture, the challenge is even greater to find and defend its rightful, relevant place in an academic context, to cherish this safe haven that allows one to wander, and to explore, without the demands of an immediate economic return on investment. An important task will be to continue to safeguard the artistic freedom, which can be seen as another form of the so-called academic freedom in research. This academic freedom is not without obligations: it is the task of researchers in interior architecture to formulate with great care the 'results' of their research (which can be artistic, experimental or otherwise), and to translate and disseminate these to a larger audience. Establishing this communication, in a written, visual, or even spatial form, in order to convince others that there are other ways to think, feel and perceive, will be an important challenge.

References

Abercrombie, S. (1990). *A philosophy of interior design*. Oxford: Westview Press.

Bachelard, G. (1964/1994). *The poetics of space*. Boston: Beacon Press.

Berg, B. (2008). Visual ethnography. In L. Given (Ed.), *The sage encyclopedia of qualitative research methods* (pp. 934–938). Los Angeles: Sage.

Bollnow, O. (2011). *Human space*. London: Hyphen Press.

Borch, C. (Ed.). (2014). *Architectural atmospheres. On the experience and politics of architecture*. Basel: Birkhäuser.
Boudon, P. (1969). *Pessac de Le Corbusier*. Paris: Dunod.
Brooker, G., & Stone, S. (2007). From organization to decoration. In J. Gigli, F. Hay, E. Hollis, A. Milligan, A. Milton, & D. Plunkett (Eds.), *Thinking inside the box. A reader in interiors for the 21st century* (pp. 125–132). London: Middlesex University Press.
Brooker, G., & Stone, S. (2010). *What is interior design?* Hove: Rotovision.
Carll White, A. (2009). What's in a name? Interior design and/or interior architecture: the discussion continues. Journal of Interior Design, 35(1), x-xviii.
Clemons, S., & Eckman, M. (2008). Toward a common language: Proposed index categories to enhance dissemination and retrieval of interior design scholarship. *Journal of Interior Design, 30*(2), 13–30.
Clemons, S., & Eckman, M. (2011). Exploring theories identified in the journal of interior design. *Journal of Interior Design, 36*(4), 31–50.
Collier, J. (1957). Photography in anthropology: A report on two experiments. *American Anthropologist, 59*, 843–859.
Constanza, R., Fisher, B., Ali, S., Beer, C., Bond, L., Boumans, R., Danigelis, N., Dickinson, J., Elliott, C., Farley, J., Elliott Gayer, D., MacDonald Glenn, L., Hudspeth, T., Mahoney, D., McCahill, L., McIntosh, B., Reed, B., Turab Rizvi, S., Rizzo, D., Simpatico, T., & Snapp, R. (2007). Quality of life: An approach integrating opportunities, human needs, and subjective well-being. *Ecological Economics, 61*(2–3), 267–276.
Corbusier, L. (2007). *Toward an architecture*. London: Getty Research Institute.
Creswell, J. (2003). *Research design. Qualitative, quantitative, and mixed methods approaches*. Thousand Oaks: Sage Publications.
Cupchik, G., & Hilscher, M. (2008). Holistic perspectives on the design of experience. In H. Schifferstein & P. Hekkert (Eds.), *Product experience* (pp. 241–255). Amsterdam: Elsevier.
Deleuze, G. (1992). Postscript on the societies of control. *October, 59*, 3–7.
Desmet, P., & Pohlmeyer, A. (2013). Positive design: An introduction to design for subjective well-being. *International Journal of Design, 7*(3), 5–19.
Dickinson, J., Anthony, L., & Marsden, J. (2009). Faculty perceptions regarding research: are we on the right track? *Journal of Interior Design, 35*(1), 1–14.
Doyle, E., Escobar-Tello, C., & Pui Ying Lo, K. (2015). *Taking a softer approach: Using photo elicitation to explore the home as a system for happiness and sustainability*. Surrey: Sustainable Innovation Conference.
Edwards, C. (2011). *Interior design. A critical introduction*. Oxford: Berg Publishers.
Fetterman, D. (2008). Ethnography. In L. Given (Ed.), *The Sage encyclopedia of qualitative research methods* (Vol. Vol. 1 & 2, pp. 288–292). Los Angeles: Sage.
Foucault, M. (1977). *Discipline and punish. The birth of the prison*. New York: Vintage Books.
Foucault, M. (1983). The subject and power. In H. Dreyfus, L. Hubert, & P. Rabinow (Eds.), *Michel Foucault, beyond structuralism and hermeneutics* (pp. 208–226). Chicago: University of Chicago Press.
Foucault, M. (1997a). Technologies of the Self. In P. Rabinow (Ed.)., 1997. Ibid. *Ethics. Subjectivity and truth* (pp. 223–252). New York: The New Press.
Foucault, M. (1997b). What is enlightment? In P. Rabinow (Ed.)., 1997. Ibid. *Ethics. Subjectivity and truth* (pp. 303–319). New York: The New Press.
Foucault, M. (2001). Interview. In J. D. Faubion (Ed.)., 2001. Ibid. *Power* (pp. 239–297). New York: The New Press.
Foucault, M. (2002). *The order of things: an archaeology of the human sciences*. New York/ London: Routledge.
Foucault, M. (2011). *The courage of the truth (The government of self and others II). Lectures at the Collège de France 1983–1984* (ed. by F. Gros). New York: Palgrave Macmillan.
Groat, L., & Wang, D. (2013). *Architectural research methods* (2nd. ed.). Hoboken: Wiley.
Han, B.-C. (2015). *The burnout society*. Stanford: Stanford University Press.

Harper, D. (2002). Talking about pictures: A case for photo-elicitation. *Visual Studies, 17*(1), 13–26.
Hassenzahl, M., Eckoldt, K., Diefenbach, S., Laschke, M., Lenz, E., & Kim, J. (2013). Designing moments of meaning and pleasure. Experience design and happiness. *International Journal of Design, 7*(3), 21–31.
Huppert, F., & So, T. (2013). Flourishing across Europe: Application of a new conceptual framework for defining well-being. *Social Indicators Research, 110*, 837–861.
Johnson, M. (2007). *The meaning of the body : Aesthetics of human understanding*. Chicago: Chicago University Press.
Lee, J., Je, H., & Byun, J. (2011). Well-being index of super tall residential buildings in Korea. *Building and Environment, 46*, 1184–1194.
Lyubomirsky, S. (2008). *The how of happiness. A scientific approach to getting the life you want*. New York: The Penguin Press.
Lyubomirsky, S., King, L., & Diener, E. (2005). The benefits of frequent positive affect : Does happiness lead to success? *Psychological Bulletin, 131*(6), 803–855.
Mariampolski, H. (1999). The power of ethnography. *Journal of the Market Research Society, 41*(1), 75–86.
May, T. (2005). Foucault's relation to phenomenology. In G. Gutting (Ed.)., 2005 *The Cambridge companion to Foucault* (pp. 284–311). Cambridge/New York: Cambridge University Press.
Merleau-Ponty, M. (1964). The film and the new psychology. In H. Dreyfus & P. Dreyfus (Eds.)., 1964 *Sense and non-sense* (pp. 48–59). Evanston: Northwestern University Press.
Morgan, D. (2007). Paradigms lost and pragmatism regained: Methodological implications of combining qualitative and quantitative methods. *Journal of Mixed Methods Research, 1*, 48–76.
Nelson, H., & Stolterman, E. (2003). *The design way: Intentional change in an unintentional world*. Englewood Cliffs: Educational Technology PublicationsInc.
Neuckermans, H. (2004). Nurture and nature of research in architecture. In L. Fontein & H. Neuckermans (Eds.)., 2004 *ARCC/EAAE Montreal conference on architectural research proceedings* (pp. 23–29). Montreal/Leuven: EAAE.
Norberg-Schulz, C. (1988). *Architecture: Meaning and place*. New York: Electa/Rizzoli.
Pallasmaa, J. (1995). Identity, intimacy and domicile: Notes on the phenomenology of home. In D. Benjamin (Ed.)., 1995 *The home: Worlds, interpretations, meanings and environments* (pp. 131–147). Aldershot/Brookfield: Avebury.
Pallasmaa, J. (2005a). *The eyes of the skin. Architecture and the senses*. West Sussex: Wiley.
Pallasmaa, J. (2005b). *Encounters - architectural essays*. Helsinki: Rakennustieto Publishing.
Pallasmaa, J. (2011). *The embodied image. Imaginations and imagery in architecture*. West Sussex: Wiley.
Petermans, A. (2012). Retail design in the experience economy: Conceptualizing and 'measuring' customer experiences in retail environments. In *Unpublished doctoral dissertation*. Hasselt: Hasselt University.
Petermans, A., & Nuyts, E. (2016). Happiness in place and space: Exploring the contribution of architecture and interior architecture to happiness. In P. M. A. Desmet, S. F. Fokkinga, G. D. S. Ludden, N. Cila, & H. Van Zuthem (Eds.), *Proceedings of the 10th international conference on design and emotion: Celebration & Contemplation* (pp. 114–122). Amsterdam: The Design & Emotion Society.
Petermans, A., & Pohlmeyer, A. E. (2014). Design for subjective well-being in interior architecture. In *Proceedings of the 6th annual architectural research symposium in Finland 2014: Designing and planning the built environment for human well-being* (pp. 206–218). Oulu: University of Oulu.
Petermans, A., & Van Cleempoel, K. (2010). Research in retail design: methodological considerations for an emerging discipline. In K. Sato, P. M. A. Desmet, P. Hekkert, G. Ludden, & A. Mathew (Eds.)., 2010 *Proceedings of the 7th international conference on Design & Emotion*. Chicago: IIT Institute of Design.

Petermans, A., Janssens, W., & Van Cleempoel, K. (2013). A holistic framework for conceptualizing customer experiences in retail environments. *International Journal of Design, 7*(2), 1–18.
Pink, S. (2015). *Doing sensory ethnography* (2nd ed.). Los Angeles: Sage.
Poldma, T. (2015). Engaging voices within a dynamic problem-based learning context. In J. A. Thompson & N. Blossom (Eds.)., 2015 *The Handbook of Interior Design* (pp. 465–477). West Sussex: Wiley.
Poldma, T., & Asher Thompson, J. (2009). Proposing a dialogue about design research in interior design: New frontiers and possibilities. In *Interior design educators council conference 2009, Proceedings of the IDEC conference 2009*, March 25–28, St. Louis, Missouri, USA.
Prosser, J. (2007). Visual methods and the visual culture of schools. *Visual Studies, 22*(6), 13–30.
Riemer, F. (2008). *Ethnography research*. Retrieved July 15, 2010, from http://media.wiley.com/product_data/excerpt/95/04701810/0470181095_2.pdf.
Ritchie, J., & Lewis, J. (2003). Preface. In J. Ritchie & J. Lewis (Eds.)., 2003 *Qualitative research practice. A guide for social science students and researchers* (pp. xiii–xixv). London: Sage Publications.
Roes, R. (2016). *Traversing the interior landscape: Five dialogues in existential space*. PhD dissertation. Hasselt: Hasselt University.
Schroeder, J. (1998). Consuming representation: A visual approach to consumer research. In B. Stern (Ed.)., 1998 *Representing consumers: voices, views, and visions* (pp. 193–230). New York: Routledge.
Schroeder, J. (2002). *Visual consumption*. New York: Routledge.
Seamon, D. (2000). A way of seeing people and place: Phenomenology in environment-behavior research. In S. Wapner, J. Demick, T. Yamamoto, & H. Minami (Eds.)., 2000 *Theoretical perspectives in environment-behavior research* (pp. 157–178). New York: Plenum.
Seamon, D. (2015). The phenomenological contribution to interior design education and research: Place, environmental embodiment, and architectural sustenance. In J. A. Thompson & N. Blossom (Eds.)., 2015 *The handbook of interior design* (pp. 417–431). West Sussex: Wiley.
Seligman, M. E. P. (2011). *Flourish*. New York: Free Press.
Sheldon, K. M., & Lyubomirsky, S. (2006). Achieving sustainable gains in happiness: Change your actions, not your circumstances. *Journal of Happiness Studies, 7*(1), 55–86.
Smith, D. (2011). Interiors can address social justice: Fact or fiction? In D. Smith, M. Lommerse, & P. Metcalfe (Eds.), *Life from the inside: Perspectives on social sustainability and interior architecture* (pp. 77–109). Perth: Paper and Pencil Press.
Smith, D., Metcalfe, P., & Lommerse, M. (2012). Interior architecture as an agent for wellbeing. *Journal of the HEIA, 19*(3), 2–9.
Snape, D., & Spencer, L. (2003). The foundations of qualitative research. In J. Ritchie & J. Lewis (Eds.)., 2003 *Qualitative research practice. A guide for social science students and researchers* (pp. 1–23). London: Sage Publications.
Veenhoven, R., Arampatzi, E., Bakker, A., Bruel, M., Burger, M., Commandeur, H., Gupta-Mannak, J., van Geest, P., van Haastrecht, J., Hendriks, M., Hessels, J., van Liemt, G., Oerlemans, W., Volberda, H., & van der Zwan, P. (2014). *Het rendement van geluk. Inzichten uit wetenschap en praktijk*. Den Haag: Stichting Maatschappij en Onderneming.
Verhaeghe, P. (2014). *What about me? The struggle for identity in a market-based society*. Melbourne/London: Scribe.
Wagner, J. (1979). *Images of information*. London: Sage.
Wang, D. (2015). An overview of phenomenology for the design disciplines. In J. A. Thompson & N. Blossom (Eds.)., 2015 *The handbook of interior design* (pp. 11–28). West Sussex: Wiley.
Warren, S. (2005). Photography and voice in critical qualitative management research. *Accounting, Auditing & Accountability Journal, 18*(6), 861–882.

The Philosophical Underpinnings of Design Theory

Anne-Françoise Schmid

Abstract This paper describes the philosophical conditions for a 'Design Theory'. A theory of design is a theory that recognises a mode of reasoning that produces the new. There is a massive gap between the traditional Philosophy and the Design Theory. Nevertheless, in the current context philosophy is modifying itself and can invent new relationships with the concepts of the design theory. The introduction of the notions of fiction, interdisciplinary site, genericity, enables opening spaces in which we can address heterogeneities. Classical scientific and philosophical disciplines are no longer able to describe these and Design can treat them with the idea of "new". These relationships between Philosophy and Design are developed in an example, the creation of Genetically Modified Organism Fish. We show that it is not a complex object, but an integrative one, which implies new methodological approaches, participating to philosophy and to design theory together.

Keywords Design · Philosophy · Science · Generic epistemology · Non-standard heterogeneity · Integrative object · Interdisciplinary site

1 Introduction: Design as a Generic Thought between Science and Art

For a long time, theories of design have lacked any philosophical support or foundation. This can be explained by the particular place design occupies, between science and art. From science's point of view, design seems to mess up and interfere with otherwise beautiful hypothetico-deductive constructions. It seems like a sort of 'epistemic engineering' (as one might speak of 'mechanical engineering' or 'electrical engineering'), like some kind of extension of what science knows onto and within the sciences themselves. Whereas from the point of view of the arts, design

Translation: Robin Mackay

A.-F. Schmid (✉)
Associate of TMCI, Théorie et Méthodes de la Conception Innovante, Ecole des Mines de Paris, Associate of the Archives Poincaré - Laboratoire de Philosophie et d'histoire des scineces (UMR 7117 du CNRS, et honorary of INSA de Lyon), INSA de Lyon, Villeurbanne, France
e-mail: anne-francoise.schmid@mines-paristech.fr

© Springer International Publishing AG 2018

P. E. Vermaas, S. Vial (eds.), *Advancements in the Philosophy of Design*, Design Research Foundations, https://doi.org/10.1007/978-3-319-73302-9_19

appears, at best, of the order of industrial creation, and thus once more involves application.

This situation is beginning to alter. On the one hand, there exist theories of design which prevent limiting design any longer to 'making' and 'practice'. On the other hand, the contemporary sciences can no longer be reduced to those elegant classical arrangements: science does not involve only the object of study, isolated, as it must be, in the name of objectivity. It has programmes and meta-programmes. It has multiple levels, and the tools it employs cannot be related in any simple way to theory or to experience alone. The field of the ingredients of science must be opened up, and its unity must be sought elsewhere than in a general theory. Likewise, the multiplicity of experiences in art no longer permits the radical separation of inspiration, inner knowledge and design.

Concepts are thus beginning to be redistributed, and the theory of design takes its cue no longer from a theory of application, but instead from a theory of the infrastructures of thought necessary for both science and art – without itself being either science or art. It depends neither on science nor on art; it is rather of the order of a *generic thought*, and it is in virtue of this that it will encounter philosophy.[1]

It is in this changing landscape that we situate ourselves – in the midst of a redistribution of the relations between classical scientific disciplines, but also those between philosophies.

2 Is it a Philosophy for a Theory of Design?

Might there be one special philosophy – a philosophy of action, a philosophy of science, a philosophy of art, or perhaps a philosophy of mathematics – that would be most appropriate to account for theories of design? We can find, dating from classical Antiquity (in Aristotle, for example) means to distinguish what is of the order of *poiesis*, and what of the order of contemplation. The philosophical tradition is rich, and we shall see that we can draw from it many useful instruments to think the theory of design.

2.1 The Classical Structure of Philosophy Is Not Compatible with Design Theory

However, there is a major obstacle to the direct use of these philosophical notions in theories of design. We must find another – more indirect – usage for them. For the twentieth century brought to light the functioning or quasi-structure of philosophy. Philosophers, as Derrida, describe philosophy as sets of tensions between

[1]Anne-Françoise Schmid & Armand Hatchuel, "On generic Epistemology", Angelaki, Journal of the Theoretical Humanities, vol.19, n° 2, 2014, pp. 131–144.

contraries, small and great, conceptual and empirical, maintained together with an invisible link, the transcendental. So, philosophy was understood as a 2/3 structure – neither merely 2 nor entirely 3, for the union of contraries does not form an independent term.[2] It is very important to take note of this work on philosophy, not as an outcome *of* philosophy, but as a key to better understand the relations of philosophy to sciences and to other bodies of knowledge. Indeed, the contraries would be philosophy and human sciences, as in Derrida. But this structure of philosophy is not necessarily best suited either to the comprehension of the sciences, or that of theories of design. Our knowledge about philosophy has also changed, and has become far more complex. Both analytic and continental philosophers have sought to distance themselves from the mechanisms of contraries, to expose them or to utilise them as means to examine that which cannot be reduced to either one of the contrary terms. For example, the Vienna Circle sought to reduce all concepts of physics to 'observational terms', so as to understand at what point theoretical terms come into play. In similar fashion, the term 'ordinary' was used (Ryle,[3] Austin[4]) to break this circulation of terms. Or again, the critique of classical oppositions: for example, that of the analytic and the synthetic (Quine[5]). This is one way of distancing oneself from the play of contraries between experience and theory. But continental philosophy, also, has sought to separate radically the immanent from the transcendent (Michel Henry[6]), to make a radical separation between the finite and the infinite and to interrupt the rules of dialogue (Levinas[7]), to divide them without a rule of unification, but with only half-rules (Derrida), to extend them on the same infinite lines (Deleuze[8]), or to seek their archaeological and interdisciplinary conditions (Foucault[9]). If we wish to forge a philosophical environment for theories of design, we must take account of such modifications. Every philosophical concept may be utilised – even antique concepts – on condition that we take account of these transformations. But a dogmatic use of philosophy, subtended by one philosophy taken in isolation, is no longer possible, because it cannot account for the type of heterogeneity that design has to deal with today. It is no longer a question of application, but of the *production of objects,* making use of widely differing disciplines and bodies of knowledge which are not synthesised by any discipline or philosophy.

[2] Jacques Derrida, *De la Grammatologie*, Paris, Minuit, 1967 et *Marges de la Philosophie*, Paris, Minuit, 1972.

[3] Gilbert Ryle, *The Concept of Mind*, Routledge, 1949.

[4] John Langshaw Austin, *How to Do Things with Words: The William James Lectures delivered at Harvard University in 1955*, 1962 (eds. J. O. Urmson and Marina Sbisà), Oxford, Clarendon Press.

[5] Willard von Orman Quine, 1951, "Two Dogmas of Empiricism". *The Philosophical Review,* **60** (1), 20–43.

[6] Michel Henry, *L'essence de la manifestation*, Paris, PUF, 1963, reed. 1990.

[7] Emmanuel Levinas, *Humanism of the Other*. Trans. Nidra Poller, Introduction by Richard A. Cohen. Urbana and Chicago, IL: Illinois University Press, 2003.

[8] Gilles Deleuze, Différence et Répétition

[9] Michel Foucault, *Archéologie du savoir, The Archaeology of Knowledge* (1969).

As for the philosophical interpretation of sciences, we know from current debates how the opposition or complementarity that has been constructed between the concepts of theory and experience, has rendered it very difficult to understand the place of models, of modelization, or of simulations. Debates on climate change are testimony to this. And an analogous phenomenon has come about in theories of design. To approach design using contradictory or complementary terms is to see it as a classical philosophical theory would; the links and the interactions between design and philosophy will be far richer and freer if we proceed otherwise. It is enough to recall that the couplets of contraries transposed into epistemology have made it impossible to think engineering sciences except by inverting the terms, in such a way that the sciences become a banal undertaking. Our stance is to affirm that there are sciences, that there is philosophy, that there is art, that there is design, and that one can theorise about all of these without taking away from one what one gives to the other. But only if we suppose another usage of contraries and of oppositions, another interpretation of complementaries and multiplicities.

2.2 *Is the Structure of Contraries Pertinent for Design? Design as an Activity*

I should like to give an example of this dysfunctionality of contraries for design. Following the work of Herbert Simon, French interpreters have opposed science to engineering, claiming that the former is analytic, the latter synthetic. This opposition was one of the philosophical bases for the establishment of the engineering sciences. Engineering had been greeted by a profound silence in epistemology, since its problem is not theory, which is the concern of the great epistemologies of Duhem, Carnap, Popper, Lakatos, Feyerabend. Design could earn its credentials by giving it its own field, that of synthesis and of the artificial. We find ourselves here precisely in an opposition between contraries. The problem is that the analytical point of view is far too meagre for an understanding of the sciences. Thus, one ends up having to simplify in the extreme what one understands by science, in order to find a place in it for design and engineering. Now, this opposition with regard to engineering was constructed at the moment when classical epistemology found itself troubled by certain scientific developments – the expansion of models, modelization, simulations – developments that were all synthetic, and which one did not know whether to class as science proper, or as pragmatic tools continuing or extending science by other means. This dilemma invariably recurs in current debates on climate change. The idea of an analytical science has created a false debate, and has prevented epistemology from conserving its place in relation to the sciences, a place that has now been usurped by Social Studies (sociology of science). Things are changing now, as epistemology returns to the sciences and helps to pacify these debates. Armand Hatchuel and myself are working on a *generic epistemology* that would refuse the submission of epistemology to the constraints of the opposition between analysis and synthesis.[10]

Finally, one can see design as an activity, as a partly ordered series that contributes to the constitution of the subject, just as science can be understood as a contemplation that constitutes the scientist. One might here make use of Foucault's philosophy, for example.

2.3 Science Creates Objects, Not Only Knowledge

Another opposition has been problematic as much for the comprehension of sciences as for design – that of the True and the False, the Verifiable and the Refutable. All of these notions are fundamental to the sciences, but not necessarily central. For, like design, the sciences cannot be reduced to a system of proof, but also *create objects*. In the sciences, just as in design, it can be necessary to verify or to refute. But this does not mean to say that all of science and all of design are universally subservient to these values.

In our world of mixtures, of intrication of different levels, of amalgamations of sciences, techniques and technologies, the visibility and the self-evidence of the continued effectiveness of contraries is ensured by the disciplines and the great domains – mathematics, physics, philosophies. No amalgam could enable the identification of great concepts with which to orient analysis. It is in the light of a disciplinary logic that one can determine the true from the false, make the difference between the analytic and the synthetic, between the scientific and the technical, distinguish that which appears as knowledge and that which appears as action, as theory or as experience, or the possible exchanges between these distinctions. But, as we can see, the new scientific objects are no longer synthesised by disciplines; instead, disciplines are their *dimensions*. We must therefore modify our way of seeing, no longer aiming directly to grasp an object, but instead fashioning it indirectly. From this perspective, design does not depend upon any discipline in particular; it makes use of them, it constructs islets of expert knowledge to surround a concept. This expert knowledge is required qua generic – as scientific rather than as originating in a disciplinary logic.

---

[10] See also Franck Varenne, Marc Silberstein, Sébastien Dutreuil, Philippe Huheman eds., *Modéliser et simuler. Epistémologies et pratiques de la modélisation et de la simulation,* tome 1 et 2, Paris, Editions Matériologiques, 2013 et 2014.

3 What Philosophical Environment for Theories of Design? The Question of the "new"

So, what can guide us in constructing a philosophical environment for the theory of design?

What defines a theory of design is that it recognises a mode of reasoning that produces the new. This seems to go without saying; however, this aspect is often the object of polemics, some seeing such a type of reasoning as a simple technique of demonstration, others as the means to create the new. What makes the difference? At what moment can we say that we are dealing with a 'new object' or an 'unknown object'? It are the disciplines that make us see distinctions between true and false, analytic and synthetic, contemplation and action; but what manifests to us the 'new' and the 'unknown', the heterogeneous and the future?

We shall thus seek the conditions of those modes of reasoning that are reputed to produce the new.

3.1 What Is "new" in Philosophy

In philosophy, what are the modes of reasoning that produce the new? Those that involve the co-construction of thought and the real. In a certain way, all philosophy produces this interrelation between thought and the real, between the one and the multiple. Thus in a certain sense one could use any philosophy to describe design reasoning, whether through classic dialectic (Plato, Aristotle, Proclus, Hegel), topological inversion (Deleuze), difference (Heidegger, Derrida and Deleuze), or the 'flexuous line' (Ravaisson). In so far as these methods are oriented not towards justification, but towards an 'X' that one cannot completely foresee, it could be said that the method is heuristic and creative. The new will be that which, in the combinations of Thought and Being, will give an unknown combination of the One and the Multiple that the old combinations were unable to envisage. We could even read the Hegelian dialectic as a way of seeing history as pure design. But again, these methods must always be transformed according to what we know about philosophy and philosophies.

If, regarding philosophies, we pay attention not to their substantive forms but to their aspects of action and of creation, we come close to something like design. The attention paid by philosophy (classical, but also contemporary) to rhetoric, a form of invention – as shown by current research such as Balthazar Gracian's on the 'agudeza' – has been one of the ways in which it has been able to assimilate something of the new and unknown. In so far as one can postulate that every philosophy has seen something that others were not able to attain, one can admit that they have some empathy with theories of design. On the other hand, the object attained is not the same in philosophy as in design. In design, one creates an object whose variation from the existent series will depend partly upon the underlying mathematics of the theory. In philosophy, it is not the object that is important, so much as the mode of seeing the object; rather than a mathematics, it is a mode of purification, as the neo-platonists say – a pedagogy that leads to the true philosophy, or a return to the things themselves, as in phenomenology. If, in certain philosophies, mathematics come into play, they do so not as structures but as a mode of pedagogical purification or as ontology (as in Badiou[11]). They accompany philosophy, imposing constraints upon it, but do not enter into its very heart. Mathematics intervene as a sort of double or doublet, in conformity to the very structures of philosophy, which sees the real as if between doublets, in the specular mode – what we have called contraries. We come back always to the same difference between design and philosophy.

Thus, the usage of philosophy in the underpinnings of design engages the latter in a logic that does not entirely belong to it, just as the philosophy of science projected upon science structures that do not belong to it. A generic, non-specular usage of philosophies, which enables us to give an account of the possible relations between philosophy and design.

4 The New Objects of Science and the State of Non-Art: New Methodologies for Philosophy and Design

If one does not wish to exclude new methods from science, then the latter must be understood not only as a system of proofs, but as a creator of objects. Such a conception represents an abandonment of positivist science, which seeks to reduce science to the facts that justify it. Moreover, current sciences do not deal so much with 'facts', as with *givens*. The 'fact' is that with which the theory or the hypothesis is confronted, in view of either confirming or invalidating it. Givens are terms that do not engage in any such confrontation; they are neutral from the point of view of theory or model. They are infinitely numerous, they are generic, sometimes disposable. We do not always know in advance what theory could account for them; sometimes we hypothesise that only an 'interdisciplinary site' in a generic mode could allow for their 'interpretation'. To understand contemporary sciences, we are obliged to open up the oppositions and complementarities. We are constrained to take account of heterogeneities that can no longer be reduced by disciplinary knowledge. Confronted with the unknown, we must introduce the future into epistemology, which supposes an opening-up and an autonomisation of its principles and its ingredients.

Confronted with this deluge of givens, this heterogeneity, this unknown, it is important that the scientist does not transmit only what he knows according to his specialism (this happens anyway, through publications and colloquia). He must also transmit what he *does not* know of these givens from the point of view of his discipline, what heterogeneities or futures he thinks his discipline incapable of accounting

[11] Alain Badiou, *L'Etre et l'Evenement*, Being and Event

for. We must regard this *non-knowledge* as equal in importance to what is known; otherwise these heterogeneities will escape any scientific treatment apart from the disciplinary. This non-knowledge creates something other than new expertise: a space in which to project these heterogeneities and to treat them in such a way as to put them in relation with already-acquired expertise. This non-knowledge creates another logic of interdisciplinarity. Specialists are linked not only by the exchange and combining of the latest disciplinary acquisitions, but by a new form of collective intimacy, for non-knowledge is not additive, it cannot be detached from interaction as a positive fact; it creates a condition, a site and a mode of exchange whose particularity lies in its ability to favour the generic. A discipline taken on its own does not, of itself, create anything of the generic. For this, an external or heterogeneous element is necessary. What we usually call a generic discipline is a discipline that develops its expertise by involving itself in all the other disciplines that encounter the object it treats of – tribology, for example. In certain cases or according to certain usages, mathematics and informatics can be treated as generic disciplines. But there is also a generalised sense, where the generic is no longer directly disciplinary and is one of the consequences of the introduction of heterogeneity into the sciences. For this, we need not just a space (too neutral), but a site. A site of minimal, scientific and human notions, that permit invention in such a way that disciplines would be just the dimensions of it.

4.1 Science and Design Around Objects, the Example of GMO Fish

Now, if we take a new object created by the sciences, for example a Genetically-Modified Organism (GMO), the product of synthetic biology, we will find problems on the very edge of science and design, problems that will change certain aspects of design's relation to philosophy.

Take a GMO fish, for example. We can treat it as a fish + a genetic manipulation. But we perceive very quickly that this decision creates a rather impoverished outlook – as witnessed by the repetitive debates about vegetable GMOs that have taken place, at least in France, for the last 20 years – and that it exhibits important shortcomings with regard to the scientific knowledge involved. If we treat this fish instead as a design object – that is, as an 'X' whose properties, following the experience of non-Art, are divided in unprecedented fashion between diverse disciplines, the effects will be very rich. The first reason for this richness is that the hierarchy of disciplines is undone: all count for one, with the same weight. The fish will no longer be thought only as a technical product of molecular biology, with the aid, afterwards, of other disciplines as necessary – for example, chemistry for traceability, quantitative genetics for the expression of genes, economics for the commercial channels and the risk of chance contamination, sociology for consumer perception, law for marketing and labelling, epistemology to understand the variety of scientific ingredients, and ethics evoked in the question of social acceptability. Non-Art places all disciplines outside of this temporal series and concentrates on the identity of the object, using the methods of design. We know that in September 2010 a GMO salmon for human consumption was presented at the FDA for authorisation to bring to market, but that it was refused because the company concerned had not thought through the identity of its object. Is it a question for veterinary science? Does it fall under the laws of food additives? Medicine? Thus, theories of design are now responsible for thinking the identity of objects.

Science now creates objects whose identity is no longer fixed by the discipline of origin, nor by given disciplinary combinations. They are no longer objects considered as 'natural', seen within one discipline alone (the 'nebula', for example), nor objects modelled within one principal discipline with various adjunctions (the 'galaxy', for example); nor even complex objects that could be articulated through a convergence of disciplinary perspectives. A true exercise in design is necessary to understand the identity of the object. But this design makes for something other than a classical, manipulable object, a site of the convergence of disciplines; something other than a complex object. We have called this object an 'integrative object'; it presupposes superpositions of knowledge and non-knowledge, the intention of a collective of researchers, a relation to the real that is not one of co-construction, a non-manipulable object, and a recourse to disciplines as dimensions, not as means of synthesis. To understand such an object, many bodies of knowledge must be brought into play, including ethics and philosophy – on condition that they be integrated in a form of simplicity that does not double the terms in presence. Disciplines are like the dimensions of the object: they are no longer at the centre, but are made use of in the construction of the object.

4.1.1 The Methods of "without"

Now, in design also, the disciplines and types of knowledge required are not hierarchised. They are neutral without supremacy, and made compatible through their relation to the (heterogeneous, unknown) object 'X'. How is this unknown thought? Not through an overdetermination, but through an 'underdetermination', like that supposed by Non-Art. One reasons with a 'without' that is not a lack. One seeks to create an object 'without' one of its supposedly natural properties, so as to redistribute knowledges in relation to the object. The properties of the object X are redistributed according to unexpected disciplines. The 'X' supposes that the object is not extracted from a series of objects transformed in a rule-governed way, as would be supposed if it were a case of a simple expansion of knowledge.

Now, this method was developed independently in different domains and on different continents: in analytic philosophy of mathematics, in C/K Design Theory[12]

[12] Armand Hatchuel et Benoît Weil eds., *Les nouveaux régimes de la conception. Langages, théories, métiers*, Paris, Vuibert et Cerisy, 2008. Hatchuel A. and Weil B., C-K design theory: An advanced formulation, Research in Engineering Design, 19(4):181–192, 2009

(which supposes relations between the two spaces of Concept and Knowledge) and in François Laruelle's Non-Standard Philosophy. In each instance, it is a question of taking an unknown object and enriching it. Admitting that we do not know what mathematics are, or that we know only things that do not allow us to understand them in a rich way, then let us ask what mathematics would be without one of its 'banal' characteristics – without object, without number, without magnitude, without proof, etc. We are then obliged to bring in new islets of knowledge and expertise to account for what mathematics are. C/K Design Theory has already amply manifested what it makes of this method, and shows in its practice that not every object is necessarily thought within a series of objects. Non-standard philosophy transforms the rules of the writing of philosophy by multiplying and generalising them, withdrawing their authority over the real. The 'without' without lack, the 'non-', releases effects of extension or expansion, all the while distancing itself in relation to existent objects. By taking a minimal concept 'without' one of its properties, one can immerse it in new knowledge 'under' this knowledge, according to the formula C × K / K (where C=Concept, K=Knowledge). It is not a matter of opposing C and K, but precisely of extending them mutually without opposing them or placing them into a philosophical-type complementarity.

Science creates 'X'-objects; non-standard philosophy universalises philosophy by removing its sufficiency; a mathematics without object gives us back a richer mathematics, eventually including music and dance.

The 'new' is manifested here as placed outside an object-series, finding the real of its identity by placing in superposition knowledge and non-knowledge, intentions, which no longer separate the researcher and his object, superpositions that are more those of an oscillation than those of geological strata. It is at once one and plural.

4.1.2 Passage through Models

But there is another way in which the new can manifest itself, and which depends on the interpretation of reasoning. Let us take an example: Poincaré said that in mathematics there are chains of reasoning that are not mere tautologies, but in fact he gives only one example: arguments by complete induction. Such arguments link the 'repetition' of '+1' to what Poincaré identifies as what is mathematically at issue in 'a priori synthetic judgment', namely that one can reason by induction only on objects defined by induction. According to Poincaré, such arguments allow the passage from the finite to the infinite, and thus create something new. This was not Russell's position – he thought, on one hand, that the idea of repetition was far more complicated than Poincaré believed, and on the other, that logic was not a tautology, nor a structure of thought, but an experimental science having mathematics as its object. According to him, somewhat as in Peano, complete induction was one of the definitions of finite numbers, and did not exhibit any new object as such. According to Russell, a mode of reasoning did not produce something new in an absolute fashion, but as a function of one's design on it, the usage one made of it.

Another type of reasoning that permits the new is negation, as we have seen for non-art. In the most classical dialectic, negation plays a central role, and it is one of the modes of reasoning that, in philosophy, has been considered as generative of the new. It is the very principle of Hegelian dialectic, where concept and the real are ultimately the same. In the negation we see the importance of 'not', of its various significations, from the classical definition, where not-not-A = A, and the 'nots' that augment truth values (*n* values between the true and the false), to the non- as pure extension. One can put side-by-side spaces not having the same properties, just as one can have non-classical logics. C/K theory is interested in Brouwer's intuitionist theory, which precisely changed the signification of 'not' by refusing the classical definition of double negation (and thus of the excluded middle), on the grounds that it could not construct mathematical beings. For it sometimes posits them 'arbitrarily' through a reasoning by the absurd, on the grounds that one can exhibit both P and not-P if one renounces such-and-such a being. Brouwer wished to modify mathematics according to his alternative acceptation of negation; at the same time, he generated a theory rather close to C/K Design Theory, as has been discussed by Akin Kazaçki.

4.1.3 Changing the Level of Epistemological Analysis

This difference of interpretation shows that one cannot remain at the first degree, and that to say simply that a procedure induces the new does not take account of our historical knowledge of scientific 'crises'. We must 'go up a level' in epistemological analysis, and accept that two interpretations are possible according to different perspectives. The mathematician may see in the methods of 'forcing' just a method to prove the independence of the axiom of choice in relation to ZF, whereas the designer might see it as a method for the creation of objects and of sets in proportions never before known, at least not since Cantor. The mathematician might see topology as a non-metric geometry, whereas the designer will see in relations of neighbourhood a way of putting objects into series. What would be the point in disputing this point, when the debate between Poincaré and Russell is well known? It would amount to believing that all epistemological problems are posited at the same level, that positions can be opposed and destroyed amongst themselves.

4.1.4 Thinking through Models

To explain this change of level, I will take a detour via models, whose multiplicity Poincaré saw so well: if one finds a mechanical model for thermodynamic or electromagnetic phenomena, then there are an infinity of others, and it is a false question which is the 'right' one. There has to be a model if one seeks a compatibility between theories and objects, but we should not seek the one that exhibits the 'mechanical' functioning of thermodynamics or electromagnetism. At this point, we change level. Epistemology does indeed bear upon an object, on condition that we cannot say which object in particular. Epistemology only describes its object indirectly, so there is no contradiction involved in accepting an epistemological plurality. Epistemological disputes arise from the illusion that one is directly describing the sciences, when in fact an historical case is being exhibited qua ideal case.

A theory of design might find itself compatible with mathematics, topology, combinatory algebra, *forcing*, without placing itself in contradiction with pure mathematical work (with regard to which, moreover, different mathematicians may have different ideas). One might interpret this in terms of a philosophy of creation (on condition that it itself does not become the object of this same illusion) and thus proceed via fiction: what is an object 'without' forcing? What is an object 'without' topology?

Thus, new objects, modes of repetitions interpreted differently according to whether one sees mathematics in its pure specificity, or as that which permits the creation of what common sense cannot of itself create. The heterogeneous created in the 'object', or the heterogeneous produced by reasoning, and subsequently by models, but rendered compatible by the reasoning itself. With current theories of design, philosophy must reckon with a new type of modes of heterogeneity, because it is no longer reducible in terms of disciplines, because the objects are no longer synthesisable, and because the number of givens may be overwhelming. One of the questions will be how to introduce heterogeneity into philosophy.

5 Heterogeneity and Future in Design and in Philosophy

5.1 New Relations between Philosophy and Design

Theories of design construct spaces, rather than making structures out of contraries, as philosophy does. Not lines in spaces, as in Deleuzian philosophy, but true spaces, allowing the thinking of concepts and objects as well as the acquisitions of knowledge. Spaces that have a structure, with or without the excluded middle or the axiom of choice, susceptible of having 'true' or 'false' values or not. Spaces, scientific spaces – and no longer foundations, philosophical foundations. These spaces are the modalities according to which design takes account of, and distances itself from, existing objects, so as to create new ones. Whether it is a combinatorial space, where properties can be decoupled, spaces of diffusion allowing the transfer of a property of one discipline into another, or logical and non-logical spaces to combine concepts and scientific expertise, these spaces can be described by mathematical structures, topologies, combinatory algebras, algebraic extension.

The notion of space or of site is more scientific, even in the effective reality of its practice (the site of the laboratory, the site of the collider, etc...). Philosophy speaks rather of foundations, which allows for a sort of doublet justifying what appears at first sight as arbitrary or contingent.

What combinations with philosophy are then possible? Kant wrote a short treatise, fundamental for the great Critiques that followed: the 1763 *Attempt to Introduce the Concept of Negative Magnitudes into Philosophy*.

The current situation urges us to introduce into philosophy and into epistemology the concepts of the heterogeneous, the future, and the unknown. The stakes are considerable, since limiting philosophy and epistemology to contraries currently forces us to deny just as many great swathes of science as did the generalisation of philosophy and its changes of usage. If one treats the sciences only through an epistemology constructed uniquely upon the past of the sciences, then one excludes anything that is unforeseeable on the basis of the historical cases from which one sets out. For this very reason, epistemology has undergone a sort of collapse, and has been replaced for a period by the sociology of sciences (Social Studies). Epistemology could no longer say anything pertinent whatsoever about the post-war explosion in modelization, and so models were seen as practices rather than as scientific ingredients, or else as intermediaries between theory and experience.

5.2 How to Introduce the Concepts of Heterogeneity, the Unknown, and Future into Philosophy?

How to introduce a future distinct from that which is to come (the continuation of the present), a non-standard heterogeneity distinct from that of the disciplines, orders of magnitude that can uphold the distinction between fact and given?

Many conditions must be fulfilled. The first is a change of level, so as not to oppose philosophies one to another, but to describe indirectly the real, the sciences and the arts, by superposing philosophical concepts drawn from various philosophies. This supposes a very strong hypothesis as to the writing of philosophy: one can no longer write without knowing that other positions have just as much value as those from which one sets out; this writing must therefore be given, not just an alterity – as we know, this has been done very well for 50 years now – but an *identity* that does not evaluate between contraries, but is an indirect description of the real.

The second is precisely the construction of a space, where the homogeneous and the heterogeneous can take the measure of each other without opposing each other. The philosophy of oppositions is constituted by supposing that the very play of these oppositions individuates it and produces its specific model of understanding. Since we seek a generic understanding, we will give a different status to the heterogeneous, and postulate that one can know philosophy not only through itself, but through the agency of an external 'element'. Philosophy could then be modelized with the help of scientific or artistic expertise in the guise of 'underdeterminations', and placed within a matrix under the constraints of science or art. This time, rather than C × K/K, we would have a generic concept of philosophy, for example the transcendental, with which it links the concept and the empirical, T × K/K, or T × A/A. And we find that this changes the links between philosophy and the sciences, between philosophies and the arts. No longer imperious philosophies that overlook or survey, but unexpected combinations that can be divided between philosophies and sciences and arts.

The third is a change of syntax: not to transform contraries one into another as in dialectics or topology, but to postulate them either as rigorously identical, or as rigorously separable. How, for example, to describe engineering, making use of the category of the synthetic without opposing it to the analytic? And here is an interesting question: how to enrich our conception of the synthetic with areas of expertise different from the usual ones employed for this? There are doubtless many ways, which demand the bringing to bear of different bodies of knowledge, not reduced to one sole discipline, but combining generic elements, without having to oppose them to a narrow view of the sciences. The visibility of these syntheses could not be conceived on the basis of disciplines and philosophies taken in isolation; instead, they are 'lived' in an interdisciplinary site. In such sites, there would be an awareness of these syntheses close to lived experience, and an expertise elaborated on the basis of them that one could relate to disciplinary expertise. Husserl, a mathematician, himself also supposed (according to other methods) a lived experience that would be both human and scientific.

What have we done? We have constructed a generic site of interdiscipline where philosophy and design might construct new interactions and interrelations. We have changed the level of the exchange between philosophies and design. Why these modifications? To be able to account for the conditions in which both philosophy and science are currently carried out. If we did not make these hypotheses, we would exclude great swathes of science, and we would also exclude certain philosophies. Philosophies, neutralised in the sense that one no longer seeks to oppose one to another, furnish a language – but not only a 'language', something closer to the object in so far as this language is connected to the real but does not co-construct it – that is very rich for design, and for interpretations such as that of 'fiction'. Design in its turn helps philosophy by highlighting what we might call 'philosophical engineering', in the sense in which one speaks of a mechanical or electrical 'engineering', without being caught up in the totality of a system. There are philosophical concepts, there are philosophical techniques, that we can combine and superpose so as to construct interrelations with science and with the arts.

Interactions suppose rules of autonomy. It is here that we can introduce the theme of the future, of futurality, or of prosence, that allows the treatment of the heterogeneous and of the unknown otherwise than in terms of the present and the past. The future is the language of the unknown. It is also a way to speak of the unknown not as a continuity from the present to what is to come; but as a *cut* between what is to come and the present. This cut is a condition for compatibility with disciplines and with the present.

The principles are as follows:

- *Principle of Independence* (from disciplines, from the present). If we do not accept these principles of independence, the links will be those given by historical circumstances.

- *Principle of the Integrative Object*. Science as creative of objects *vs* science as system of proofs. If we do not accept the integrative object, what we can say of the sciences is impoverished to the level of disciplinary structures and the epistemology of theories. From the point of view of the relations between philosophy and design, this signifies that philosophy stands in a position of survey or of foundation in relation to design.
- *Principle of Genericity* (transformation of disciplinary elements: one does not bring together two disciplines without some transformations). If we do not accept genericity, interdisciplinarity is a matter of the combination and transfer of disciplinary expertise. From the point of view of the relations between philosophy and design, this means that the totality of the system must articulate itself with a mathematical space – a ponderous, unwieldy machine.
- *Principle of Futurality, of Prosence*. Treatment of the heterogeneous and the unknown otherwise than through present characteristics. One no longer passes from the present to what is to come in continuous fashion; one reverses the arrow of time in seeking rules of compatibility.
- *Principle of Design Reason vs Critical Reason*. From the point of view of the relations between philosophy and design, this supposes an opposition: philosophy would be critical (a system of proof for a particular philosophy), whereas design would create new objects.

These principles enable a profound modification of the current treatment of scientific and philosophical questions, and also allow us to see how emergent disciplines and new objects can be received. For this we must transform the relations between disciplines by thinking their 'common site' otherwise than as combinatorials of positive fragments of sciences, by transforming disciplinary concepts into generic concepts. Disciplines become like 'dimensions' of new objects, but without 'covering' them completely (respecting heterogeneity).

Now, between the conjunctions that we have found around the theme of fiction as extensive under-determination, and the transformations of philosophy and the knowledge of its structures that emerged in the last half-century, the development of design theories have allowed us to find links that are not authoritarian or reductive ones. From a critical reason, which always exists, we pass progressively and without crisis to a design reason, which permits the articulation of philosophy and scientific expertise in the same construction. The condition of this is the construction of spaces, the 'detaching' of notions from one another, the decentring of disciplines and theories in relation to integrative objects, and the decentering of philosophies in relation to the real.

6 Conclusion

The links between design, science, art and philosophy are forged around integrative objects. Their coherence owes not to a synthesis, nor to a recovery of disciplinary perspectives, but to the compatibility that can be constructed between the diverse dimensions of the object – something that Poincaré, once a student at the Ecole des Mines, might well understand. For all his inventions are those of a thinker who poses the question, with regard to every phenomenon, every generalisation, of its *compatibility* with *other* languages and *other* knowledges. It is in this way that Poincaré could have had, for example, the idea of an algebraic geometry. I call this 'Poincaré's criteria', because it endows interdisciplinary approaches with a certain scientificity. A spontaneous generalisation that is not 'compatibilised' with all the disciplinary knowledge that it puts into play, is not scientific. Thus, if one constructs rules of compatibility, then ethical, artistic, and philosophical aspects will also be dimensions of the integrative objects. In the same way, these disciplines will have to be rethought in a non-isolated way. For example, ethics will no longer be the child of philosophy, but a complex of which many disciplines – including philosophy – could be dimensions. And its function will change – it will be a generic science of disciplinary frontiers in the engineering of design.

In this capacity, philosophy truly provides an underpinning for design; it broadens design, giving its activity a meaning beyond that of engineering. It is one discipline among others and, through fiction, through neutralised philosophies as 'materials', it is capable of enriching design, and makign this most silent of disciplines 'speak' through the countless concepts of the tradition.

Part V
Design, Sustainability, and Ecology

Effects of Design and Sustainable Design of Technical Artefacts

Karina Vissonova

Abstract The aim with this chapter is to explicate sustainable design of technical artefacts. Given the increasing design efforts to respond to the issues of sustainability, an explication is needed as to clarify what makes a design the 'sustainable' kind. I propose to form a consistent understanding of what should fall under the 'sustainable design' kind and what should not. Such an understanding may be formed by looking at how the design practice have adopted the notion of 'sustainability' in the many design solutions. I state that in seeking to design for sustainability, the design is aimed at reconciling industrial and natural processes, and the act of designing encompasses a broad scope of considerations towards undesirable side effects. As the design aims for such a reconciliation and for resolving side effects, the specific materials are selected for their dispositions to carry the value of sustainability; as well as their dispositions not to pose hazards. I come to argue that technical artefacts are designed as sustainable based on the extent to which side effects are addressed with the design. I conclude with presenting necessary and sufficient conditions in the presence of which the design falls under the concept of sustainable design of technical artefacts.

Keywords Sustainable design · Technical intervention · Dispositions · Side effects

1 Introduction[1]

We tend to refer to 'sustainable design' as if it were discernible as a kind of design. Yet, there are no clear and shared criteria by which we may discern the design as the sustainable kind. In consequence of which, we do not share conceptions as to what are the permissible solutions when it comes to the design of technical artefacts we may qualify as 'sustainable'. Thus, I find that our conceptions on what constitutes

[1] Some of the ideas developed in this chapter were presented in preliminary form in a conference paper (Vissonova 2015), on which parts of the present exposition are based.

K. Vissonova (✉)
The Royal Danish Academy of Fine Arts, Schools of Architecture, Design and Conservation, Copenhagen, Denmark
e-mail: vissonova@icloud.com

© Springer International Publishing AG 2018

P. E. Vermaas, S. Vial (eds.), *Advancements in the Philosophy of Design*, Design Research Foundations, https://doi.org/10.1007/978-3-319-73302-9_20

the 'sustainable design' are fairly informal, as we lack the criteria by which we may discern the design as the kind.

Our evaluations about which design solutions might be permissible and which not are somewhat simplified if we content ourselves with asking whether or not the design exhibits even the slightest adaptation to an environmental discourse or to the principles of sustainable development. Thus, any extent to which some positive value is delivered by design to a selected eco-system, relative to an industrial activity posing a hazard to that system, is considered to satisfy the sufficient condition for the design to count as 'sustainable'. My concern is that such a condition may not be as sufficient as generally assumed. While it certainly seems necessary for design to deliver such positive values in order to count as sustainable, that in itself is not sufficient, as I shall argue in this chapter.

I claim that the criteria, which help to qualify the design of technical artefacts as sustainable, concern the effects of design. I construct my proposition based on the materials selected for a structure of a technical artefact having certain *dispositions* which afford effects. Some of these effects are unintended and are also undesirable, making up the *side effects* of design. I come to argue that the designs we should regard as sustainable are those where various side effects are addressed.

In addition, I propose that, by being an industrial activity with a likely production of by-products and side effects, the design of technical artefacts gives rise to an intervention in the natural processes. I call such an intervention a *technical intervention*. The technical intervention is where the effects of design ramify and where the undesirable side effects of design can be resolved.

My proposition results in an *explication* of *the sustainable design of technical artefacts*, where I identify the necessary and sufficient conditions by which we may qualify the design as the sustainable kind. An explication, as defined by Carnap (1950), takes place when we give more exact terms based on logic or empirical explanations to an imprecise and a pre-scientific concept.

The chapter offers a critical view on the notion of sustainability in design. It proposes a set of individually necessary and jointly sufficient conditions to define the sustainable design of technical artefacts, which may give rise to sustainable technical interventions. The chapter is a conceptual enquiry and thus intended as a contribution to the emerging philosophy of design.

I begin the chapter by introducing a broad overview of the notion of 'sustainability' with reference to design. I introduce the concept of a *technical intervention* to conceptualise where the effects of design manifest themselves, and I suggest the concept as a basis for more adequate evaluations of the effects of design. I continue with an explanation of the effects being afforded by dispositions of materials selected for the structure of a technical artefact. I make a distinction between desirable effects and undesirable side effects, which can be known and unknown at the time of designing. Furthermore, I explain how values and hazards may be rendered by the dispositions, and how they relate to an attainment of sustainability by design. I conclude the chapter with classifying empirical examples of sustainable design with a clear correlation to technical interventions. I then offer a definition of sustainable design of technical artefacts by conditions of which technical interventions may count as sustainable.

2 Designing for Sustainability

In design research as well as in practice, we tend to think of 'sustainability' partly as a concept of environmental conservation in ecology, and partly as principles of sustainable development (Dusch et al. 2010; Hopwood et al. 2005; Brumsen 2011; Thorpe 2010; Lovins et al. 1999; Stebbing 2015; Tischner 2015; Reller and Diesenbacher 2015; Tonkinwise 2015; Tukker 2015; Rockström et al. 2009; Manzini 2006, 2008). As such, the notion is adopted for the purpose of formulating the design requirements, where 'sustainability' is translated into the requirements as a value (van de Poel forthcoming). To design for environmental conservation and for meeting the principles of sustainable development, is typically referred to as to 'design for sustainability'. It can be said that to design for sustainability is to counter the issues arising from the climate change, loss of biodiversity, and other forms of environmental degradation, as well as to counter the negative effects of lifestyles on the well-being of people and eco-systems.

However, one of the common misconceptions about the design for sustainability, I find, is that it is presumed to offer an overall reconciliation of imbalances between industrial and natural processes. The challenge to such a conception is that the notion of 'sustainability' expresses the coping capacity of a particular eco-system, or a social system for that matter, when faced with a particular industrial activity involving that system. As proposed by Tonkinwise (2015), sustainability is "… a measure of the capacity of a system to respond to a series of more or less likely impacts". Consequently, to address the issues of sustainability is a design task selecting some considerations towards what ought to be sustained. For instance, whether it might be access to fresh water, renewable energy, transport with low carbon emissions, and so on. This, furthermore, is accompanied by a belief that by doing little by little in the selected eco-systems, many a system will be made sustainable (McDonough and Braungart 2013). Tonkinwise (2015) calls this the 'cumulative' effect. The issue with such a design approach, or we may call it a strategy, is that 'sustainability' as a notion has many expressions of its meaning. As such, it provokes many different and sometimes conflicting responses the solutions of which are difficult to evaluate as right or wrong. The well-recognised conflicting ideas of sustainability are the historic formations of two pulls dividing the notion - biocentrism and anthropocentrism (Hopwood et al. 2005; Dusch et al. 2010; van de Poel forthcoming; Brumsen 2011). The biocentrism expresses the intrinsic value of the environment, requiring its conservation for its own good. The anthropocentrism, on the other hand, expresses the preservation of the environment to benefit the people. When designing for sustainability, the one or the other perspective is adopted, although most commonly selected design approach follows the sustainable development perspective, which claims to restore a balance between the two conflicting pulls (Hopwood et al. 2005).

Yet, this is not where the conflicts of the notion of sustainability end. Once one adopts the sustainable development perspective, one is met with the issues of equal access to the natural resources, and equal distribution of the benefits delivered by the resources, both across the world and across generations (van de Poel forthcoming). These conflicts are often not resolved by design we refer to as sustainable, because the demand for goods today, which are presumed to better our lives, requires mining of resources, many of which are simply not sufficient to service also the demands of the very near future. As for instance, photovoltaic solar panels provide us with renewable energy thus weaning us off the fossil fuels, while requiring silver for generating the electricity. According to research by Reller and Diesenbacher (2015), silver is a finite resource and is expected to be exhausted within more or less 22 years. A somewhat similar issue of problematic resources we face with rare earth minerals, required for the function of electric car motors, as well as for rechargeable batteries and the wind turbine motor function (Reller and Diesenbacher 2015). In addition, the design we consider sustainable also produces waste, which to some extent we are leaving to the innovation capabilities of the future generations.

These aforementioned examples represent another less discussed issue with the notion of sustainability. That is, by designing in order to conserve one eco-system faced with hazards posed by industrial activities, it is permissible to disturb or even exhaust the coping capacities of another eco-system. In terms of the resource depletion, Reller and Diesenbacher (2015) contend: "we are practically shifting from a dependence on carbon compounds (gas) to a dependence on metals and from carbon dioxide emission to a type of mining that may well have greater effect on humans and the environment". Therefore, I would like to point out that the design for sustainability is not free from a production of side effects, which come to ramify in eco-systems irrespective of the good intentions behind the design. Additionally, the 'cumulative' effect lends a false conviction that 'sustainable design' is free of side effects. Keeping the above considerations in mind, and putting the 'cumulative' effect of the design efforts into a perspective, it indeed becomes challenging to envision as to what we are striving to attain when designing for sustainability. In consequence of which, we are challenged in claiming that what is being attained is in fact part of the solution and not still part of the problem.

Despite the different views we appear to hold towards the notion of sustainability, leading to somewhat conflicting responses by design, it makes sense to recognise sustainability as a value, as suggested by van de Poel (forthcoming). Meaning, an eco-system being preserved and industrial and natural processes reconciled represent conditions of 'goodness' on which we commonly agree and which we consider desirable. The conditions can be perceived objectively, such as human welfare, air and oceans free of pollution, biodiversity, and so forth. By saying that these conditions may be perceived objectively, I mean that irrespective of our individual preferences and aspirations, the conditions of goodness are not disputable due to us being able to share the foundations for judgment whether the conditions represent the goodness or not (see Rittel's discussion on 'objectification', 1972). The value we associate with 'sustainability' and as such translate into design requirements, relates to this 'goodness'.

Furthermore, an aspect of designing for sustainability, which deserves a place in the discussion, is that the notion of sustainability implies an acknowledgement of restrictions to our ways of designing technical artefacts and how we are accustomed to use them. Tonkinwise (2015) claims that we have certain existing notions of freedom to which sustainability poses a challenge. Sustainability, he argues, "... is about acknowledging limits; it is about accepting responsibility for longer term and wider afield consequences". Designing for sustainability entails a much broader scope of considerations, such as toxicity of materials and their biodegradability. Undoubtedly, it is a strenuous task to find alternative resources replacing ones with higher environmental impacts, and then, by design to mitigate the negative impacts in the artefact's use phases, while planning for a safe and environmentally friendly way to dispose of the artefact. On top of these challenges, our methods for sourcing of the materials, our setups for the manufacturing, and our custom of consumerism are not particularly well suited to accommodate the alternative proposals shaped by the design for sustainability. In this regard, an idea promoted by Manzini (2006), Thorpe (2010), and Chapman (2009), expresses a lesser concern with sustainable design as a design of 'objects', but rather designs for psychological well-being through formed relationships with technical artefacts, or particularly well-being in their absence. Thus, the design emphasises the human experience, essentially aiming at replacing consumerism and thus slowing down the production of goods and of waste.

In the light of the diverse views we hold towards 'sustainability' leading to diverse responses by design, as discussed above, how may we evaluate what are the right or the wrong solutions of design we may qualify as sustainable? What might appear as the 'right' solution by applying one of the sustainable design methods, might prove to be entirely 'wrong' in a view of another such a design method. For instance, the well recognised Cradle to Cradle principles, developed by McDonough and Braungart (2002), advocate an idea that we may maintain extensive production and consumption cycles, as long as we contain the natural and 'technical nutrients', as the authors call all artificially produced materials, while eliminating waste. Yet, whether we are able to contain the nutrients in the manufacturing processes, ensuring their continuous return in the production-consumption loops, while using renewable energy for these processes; and whether we are able to eliminate the waste, including pollution during the material processing and manufacturing, is one question towards our ability to attain the value of sustainability as advocated through this method. Another question is whether such a social setup of production-consumption is actually enhancing our welfare, as argued just earlier, even if it poses no cost to the environment. Therefore, although we may agree on the desirability of a certain goodness to be delivered by a design of a sustainable kind, the design methods and strategies may lead to somewhat conflicting solutions which are difficult, if not impossible, to evaluate as being right or wrong.

In consequence to the above, I find it reasonable to ask, whether the sustainable design of technical artefacts is discernible as a kind of design. Perhaps, rather, it is a design that deals with 'wicked problems' (Rittel 1972) occurring in eco-systems and in social setups that demand equal distribution of risks and benefits of resources. Rittel and Weber (1973) argue that wicked problems are the type of problems for solving of which we may not have an objective procedure, but where some resolutions may be delivered. Such problems imply some uniqueness and therefore the resolutions may not be replicable when a new situation obtains. Furthermore, as argued by the authors, the wicked problems typically would have many parties "… equally equipped, interested, and/or entitled to judge the solutions, although none has the power to set formal decision rules to determine correctness". The authors further argue that the judgments vary according to varying interests and value-sets, and hence any evaluations of proposed solutions may be expressed only as 'good or bad', or 'better or worse', or perhaps just 'good enough' (Rittel and Weber 1973). Considering the incoherencies of the notion of sustainability, the selective way of the design responding to its matters with the diverse design methods, and hence the difficulties in evaluating what are the permissible and impermissible solutions when it comes to the 'right' sustainable design, it may indeed indicate that 'sustainability' is a wicked problem for the design to resolve. Arguably, and considering the aforementioned 'cumulative' effect, it might be thus acceptable to do 'something' for the sustainability and hence to design technical artefacts which may be considered as 'better' (than unsustainable) or perhaps just 'good enough' resolutions.

The issue I raise against such an assertion concerns the 'sustainability' being a particular "evaluation of risks" (Tonkinwise 2015) in a particular situation where industrial and natural processes may be out of balance. The notion carries diverse meanings related to the set of risks in question, and hence no contextual shared criteria may be attached to the design dealing with the type of 'wicked problems'. As we may not attach any criteria to the design, we are consequently limited in evaluating whether the design is sustainable or not, and whether it is part of the solution or still part of the problem. In other words, for us to justify the claims that what is being attained by design is sustainable, the criteria for the design must concern the kind of design and not the kind of wicked problems, and the good intentions behind addressing these. I suggest that the ample empirical examples developed by the design practice offer some more definitive criteria by which we may discern the design as the sustainable kind. Therefore, the task is to offer an *explication* (Carnap 1950) of the sustainable design of technical artefacts.

2.1 Technical Intervention

The task of explicating the sustainable design of technical artefacts entails an explication of a notion of the design giving rise to interventions in natural processes, thus, exhausting its coping mechanisms when faced with industrial processes. Since the design of technical artefacts is inherently dependent on the natural resources, and since it is disposed to produce side effects, for instance in the form of emissions, waste and pollution, I believe it is necessary to recognise this act of design as a certain intervention in the natural processes. The outcome of such an intervention, to some extent, is still 'natural', as the natural processes continue their course; but they are mixed with the 'artificial', as the artificially produced materials enter the natural processes as waste and other pollutants, more often than not posing hazards to these processes. These resulting by-products and side effects of making a technical artefact constitute a somewhat undeclared element of the design. I propose to call this element of the design a *technical intervention*.

By introducing the concept of a technical intervention, I argue that there is an element of the design of technical artefacts where the side effects ramify and where they can be evaluated and hence potentially resolved. Technical interventions occur as a result of the design as it requires mining of materials, factories and transport, while it produces polluting emissions and waste which is, albeit unintentionally, distributed in eco-systems as larger and smaller abiotic bits and parts. In this regard, a role in an intervention by design of a technical artefact made of locally sourced materials, which biodegrades after its use cycle and requires little processing, is rather insignificant, as for instance in the case of a wooden spoon.

One of the studies which, perhaps somewhat indirectly, exemplifies the full range of effects of design, is by Rockström et al. (2009; 2015). It illustrates very well the technical interventions arising from the design of technical artefacts. The authors have developed what they call "Planetary Boundaries", or some measurable safe operating limits, suggesting that exceeding these limits, the nature's coping mechanisms with the industrial processes may be irreversibly exhausted. So for instance, the natural mechanisms for resilience of nitrogen and phosphorus flows, biodiversity and climate change are nearly or entirely exhausted. As the technical interventions lack a balance between industrial and natural processes, and the environment's abilities to cope with our interventions are being exhausted, a need increasingly arises for the design to include a focus on the technical interventions and so to reconcile the industrial and natural processes.

As I see it, technical intervention can be conducted in such a way that it maintains a certain kind of goodness, or a good life, and that is the value desired by society. As argued by van de Poel (2009), values attained with technical artefacts can be contributory or instrumental in attaining a good life, and sustainability is one of such values along with privacy, freedom and health. What then can be said is that if a technical artefact is designed to deliver a value of benefiting the natural environment and the well-being of people, the technical intervention is considered sustainable. Alternatively, design that harbours negative effects to the environment and well-being is seen as an unsustainable technical intervention. Thus sustainability, in essence, is a contributory value in the way it benefits society and the environment, and is delivered through technical interventions.

In the light of the above, to judge whether or not the design of a technical artefact is sustainable, one must consider the entire technical intervention to which the design gives rise. I shall consider the design of a technical artefact sustainable to the extent that the associated technical intervention is sustainable. The task of explicating sustainable design, therefore, involves explicating sustainable technical interventions.

3 Dispositions, Effects and Values of Design

A technical artefact is significant to us due to its properties the effects of which help us to realise our goals. Based on the properties, materials are selected for the structure of the technical artefact, and the structure of the artefact, typically, helps to realise its function (Kroes 2009). The properties that constitute the technical artefact are constantly present in the artefact. However, some of the properties may manifest themselves only when certain events take place. Such properties are *dispositional*. For example, a raincoat will display its hydrophobic property when exposed to water. In other instances it may appear to an unknowing observer as a regular coat. The hydrophobicity is a *disposition*. My use of the concept of dispositions is with reference to Mumford's (1998) theory of dispositions.

By regarding a property of a technical artefact as a disposition, we grant the artefact an ability to partake in certain causal processes associated with just that property (Ellis 2002). The causal production, as I further refer to, is the specific mechanism of cause and effect in which dispositions participate. It is also the relation that defines the properties of materials based on which the design of technical artefacts is undertaken. Fragility, durability, transparency, conductivity, solubility, biodegradability, and so forth - these are dispositions due to which we select the materials and determine the design of a technical artefact. The knowledge of the dispositions also enables us to avoid the use of materials in our designs, if by doing so possible undesirable effects are prevented from ramifying through a technical intervention to which the design gives rise.

In philosophy, dispositions may be referred to as analogous to causal powers (Witt 2008). Ellis (2002) suggests a view of a causal power according to which it constitutes a driving force in a causal relation. For the technical artefact to come about, the causal relation that will lead to the desired effect is identified and the involved principles are, presumably, understood fairly well (Kleinberg 2013). Therefore, the act of designing to a high degree pertains to formation of knowledge about causal relations and the principles that govern them, and hence the mechanism by which a single act of designing extends to a technical intervention.

It can thus be said that the design of technical artefacts is aimed at producing some desired effect. For instance, the effect of wearing a shoe is protection of the foot from a roughness of ground surface, at a bare minimum of the possible effects afforded by dispositions of the selected materials. The effect of packaging is preservation and containment of goods for us, if the dispositions of the packaging materials are suitably selected. We often refer to the design as delivering an improvement in existing conditions, as proposed by Franssen et al. (2009), or as defined by Simon (1996), according to whom the design aims at "changing existing situations into *preferred* ones" (emphasis added).

Furthermore, being dispositional, the properties display their causal production in given circumstances, thus enabling the particular structural compositions and hence the functions of technical artefacts. Therefore, the constitutive properties are central to reaching the desired effects of the design of technical artefacts. As the materials are selected due to their specific dispositions affording specific effects, I call these the *known desirable effects*.[2]

The known desirable effects are always intended by the design of technical artefacts. These are the effects by which we may characterise the artefact, but which are not necessarily essential for the artefact to perform its function. Crilly (2013) argues that the desired effects inherently are functional when it comes to technical artefacts. However, my argument is that a function of a technical artefact pertains to the many desirable effects afforded by the dispositions of the artefact's structure, rather than all the effects pertaining to its function. Meaning, during the design phases of selecting materials for a structure of a technical artefact, dispositions may be sought after, which afford certain effects carrying some values other than the function of that particular artefact. These dispositions might be certain thermal conductivity of materials used in construction of a window frame, adding to its insulation qualities, for instance. That is to say, it is the dispositions that may make a technical artefact valuable for a particular reason, relative to other artefacts with the same function.

The values carried by the known desirable effects are *utility values* based on which we select particularly these materials over other alternatives. The utility values can be said to be deliberately accommodated in the technical artefact's structure, and, depending on the materials selected, an artefact may accommodate several utility values in addition to its function. With an attainment of utility values, the effects manifesting themselves through a technical intervention can be said to be desirable or undesirable, also determining if the nature of the intervention might be sustainable.

In discerning the dispositions of the chosen materials, we ascribe a value to certain dispositions. We do so, according to Harre and Madden (1975), by grasping the causal production characteristic of these dispositions, i.e. the specifics of the cause and the resulting effects, and thus we are led to claiming the necessity of these effects. Moreover, the authors further argue, legitimacy is created by just this causal production resulting in the particular effects. The implications are, firstly, that we thus justify the dispositions of chosen materials over other alternatives. Secondly, the implications are that our intentions are concentrated on the dispositions and their causal mechanisms which best benefit our goals in question, as defined in design requirements. Understandably so, as the dispositions render values for which we find the material employable in designing the technical artefact, which in turn makes the artefact useful to us - the effects are desirable due to the value they carry. However, a further connotation is such that our regard for dispositions is primarily for the known desirable effects, including those that contribute to fulfilling the artefact's function. Meanwhile, the dispositions may afford other effects than those desired with the design of technical artefacts, yet these effects are accepted rather than intended with the design. The unintended effects that occur along with the desirable effects are the *side effects* of the design of that artefact.

[2] Alternatively, the desired effects might also be unknown at the time of designing. For instance, some uses of technical artefacts are desirable but were not intended at the time of designing. These may be conceived as the unknown desirable effects.

A side effect implies an unintended effect of bringing about some intended end (McIntyre 2014). Although side effects may also be desirable, my focus is on the *undesirable side effects*, due to their direct correlation with how the design of technical artefacts is resolving the issues of sustainability.

3.1 Undesirable Side Effects of the Design of Technical Artefacts

Irrespective of what is intended by the design of technical artefacts, dispositions inherent in the structure of a technical artefact can harbour side effects. Side effects are undesirable due to their capacity to carry hazards. As dispositions may afford the effects only when certain conditions obtain, so the hazards posed by the dispositions may be expressed as risks. In this regard, Hansson (2009) suggests that a hazard may be a risk potential, as well as it may be a negative value carried by an undesirable side effect which has manifested itself. Either way, materials may pose hazards to human well-being and eco-systems during the mining, processing and manufacturing needed for them to become components for a structure of technical artefacts, or when they are used, or when discarded. So for instance, pollutants such as nitrogen oxides occur as a side effect of the function of combustion engines. The phthalate substance, which is a type of plasticiser, is contained in numerous technical artefacts including electronics hardware, packaging, detergents, as well as cosmetics and toys. Phthalates are compounds with a disposition that poses varying degrees of hazard to human health, according to EurActive Network (Jacobsen 2012). Due to this disposition, some phthalates are restricted or banned under the EU's REACH regulations (2007).

Hazards are also posed to the eco-systems by materials having dispositions that continue to afford effects when the technical artefact is no longer functional and is discarded. A particularly relevant example of such a material with respect to designing for sustainability is a wind turbine blade. The wind turbine counters the use of fossil fuels and significantly reduces the carbon emissions - the side effect of the production and use of the fossil fuels. The technical artefact, it can be said, is thus instrumental in attaining sustainability by its function. However, while the dispositions in the structural composition of the blades render certain utility values, they also afford undesirable side effects, which carry hazards when the wind turbine is no longer functional. This is because the blades are made of composite materials, which are particularly unfriendly towards reprocessing and hence the reuse of the materials. Due to the large size of the blades (some are 60 metres long) and their sizeable volumes of consumed raw materials, the concern is for our abilities to reprocess these colossal artefacts so that we avoid them being discarded in landfills. The dispositions of the blades pose hazards to the environment when the blades are discarded, as well as when incinerated for energy recovery, producing high amounts of abiotic ash. Currently, there are still too few and too inefficient methods for processing the blades, while applications of a value retrieval of the blade materials, through reuse of the composite waste, are yet not sufficiently developed. Hence, due to the significant side effects affecting the environment, the technical intervention, to which the design of the wind turbine gives rise, is unsustainable.

In consequence to the above, as I argue earlier in this chapter, the view that the permissible solutions in the design of a sustainable kind are those that reconcile natural and industrial processes is misleading. Sustainable design is not design free of side effects, but rather design where the undesirable side effects are addressed to a certain degree in the subsequent technical intervention, or in a technical intervention resulting from other designs.

Continuing on the subject of dispositions affording side effects, while the production of undesirable side effects may be permissible as a result of a production of known desirable effects, as in the several examples above, some side effects may occur unbeknown to us due to us having no knowledge of certain dispositions and the nature of their causal production (Mumford 1998). As Sellars (2008) notes, things exist and possess properties "…independently of our knowing them". A range of substances and toxic compounds have dispositions that pose hazards which we may learn once the undesirable side effect manifests itself and the extent of its harm is rated as inadequate. For example, the freon gas in the earlier refrigerator designs, lead in earlier PVC materials, as well as the previously common asbestos containing materials - before these toxic substances were banned. In this regard, Boodin (1911) notes that we do not know all the properties of artefacts, although we may apply the best instruments, artificial and our own intuitive ones. Nevertheless, he argues, however incomplete our knowledge about the properties and their causal production might be, we discern the artefact for the properties we do know.

3.2 *Conceptual Machinery of Dispositions*

To draw some preliminary conclusions from the arguments discussed thus far, materials are selected for the structure of technical artefacts based on certain inherent dispositions. They are selected over other alternatives due to the dispositions affording certain desirable effects, which in their turn carry utility values. Such a process, more often than not, involves a production of side effects, which occur as unintended effects of designing for some desirable effects. The unintended side effects, irrespective of whether they are known or not at the time of the designing, are undesirable due to their capacity to pose hazards. That being so, the materials are also selected to compose technical artefacts with lesser side effects. So for instance, a disposition that poses a hazard, which might be deemed greater than a utility value rendered by the same disposition, makes the material an unlikely candidate for the structure of a technical artefact. While a disposition rendering a utility value while posing a hazard which is rated as a slight risk is permissible (see discussion of double-effect in McIntyre 2014) .

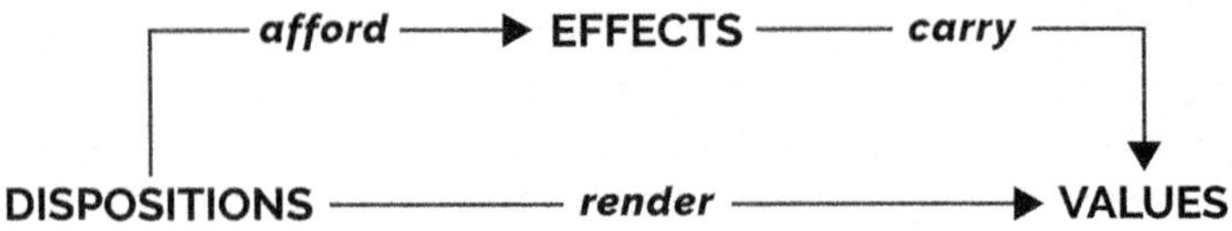

Fig. 1 The conceptual machinery of dispositions and effects

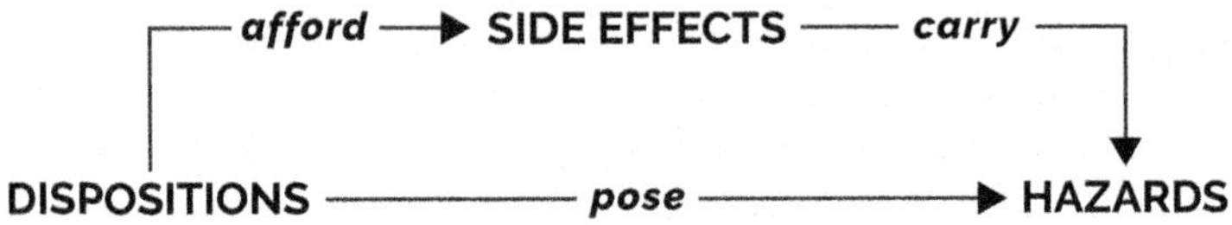

Fig. 2 The conceptual machinery of dispositions and side effects

I summarise the *conceptual machinery* of the dispositions rendering the positive values in Fig. 1, and the dispositions posing hazards in Fig. 2.

The dispositions affording effects are intrinsic to the structure of a technical artefact, for which a set of materials are selected to support its function.

4 Explicating the Sustainable Design of Technical Artefacts

Sustainability being a value, as proposed earlier in this chapter, is carried by the known desirable effects afforded by dispositions intrinsic to the structure of a technical artefact. In attaining the value of sustainability, the 'good' effects are obtained by resolving the undesirable side effects ramifying through technical interventions. The act of designing for sustainability generally includes a broader scope of considerations towards the undesirable side effects. The scope is not only broader than in a 'conventional design', as it were, but is also carried out with a particular focus on a reconciliation of industrial and natural processes. As the design aims for such reconciliation, only those materials are selected with the particular dispositions affording the effects that carry the value of sustainability; as well as the dispositions not posing hazards. In this way sustainability is translated into the design requirements.

However, one may assert that sustainability is attainable by addressing a design problem just the same as we may address safety in a car design, for instance. It appears not to be so due to safety relying on properties which explicitly render safety, relative to our conceptions of what a 'safe' car might be (Hansson 2012). Design solutions pertaining to sustainable design, on the other hand, are more difficult to evaluate and are only comparable to their alternatives which we find impermissible due to them involving a higher degree of undesirable side effects. That is to say, the 'goodness' attained through the act of resolving 'badness' is a process which is not necessarily free from the occurrence of undesirable side effects.

Consequently, the common misconception concerning the sustainable design, I believe, is that the impermissible solutions are those lacking a balance between industrial and natural processes. However, we ought to recognise that the design's concerns for the reconciliation of the two processes do not guarantee a resolved balance. Rather, the design of technical artefacts is aiming at resolving a selected issue in a selected eco-system or a social structure. It thus follows that the design is instrumental in bringing about sustainability in a selected eco-system by preserving its goodness, and it may do so at the expense of exhausting coping capacities of other eco-systems through unsustainable interventions to which it may give rise.

4.1 The Four Classes of Sustainable Design of Technical Artefacts

In discerning the criteria for a sustainable design as a kind of design, and not merely a design dealing with the kind of 'wicked problems' occurring in eco-systems, as discussed earlier in this chapter, I come to proposing that sustainable design is distinctive in four different ways. I classify the distinctive ways by proposing that some technical artefacts are designed to attain sustainability 'by function' of the technical artefact in question, and some 'by structure' of the artefact. I do so on the basis of ample empirical examples of the design we currently conceive to be sustainable. On the foundations of the sustainable design so classified, I propose that there are certain conditions by which we may qualify design as the sustainable kind.

The first class of the design we refer to as sustainable is a design where an emphasis is placed on that artefact's function in attaining sustainability. The technical artefacts in this class are designed as a more sustainable alternative to artefacts with a similar or the same function but by which undesirable side effects are produced that we consider impermissible. These are the technical artefacts designed to address carbon emissions, such as renewable energy sourcing technologies, electric vehicles, also designs to reduce the consumption of energy, such as LED lights, and various insulation improvements wherever thermal energy might escape into the atmosphere. Furthermore in this class of sustainable technical artefacts, we may place the artefacts designed to address pollution and saturation of bio-capacity. These are various reprocessing plants for chemicals and other hazardous substances, as well as all types of waste reprocessing machinery. Also in this class are technical artefacts designed to address various forms of resource depletion, such as fresh water delivery and water saving technologies. These examples illustrate the purposes of this class of sustainable design, which are to either replace more polluting artefacts, or to resolve the resource depletion by design. Nonetheless, the designs in this class have structural compositions with little or no concern for the technical intervention they give rise to. The technical artefacts contain rare minerals, metals and other non-renewable resources demanding their mining. They are also often non-processable for reuse, and their waste contains large amounts of abiotic compounds.

It seems that in this class, the emphasis is placed on the function of the technical artefact, and hence, the sustainability attained is relative to not attaining the specific goals at all. Moreover, the design of technical artefacts pertaining to this class is aimed at curbing undesirable side effects resulting from other designs and the technical interventions to which they have given rise. Meanwhile, the technical intervention subsequent to the design in question is fairly unsustainable, and is dependent on other designs, perhaps those of the future, to resolve the therein ramified side effects.

As for the second class of the sustainable design, I find that specific materials are selected which possess dispositions rendering specific utility values beneficial for reconciling industrial and natural processes. That is to say, a technical artefact may be designed so that its structure accommodates utility values by which a sustainability issue is resolved by the design.

One such an artefact is a roofing membrane, "Noxite", developed by Icopal, which purifies air from certain pollutants. The product is a bituminous waterproofing sheet covering a roof surface. Besides this function, the membrane purifies air by ridding it of nitrogen oxide pollutants called NOx, which are atmospheric pollutants and are born as a side effect of fossil fuel combustion from cars, incinerators, manufacturing etc. The air purification works by the membrane containing on its surface granules coated with titanium dioxide, which through exposure to the sun alters the NOx molecules converting them into harmless levels of nitrates, carbon dioxide and water. The converted pollutants are washed away with the rain. The utility values accommodated in the structure of "Noxite" benefit the reconciliation of industrial and natural processes while the function of this artefact, so to say, serves its regular purpose of protecting the roof from a premature deterioration. A rather similar example of such added utility values accommodated in the structure of an artefact is a carpet, "Airmaster", developed by Desso. The carpet purifies indoor air by absorbing various pollutants and hence preventing them from escaping into the air. In addition to these two examples, both of which are recyclable, what we may also consider as belonging to this class, are designs which address waste by optimising the technical artefacts for transportation, such as flatpack, or liquid concentrate packaging.

Overall, the technical artefacts in the second class have substantially altered structural compositions in order to deliberately accommodate certain utility values. The design here may not necessarily be 'sustainable' by function, while the structural compositions and the subsequent technical interventions are not necessarily free from side effects. However, the utility values rendered by the dispositions make the design of a technical artefact instrumental in attaining a sustainable technical intervention. Subsequently, as the attention is given to the specific dispositions inherent in the artefact's structure, the technical intervention can be said to be sustainable to the extent the undesirable side effects are being resolved.

In the third class of sustainable design, the design appears to focus particularly on dispositions inherent in structural compositions of artefacts. A quality emphasised within this class is that utility values of materials may be retrievable when the artefact is no longer in use. Such a notion of value retrieval pertains to various materials, substances and entire component looping methods where a utility value is somehow retrieved and the material in question is reused in the same loop or in a new design. The premises for the value retrieval are that the material with the value is separable from other materials, and processable for reuse or other adaptations of its value. Design for 'disassembly' ensures the retrievability of the utility values. However, recovering materials so that their value may be retrieved is still a challenge, as appropriate technologies are not readily available. Value retrieval to a lesser extent also applies to energy recovery in incineration plants, where waste is processed for energy production. As a general rule, the materials used for the structural compositions of technical artefacts in this class are biodegradable, modifiable, upgradable, recyclable and reusable. The technical artefacts may be designed so as to extend their life, or to extend the life of their components or just materials. Hence, the components might be designed as modular to facilitate their replacement. Thus waste and depletion of resources are being addressed.

Also in this class, we will find designs reducing the use of virgin materials, such as wood, metals and rare minerals and hence helping in managing the resource flows. The sourcing of virgin resources and a need for the production of new materials is minimised as we extend the life of the technical artefacts and the materials they are composed of. Hence the technical interventions resulting from the design belonging to this class are relatively more sustainable, as unsustainable mining and manufacturing practices are being addressed by freeing the interventions from some of the undesirable side effects of the design.

The fourth class of sustainable design contains solutions with the most concern for the technical interventions to which their design gives rise. The technical artefacts are 'sustainable' by their function as well as by their structure, to the extent side effects are resolved in the subsequent technical intervention. In other words, it could be said that this class of sustainable design deliberately limits the technical interventions to which they may give rise. The designs in this class are focused on local sourcing of materials, local use, and generally committed to enhancing welfare while maximally reducing the risk of posing hazards. One such example is WarkaWater, developed by an organisation under the same name. The WarkaWater is a construction for collecting condensation, providing fresh water to communities where water is scarce. The construction is made from bamboo, metal pins, hemp, bioplastic netting and polyester ropes, which may be repurposed. The technical intervention subsequent to the design of WarkaWater is minimal, as no undesirable side effects are likely to ramify as a result of the design. Therefore, perhaps, this last class may be an indication of a design most sustainable as the associated technical intervention is most free of side effects.

I classify the design of technical artefacts in the four fairly distinctive classes for the purpose of evaluating what is being attained by the design of technical artefacts we already qualify as 'sustainable'. With the reference to the four classes as described above, the designs instrumental in attaining sustainability 'by function' address undesirable side effects of other designs. The designs instrumental in attaining sustainability 'by structure' address their own undesirable side effects, making the subsequent technical intervention sustainable. Irrespective of the one or the other way of addressing the undesirable side effects, the resulting benefit may be evaluated against the sustainability of the technical interventions, rather than certain isolated values attained with technical artefacts.

Furthermore, an evaluation of effects, which are afforded by dispositions of materials, is how the sustainable design kind obtains sustainable technical interventions. This is so as besides the design of technical artefacts bringing about sustainability through specific utility values beneficial to reconciliation of industrial and natural processes, specific materials are also selected which have the least dispositions affording undesirable side effects with potentiality of hazards. While it may not be possible to design technical artefacts without setting off undesirable side effects, all design of technical artefacts intended to qualify as sustainable, irrespective of the method applied, only succeeds to the extent to which the side effects are addressed. Consequently, I propose to replace an evaluation of 'sustainable design' against the various ideas about what is desirable to attain, with an evaluation of the specific utility values of materials in the structure of technical artefacts, *and* the specific resolving of undesirable side effects of design.

One last point, which I would like to offer in discerning the sustainable design kind, is that no implications can be made of the design kind being free from undesirable side effects, although, presumably a freedom from the undesirable side effects may be stated as the goal of the design kind. My proposition is that the design can be qualified as sustainable *to the extent* the side effects are resolved in the technical interventions to which the design gives rise, starting from mining, sourcing, manufacturing and transporting the materials, and ending with their disposal and the value retrieval mechanisms so designed.

4.2 The Sustainable Design of Technical Artefacts - the Explicatum

The task of explicating the concept of sustainable design of technical artefacts can now be completed. My proposition consists of the following two individually necessary and jointly sufficient conditions, which, when satisfied to a high degree, qualify the design of technical artefacts as the sustainable kind.

The design of technical artefacts is *sustainable* to the extent that:

1) *side effects are resolved in the technical intervention to which the design gives rise,* and

2) *dispositions inherent in the structure of the artefact render certain utility values which benefit the reconciliation of industrial and natural processes.*

5 Conclusion

Based on the considerations presented in this chapter, I have proposed two individually necessary and jointly sufficient conditions for the design of a technical artefact and the subsequent technical intervention to count as sustainable. I have claimed that when the design of technical artefacts reconciles a given imbalance between industrial and natural processes, yet does so by ramifying side effects and upsetting the coping mechanisms of other eco-systems, the design is unsustainable since the technical intervention to which it gives rise is itself unsustainable. This is due to the condition being necessary yet insufficient, as for the design to qualify as the sustainable kind, the technical intervention ought to also be sustainable.

In this chapter I have presented a way in which the sustainable design of technical artefacts may be explicated. The new concept reflects the instrumental nature of the design of technical artefacts that concern sustainability, as unsustainable interventions are problems of design yet to be resolved. My aim was to offer a simple definition that is consistent with the empirical evidence of sustainable design, and which, I hope, will be useful to practitioners and researchers in design and other fields, offering, as I believe it does, an accurate and operational demarcation of the concept of sustainable design.

Acknowledgments I would like to thank Dr. Maarten Franssen and Dr. Peter Kroes from Delft University of Technology, Philosophy, Values, Technology and Innovation Department, Ethics and Philosophy of Technology Section, and Dr. Per Galle from The Royal Danish Academy of Fine Arts, School of Design, for their expert advice and the invaluable discussions during the writing of this chapter.

References

Brumsen, M. (2011). Sustainability, ethics and technology. In I. van de Poel & L. Royakkers (Eds.), *Ethics, technology, and engineering* (pp. 277–300). Oxford: Wiley-Blackwell.

Carnap, R. (1950). *Logical foundations of probability*. London: Routledge and Kegan Paul.

Chapman, J. (2009). Design for (Emotional) Durability. *Massachusetts Institute of Technology Design Issues, 25*(4), 29–35.

Crilly, N. (2013). Function propagation through nested systems. *Design Studies, 34*, 216–242. https://doi.org/10.1016/j.destud.2012.10.003.

Dusch, B., Crilly, N., & Moultrie, J. (2010). Developing a framework for mapping sustainable design activities. Design Research Society international conference, Montreal, 7–9 July 2010, Canada.

Ellis, B. (2002). Powers and dispositions. In R. Groff (Ed.), *Revitalising causality: Realism about causality in philosophy and social science (2008)* (pp. 76–92). Oxon: Routledge.

EU's REACH Regulations. (2007). Official Journal of the European Union. http://eur-lex.europa.eu/LexUriServ/LexUriServ.do?uri=OJ:L:2007:136:0003:0280:en:PDF. Accessed September 2015.

Franssen, M., Lokhorst, G.-J. & van de Poel, I. (2009). "Philosophy of Technology", The Stanford Encyclopedia of Philosophy (Fall 2015 Edition), Edward N. Zalta (Ed.). http://plato.stanford.edu/archives/fall2015/entries/technology/. Originally published in 2009. Accessed June 2015.

Hansson, S. O. (2009). Risk and safety in technology. In A. W. Meijers (Ed.), *Philosophy of technology and engineering sciences* (pp. 1069–1102). Amsterdam: Elsevier.
Hansson, S. O. (2012). Safety is an inherently inconsistent concept. *Safety Science, 50*, 1522–1527.
Harre, R., & Madden, E. H. (1975). Conceptual and natural necessity. In R. Groff (Ed.), *Revitalising causality: Realism about causality in philosophy and social science (2008)* (pp. 56–76). Oxon: Routledge.
Hopwood, B., Mellor, M., & O'Brien, G. (2005). Sustainable development: Mapping different approaches. *Sustainable Development, 13*, 38–52. https://doi.org/10.1002/sd.244.
Jacobsen, H. (2012). EurActive Network. http://www.euractiv.com/consumers/danish-minister-bans-endocrine-d-news-514424. Accessed September 2015.
Kleinberg, S. (2013). *Causality, probability and time*. New York: Cambridge University Press.
Kroes, P. (2009). Foundational issues of engineering design. In A. W. Meijers (Ed.), *Philosophy of technology and engineering sciences* (pp. 513–543). Amsterdam: Elsevier.
Lovins, A. B., Lovins, L. H., & Hawken, P. (1999). A road map for natural capitalism. *Harvard Business Review, May-Jun*(1999), 145–158.
Manzini, E. (2006). Design, ethics and sustainability. Guidelines for a transition phase. Cumulus working papers. Nantes 16/06. Publication Series G. University of Art and Design Helsinki, 9–15.
Manzini, E. (2008). Viewpoint new design knowledge conference paper, Changing the change, Turin, Italy, July 2008. *Design Studies, 30*(2009), 4–12.
McDonough, W., & Braungart, M. (2002). *Cradle to cradle: Remaking the way we make things*. NewYork: North Point Press.
McDonough, W., & Braungart, M. (2013). *The upcycle. Beyond sustainability – designing for abundance*. NewYork: North Point Press.
McIntyre, A., "Doctrine of Double Effect", The Stanford Encyclopedia of Philosophy (Winter 2014 Edition), Edward N. Zalta (Ed.). http://plato.stanford.edu/archives/win2014/entries/double-effect/. Accessed February 2016.
Mumford, S. (1998). *Dispositions*. New York: Oxford University Press.
van de Poel, I. (2009). Values in engineering design. In A. W. Meijers (Ed.), *Philosophy of technology and engineering sciences* (pp. 973–1007). Amsterdam: Elsevier.
van de Poel, I. (forthcoming). Design for sustainability. In P. K. Brey, D. M. Callicott, & J. Baird (Eds.), *Technology and the environment*. Cambridge, MA: MIT Press.
Reller, A., and Diesenbacher, J. (2015). Are there enough resources for our lifestyle? How resource strategy leads from wasting materials to using them. In P. Stebbing & U. Tischner (Eds.), *Changing paradigms: Designing for a sustainable future*. Publication No. 1 of the Think Tank Series from the Cumulus International Association of Universities and Colleges of Art, Design and Media. (pp. 154–166). Aalto University School of Arts, Design and Architecture. Mumbai: Vedanta Arts.
Rittel, H. (1972). On the planning crisis: systems analysis of the 'First and Second Generations'. *Bedriftsøkonomen, 8*, 390–396.
Rittel, H. W. J., & Weber, M. M. (1973). Dilemmas in a general theory of planning. *Policy Sciences, 4*(1973), 155–169.
Rockström, J. (2015). Planetary boundaries. A safe operating space for humanity. Published online by the Stockholm Resilience Centre as SOS for Business. http://www.stockholmresilience.org/download/18.6d8f5d4d14b32b2493577/1422535795423/SOS+for+Business+2015.pdf. Accessed April 2015.
Rockström, J., Steffen, W., Noone, K., Persson, Å., Chapin, F. S., III, Lambin, E., Lenton, T. M., Scheffer, M., Folke, C., Schellnhuber, H., Nykvist, B., De Wit, C. A., Hughes, T., van der Leeuw, S., Rodhe, H., Sörlin, S., Snyder, P. K., Costanza, R., Svedin, U., Falkenmark, M., Karlberg, L., Corell, R. W., Fabry, V. J., Hansen, J., Walker, B., Liverman, D., Richardson, K., Crutzen, P., & Foley, J. (2009). Planetary boundaries: exploring the safe operating space for humanity. *Ecology and Society, 14*(2), 32. http:// www.ecologyandsociety.org/vol14/iss2/art32/. Accessed 18 November 2015.

Sellars, R. W. (2008). Critical realism and substance. In R. Groff (Ed.), *Revitalising causality: Realism about causality in philosophy and social science* (pp. 13–26). Oxon: Routledge.
Simon, A. H. (1996). *The sciences of the artificial*. Cambridge, MA: Massachusetts Institute of Technology.
Stebbing, P. (2015). 1. Raison d'etre. 3. Why we have to design for sustainability - the new paradigm, schesiological links and externalities. 10. Pollution, poisons and profits: Toxicity for designers. In P. Stebbing & U. Tischner (Eds.), *Changing paradigms: Designing for a sustainable future*. Publication No. 1 of the Think Tank Series from the Cumulus International Association of Universities and Colleges of Art, Design and Media (pp. 6–21), (pp. 37–53), (pp. 167–199). Aalto University School of Arts, Design and Architecture. Mumbai: Vedanta Arts.
Thorpe, A. (2010). Design's role in sustainable consumption. *Massachusetts Institute of Technology Design Issues, 26*(2), 3–16.
Tischner, U. (2015). Design for sustainability, strategies, methods and tools. In P. Stebbing & U. Tischner (Eds.), *Changing paradigms: Designing for a sustainable future*. Publication No. 1 of the Think Tank Series from the Cumulus International Association of Universities and Colleges of Art, Design and Media (pp. 302–316). Aalto University School of Arts, Design and Architecture. Mumbai: Vedanta Arts.
Tonkinwise, C. (2015). Radical sustainable innovation. In P. Stebbing & U. Tischner (Eds.), *Changing paradigms: Designing for a sustainable future*. Publication No. 1 of the Think Tank Series from the Cumulus International Association of Universities and Colleges of Art, Design and Media (pp. 284–295). Aalto University School of Arts, Design and Architecture. Mumbai: Vedanta Arts.
Tukker, A. (2015). Sustainable consumption and production, the SCP framework. In P. Stebbing & U. Tischner (Eds.), *Changing paradigms: Designing for a sustainable future*. Publication No. 1 of the Think Tank Series from the Cumulus International Association of Universities and Colleges of Art, Design and Media (pp. 272–283). Aalto University School of Arts, Design and Architecture. Mumbai: Vedanta Arts.
Vissonova, K. (2015). Insidious side effects of design: and how to turn them into values of sustainability in design. Conference paper Cumulus Mumbai 2015. Mumbai: IDC IIT Bombay Publications.
Witt, C. (2008). Aristotelian powers. In R. Groff (Ed.), *Revitalising causality: Realism about causality in philosophy and social science* (pp. 129–139). Oxon: Routledge.

Ecological Design as an Ecology of Love: Epistemological and Ethical Implications

Gonzalo Salazar and Seaton Baxter

Abstract Based on the argument that the complex environmental crisis is essentially an epistemological and ethical crisis, the intention in this chapter is double: first, to synthesise a new epistemology of design—one that we call *an ecology of design* by attending to the problem of how are we to understand the systemic relationship between individuals and their environment and comprehend the praxis of design as an integral part of it. And second, to synthesise the essential element for design to become ecological, which it is argued that only occurs when its praxis is mainly commanded by the emotion and ecology of love. Love is described as the biological and ecological foundation of what makes us human beings and therefore as the main human disposition from which a truly ecological ethics and ecological consciousness in design praxis may emerge. First, the chapter examines how design is part of an ecology of living which is epistemologically constructed as a reaction to modern rationale. Then, based on Maturana's notion of human existence in conversation, the chapter suggests that design is a human form of conversing and synthesises four implications that are constitutive of this condition. After dealing with an epistemological dimension of design as conversation, the chapter synthesises the notion of an ecology of love. Based on the exploration of several philosophical and scientific accounts, the article examines some essential aspects of an ecology of love that informs an ethical and collaborative form of designing.

Keywords Ecological design · Ecology of love · Emotion · Sustainability

---

G. Salazar (✉)
UC Centre for Local Development, Pontificia Universidad Católica de Chile, Villarrica, Chile
e-mail: gonzalosalazar@uc.cl

S. Baxter
Ecological Design Thinking Schumacher College, Fife, Scotland, UK

© Springer International Publishing AG 2018

P. E. Vermaas, S. Vial (eds.), *Advancements in the Philosophy of Design*, Design Research Foundations, https://doi.org/10.1007/978-3-319-73302-9_21

1 Introduction: The Problem of Ecological Design

If the world we live in is not the way we want it to be, then this is the fault of humanity. If, as Bill McKibben (1989) has suggested, we are seeing the "End of Nature", then this too is our fault. But, we have come to realise that we cannot survive and flourish as a species without some symbiotic lasting partnership with the best version of "Nature" that we can create together. "Nature no longer exists apart from humanity and henceforth the world we will inhabit is the one we have made" (Purdy 2015) and conservation in the Anthropocene, as envisioned by Jamie Lorimer (2015), will have to become a "dynamic cosmopolitics for wildlife". We need therefore to design and implement new cooperative partnerships and a new vision.

The notion of 'ecological design' has emerged, during the last few decades as a reaction to a complex environmental crisis and as a line of thought and practice that not only questions the current worldviews of Western culture, but also calls for practical changes. In David Orr's words (2002, p. 20), 'it is the careful meshing of human purposes with the larger patterns and flows of the natural world… 'Thinking ecologically about design is a way of strengthening the weave that links nature and culture' (Van der Ryn and Cowan 1996, p. 9). In this way, ecological design is presented as an activity that contributes to fixing the 'miscalibrations' between man and the rest of nature. 'The environmental crisis is a design crisis. It is a consequence of how things are made... [Yet] design manifests culture, and culture rests firmly on the foundation of what we believe to be true about the world' (Van der Ryn and Cowan 1996, p. 9). The way we design therefore, is part of a major cultural network of beliefs which, among other things, have lead us to the current complex crisis we are now facing. As Capra (1996, p. 4) suggests, our systems of crises may be 'just different facets of one single crisis, which is largely a crisis of perception'.

In other words, the problem of ecological design is, in this sense both epistemological (which may be illustrated through the question of how we *understand* our perception of and relationship with the other, the rest of nature and with the future generations) and *ethical* (which may be illustrated through the question of how we *should* relate with the other, the rest of nature and with the future generations).

Regarding these foundationally philosophical questions, the intention in this chapter is double: first, to synthesise a new epistemology of design—one that we call *an ecology of design* by attending to the problem of how are we to understand the systemic relationship between individuals and their environment – the human ecology of living – and comprehend the praxis of design as an integral part of it. And second, to synthesise the essential element for design to become ecological, which we will argue only occurs when its praxis is mainly commanded by the emotion and ecology of love. That is, love is described as the biological and ecological foundation of what makes us human beings and therefore as the main human disposition from which a truly ecological ethics and ecological consciousness in design praxis may emerge.

First, the chapter examines how design is part of an ecology of living which is epistemologically constructed as a reaction to modern rationale. Then, based on

Maturana's notion of human existence in conversation, the chapter suggests that design is a human form of conversing and synthesises four implications that are constitutive of this condition (Salazar and Baxter 2015). After dealing with an epistemological dimension of design as conversation, the chapter synthesises the notion of an ecology of love. Based on the exploration of several philosophical and scientific accounts, the article examines some essential aspects of an ecology of love that informs an ethical and collaborative form of designing.

2 Epistemology, Relationship and the Ecology of Design as Part of an Ecology of Living

Ecological design is a reaction to a Western epistemology of the human-nature relationship which has at least two profound implications for the understanding and practice of design:

1. The dichotomy between the artificial environment and the natural environment.
2. The negation of embodied emotion and the emergence of technocratic design.

In relation to the first of these implications, modern design has assumed that there is a sort of 'artificial environment' –the object, the building, the city, – isolated from and positioned over a 'natural' one. This is what Ingold (2000, p. 179) refers to as 'the building perspective', which believes that human beings 'inhabit the various houses of culture, pre-erected upon the universal ground of nature – including the universal of *human* nature'. Encapsulated in this artificial-cultural environment, the designer appears in opposition to nature. Thus, not only is the so-called 'natural environment' thought of as a sort of Garden of Eden for the elaboration of an artificial environment –which may be one of the deepest sources of the current environmental crisis – but also, the designer has become ecologically illiterate, and therefore unable to deal with the real challenge of sustainability. In the case of the second implication, there is a basic epistemological assumption that human intelligence (or reason) inexorably occurs dislocated from the body, emotion and the rest of nature. Freeman (2000, p. 211) synthesises this modern vision by alluding to Descartes's account of perception: 'the animal machine in man was guided by the soul as his "pilot", which sought knowledge through reasoning about the passive imprints of sensations, in order to arrive at absolute mathematical truth. Fantasy, intention and emotion were dismissed along with imagination as being nonmathematical and therefore unscientific'. As Solomon (2000, p. 3) comments, 'one of the most enduring metaphors of reason and emotion has been the metaphor of master and slave. Such a metaphor, not only suggests that emotion is 'inferior', 'more bestial', and less intelligent' than reason, so that it 'must be controlled', but it also implies a clear distinction between them 'as if we were dealing with two different natural kinds'.

In this Cartesian context, since the Enlightenment, design has been mostly understood as a process of technological invention, as the logical application of modern scientific knowledge for practical purposes,[1] and therefore as the primary means for human satisfaction and well-being. It was particularly in the Industrial Revolution when this epistemology of design based on a logic and mechanical way of thinking was deeply put into praxis. The emergence of industrial design in the industrial revolution not only was a major transformative and globalizing force, but also a key factor for the cultivation and expansion of a culture very much attached to a techno-cratic epistemology. Although the notion and practice of industrial design in the mid nineteenth century – and then in a *Fordist* fashion in the twentieth century – democ-ratized the access to a new modern and global era, it was a key element for the materialization of a Cartesian way of thinking in a global scale and a new economy based on mass consumption and exploitation of natural resources. As such, techno-cratic design has been thought of as a 'fundamental way humans take up with the world. It is a worldview that sees everything out there as having value only or pre-dominantly through its functional (instrumental) purposes' (Ehrenfeld 2008, p. 30).

However, after several hundred years, this form of designing has proved to be highly dislocative of human needs and ecological limits and is therefore highly unsustainable. Its main mode of operating is based mainly on a shallow form of enquiring and its way of dealing with complex problems, for example, has been to design new, more efficient artefacts that would eventually replace 'obsolete' ones without really asking any deep epistemological and ethical questions. This is what Orr (2002, p. 63) calls technological fundamentalism: '[designers] who fail to ask hard questions about why we do what we do, how we do it, or how these things affect long-term prospects…to question our basic assumptions about how our tools relate to our larger purposes and prospects'.

Since the second half of the last century, and as part of a systemic and holistic paradigm, two lines of research have made an important contribution to questioning and transcending the modern mind-body and culture-nature dichotomies. These are the phenomenology of perception, as developed by Merleau-Ponty, and the biology of cognition, as developed by Maturana. This has led to an ecology of living dynam-ically formed by the relational interdependency of the individual and its medium and to an epistemological change about human existence and cognition essential to the emergence of a new epistemology of design.

According to Merleau-Ponty (1962), the lived-body is the foundational basis of human experience and explanation and that every human activity, from walking in the park to theoretical reflection, is done by one's *embodied existence*. Neither the "body-subject" nor the world which it inhabits can be defined and characterized independently from each other: 'the world is inseparable from the subject, but from a subject which is nothing but a project of the world, and the subject is inseparable from the world, but from a world that it projects itself'(Merleau-Ponty 1962, p. 430).

In the biological field, the work of the Humberto Maturana (1980; 1988) arrived at a similar epistemological understanding of the individual-medium relationship and concluded that the phenomenon of perception is not the passive representation

[1] "technology". Oxford Dictionaries. April 2010. Oxford University Press. http://english.oxford-dictionaries.com (accessed March 14, 2011).

of a pre-existent world out there, but on the contrary, is determined by the physical structure of the organism's body that is itself organized in a circular fashion, as a 'closed network of interactions' of components. This circular organization, he discovered, is actually the general form that embraces all living beings – later defined as autopoiesis—literally a network of components that continually regenerates itself and constitutes its own topological domain for its realization as a network (Maturana and Varela 1987). Since an *autopoietic* system is dynamically constituted by a closed network of particular components, its perception is necessarily determined by it. That is, all the continuous changes of a living system, as a consequence of either its internal dynamics or its interaction with the medium, are specified by its actual embodied components. This implies that any external agent that interacts with a living being only *triggers* structural changes in the system, but does not determine them (Maturana and Varela 1987). Moreover, a 'living being only encounters the structural features of the medium that its own structure specifies' (Maturana and Mpodozis 2000, p. 5). In this sense, 'perception', Maturana noticed, 'should not be viewed as a grasping of an external reality, but rather as the *specification* of one' (Maturana, in Maturana and Varela 1980, pp. xv–xvi) and is not about a passive representation of an external world 'out there', but rather 'the bringing forth of a world' (Maturana and Varela 1987, p. 174). So, the biological mechanism of the organism-medium relationship in which both are dynamically formed by their structural coupling is a continuous process of recurrent interaction in which organism and medium mutually trigger structural changes *(*Maturana and Varela 1987, *p. 75).* Ultimately, a living being's ongoing capacity to maintain its autopoiesis through an enactive structural coupling with a world is an act of cognition, 'the throbbing of all life' (Maturana and Varela 1987, p. 100), 'the breath of life' *(*Capra 1996, *p. 257).*

These accounts by Merleau-Ponty and Maturana imply an epistemological change in which form is inescapably related to the process (Capra 1996; Ingold 2000). It is a change of the understanding of life and cognition from the *static* to the *active*, 'from being to doing'(Maturana and Poerksen 2004), from self to 'self-in-process'(Guidano 1991) and from the universal to the experiential.

Thus, nothing is static in life; nothing is totally fixed or programmed before the realization of living in relationships. Thus, neither the individual nor the medium can be defined as pure facts before their interaction. In contrast, they form one indissoluble totality, a being-in-the-world, which biologically occurs as a recurrent structural coupling. Thus, every human experience depends on being in a world that is inseparable from its own 'embodiment'(Varela et al. 1991). This is the cornerstone for the ecology of the living, and what makes us living beings a part of the phenomenon of life.

Human design is no exception to this ecology. Every act of design, good or bad, beautiful or ugly, useful or useless, both biologically and phenomenologically, belongs to our embodied inhabitation of a world. It is a systemic part of our ecology of living. Design does not create an artificial or cultural world separated from the environment, and this does not happen outside the embodiment of cognition. Design affects how we continuously create ourselves and the web of life in which we exist.

Every action of design emerges from, or is determined by, this embodied socio-ecological existence. It is in this circular causality that lies the power of the ecology of design. It is not ecologically passive, but active. In this sense, it can be said that, although every design is creative or active, only when it contributes to maintaining our bio-ecological existence in the pleasure of wellbeing, is it a constructive act.

3 Conversation: The Human Way of Existence

We have now reached the stage at which we can suggest how design actually happens in cognitive terms. We suggest this occurs in a further more complex level of cognition that is a fundamental aspect of our humanness, our *eco-cultural existence in conversation*.

To explain the notion of conversation, it is necessary to note three things about the phenomenon of emotion from a systemic perspective.

First, Emotion is an embodied cognitive event. The latest neurobiological views explain emotion as a 'prototype whole brain event', that is, as a complex phenomenon that embraces many regions and subsystems of the brain (Watt, quoted in Thompson 2007, p. 362). As Damasio (1995, p. 128) asserts, the neocortex area of the brain (traditionally linked to reason and 'will-power') is highly interconnected with 'downstairs', the subcortex, (traditionally linked to emotions) which plays a key role in 'biological regulations' within the system. 'Nature´, he suggests, ´appears to have built the apparatus of rationality not just on top of the apparatus of biological regulation, but also from it and with it. The neocortex becomes engaged along with the older brain core, and rationality results from their concerted activity'. Emotion can be seen as a 'prototype whole-organism event, for it mobilizes and coordinates virtually every aspect of the organism' (Thompson 2007, p. 362) and so it plays a key cognitive role in the dynamic homeostasis of an individual in recurrent interaction with its medium. As Damasio (1999, p. 50) concludes, 'emotions are *about* the life of an organism, its body to be precise, and their role is to assist the organism in maintaining life'.

Second, emotion also commands and constricts human intention. In behavioural terms, emotion is an embodied disposition to action, and therefore, always defines a kind of behaviour (Maturana 1988, pp. 48–49). As Thompson (2007, p. 365) synthesises, 'emotion is not a function in the input-output sense, but rather a feature of the action-perception cycle—namely, the endogenous initiation and direction of behaviour outward into the world.'

Third, emotion also takes place in a phenomenological domain, as a systemic part of the ecology in which the individual exists. This means that, through embodied and ever-changing emotional-appraisal behaviours, the individual continuously participates in and makes a particular eco-cultural network of interactions, while the networks that emerge from these interactions also constrain (cognitively trigger) the self-organizing embodied emotion. In this way, we can talk about an eco-enculturation of emotion. It is the ecological domain from which emotion and behaviour develop in particular ways. So, the phenomenon of human eco-cultural living is absolutely inseparable from the process of *how to live*, and emotion is at its core.

Armed with this explanation of emotion, we can now return to the notion of conversation, which, as Maturana suggests, is the human cognitive way of existence—namely, that humanness emerged when conversation started to be used in a recurrent and intergenerational fashion. Conversation, Maturana points out, is the systemic interweaving of our coexistence in language and our emotions (Maturana and Verden-Zöller 2003, p. 30). In opposition to a modern epistemology in which the rationality of language is separated from the body and emotion, here language and emotion form a unique embodied dynamic in which a higher order of *cognition as conversation* emerges. Every appraisal action such as thinking, talking, dreaming, reflecting and walking, occur through language as a central cognitive aspect of human intelligence. Yet, as embodied actions, they are inevitably emotional actions, and as such, they are emotionally driven. Cognitively, this is what Lewis (2005, 2000) calls 'appraisal-emotion amalgams' which, in a systemic causation, form the basis of human 'intentions', 'moods', and, ultimately, a 'sense of self'. Conversation is therefore not just the description of a world (as believed in modern theories of language) but a continuous bringing forth of it. It is the form in which humans coordinate their embodied-ecological existence. Thus, it is through conversing, as a cognitive and relational phenomenon, that humans *inhabit in and create an ecology of living*—namely, an eco-cultural network of conversation, and through it, make themselves in the world.

4 Design: A Human Manner of Conversing

So, by understanding human existence as a conversation, this allows us to suggest a genuine ecology of design in cognitive terms and to propose that *to design is to converse (see* Salazar and Baxter 2015*)*. This then becomes the primary way of understanding design as a biological and eco-cultural phenomenon and it allows us to synthesise the experiential, relational, embodied, emotional and linguistic constituents of the ecology of design.

Other design theorists such as Klaus Krippendorff and Ranulph Glanville, have also explored, either explicitly or implicitly, a notion of design as conversation. Arguing from a "second-order cybernetic" approach (Von Foerster 1974), they have highlighted the indispensable circular interaction or coordination between the observer and the observed – or the individual and the artefact – implicit in any design process.

Krippendorff (2005) for example, proposes a new foundation for design that emphasises the relational and experiential dynamics between users and artefacts and suggests that artefacts do not exist and have meaning by themselves (as modern design would assume) but are "manifest in the form of interfaces"—i.e. "sequences of ideally meaningful interactions". Artefacts "are constructed by those involved and account for their experiences under conditions of recursively stable and hence reliable interactions". This helps to explain why people may not only interface in different ways with the same artefact, but also act independently and responsively on what they mean to them. So he claims, that in order to design artefacts, it is important to understand the second-order cybernetics—that is, "the cybernetics of participating in systems under continuous reconstruction by their constituents" (Krippendorff 2007). This cybernetics of participatory design occurs in language.

Glanville (1999, p. 88) also defines design as a conversation, but for him, "conversation is a circular form of communication, in which understandings are exchanged. So, in a conversation, participants build meaning through the conversational form, rather than trying to communicate a predetermined meaning through coding and "words do not hold meaning—we do." For Glanville, design as conversation appears as "a means of exercising our creativity"—one that emerges from the circularity of conversation.

Both Krippendorff and Glanville contribute to our understanding of the design process in relational and experiential terms as an interactive and participatory phenomenon and as "a way of realizing oneself in coordination with others" (Krippendorff 2007). However, when design is understood as conversation, it is also necessary to pay attention to the embodied and emotional aspects of this ecology. Conversation must be understood not only as a relational and linguistic phenomenon but also as a cognitive mode of existence continuously determined by emotional dispositions. From this *Maturanean* understanding of conversation, we have offered important implications for design praxis (Salazar and Baxter 2015) synthesised in the following five points:

First: *the praxis of designing is always commanded by emotions*. To design is to apply several appraisal dynamics such as reflection, evaluation and critical observation. They are the key aspects of any design process. However, in our understanding of design as conversation, these appraisal dynamics are inevitably interwoven with emotions. So, if we want to understand what happens when a designing process is carried out, we have to focus on the emotions that specify the praxis of design and we have to pay attention to how the appraisal aspect of a design process is interwoven with the *emotioning* that it implies. 'What' we design is ineluctably attached to 'how' we design it. Clearly, design is not a purely technological method. In fact, technocratic design does not imply that design is just a logical and objective action either, but that it is only an explanation developed from a manner of conversing that leads us to negate our embodied and ecological existence, and to think of design solely in utilitarian terms. Design is always defined by the unique embodied and ecological experience of the designer. As such, there is no objective and value free process of design. In contrast, when the designer becomes conscious of the importance of emotions, a responsive disposition may also emerge. This disposition will facilitate reflecting on which kind of emotions contribute to the emergence and conservation of more sustainable relations both socially and ecologically.

Second: *designing is a kind of conversation about facilitating human existence in conversations.* The prime objective of design in a materialistically oriented society like the present Western-European culture is generally understood as the planning and producing of things. Designers, it is believed, create books, houses, cities, parks, etc. The problem with this vision is that, it not only construes the designer (subject) and the designed (object) in opposite Cartesian domains, but also, it elevates the created thing conceptually above what design truly does. In this way, a universalistic and domineering design is constructed. In contrast, the understanding that design does not primarily create specific things, but is a human conversation that offers *platforms that facilitate particular modes of conversing* is the cognitive substance of what Krippendorff (2007) has termed 'human-centred design'. This 'calls on designers to conceive of their jobs not as designing particular products, but to design affordances for users to engage in the interfaces that are meaningful to them'.

Third: *designing is a systemic part of an eco-cultural network of conversations.* As design is part of a whole process of inhabitation, it is necessarily entwined with many other forms of conversations and other non-human beings and eco-systemic flows and cycles and this means that every design process and designed product is eco-culturally shaped. It is an eco-cultural manifestation of a specific form of living. Design also *makes* the eco-cultural domain in which it takes place and it does so by facilitating modes of conversations. Although the designer cannot design conversations, nor specify emotional changes, design always *triggers* emotional changes in an agent, who may either conserve or change certain behavioural patterns. Design therefore 'encourages' a certain manner of conversing that may result either in the conservation of a particular pattern of acting or in the changing of it. Moreover, as conscious beings who are able to feel our emotions and associate them with certain actions, we can deliberately design platforms that try to control the way that people converse thereby triggering emotions such as fear and aggression, or we can design platforms that invite people to converse in a more participatory way, thereby triggering a sense of well-being. So, design ineluctably has an effect on the quality of life of other human and non-human beings.

Fourth: *every human being is a designer*, simply because human beings live in conversation and design is a kind of conversation. This implies a reconsideration of the importance of the interdisciplinary and trans-disciplinary (Max-Neef 2005) process of co-designing and entails a radical change of the designing process, from objective and mandatory design, to an inter-subjective and cooperative one. Ultimately, every human being participates in the creation of a world experienced as a multiverse of diverse others. Listening to the other with respect and cooperatively finding ways to afford diversity now become the essential ingredients of creativity by designing ecologically. In this form of creation, the professional designer's contribution need not be suppressed, but needs to be redefined as a *facilitation* of socio-ecological processes. As Wahl and Baxter (2008) have pointed out, 'as facilitators of trans-disciplinary integration, designers can help to change culturally dominant worldviews and value systems'.

Fifth: *learning to pay attention to emotions is a first step towards a more ecological form of designing.* We need to be given the opportunity to pay attention to emotions and to learn from them—that is to converse with ourselves, with the bodies in which we exist, and to create meaning from there. In the impersonal and hurried globalized world of today, this awareness of connection to the body has been critically constrained, but design as a process of facilitating conversation can improve this situation. This however suggests radical changes in the design education. For example, it will require a move away from a functional characterization of design such as 'industrial design', 'graphic design' or 'media design', to a qualitative one, which explicitly elucidates its emotional intentions, and encourages the study of emotion as a central aspect of understanding the world in which the designer lives.

By adopting this position, designers will be able to (1) understand which kind of emotions in general and conversations in particular have led us, as a society, to generate so much ecological destruction—i.e., to understand the roots of the social and environmental crisis we are facing; and (2) to ask and learn which emotions – which forms of conversing with the other and the rest of nature – are the ones from which more sustainable or harmonious patterns of human inhabitation may emerge.

5 Ecological Design as Part of an Ecology of Love

When we become aware that through design as conversation a world is continually made in a certain way, we are compelled to look for those forms of conversing by design that would not only cultivate the consciousness of our embodied existence in an ongoing ecology of coexistence, but also to look for ways that would facilitate the conservation of this in a harmonious fashion. Based on this, the fundamentally ethical challenge of ecological design may be rephrased as follows: To facilitate the emergence and conservation of a human existence in conversation in which ourselves, our contemporaries, future generations and the rest of nature are seen as legitimate participants of the web of life. Following the accounts of several philosophers and scientists, we suggest that both the ethical disposition and cooperative praxis of ecological design really flourish as part of an ecology of love, which is a fundamental emotional and relational dynamic for human flourishing and survival. When Erich Fromm (1995, p. 6) says that 'a theory of love must begin with a theory of man' he implicitly suggests that humanness and love are part of a circular causation – that is, that a theory of man should also begin with a theory of love. Similarly, Pitirim Sorokin (2002, p. 6) defines love as a fundamental 'energy' for human endurance, a 'vital power'. He says: 'ontologically love is…a unifying, integrating, harmonizing, creative energy or power… …Without the operation of love energy the physical, the biological and the socio-cultural cosmos would have fallen apart'. This energy, he explains, is not only emotional and biological but also a basic human need.

First we would like to introduce the term of ecology of love as a comprehensive concept that unifies in a relational and indivisible dynamics the social love, the love of nature and the love of place. We do this by respectively examining and then integrating the accounts of Maturana, Wilson and Tuan related to a human's loving existence. What unites these three authors is that they generate crossbred explanations of biological (innate) origins with cultural processes of learning and experience of human existence in love. Nevertheless, we argue that there is still need for a comprehensive vision of love that unites all of this into of what we call an ecology of love. From this synthesis we then propose what we consider are key implications for the thriving and praxis of ecological design in a global era.

From a social dimension, Humberto Maturana asserts that love is the only emotional disposition (or 'domain of relational behaviours'[2]) that allows human recurrent interactions in which 'the other' is accepted as an authentic other 'without expectation, and that can be amplified and stabilized'. As such, he argues that love has biologically been an essential emotion for the emergence of the human social and language domains (Maturana and Verden-Zöller 2008, p. 223; Maturana 2005, pp. 45–46). Maturana and Verden-Zöller suggest that our loving existence is the result of a particular epigenetic trans-generational manner of living. This manner of living was mainly constituted, conserved and extended through a *social* and *recurrent* trans-generational phenomenon. They point out that the human lineage emerged through the conservation of the progressive expansion of the mother/child relationship of mutual body acceptance, nearness, and mutual care in playfulness and total trust, in a manner that also involved the male, and progressively extended beyond the age of reproduction into the adult life in a *neotenic* evolutionary trend. Eventually, living through loving conversations became a central manner of living through every developmental period of a human being. Thus, Maturana and Verden-Zöller claim that the conservation and trend beyond neoteny of this manner of living in intimate playing in love, generated the possibility for the spontaneous arising of language as an intergenerational phenomenon learned by children in the interaction with their mothers. They concluded that this 'constituted humanness as the basic loving manner of living that we live now' (Maturana and Verden-Zöller 2008, pp. 3–4).

From a more ecological standpoint, and transcending the primarily social dimension of love examined by Maturana, Wilson (1984, p. 31; Kellert and Wilson 1993) introduces the term biophilia, as 'the innate tendency to focus on life and life like

---

[2] Maturana argues that, because emotions are internal body dispositions, that is, they are dynamic body changes that belong to a domain different from the domain of the observer, we cannot see them directly. However, considering that the observer has access to the behavioural domain of a living being, what the observer connotes when he distinguishes emotions, 'is a domain of relational behaviours'. In other words, when we talk about emotions, we always refer to some domain of behaviours (such as, seeing, hearing, moving, thinking, reflecting, etc.) that an animal or person may do, and we speak in terms of the 'kinds of doings that it may generate'. So, as Maturana explains, 'the different emotions or moods can be fully characterized in terms of the kinds of relational behaviours that they entail as a domain of actions'.

processes'. From an Aristotelian perspective, biophilia not only means *love*, but also embraces, through the concept of friendship (*philia*), the notion of *relationship with others*. This is a central point in the idea behind the notion of biophilia. Indeed, both Wilson and others who have contributed to the formulation of this hypothesis, propose that love of other living beings is materialized in a human 'tendency' or a 'need' to 'affiliate' with Nature. 'From infancy', Wilson says, 'we concentrate happily on ourselves and other organisms. We learn to distinguish life from the inanimate and move toward it like moths to a porch light'(Wilson 1984, p. 1). As he argues, "for more than 99% of human history people have lived in hunter-gatherer bands totally and intimately involved with other organisms. During this period of deep history, and still farther back, into paleohominid times, they depended on an exact *learned knowledge* of crucial aspects of natural history. ...In short, the brain evolved in a *biocentric* world, not in a machine-regulated world". From this argument Wilson concludes that only by understanding and establishing deep relationships with other organisms, "we will place greater value on them, and on ourselves' (Wilson 1984, p. 1).

Finally, as one of several environmental geographers, anthropologists and psychologists who have brought human relationship to place as an essential element of human experience, attitude and behaviour, Tuan (1974) introduces the notion of *topophilia.* In a similar, but also broader fashion than Wilson's biophilia (Berry et al. 2015), Tuan asserts that topophilia includes every affective bond of human beings with their material environment or place. Examining Tuan's topophilia, Sampson (2012, p. 26) has described it as "an innate bias to bond with local place, including both living and non-living components". In this regard, Sampson asserts that the notion of topophilia recognises the innate human attachment to other living beings as proposed by Wilson, but has expanded it to include every physical and non-living element of a place. Furthermore, Tuan argues that this bond is essentially aesthetic and experience-based, so topophilia is often related to a particular place or environment. Thus the aesthetic experience of topophilia always varies in its intensity, subtlety and mode of expression. Yet, when topophilia becomes a strong emotion, he points out, we can be sure that the place we inhabit is replete with intimate events, symbols and memories; a place that we feel as home.

What unites Maturana, Wilson and Tuan, in addition to the common bio-cultural essence of their respective arguments, is a relational vision of human love. A key concept that unites these relational accounts is *intimacy.* By synthesising these account we can argue that social and ecological intimacy are actually one loving phenomenon that was biologically essential for the emergence of humanness. Humanness emerged immersed in an intimate socio-ecological network of conviviality. We can observe in the ancestral family, a full interdependence between social phenomena and intimacy with nature (Shepard 1998). The main cooperative social activities such as, gathering of food, hunting of animals, obtaining necessary timber for a fire, cooking and sharing of food, orally sharing and teaching the myth, the spirits and knowledge of nature, walking in a wet forest, and resting near a river after a long journey, necessarily happened in a deeply congruent and intimate coupling with eco-systemic processes that were ultimately felt as home. In this sense, every social activity arose from and in deep relations with a place. Therefore, the trend from loving mother-child neoteny to every human developmental phase examined by Maturana as a fundamental aspect for the emergence of humanness as a *languaging* and social being, may be rephrased as a trend to *philic* relationships with other living beings and ultimately to a whole place. This synthesis allows us to: (1) understand humanness not only as a social phenomenon but more broadly as an ecological one that takes shape in recurrent relationship with a whole ecosystem; (2) comprehend that humanness, as a relational being, is dependent on a socio-ecological love, since this is the only emotion that allows the emergence and conservation of recurrent co-operative relationships; (3) recognise that the love for other human beings, the love for other non-human beings and the love of the places we inhabit are essentially one indivisible dynamic dependent on human experience – i.e. an ecology of love; (4) realize that through the cultivation of an intimate relationship with the eco-cultural places which we inhabit, we come to know ourselves and our wildness; it is through this knowledge that we become aware that we are part of a major ecosystem (or nature); and (5) to understand that, as Maturana, Wilson and Tuan warn us in their respective terms, the loss of love as an intimate relational phenomenon in modern society not only jeopardizes human's ability to flourish but it is also an essential factor in socio-ecological diseases and destruction.

Equipped with a broad synthesis of an ecology of love, we suggest that ecological design, as a conversing phenomenon that has reacted to the socio-ecological crisis we are now facing, is not only dependent on love, but also it is a platform that enhances the human capacity to intimately affiliate with the other, the rest of nature and the particular places we inhabit. As such, by linking ecological design with an ecology of love, there are at least four implications that we would like to suggest:

1. Ecological Design is about listening to the other in its own legitimacy.

Love is an emotion that defines an intention of openness to oneself and the other, one in which the other is brought forth as a legitimate other in coexistence with oneself (Maturana and Verden-Zöller 2008). Therefore the disposition of love is one that 'disarms our emotional defences' and makes us 'vulnerable to the other'', it is a 'matter of "really looking"' at the beloved (Velleman 1999, p. 361). Similarly, Badhwar (2003, p. 43) suggests that "love is a perceptive look, a look that seems really to *see* the loved object, not a falsifying look of projection and fantasy, or a self-centred look of appropriation. And, seeing the loved object as it is, the look of love seems to affirm the object's value in its own right". Therefore, the look of love is the essence of the ecological look, it is the point of departure of the encounter with the other – one in which the other emerges in its legitimacy in relationship with oneself.

As such, the disposition or look of love is crucial for ecological design. Ecological design emerges when it is realized that the *first* and most important action in an ecological conversation is not about giving speeches (Irigaray 2002), teaching the other, proposing new ideas, new technologies, or making or creating new solutions, but rather, it is about listening to oneself and to the other in their legitimacy.

Listening is the primordial, most important, action of ecological design. It is its point of departure. Without listening, the action of designing becomes a noise and is ecologically displaced. The disposition of love is the only emotion that allows us to listen to oneself, and from there, spontaneously to start listening to the other.

2. Ecological design is a collaborative learning process.

Global design praxis, happening in a competitive and technocratic disposition and triggered by a purely neo-economic vision of development, is based on the belief that the designer defines and creates 'solutions' for the other—globally, for other places that many times are distant, almost foreign to him. Based on a Cartesian subject-object separation, Western design then has reinforced a profound division between the designer and the consumer that, in a complex global era, has readily been translated into a dichotomy of designator and designated dimensions. Thus, design praxis has been dominated by the belief that one designs *for* people, *for* the user, *for* the mass but not *with* the other. As such it has become epistemologically and practically anti-ecological, hierarchical and often, overbearing, and therefore has faced severe constraints and thresholds that prevent it from adequately informing itself in ecological and socio-cultural terms.

Ecological design demands that we pass from the competitive vision of development to a cooperative one. Epistemologically, ecological design is about exploring the significance of working *with* the other and not *for* it. By doing this, design praxis not only frees itself from the designer-consumer Cartesian dichotomy but also enters into the ethical and cooperative domain of relationship that – we have argued - only comes through the ecology of love. Working *with* the other is a conversation that happens – in its most basic form and smaller scale – with a particular and irreplaceable other (or others) and is concerned with the other for its own sake (the essence of the Aristotelian complete state of *philia*). As Frankfurt (2004, p. 42) has synthesised, "love consists most basically in a disinterested concern for the wellbeing or flourishing of the person who is loved. It is not driven by any ulterior purpose but seeks the good of the beloved as something that is desired for his own sake". As such, following Singer's account, designing with the other and for the other's sake implies passing from a merely appraisal form of relationship in which "we are all commodities for each other" to one that also incorporates a bestowal dimension: "an engendering of value by means of one's appreciative attitude towards the person, thing or ideal, to which we attend" (Singer 2009, p. 53).

Therefore, designing *with* a particular other and *for* the other's own sake has at least two major implications: first, it implies an attentive opening and intimate relationship that informs design about the particularity of the other, its constitutive structure and conditions, and everything that is needed to contribute to its conservation and well-being. In this way, the designer is occupied with the other as a whole being, as an end in itself. But also, designing with the other for its own sake is about being aware that the existence of the other is ineluctably interwoven with the medium in which it exists. That is, the designer's loving attention also brings to the conversation (or design process) environmental factors that are fundamental to the wellbeing of the beloved. Briefly, therefore the designer gets to know the other and its medium, and acts ethically and ecologically in accordance with that knowledge.

Second, as a conversation, ecological design is fundamentally an inclusive and intimate co-facilitation and co-creation process. It is a dynamic that incorporates the other, that invites the other to give its point of view, to create the world in which it 'wants' to live. In such a scenario, ecological design questions and tends to transcend the globalized separation between designer and consumer. So, "*designing with*" enhances a self-producing order. Ecological design implies – in Princen's terms (2010, p. 93) - to pass from a mining and consuming economy to a home economy in a global era. While the first two - which have commanded the neo-liberal and pro-growth patterns of globalization – have respectively fostered "humanity's excess throughput of material and energy" and "the construction of a ´sovereign consumer` [that] is all too convenient excuse for powerful actors to evade their societal and environmental responsibilities", a home economy recalls the notions of co-production and self-reliance of the places where we live. Consequently, Princen argues that "a useful starting point would be a consumer's apparent polar opposite, a ´producer economy`", in which every one is seen as a designer in a sense of creating, making, using, recycling and caring for.

3. Ecological design is about establishing an intimate and aesthetic encounter with the place we inhabit.

Globalization may have brought positive things to many people around the world, but local communities have been one of the losers of globalization (Shuman 2000). The main negative effect is that people have lost the capacity to produce for themselves, becoming reliant on a single, highly unstable and complex economic system (Douthwaite 1996). Being reliant on global factors that are socially and environmentally unsustainable has had serious social consequences in the local domain, such as, the breakdown in family structures and extended communal groups (Laszlo 2006), the jeopardizing of socio-ecological dynamics in local communities and neighbourhoods (Barton 2000; Shuman 2000) and the loss of administrative power in local governments (Hess 2009).

In this scenario, ecological design must lead the way to a localization process in a global era. This is not an anti-global discourse or the enhancement of a local-global dichotomy. On the contrary, it is the inevitable exploration of localizing our life while thinking of and appropriately participating in a global age. This is what Schumacher meant by developing a new economic process in 'a direction that shall lead it back to the real needs of man, and that also means: *to the actual size of man.*' Similarly, bioregionalism has also called for a local living in a global era, materialized through the holistic process of getting back into place. As Berg and Dasmann (quoted in Aberley 1999, p. 23) put it: "Living-in-place means following the necessities and pleasures of life as they are uniquely presented by a particular site, and evolving ways to ensure long-term occupancy of that site. A society which practices living-in-place keeps a balance with its region of support through links between human lives, other things, and the process of the planet – seasons, weather, water cycles – as revealed by the place itself".

This localization of human living through design is essentially a *topophilic* process. As such, there are two main implications for ecological design: first, it implies a design praxis intimately and aesthetically related with the place. This reciprocal intimacy not only brings the emotional, perceptive and relational aspects of design praxis together in epistemological terms, but is also the bedrock of an ecoliteracy process of design. Only through a *topophilic* relationship can the designer opens himself to really look at the place, to learn about its eco-systemic components and processes and design in coherence with it.

Second, it implies the emergence of an ecological consciousness—the feeling that one's identity, health and wellbeing are deeply rooted and interdependent with an inhabited place and the major biosphere. Snyder illustrates this by describing that 'after twenty years of walking right past it on my way to chores in the meadow, I actually paid attention to a certain gnarly canyon live oak one day. Or maybe it was ready to show itself to me. I felt its oldness, suchness, inwardness, oakness as if it were my own. Such intimacy makes you totally at home in life and in yourself'(Gary Snyder in, Nicholsen 2003, p. 65) In this way, acting from an ecological consciousness, the participants of an intimate process of co-design would understand the scale-linking essence of ecological design and would apply an ethical framework to it. Through the emergence of ecological consciousness, designers eventually become aware that their actions inevitably take part in the construction and conservation of the major biosphere. They become aware that their actions involve ethically respecting their co-existence with uncountable other beings, things and flows, and therefore they act in a way that, appraising from their own ethical concerns, does not unnecessarily reduce the possibility for other beings to develop their own lifes.

4. Ecological design is about learning and acting in the flow of the present.

As the heirs of a platonic epistemology, Western culture is firmly attached to the understanding of wellbeing as a final stage, as a transcendental value. Based on the idea of endless progress and growth, we do not only initiate a process of boundless accumulation, but also hope that at a certain point we will reach a level of satisfaction. This is what Fromm (1995) refers to when he points out that our culture has moved from a "Being" mode of life to a "Having" mode of life. In a globalized scenario, governed by this progressive epistemology, we are now realizing that infinite growth (or what now some call sustainable growth) is an oxymoron, especially when we are facing the environmental breakdown of a very finite planet. Ultimately, the incessant progressive pilgrimage to attain a predefined and transcendental state of wellbeing (or beauty in ancient philosophy) has generated cultural patterns of lives that are anxious, painful, and ultimately socially and ecologically destructive. This is a basic factor in the global ecological crisis and a main obstacle to understanding and living in the pleasure that love brings.

In contrast, as Maturana (2008, p. 98,99) asserts, the emotion of loving happens in the flow-of-living-in-the-present in the legitimacy of everything. Loving, he suggests, is living in the spontaneous unity of everything. It is living in the corporeal wellbeing in the flow of social conviviality. In this sense, Maturana links the process of loving with the ancestral notion of 'the path of Tao'—namely, the oriental idea of living in the wellbeing that emerges through the process of living without the suffering that comes with the attachment to anything that is declared transcendental, without looking for the permanent (Maturana 2008, p. 97). Similarly, Fromm (1995, p. 17) proposes that loving is an art—'the art of loving'. As an art, he says that love 'is an action', 'an activity', a 'personal experience', so there are no 'prescriptions' to its practice. Yet he proposes that, as with any art, love requires discipline, concentration and patience. As such, the art of loving is about continuously co-creating and conserving the places where we live by accepting one's own and the other's legitimate participation in it.

With a progressive focus, the possibility of an ecological design becomes an impossibility. The relational aspect of its ecology becomes instrumental, a secondary phenomenon in which the attention is reduced to the attendance on a final product. In contrast, as a cooperative conversation, ecological design brings back the importance of daily life in order to change the current unsustainable form of our global era. It is in the flow of the present where real actions occur, and from where a more sustainable form of organization of our institutions and socio-ecological living can emerge. Unfortunately however, the mainstream visions of sustainable development and ecological design have become yet another representation of a progressive form of thinking that ultimately negates our inevitable existence in a complex interconnected present and unpredictable future (Foster 2014, see 2012). In contrast, ecological design must be understood as a platform that facilitates the continuous emergence and cultivation of a more sustainable world, not as a fixed point in the future, but through the interconnection of an essentially diverse and in-placed dynamics that occur in our on-going present for which we are responsible.

6 Conclusion

When we realize that phenomenologically we are participants and creators of an ecosphere that is at the edge of collapsing, this means that, among many other things, in our hands rests the co-creation of a world that we would conceive spiritually, socially and ecologically sustainable. In this sense, the global environmental crises we are facing become an opportunity for deep reflection, imagination and creativity by design.

In this Chapter we have elucidated an ecology of design and explained its cognitive dynamics as a form of conversing from a bio-cognitive and phenomenological point of view. This may contribute to an epistemological change in which the embodied, emotional and ecological dimensions of design praxis are reinvigorated. Design, we have suggested, is a human conversation about *facilitating* our existence in conversation. This implies, among other things, that every action of design is commanded by the embodied, self-organizing *emotioning* of the designer. It also implies that the practice of design unfailingly defines and is defined by the eco-cultural medium in which the designer exists. However, this epistemological realization also opens up the essentially ethical question of ecological design. As an ethical phenomenon, we have suggested that design only becomes ecological when it is primarily guided by the emotion and ecology of love. It is through this emotion that design becomes a platform of conversation that facilitates and encourages the emergence of ecological consciousness. Essentially, we conclude that the platforms of ecological design are systems (of every kind) that treat the other as an authentic being and invite it to co-create a cooperative form of being at home in the world.

Acknowledgments This research was supported by Fondecyt Iniciación Project N° 11130519; by Project CONICYT/FONDAP N°15110020 (CEDEUS); and by Project CONICYT/FONDAP N° 15110006 (CIIR).

References

Aberley, D. (1999). Interpreting bio-regionalism: A story from many voices. In M. V. McGinnis (Ed.), *Bioregionalism*. Routledge.
Badhwar, N. K. (2003). Love. In H. LaFollette (Ed.), *The Oxford handbook of practical ethics*. Oxford: Oxford University Press.
Barton, H. (Ed.). (2000). Sustainable communities: The potential for eco-Neighbourhoods. Earthscan.
Berry, T., Jönsson, K. I., & Elmberg, J. (2015). From environmental connectedness to sustainable futures: Topophilia and human affiliation with nature. *Sustainability*, 8837–8854.
Capra, F. (1996). *The web of life: A new synthesis of mind and matter*. London: HarperCollins.
Damasio, A. R. (1999). In W. Heinemann (Ed.), *The feeling of what happens: Body and emotion in the making of consciousness* (1st ed.). London.
Damasio, A. R. (1995). *Descartes' error: Emotion, Reason and the Human Brain*. Picador.
Douthwaite, R. J. (1996). *Short circuit: Strengthening local economies for security in an unstable world*. Dublin: Green Books, Lilliput Press.
Ehrenfeld, J. R. (2008). *Sustainability by design: A subversive strategy for transforming our consumer culture*. Yale University Press.
Foster, J. (2014). *After sustainability*. Abingdon: Routledge.
Foster, J. (2012). *The sustainability mirage: Illusion and reality in the coming war on climate change*. Routledge.
Frankfurt, H. G. (2004). *The reasons of love*. Princeton University Press.
Freeman, W. (2000). Emotion is essential to all intentional behaviors. In M. D. Lewis & I. Granic (Eds.), *Emotion, development, and self-organization: Dynamic systems approaches to emotional development* (pp. 209–235). Cambridge: Cambridge University Press.
Fromm, E. (1995). *The art of loving*. Thorsons.
Glanville, R. (1999). Reserching design and designing research. *Design Issues, 15*, 80–91.
Guidano, V. F. (1991). *The self in process: Toward a post-rationalist cognitive therapy*. New York: Guilford Press.
Hess, D. J. (2009). *Localist movements in a global economy: Sustainability, justice, and urban development in the United States*. Massachusetts: MIT Press.
Ingold, T. (2000). *The perception of the environment: Essays on livelihood, dwelling and skill*. London: Routledge.
Irigaray, L. (2002). *The way of love*. Continuum International Publishing Group.
Kellert, S. R., & Wilson, E. O. (Eds.). (1993). *The Biophilia hypothesis*. Island Press.
Krippendorff, K. (2007). The cybernetics of design and the Design of Cybernetics. *Kybernetes, 36*, 1381–1382.

Krippendorff, K. (2005). *Semantic turn: New foundations for design*. Boca Raton: CRC.
Laszlo, E. (2006). *The chaos point: The world at the crossroads*. London: Piatkus Books Ltda.
Lewis, M. D. (2005). Bridging emotion theory and neurobiology through dynamic systems modeling. *Behavioral and Brain Sciences, 28*, 169–194.
Lewis, M. D. (2000). Emotional self-organization at three time scales. In M. D. Lewis & I. Granic (Eds.), *Emotion, development, and self-organization: Dynamic systems approaches to emotional development* (pp. 37–69). Cambridge: Cambridge University Press.
Lorimer, J. (2015). *Wildlife in the anthropocene: Conservation after nature*. London: University of Minnesota Press.
Maturana, H. (1980). Biology of cognition. In H. Maturana & F. J. Varela (Eds.), *Autopoiesis and cognition: The realization of the living* (pp. 1–58). Dordrecht: Reidel.
Maturana, H. (1988). Reality: The search for objectivity, or the quest for a compelling argument. *The Irish Journal of Psychology, 9*, 25–82.
Maturana, H. (2005). El Sentido de lo Humano. J. C. Sáez.
Maturana, H. (2008). Biología del Tao o el Camino del Amar, in: Dávila, X.Y., Maturana, H. (Eds.), Habitar Humano: En Seis Ensayos de Biología-Cultural. J. C. Sáez.
Maturana, H., & Mpodozis, J. (2000). The origin of species by means of natural drift. *Revista Chilena de Historia Natural*, 261–310.
Maturana, H., Poerksen, B., 2004. From being to doing: The origins of the biology of cognition. Carl Auer Verlag.
Maturana, H., & Varela, F. J. (1987). *The tree of knowledge: The biological roots of human understanding*. Boston: Shambhala Publications.
Maturana, H., & Varela, F. J. (1980). *Autopoiesis and cognition: The realization of the living*. Dordrecht/Holland: Reidel.
Maturana, H., Verden-Zöller, G. (2008). The origin of humanness in the biology of love. Imprint Academic, Exeter.
Maturana, H., & Verden-Zöller, G. (2003). In J. C. Sáez (Ed.), *Amor y Juego: Fundamentos Olvidados de lo Humano* (6th ed.).
Max-Neef, M. (2005). Foundations of transdisciplinarity. *Ecological Economics*, 5–16.
McKibben, B. (1989). *The end of nature*. New York: Random House.
Merleau-Ponty, M. (1962). *Phenomenology of perception*. London: Routledge & Kegan Paul.
Nicholsen, S. W. (2003). *The love of nature and the end of the world: The unspoken dimensions of environmental concern*. MIT Press.
Orr, D. W. (2002). *The nature of design: Ecology, culture, and human intention*. New York: Oxford University Press.
Princen, T. (2010). *Treading softly: Paths to ecological order*. Massachusetts: MIT Press.
Purdy, J. (2015). *After nature: A politics for the anthropocene*. London: Harvard University Press.
Salazar, G., & Baxter, S. (2015). Towards an ecology of design praxis. *The Design Journal, 18*, 421–438.
Sampson, S. (2012). The topophilia hypothesis: Ecopsychology meets evolutionary psychology. In P. Kahn & P. Hasbach (Eds.), *Ecopsychology: Science, totems, and the technological species* (pp. 23–54). Cambridge: MIT Press.
Shepard, P. (1998). *Coming home to the Pleistocene*. Island Press.
Shuman, M. (2000). *Going local: Creating self-reliant communities in a global age*. New York: Routledge.
Singer, I. (2009). *The nature of love, volume 3: The modern world*. MIT Press.
Solomon, R. C. (2000). The philosophy of emotions. In M. Lewis & J. M. Haviland-Jones (Eds.), *Handbook of emotions* (pp. 3–15). New York, US: Guilford Press.
Sorokin, P. A. (2002). *The ways and power of love: Types, factors, and techniques of moral transformation*. Templeton Foundation Press.
Thompson, E. (2007). *Mind in life: Biology, phenomenology, and the sciences of mind*. Belknap Press of Harvard University Press.

Tuan, Y. (1974). *Topophilia: A study of environmental perception, attitudes, and values.* COLUMBIA University Press.
Van der Ryn, S., & Cowan, S. (1996). *Ecological Design* (1st ed.). Washington: Island Press.
Varela, F.J., Thompson, E., Rosch, E., (1991). The embodied mind : Cognitive science and human experience, first MIT press paperback edition, 1993. ed. MIT Press.
Velleman, J. D. (1999). Love as a moral emotion. *Ethics, 109*, 338–374.
Von Foerster, H. (1974). *Cybernetics of cybernetics or the control of control and the communication of communication.* Urbana: Biological Computer Laboratory, University of Illinois.
Wahl, D. C., & Baxter, S. (2008). The Designer's role in facilitating sustainable solutions. *Design Issues, 24*, 72–83.
Wilson, E. O. (1984). *Biophilia.* Harvard University Press.

Scales of Design: Ecodesign and the Anthropocene

Victor Petit and Bertrand Guillaume

Abstract In this chapter, we provide with a brief historical overview of the encounter between design and the global environment in the Anthropocene, investigating the moment when the traditional figures of conception (namely the engineer and the architect) merged under the single term "design". We then offer a philosophical inquiry into the meaning of design today, and the crucial role of designers next to engineers. We more particularly look at the issue of scales (space scale and time scale) in the context of the ecological crisis, and at the significance and the usefulness of an approach in terms of the "milieu" rather than of the environment. We show that the articulation of global and local is possible, as is possible a transition both digital and ecological, and that the important distinction is not between global and local design, but between two ideas of eco-design.

Keywords Anthropocene · Ecodesign · Milieu · Scale

1 Introduction

It is often claimed that industrial revolutions have set humans free from the limits of space and time, that is to say that technology modifies spatial and temporal scales. The space-time of technology is, however, no longer commensurate with ours. As Gunther Anders wrote: "We humans are smaller than ourselves" (Anders 1961: 11).

The question of the scale change in the history of design is much wider than the following paragraphs. Would not it be feasible to see a great deal of discontinuities in the history of art, or more generally in the history of technology? Is the history of industrial design in the twentieth century the passage from objects to homes, and from homes to towns? When exactly design became global? When design meets the

V. Petit
University of Technology, Troyes, France

B. Guillaume (✉)
Max-Planck Institute for the History of Science, Berlin, Germany
e-mail: bertrand.guillaume@utt.fr

© Springer International Publishing AG 2018

P. E. Vermaas, S. Vial (eds.), *Advancements in the Philosophy of Design*, Design Research Foundations, https://doi.org/10.1007/978-3-319-73302-9_22

Anthropocene, which is not accidental, contrariwise to what the naturalistic discourse assumes (Bonneuil and Fressoz 2013)?

The Anthropocene, like design, is a much-debated idea regarding its nature, its origin, or its name. It refers either to the Great Acceleration or Gaïa. While Italian futurism is a candidate for the design of the Anthropocene from the point of view of the former, the design science of Buckminster Fuller is a candidate from the point of view of the latter, and one can actually find parallels between Fuller and James Lovelöck (Grevsmühl 2014).

A brief historical overview of the encounter between design and the global environment will lead us to a more philosophical inquiry into the meaning of design today, and the crucial role of designers next to engineers. We will show that the articulation of global and local is possible, as is possible a transition both digital and ecological, and that the real distinction is not between global and local design, but between two ideas of eco-design.

2 From Architecture to Design

Studying the relationship between design and the Anthropocene is in a sense investigating the moment when the traditional figures of conception (namely the engineer and the architect) merged under the single term "design". It is about questioning the time when everything, at all scales, became epithet of design. Is the Anthropocene so different from climate design? Design is about art, as much as about technology, and its scales range from object to Earth through the city.

2.1 From the Spoon to the City

The design changes scale when it becomes mixed with the social planning, when it tries to answer what Horst Rittel called wicked problems, such as the environmental problems (Rittel and Webber 1969). The idea of "large-scale design", which echoes design for natural environments as much as design for social organization, is inherited from what James Scott (1998) labeled "high modernism". Le Corbusier, who could have said: "Big is beautiful", is its incarnation (Scott 1998: 104). His book entitled *La ville radieuse. Eléments d'une doctrine d'urbanisme pour l'équipement de la civilisation machiniste* (1933) illustrates such a scale change and his totalizing (totalitarian?) approach disregarding any context. An exhibition held at the Centre Pompidou, Paris (*Le Corbusier. Mesures de l'Homme*, 2015) has recently reopened the debate on Le Corbusier's fascist ideas (De Jarcy 2015). Leaving this debate aside, let us observe that the recognition of Le Corbusier is first and foremost a scale issue: at the scale of furniture (as a designer), Le Corbusier is acclaimed; at the scale of housing (as an architect), le Corbusier is admired; but at the scale of the city (as a planner), Le Corbusier is, at best, contested. If his *Modulor* (1949) commands to systematically take into account the human scale, he sketches what appears to be an "ethereal" man, for whom the scale change is exclusively an arithmetic operation! Meanwhile, he overlooks that scale changes are not only quantitative ones, but also qualitative ones, and forgets that the scales of man are not the measures of man. At least two contradictions fall within the scope of our inquiry. Firstly, Le Corbusier uses the very notion of "milieu", but his philosophy is in reality more about removing the variability of the milieu.[1] Secondly, he evokes scale, but he seems caught in an inescapable contradiction between scale and proportion. Indeed, the *Modulor* aporia lies in the assumption that scale is unique, and also in the confusion between proportion and scale (Boudon 2002). Philippe Boudon uses scale as the central notion of his "architecturology", and defines it as "the relevance of measure": one has to measure in a certain scale, but this scale does not result from the measure (Boudon 1991). Contrariwise to geometric proportion, architectural scale only makes sense in its connection with the human body. However, scale is precisely not just a matter of measure and size relatively to the human body.

Le Corbusier is, of course, not the only architect who dreams of expanding scales. There are also, for instance, the architects of the "Megatructure" (Deyong 2002). The dream is not only to change scale, but also to eliminate the differences between different scales. It was aptly summarized by Milanese architect Ernesto Rogers in 1952: *Dal cucchiaio alla città*. "From the spoon to the town" is the slogan, which characterizes the spirit of the 1943 Athens Charter, but it has nearly become an equivalent to the word "design". Such a motto reveals the dream of modern design, namely to abolish all scale, in a complete identification between the spoon and the town, the inside and the outside, the object and its blueprint. For that reason, the term "design" was quite contested in France. For instance, French architect Claude Parent recused the Anglo-Saxon definition, which covered all creation of the built environment, from the lighter to the city (Parent 1971: 19). Jacques Viénot, before him, disputed the Anglo-Saxon term of design (in particular the industrial design of Loewy) to prefer the term of industrial aesthetic; but he also admitted that the industrial aesthetic had to extend "from the dam to the pen", and he was able to say "*it is one or the other: either the planet explodes, either we organize it*" (Viénot 1948). Industrial aesthetics designs a "society" of objects before designing individual items, a dwelling environment before isolated artefacts: "*move from the small problem to the big one, which is truly the environment of Man*" (Patrix 1962: 101).

---

[1] By contrast, one could quote the architect and designer Alvar Aalto, who never ceased to emphasize that art, contrariwise to technology, has to be anchored in that variability of the *milieu* (Aalto 2012 [1997], p.162).

2.2 *Post-Sputnik Design. Design by Environment*

The genuine discovery of space conquest is not outer space, but our very planet. The space age fully participates in the Anthropocene, for it has exposed that, in a sense, the techno-sphere had exceeded the bio-sphere. McLuhan was not wrong when he claimed that October 17th, 1957 the Earth itself became, with the first man-made satellite, an artifact, a design form, an anti-environment aware of itself (McLuhan 1966 2005: 10, McLuhan 1971 2005: 22). From then on, ecology has been about the technical *milieu*, both a medium and an environment. In 1959, Nixon and Khrouchtchev inaugurated the *American National Exhibition*. According to art historian Beatriz Colomina, the post-Sputnik architecture began with this first USSR-USA cultural exchange, when Ray and Charles Eames projected the film "Glimpses of the USA" onto screens under a dome designed by Fuller in Sokolniki Park. In 1968, the designers released "Powers of Ten", a documentary offering a 9-min journey from the human scale to the infinitely large and to the infinitely small. This amazing film encourages a typical epistemological mistake of global design, namely the idea that one can understand different scales on the same level. "The logic of both films is the same. Intimate domesticity is suspended within an entirely new spatial system" (Colomina 2001: 12). With the two of them, the space of architecture and the space of media, and the global village and the Earthship overlay, and never separate again.

Designs concerned about environmental issues have been trying to be both *economical* and *ecological* since at least the 1960s. This topic of the *oïkos* has been too much disregarded by design, mainly maybe because the entire world adopted the term "environment" to label ecological questions. The French designer and colorist Georges Patrix noted that: "The year 1969 saw the introduction of the word 'design', while in 1970 the word 'environment' was introduced." (Patrix 1973: 30).[2] Jean Baudrillard early understood that: "to this theoretical concept of 'environment' corresponds the practical concept of 'design'" (Baudrillard 1972: 251). The control of the environment, under the sign of protection (Fuller), is also the control of men, who have to be protected from themselves (Baudrillard).

2.3 *Bigness*

In architecture, "Big Design" might evoke Rem Koolhaas, whose analysis seems to be restricted to a simple matter of size: "S, M, L, XL organizes architectural material according to size; there is no connective issues" (Koolhaas and Mau 1997: xix). It takes just a few steps from his *Delirious New-York* (1978) to "Bigness", which means that beyond some size, scales are abolished, and architecture loses any

[2] In France we used to speak of "industrial estheticism" and "milieu", provided that estheticism is about the insertion into the milieu inasmuch as about beauty, as Gilbert Simondon exemplifies.

context (Koolhaas and Mau 1997: 494–517). "JunkSpace", the encounter of the escalator and the air-conditioning in urban systems, is like a child of the *Bigness*: no more scale, the maximum and the minimum blend imperceptibly into one another. It would be hard to drawn any theory of scales from Koolhaas's writings, for his theory of design is quite rhetorical: "Junkspace thrives from design, but design dies in the Junkspace" (Koolhaas 2002: 175). Between Le Corbusier and Koolhaas, did we bluntly move from high modernism to post-modernism? So thinks Latour, who advises: "no longer be a modernist, and change scale" (Latour 2005: 73). Our interpretation rather advocates that continuity prevails over differences. Both of them work for the Bigness, and the abolishment of scales. Applying the laboratory to the planetary level is, as Bruno Latour noted, renouncing to the very idea of scale: is to renounce to the dichotomy of the inside and the outside, and to the dichotomy of the micro- and the macro-level (Latour 1993). In the Actor-Network Theory, the difference in scale between actors (local/global) is fairly irrelevant (Latour 1994).

Tim Morton speaks of "Hyperobjects" to refer to things that are massively distributed in time and space relatively to humans. Hyperobjects are viscous, nonlocal, a-temporal, and exhibit their effects inter-objectively. For Morton, inter-objectivity includes "subjects", for we are ourselves effects of hyperobjects, which are above all scalar dilemmas exceeding our reality, our medium-sized objects. Hyperobjects "cannot be thought as occupying a series of now-points 'in' time or space" (Morton 2013: 47). As hyperobjects have no more scale, Morton's philosophy does not favor any scale for the proper design of hyperobjects. He draws on an example, namely the project entitled "Dusty Relief" (2002) led by architect François Roche,[3] a building that would attract dirt rather than redistributing it, which is misleading on a planetary scale. "It would make more sense", he wrote provokingly, "to design in a dark ecological way, admitting our coexistence with toxic substances we have created and exploited" as the electrostatic building suggests. In the time of "hyperobjects", design has no other choice but to include them straight from the start (Morton 2013: 109–111).

Peter Sloterdijk finds in architecture the more fundamental intuitions of his "spherology", which assumes in relation with technology that the "environment" has become internal and artefactual.[4] But we could also say, with Latour (2008), that Sloterdijk is *the* philosopher of design, because what was previously called Nature has now become a matter of Design. The crystal palace of capitalism, this greenhouse, which has no outside (Sloterdijk 2008), is the dome of Fuller.

---

[3] New Territories/ R& Sie, Dusty Relief (2002). By François Roche, Stephanie Lavaux, and Jean Navarro.

[4] In an interview with Heinrich, Sloterdijk acknowledges his intellectual debt to architects, pioneers and visionaries of spherology, and quotes: Bruno Taut, Vladimir Tatline, Constantin Melkinov and Hermann Finsterlin (Sloterdijk 2011).

3 Buckminster Fuller and Global Design

While design has long been related to environment (e.g. involving Walter Gropius at Harvard Universty, Lászl6 Moholy-Nagy at the Chicago Institute of Design, and Herbert Bayer in Aspen, Colorado), as Peder Anker (2010) has shown in his book collecting papers design truly meets the Anthropocene with Fuller: what was required was the inspiration from the space ecological engineering; what was needed was to think like an astronaut.[5]

> The imagined life of astronauts became the model for how to live in harmony with nature. The ecologically engineered machinery of the space cabin, such as bio-toilets, solar cells, and recirculation devices, became essential devices for ecological architects such as Fuller, John Todd, and Kenneth Yeang. (Anker 2010: 127).

Buckminster Fuller is the perfect "Technocrat for the Counterculture" in Fred Turner's term (Turner 2009). He is, problematically, a technocratic promoter of democracy through design.[6] Nikola Jankovic managed to describe him in just a few lines:

> Willingly collaborating with the US Army, and agent of propaganda for a 'fluid' world which would safeguard freedom and democracy, Buckminster Fuller became a messianic Captain Pax Americana. Steward of the spaceship Earth, and based on evidence, he believed simultaneously: that there is always a better way to manage the planetary stock (dymaxion-tensegrity-synergy); that economic growth can be optimized in a finite world (do more with less); and eventually in a sharing for all from the top (strict maximum). (Jankovic 2012a: 15).

His *Dymaxion* philosophy and domes are distributed among very different scales, which is the direct legacy of the "general theory of systems" (Bertalanffy 1968). Fuller is placed in the context of cold-war technology: atomic bombs, space vehicles, and computers. The meeting of these technologies called for a conception of the Earth as a controllable entity, a "closed world" (Edwards 1996). In the words of Pang (1997), Fuller is "a superb example of a Cold War designer". One can recall that he was first and foremost a son of the US Navy. He has learnt his favorite maxima ("to do more with less") from the military thought, and this is certainly where his faith in the capacity of design to solve anything came from, as well as the metaphor of the Earthship.

---

[5]The history of ecodesign proposed by Peder Anker (2010) takes place in America. It ignores divergences within the field, one effect of this view being to reduce ecology to the environment.

[6]According to Jonathan Massey "*Fuller conceptualized design as the art of reconciling systemic rationalization with individual initiative*" in order to solve the contradiction between his "*technocratic conviction that there existed only 'one best way'*" and his belief in the "*superiority of democratic governance and market economies*" (Massey 2009: 191).

3.1 *Dome at all Scales*

In his history of the Earth seen from above, Sébastian Grevsmühl (2014) interprets geodesic and pneumatic domes (after Buckminster Fuller and Frei Otto, respectively), as tangible achievements of the "spaceship Earth" idea. The dream of a dome town is fashionable; speaking of the Rockefeller Center, its designer referred to a "city under a single roof" (Hood 1929). And while Buckminster Fuller dreamed about guaranteeing complete environmental control (first in any place, then at any scale), Frei Otto imagined a mega-structure for controlling the climate, for under any radically hostile climate "*man has no choice, but to trust technology*" (Otto 1954: 116). As Antoine Picon (2008) has noted, the 'more with less' of Fuller's dome is big but light, so it could lead to 'less through more'.

Fuller's dome, once the American Pavilion at the 1967 Montréal World Fair, is today a museum dedicated to the environment. It was part of an earlier and broader project, also known as the *Geoscope* (1962), which consisted, like a premonitory dream of *Google Earth*, of including views of the universe in computers. *"We will be in 'one town world' in a realistic way"* (Fuller 1964: 28). Fuller's dome is the metaphor of global and multi-scale design, for instance for a city, as in his *Manhattan Island Dome* exhibited during the 1959 exhibition "Three Structures" at the MoMa, under the form of a photomontage (Gough 2009). A project of such a kind, namely to have an entire town under an artificial dome, is intrinsically a project of full control over the environment, of complete air conditioning, of a pure "endosphere". Buckminster Fuller "regards domes as basic environment valves, differentiating human ecological patterns form all other patterns, microcosm from macrocosm [as such] not limited in size" (Zung 2001: 39). They constitute architectures of the Cold War *par excellence*, for they are at the same time means of military expansion (towards polar regions) and means of protection to threats resulting from that expansion (especially nuclear risks). Implying a sharp distinction between the inside and the outside, the concept of the dome contradicts the very philosophy of ecology, which refuses the idea that the environment is merely external. Contrariwise, the Anthropocene is illustrated by the awareness that it is no longer an option to 'externalize' ecological impacts. Any dome would have to be global, and master planetary feedbacks: there is nothing external to design anymore. This nowadays echoes geoengineering prospects, namely the alleged control of the planet and the endless management of its data under the benevolent cupola of advanced techno-science. We thus agree with the sentence of Maldonado:

> The idea [of the 'Dome Over Manhattan'] demonstrates once again the admirable imagination of Buckminster Fuller as an engineer, but it also shows his dangerous ingenuity as an ex-temporaneous ecologist. (Maldonado 1972: 47).

3.2 Spaceship Earth and World Game

It seems that Fuller formulated for the first time the Earthship metaphor in 1951 (Krausse and Lichtenstein 1999: 279). Sabine Höhler has offered a history of the Earthship, starting from the Foucaldian statement of a 'spatial turn' in our time, namely the idea "*that the anxiety of our era has to do fundamentally with space*" (Höhler 2014; Foucault 1986: 23). With Fuller's book *Operating manual for spaceship Earth* (1969), in contrast to Ward's and to Boulding's references to Spaceship Earth, the metaphor is not used as suggesting vulnerability and community but as an operator granting the transformation of the Earth into a design object (Höhler 2014). Fuller invites us to understand Spaceship Earth as an "integrally-designed machine" (Fuller 1969: 52). With widespread, anticipatory design science, Fuller stands in the way of the growing success of Neo-Malthusianism.

The (at first unsuccessful) "World Game" he proposed for the World Fair held in Montreal in 1967, is a perfect example of the designer operating the Earthship. Initially conceived of as a game for students to model (and optimize) the worldwide circulation of resources and information, the spirit of the *World Game* exemplifies how design can be used instead of politics. When Gene Youngblood introduced the game, he explicitly announced "a concrete scientific alternative to politics" (Youngblood 1970: 30), in so far as it would "gradually force world politics to yield to the computer-indicated, mutually-beneficial world programs" (Fuller 1973: 116). The first session of the *World Game* was held in 1971, just before the release of the Meadows et al. (1972) report to the Club of Rome. Both exercises use scenarios in a context of a looming ecological crisis. Yet, their conclusions sharply diverge (Jankovic 2012b). In the first paragraphs of the fourth chapter of the *The Limits to Growth* (Meadows et al. 1972), Fuller's optimism (World Resources Inventory, 1967) is referred to (and mocked). The sessions of the World Game resulted in the publication of *Energy, Earth and Everyone Energy Strategies for Spaceship Earth* (Gabel 1974). The editor of the book, Medard Gabel, a designer and co-founder of the *World Game Institute* in 1972, today continues designing a world that works for all, through the Design Science/Global Solutions Lab (Gabel 2010). Would he maintain, as Buckminster Fuller said in the forewords of the 1974 volume, that the energy question is about to be resolved?

The dome, the Earthship and the World Game all exemplify two dominating attitudes towards the environment: firstly as *metaphors*, or ways to grasp the whole; secondly as *models*, or ways to quantify and to parameterize our planetary lifeboat. The idea is nothing less than to replace the Earth, our primary biosphere, by *design*, as some tried, twice, with *Biosphere II*, a heterotopic experiment conducted in the desert of Arizona in the late 1980s (Höhler 2014).

Yet, the advent of neoliberalism with the election of Ronald Reagan was not the only cause for the premature death of the movement for alternative technologies. For sure, another reason was that the *Whole Earth Catalogue* (1968), much inspired by Fuller's legacy, avoided feature articles on political, social, or ecological questions (Winner 1986, ch.4). It is naturally part of "global design" to be based on a global knowledge, essentially statistical and quantitative, purified of local contingencies, thus giving the misleading impression that the problem is fundamentally about nature (or science) and not society (or politics). We have incidentally mentioned *Google Earth*. Fuller, in the past, and Google, today, have the very same ideology (which is the ideology of the death of ideologies), and they have the very same fantasy as well: to substitute the map to the territory.[7] Fuller is like Mr. Jourdain in Molière's play: he has been doing politics all his life without knowing it!

3.3 Earth Design

Historical inquiries into the US counterculture movement and the role of Richard Buckminster Fuller or Steward Brand made clear their surprising connection to the military culture (Turner 2006; Kirk 2001; Kirk 2007). Felicity Scott has stressed the paradoxical relationships between: counterculture's architecture and Fuller's technocracy ("Revolutionaries or dropouts"); psychedelic and intermedia cultures ("Acid Visions"); and the politics of ecology movement and increasingly militarized environments ("Shouting Apocalypse"; Scott 2007: Ch.6–8). The pans of the scale between computers and hippies, nuclear mushrooms and hallucinogenic ones, technological fixes and natural life of the 1960–70s, finally tip in one direction. The Web made the *Whole Earth Catalog*, but did not solve its contradictions. Intellectuals and artists still debate about such a "California dreaming", as suggested the exhibition *The Whole Earth California and the Disappearance of the Outside* (2013) in Berlin, telling the story of that shift from eco-psychedelia and countercultural communality to the networked neoliberalism and the Anthropocene Project (Diederichsen and Franke 2013).

The life of Steward Brand is maybe sufficient to exemplify how 'ecodesign' lost its subversive impact: we moved from the desire to create soft technologies likely to promote autonomous communities (it was one of the objective of the *Whole Earth Catalog*) to the wish to enter an era of the large-scale engineering of ecosystems (it is the aim of the *Whole Earth Discipline*). Brand (2010) indeed intends to show, as the subtitle of the later book suggests: "why dense cities, nuclear power, transgenic crops, restored wildlands, radical science and geoengineering are necessary."

How big is Big Design? *Earth Metabolic Design* once sponsored the 'World Game Studies Workshop'. It is precisely around this notion of *metabolism* that the scale expanded from the individual, to the town, then to the region and the world. It is fascinating to see how easy it was (and how easy it is now) to some to move from the regional scale to the planetary scale, even though no (or little) proof of success exists (for industrial ecology for instance) at the scale of cities or regional level. As conceived of by Brad Allenby (Allenby 1998; Allenby 2005), industrial ecology

[7] One could make another bold comparison: in the *World Game* as in *Facebook*, the world has only friends.

should, and will, result in earth systems engineering as opposed as environmental engineering. Such an opposition, however, is first and foremost an issue of scale, rather than an issue of method, or epistemology. Allenby insists he is not a reductionist, but the way he deals with scale change make us believe, indeed, that the whole is nothing more than the sum of the parts. With its Fullerian-like vision of design as pure engineering, he pretends to solve industrial problems without touching society. Industrial ecology, however, can follow different paths, as design in general: one is focused on the production sector, and could be summarized as the engineering of (optimized) material and energy flows in industry; another embraces the whole economic system, and rather deals with so-called "metabolism" of society as a whole (Fischer-Kowalski and Hütler 1999; Fischer-Kowalski and Haberl 2007). An opposition of that kind is valid at all scale, and to us only the latter reading can really be said "multiscale".

4 Conflicting Scales?

As a matter of fact, ecology has been torn between "techno-skeptics" and "techno-optimists" since the 1970s, between those who refuse the Earthship metaphor and those who subscribe to it, or between big design and small design. The Earthship metaphor carries the dualisms of modernity (the soul and the body, nature and culture). Man is on Earth like the Cartesian soul in the body, namely like a pilot in his ship, mastering and possessing the world and himself. The new representation of the Earth refers both to a limited whole to be preserved and to the possibility of excessing it, namely the principle of non-conservation which characterizes the general economy of Georges Bataille as well as the essence of triumphant technoscience and its "boundless space of technical possibility" (Schwarz and Nordmann 2011).

4.1 Small Design

Small Design refers to the adage *Small is Beautiful* of "degrowth" philosopher Leopold Kohr, picked up by his disciple and friend E. Schumacher in his 1973 best-seller. Kohr's main idea is that "there is only one cause behind all forms of social misery: bigness" (Kohr 2001: 21).

The Papanek's *Tin Car Radio* designed in 1965 (Papanek 2011: 224–228) appears as a revolution in the history of design, neither for it was affordable for the many, nor for it was made of recovered materials, not even because it assumed a smart and creative user (it both opened the black box of technology, and challenged aesthetic uniformity), but rather because, by bringing an artisanal dimension to an industrial object and a mass media, it constituted a manifesto for small design. Buckminster Fuller was not without influence on Papanek, for instance in his approach of a global, non-specialized, multidisciplinary design. Victor Papanek, however, took a different road, probably because his encounter with the Third World. His options to revolutionize design are various, and sounds familiar today (tackling planned obsolescence, being inspired by biological models –in other words "biomimicry"–, regarding "environmental design" as a fight against pollution, etc.). Like Fuller, he sometimes seems to believe that design can supplant politics, but unlike Fuller he understood that one cannot dream of a sustainable world based on the *American way of life*. His opinion regarding the scale of design is nonetheless quite fluctuating, even if he wrote: "My primary conviction as a human being, a designer and an ecologist is *Nothing Big Works – Ever!*" (Papanek 1995: 24). Looking for a middle path, neither techno-fix nor techno-critics, he still thought that technoscience (like satellites) was mandatory for a proper knowledge of the ecological crisis (like the evolution of desertification). Papanek's contradictions are ours: how to build a human scale with non-human scale technologies?

Speaking of designer Ken Isaacs, Victor Margolin writes:

> Though modernist in his design vocabulary, he has never followed the dictum that the same design philosophy should operate at all scales. Ken's projects do not move from a spoon to a town, as the Italian architect Ernesto Rogers once envisioned, but rather from a spoon to a spoon as Andrea Branzi once remarked. (Margolin 2002: 75).

Departing from the modernist project would so be admitting that a new theory of design should correspond to every new scale. Fuller and Isaacs belong to the same American counterculture, which circle around the *Whole Earth Catalog*. Yet the difference in scale of their approach appears to separate their paths. Fuller (and his *Operating Manual for Spaceship Earth*) is heading big; Isaacs (and his *How to Build your own Living Structures*, 1974) is heading small. In both cases, however, it is a question of designing the whole, the matrix. There is thus a confusion, which should be avoided regarding global design, namely to believe that it is necessarily big.

Small design has fundamentally a negative definition: it argues against big design and "large scale research" (Galison 1992). By contrast, one speaks today of "small science" or, in design, of *Jugaad Innovation* (Radjou et al. 2012), for instance, which means basically the same: innovation in science or technology is not always to be found in the word of big investments. Small design cannot mean very small scale, unless to keep the expression for Feynmann's "*plenty of room at the bottom*", namely the scale of nanotechnologies. In fact, Small Design is above all a reaction to technocracy, and to what Juliet Schor (2013) has called Fast-Fashion. Small design is not an expression of common use, but one speaks of slow design, which has its urban correspondences (*Cittaslow* and *transition town*). With Hopkins and the context of resilience (to the growth crisis), "small is beautiful" has become "small is inevitable" (Hopkins 2008; Hopkins 2013). It is worth stressing that these local movements have become global; likewise, they grow in networks.

In a reference book, which acts against technological fixes and thus explains why technology will not save us (or the environment), one can read that to make sure that future technologies are "socially appropriate" their design has to be established following three rules:

> First, they should be local and decentralized; second, they should be human-scale and simple to operate; and third, they should provide an environment for humane, satisfying and meaningful work. (Huesemann 2011: 300).

Do It Yourself; *Small is Beautiful*; "simple living", and "convivial de-growth" are now forty-years-old schools of thought. They may be rediscovered under a series of names: *Incredible Edible*, "grassroots innovations"; *Slow Food*; *buen vivir*; etc. They offer solutions, which are admittedly local yet global as well, in the sense they are polycentric and systemic. The human scale does not entail the absence of global perspectives, but it assumes a specific approach of the global, as notably defended by Elinor Ostrom: "*An important lesson is that simply recommending a single governmental unit to solve global collective action problems—because of global impacts—needs to be seriously rethought and the important role of smaller-scale effects recognized*" (Ostrom 2009: 35).

4.2 Milieu Design

In the early 1970s, design was entrenched, whether one is under Buckminster Fuller's dome or not. One of the more virulent critics of such a global design is Maldonado. Having understood that global design encompasses everything, including philosophy and politics, he strongly contested it:

> "Evidently, Buckminster Fuller thinks design and planning would resolve the problems that politics has left unsolved for centuries. [...] he considers the 'Revolution by Design' to be exclusively an act of technical imagination: a position typical of technocratic utopianism." (Maldonado 1972: 29).

> Yet "the very high complexity of environmental problems obliges us to try to solve them technically; and we have already seen, when discussing Buckminster Fuller's approach, to what degree of abstraction technological imagination can lead when its infinite possibilities are given free play—especially when it functions without the aid of the sociological imagination." (Maldonado 1972: 40)

In a letter to the students in architecture of all countries, Fuller explains that everything, including youth crime, can be resolved by design rather than reforms. In a nutshell, this can be enunciated as follows: "*Reshape environment; don't try to reshape man*" (Krausse and Lichtenstein 1999: 253). We would like to advocate that alternative approaches of design exist, in which it is not possible to change the environment without changing man. We call such a vision of design "design of the *milieu*". In the early 1970s, the relation between *design* and *environment* bifurcated into two different branches (Petit 2015). The first would determine the environment as an objective and a quantifiable entity, in the manner suggested by Buckminster Fuller; the second one would identify the environment as a qualitative *milieu*, relatively to its dweller, in the manner developed by Maldonado (1972). As Papanek completed his manifesto book (1971), Tomás Maldonado published a lucid book on the ecological crisis (*La Speranza Progettuale. Ambiente e societa*, 1970). As we have seen, it was about responding to Fuller's design science, among other things. Maldonado's theory is a critical recasting of Fuller: "A 'Revolution by Design' has real meaning only if it is supported by 'Design by Revolution'", he wrote (Maldonado 1972: 29). His (philosophical) book, which aimed at a general theory of design (that is to say a project *praxis*, according to the author), is a book on the human milieu, which is situated, Maldonado says, at the scale of the "meso-cosme". While according to Fuller the human environment is about technology (a spaceship, which is not, as such, integrated in a larger environment), things are different for Tomás Maldonado, who holds the human milieu as a part of the global natural ecosystem.

This philosophical disagreement resulted in a disjunction regarding the ecological approach of design. The designer must be suspicious of the tendency to entirely artificializing the physical environment of man, namely the idea that one could produce a completely technical milieu. Rather, he should realize that the only way the heal Nature is to heal Society, and he has to look out for the ecological fashion, which weakens an "essentially critical ecological conscience–critical toward the scandal of society" (Maldonado 1972: 77).

We believe that genuine ecodesign is therefore on the side of Madonado, rather that on the side of Fuller, for the design of the *milieu*, unlike the design of the environment, involves the entire being of man. Maldonado fully understood that, in order to reach the human scale, one should stop thinking like an astronaut. He also perfectly understood that questioning the scale of design was in fact questioning the way one articulates design to the other theoretical spheres, and to the other practical dimensions.

4.3 The two Ecologies

Since André Gorz, at least, two ecologies have been opposed: one being technocratic, and the other democratic (Gorz 1975). This clash of visions has come in many ecological flavors: top-down or bottom-up, global or local, high-tech or low-tech, industrial or political, artist-authentic or social-fair, and so on. In fact, these oppositions are about the signification and the scope of the ecological crisis. To say it with our words and concepts, we believe that the main difference has to do with the discrepancy between an ecology of the environment (*Umgebung*) and an ecology of the milieu (*Umwelt*). The former focuses on environmental impacts, and aims at modifying our technologies to make them more "eco-compatible", while the latter deals with our living milieu, and tries to change our relationship to technology, that is to say our consumption-production mode. The environment, as the epistemology suggests, is a state of being environed. The milieu is, likewise, a middle place. The environment is absolute, while the milieu is relative (relative to the living which inhabits it, between the inside and the outside, between biology and society). The ecology of the environment is the same everywhere; it is reproducible. Contrariwise, the ecology of the milieu is specific to each place, is related to actors, to the *commoners*.

In French, the signification of the scale of the *milieu* has to do with both technology and ethics. Labeling the (physical, then biological) environment, before it became sociological, then geographical, the word *milieu* has a social sense (the eco-nomy) together with a natural sense (the eco-logy). It also suggests the intermediate, the excluded middle between the inside and the outside, and for our purpose, between the small and the big.

While the distinction between ecologies of the environment and of the milieu is theoretical and rather abstract, it has practical consequences. It allows, as we will see, a better understanding of what, in the world of design, can today distinguish *Design for Environment* (*DfE*, which is merely about reducing the environmental impacts of our technologies) from *Design for Sustainability* (*DfS*, which insists as well on the need to change our milieu). What matters in new models such as WikiSpeed or WikiHouse is based on this distinction between lower impact on the environment and a higher impact on society.

5 What Scale for Design?

5.1 When Global Meets Local

In fact, the philosophical conflict of the 1970s has not changed so much. Regarding the *space* scale, the local movement (in line with Murray Bookchin) opposes global engineering (in line with Fuller). Regarding *time* scale, the economics of free time (in line with André Gorz) opposes the tenants of accelerationism (in line with Félix Guattari). It seems like nothing has profoundly changed in the gap between a Steward-Brand-like ecology of "global technological optimization" and a Rob-Hopkins-like ecology of "local social solutions".

When Peder Anker and his colleagues of GLOBAL Design NYU propose to rethink global design from the local level, they present as a new approach what has in fact long been an old problem (Anker et al. 2014). Having said that designers are not astronauts, much remains to be done. The Anthropocene says something about the scale of the problem, but nothing about the proper level of action to solve it. For instance, although they are in keeping with a work that reflects on the Anthropocene, all projects put forward by Etienne Turpin are not a matter of Big Design, in the sense of Big TechnoScience (Turpin 2013). The series of divides we mentioned before (between DfE and DfS) are not necessarily divides in the scales of design. One example offered by design activist Ann Thorpe in her book (2012) exemplifies how "slow design" does not mean "small scale" or "low-tech": the Luna-resonant street light (designed by Civil Twilight LLC) is not only an efficient street-lamp; it is a connection to the Moon with high-tech! The opposition between local design and global design has no real significance, at least not the one of the above-mentioned divide. The D-Lab, founded by Amy Smith at MIT, is for example typically invested in small design, in the very sense that it argues for soft technologies.

However, its conception of the world and its field of action is global. Suggestively, one of the D-Lab programs, which aim at bringing local solutions to the global market, is labeled *Scale-Ups*.

As soon as we live in a digital environment, resorting to "small design" is dishonest. While the global move us away from the local, however, the hyper-global (like the Web) move us back to it.

5.2 From DfE to DfS

Authors of a famous ecodesign book say they found their inspiration in E.F Schumacher, Murray Boockchin, William Irvin Thompson, Margaret Mead, Steward Brand, Gregory Bateson and Buckminster Fuller (Todd and Todd 1994: 11). They seem to attempt to reconcile the irreconcilable! Is stewardship relevant to permaculture or bioregionalism? Certainly, they distinguish between environment and ecology. However, it is not sure that they draw the philosophical and political conclusions from such a distinction.[8]

After the "Small/Big" debate in the 1970s and the achievements of the "Design for Environment" in the 1980s–1990s, the time has come for the "Design for Sustainability". Whether one reads Pauline Madge and her three stages of ecodesign (Magde 1997) or Martina Keitsch and her three approaches to eco-design (Keitsch 2012), the work of Ezio Manzini comes up as a genuine contribution to mature sustainable design. In Manzini's perspective "the link between the environmental and social dimensions of sustainability appears clearly" (Manzini 2007). While *DfE* merely draws upon technical innovation, *DfS* draws upon both technical and social innovation. While DfE "assesses the short and medium term environmental and economic impacts for all stages of the life cycle product or service; DfS assessment of long term and global impacts based on the four dimensions of sustainable development for all stages in the life cycle of a product or service" (Spangenberg et al. 2010: 1488).[9]

To continue with the distinction, *DfE* seems to belong to the *design of the environment* (inherited from Fuller), while *DfS* seems to belong to the *design of the milieu* (inherited from Maldonado). In the first case, it is possible to modify the environment without changing the lifestyle of its dweller; in the second case, the milieu is to be modified, starting from the inhabitants' lifestyle. This is undoubtedly why networks of designers (like DESIS: *Design for Social innovation and*

[8] "*Unlike the term environment, which denote one's surroundings in a objectified sense, ecology by its very inclusiveness implied interconnectedness*" (Todd and Todd 1994:3). However, the "*New Alchemy Institute [1969–1991] to restore the Land, protect the seas, and inform the Earth's stewards*" (Todd and Todd 1994, p.172–174) is a good example of biological design, which opposes in every respect the Living Machines and the Conventional technologies. It is also a good example of environmentalism without political ecology…

[9] It is instructive that one of the guidelines for action of the DEsign EDucation & Sustainability (DEED) Project led (among others) by Alastair Fuad-Luke was entitled 'SCALES'.

Sustainability or SEP: *Sustainable Everyday Project*, or SDS: *Strategic Design Scenarios*) so much insist on "creative communities", namely people forming groups according to some contributive and bottom-up approach, in order to build *commons* (Ostrom 1990).

The metaphorical use of the ecosystem concept leads to (or often aims at) blurring even more the boundaries between the natural and the artificial environments, or between biological and information systems. The metaphor of the "ecosystem" of an object, still, is crucial provided that it allows to thinking in terms of processes and cycles (life-cycle assessment), rather than in terms of objects and products. But, while ecodesign aims at assessing life cycles, it is focused on the product, and therefore distinguishes between the product and its "ecosystem". Such a partition is highly problematic, for in reality the object is not separated from its *milieu*. This is why eco-friendly products have no significance in systems of practices that are not. In order to modify the object, one has to modify the system.

It is perhaps because the design practice has been mainly focused on technology that the ecodesign approach has come to be a practice of engineers, rather than a practice of designers. The same applies to the field of industrial ecology. They both consider the environment as a receptacle for impacts only, and disregard the living systems of human practices. It's why we call for an ecodesign of our practices, our social milieu, in order to remedy to this situation. In other words, ecodesign has surely been too much focused on *poïesis* and not enough on *praxis*. Rethinking praxis, however, is not an easy thing, because, for instance, "there is a fundamental difference between designing things to be used and trying to design use or the user experience" (Redstrom 2006: 135).

Contrariwise to engineers, designers should be more particularly focuses on uses. For instance, more efficient cars are projects typical of engineering, but a car sharing service is an innovation of design. Another example is the "circular economy" (Webster 2015), which often sounds like a project of engineers rooted in industrial ecology, and the so-called "performance economy" (Stahel 2010), which has in our view much more to do with design. We need to faces our incapacity to loop flows at all scales. The spatial scale is problematic: one can probably loop material or energy flows at some scale, but can definitely not at some other. Imagine for example a smartphone, which is completely recyclable. This does not mean that it is "unplugged" from increasingly energy-consuming data centers! The temporal scale is problematic as well, because the time of our resource and waste is not akin to the time of our social life. Fossil fuels and nuclear waste are the best (or the worst) examples. What engineering sometimes forgets is that reality rarely matches plans. This is another way to say that designer should be situated between handymen, and engineers, the two different modes of thinking once highlighted by Claude Levi-Strauss (1966).

5.3 *Ecological and Digital Transition: Reconciling Scales*

As we said before, the encounter of computers and ecological issues can lead either to pessimism (the Meadows report) or the optimism (the World Game). This is still ambivalent today: the digital network, as once the computer, serves to both counter-culture and mainstream lifestyles. Regarding counterculture, one can think of various manifestos of *hackers* and *fixers*, which seem to reuse the "DIY" idea. Regarding mainstream thinking, it seems to lead toward "smart cities", as Masdar. Let us remind that what is at stake with smart grids is not only to reduce our energy consumption, but also to change our relationship to energy through more involvement, and to make smart citizens.

ITC is not environmentally friendly in itself[10], ITC is "green" only if it is part of the pooling of a know-how directed to sustainable design. Specialized literature distinguishes between three orders of ecological effects for ITC: direct effects of technologies (*Life Cycle Impact*), indirect effects of applications (*Enabling Impact*), and systemic effects involving behavioral and structural changes (*Structural Impact*) (Hilty and Aebisher 2015). The possible encounter between the digital transition and the ecological transition, between open design and eco-design, lies in this last framework. But, again, two sides support the dream of convergence between open design and eco-design. One, exemplified by Jeremy Rifkin (2014), is close to the mentality of engineers and looks at technology. The other, closer to the design culture, looks at society. The designers of the POC21 are its troops[11]. The promising ecology of our new technical milieu is driven by open source culture and open hardware. It leads us toward the ideal of a recovered collective autonomy, of which energy self-sufficiency is just an aspect.

Hence, one should go beyond the opposition between Papanek's low tech and Buckminster Fuller's high-tech (Margolin 1998: 84), or between small design and big design. There is a third way, we argue, which mixes digital technology and ecological transition. Designer Cesar Harada, who defended a dissertation entitled "*Open Hardware for the environment*" and who is well known for his open source cleaner drone, perfectly exemplifies such a synergy that goes along with a contributive view of design. This "open design" or "design of the commons" is not so much about saving the Earth, or Nature. Rather it is about defending a technical *milieu* deprived of its culture, its know-how, and its know-how-to-live (Gorz 2008). Like the open community for digital empowerment "Jerry, Do it Together", this new culture of the "makers" is indissolubly linked to a democratic and an ecological revival. That is exactly what the ecological crisis requires to be overcome: an

---

[10] As Professor Lorenz Hilty underlined in his conference at Lift in 2016: 1) Despite Moore's Law, we are using more material for ICT hardware, 2) Despite Koomey's Law, we are using more and more energy for ICT services. 3) Despite increasing service-sector outputs, total requirements are not decreasing.

[11] *Ouishare* and *Openstate* organized this eco-hacking event from August 15th to September 20th, 2015. In reference to the 21st "Conference of the Parties" (COP) of the UNFCCC, it was named "Proof of Concept" (POC).

eco-engineering which implies the life cycle of things, but more importantly an eco-design which considers communities that reinvent their use value.

The risk of "small design" is to withdraw into simple nostalgia for a lost artisanal world. Simondon's philosophy is precious for the design thought, because it attempts to reconcile production and use in the industrial context without relying on nostalgia for craftsmanship:

> It is in this very emphasis on industrial production, in the deepening of its characteristics that an overcoming of the antithesis between the artisanal modality and the industrial one can be studied with a greater likelihood of success. (Simondon 2014: 305).

One of the characteristics of this industrial reality is its reticular dimension. Simondon invites us to change the emphasis from the machine to the infra-individual technical elements and to networks, considering in particular the modularity of objects. The distinction he operates between "closed objects" and "open objects" turns out to be illuminating. A closed object is sold fully constituted: its vocation is wear and degradation. On the contrary, an open object is *neotenic*: it is to some extent always in progress, made to be remade, and produced to last. The object opened to its milieu (and to the user-repairer) is, according to Simondon, "the essence of what might be called the crusade for the salvation of technology" (Simondon 2014: 401). *Fab Labs*, which favor open model, seem today to pursue this type of "technical culture". As with Simondon yesterday, new "amateurs" of technology strive to overcome the pair industry-craft through the deepening of the former. One may think here to the network of "Open Source Ecology" initiated by Marcin Jakubowski, for instance, or to the French cooperative "L'Atelier Paysan" for the auto-construction of open farm equipment. This renewal of open design, all at once rural and urban, local and global, ecological and digital, seems well equipped and promising to face the challenges of the Anthropocene.

6 Conclusion

It is widely acknowledge that the ecological crisis is a matter of scale. Rooted in trade globalization, which creates and outsources massive artificial flows, which disrupt natural biogeochemical flows and ecosystems, the ecological crisis has indeed to do with some upheaval or dislocation in scales. However, the concept of "scale" is underexplored in science and philosophy. A key review paper on this question ends up like this:

> In this paper, we survey one of the most important conceptual challenges to that union [between the physical sciences and the social sciences]— the concept of scale. We argue that common definitions do not exist for scale — even within disciplines — and especially in the social sciences. (Gibson et al. 2000: 236).

The concept of scale is of a colossal scientific and philosophical difficulty, and we never pretended to solve such a difficulty.

We wanted to show why making sense of the different "scales of design" seems more relevant in terms of the "milieu", how such an approach can go beyond the opposition between scales in itself, and unveil a more fundamental tension related to the very signification of design. As we have argued, all the stakeholders of a design trying to think simultaneously the digital transition and the ecological transition are likely not only to make ponder how local and global articulate, but also to offer some practical solutions in the face of the Anthropocene.

References

Aalto, A. (2012). *La table blanche et autres textes*. Marseille: Parenthèses. [1st Ed. 1997].
Allenby, B. R. (1998). Earth systems engineering: The role of industrial ecology in an engineered world. *Journal of Industrial Ecology, 2*(3), 73–93.
Allenby, B. (2005). *Reconstructing earth: Technology and environment in the age of humans*. Washington, DC: Island Press.
Anders, G. (1961). Commandment in the atomic age. In *Burning Conscience: The Case of the Hiroshima Pilot Claude Eatherly, told in his Letters to Gunter Anders*. New York: Monthly Review Press.
Anker, P. (2010). *From Bauhaus to Ecohouse: A history of ecological design*. Baton Rouge: Louisiana State University Press.
Anker, P., Harpman, L., & Joachim, M. (2014). *Global design: Elsewhere envisioned*. Munich: Prestel.
Baudrillard, J. (1972). *Design et environnement ou l'escalade de l'économie politique In Pour une critique de l'économie politique du signe* (pp. 229–255). Paris: Gallimard.
Bertalanffy, L. V. (1968). *General system theory: Foundations, development, applications*. New York: George Braziller.
Bonneuil, C., & Fressoz, J.-B. (2013). *L'événement Anthropocène*. Paris: Seuil.
Boudon, Ph. (Dir.). (1991). De l'architecture à l'épistémologie. La question de l'échelle, Paris: Presses Universitaires de France.
Boudon, P. (2002). *Échelle(s)*. Paris: Economica.
Colomina, B. (2001). Enclosed by images: The Eameses' multiscreen architecture. *GreyRoom, 2*, 6–29.
Corbusier, L. (1949). *Le Modulor, essai sur une mesure harmonique à l'échelle humaine applicable universellement à l'Architecture et à la mécanique*. Boulogne: Éditions de l'Architecture d'Aujourd'hui.
De Jarcy, X. (2015). *Le Corbusier, un fascisme français*. Paris: Albin Michel.
Deyong, S. (2002). The rise and fall of the Megastructure: Memories of the urban future. In T. Riley (dir.), *The changing of the Avant-Garde: Visionary architectural drawings from the Howard Gilman collection* (pp. 23–36). New York: MoMA.
Diederichsen, D., & Franke, A. (Eds.). (2013). *The whole earth California and the disappearance of the outside*. Berlin: Sternberg Press.
Edwards, P. (1996). *The closed world: Computers and the politics of discourse in cold war America*. Cambridge, MA: MIT Press.
Fischer-Kowalski, M., & Hütler, W. (1999). Society's metabolism. The intellectual history of material flow analysis. Part II.1970-1988. *Journal of Industrial Ecology, 2*(4), 107–135.
Fisher-Kowalski, M., & Haberl, E. (2007). *Socioecological transitions and global change. Trajectories of social metabolism and land use*. Cheltenham: Edward Elgar Publishing.
Foucault, M. (1986). Of other spaces. *Diacritics, 16*, 22–27.
Fuller, R. B. (1964). *Education automation*. Carbondale: Southern Illinois University Press.

Fuller, R. B. (1969). *Operating manual for spaceship earth*. Carbondale: Southern Illinois University Press.
Fuller, R. B. (1973). *Earth, Inc*. Garden City: Anchor.
Gabel, M. (1974). *Energy, earth and everyone: Energy strategies for spaceship earth, with the world game laboratory*. New York: Doubleday. (1980).
Gabel, M. (2010). *Designing a world that works for all*. http://designsciencelab.com/resources/dsl_full%20book.pdf
Galison, P. (1992). *Big science: The growth of large scale research*. Stanford: Stanford University Press.
Gibson, C. C., Ostrom, E., & Ahn, T. K. (2000). The concept of scale and the human dimensions of global change: A survey. *Ecological Economics, 32*(2), 217–239
Gorz, A. (1975). *Ecologie et politique*. Paris: Galilée.
Gorz, A. (2008). *Ecologica*. Paris: Galilée.
Gough, M. (2009). Backyard landing: Three structures by Buckminster fuller at the Museum of Modern art, 1959. In H.-Y. Chu & R. G. Trujillo (Eds.), *New Views on R. Buckminster Fuller* (pp. 124–145). Stanford: Stanford University Press.
Grevsmühl, S. (2014). *La Terre vue d'en haut: l'invention de l'environnement global*. Paris: Seuil.
Hilty, L. M., & Aebischer, B. (Eds.). (2015). *ICT innovations for sustainability*. International Publishing Switzerland: Springer.
Höhler, S. (2014). *Spaceship earth in the environmental age, 1960–1990*. New York: Routledge.
Hood, R. (1929). A city under a single roof. *Nation's Business, 17*(12), 19–29.
Hopkins, R. (2008). *The transition handbook: From oil dependency to local resilience* (p. 2008). Chelsea: Green Publishing.
Hopkins, R. (2013). *The power of just doing stuff: How local action can change the world*. Totnes: Green Books.
Ingold, T. (1993). Globes and spheres: The topology of environmentalism. In K. Milton (Ed.), *Environmentalism: The view from anthropology* (pp. 29–40). London/New York: Routledge.
Jankovic, N. (2012a). World game versus Club de Rome? 1972-2012. In *E3 – Energy, Earth and Everyone. Une stratégie énergétique globale pour le vaisseau spatial Terre? World Game, 1969-1977* (pp. 101–123). Paris: Éditions B2.
Jankovic, N. (2012b). Revolution by global design: croire en la croissance ? In *E3 – Energy, Earth and Everyone. Une stratégie énergétique globale pour le vaisseau spatial Terre? World Game, 1969-1977* (pp. 7–31). Paris: Éditions B2.
Keitsch, M. (2012). Sustainable design: A brief appraisal of its main concepts. *Sustainable Development, 20*, 180–188.
Khor, L. (2001). *The Breakdown of Nations (1957)*. Foxhole/Dartington/Totnes: Green Press.
Kirk, A. (2001). Appropriating technology. The whole earth catalog and counterculture environmental politics. *Environmental History, 6*(3), 374–394.
Kirk, A. (2007). *Counterculture green*. Kansas: University Press of Kansas.
Koolhaas, R. (2002). Junkspace. *October, 100*, 175–190.
Koolhaas, R., & Mau, B. (1997). *S,M,L,XL*. New York: Monacceli Press.
Krausse, J., & Lichtenstein, C. (Eds.). (1999). *Your private sky: Richard Buckminster fuller, the art of design science*. Baden: Lars Müller.
Latour, B. (1993). Give me a laboratory and I will raise the world. In K. D. Knorr-Cetina & M. Mulkay (Eds.), *Science observed* (pp. 141–170). London: Sage.
Latour, B. (1994). Une sociologie sans objet. Note théorique sur l'interobjectivité. *Sociologie du travail, 36*(4), 587–607.
Latour, B. (2005). En tapotant sur Rem Koolhaas avec un bâton d'aveugle. *Architecture d'aujourd'hui, 361*, 70–79.
Latour, B. (2008). A Cautious Prometheus? A Few Steps Toward a Philosophy of Design (with Special Attention to Peter Sloterdijk). In F. Hackne, J. Glynne and V. Minto (Eds.) *Proceedings of the* 2008 *Annual International Conference of the Design History Society* (pp. 2–10). Falmouth: 3–6 September 2009, e-books, Universal Publishers.

Le Corbusier (Charles-Edouard Jeanneret). (1964). *The radiant City: Elements of a doctrine of urbanism to be used as the basis of our machine-age civilization* (P. Knight, Trans.). New York: Orion Press. The original French edition is La ville radieuse: Eléments d'une doctrine d'urbanisme pour l'équipement de la civilisation machiniste (Boulogne: Editions de l'Architecture d'Aujourd'hui, 1933).
Levi-Strauss, C. (1966). *The savage mind*. Chicago: The University of Chicago Press.
Madge, P. (1997). Ecological design: a new critique. *Design Issues, 13*(2), 44–54.
Maldonado, T. (1972). *Design, nature, and revolution: Toward a critical ecology*. New York: Harper & Row.
Manzini, E. (2007). Design research for sustainable social innovation. In R. Michel (Ed.), *Design research now. Essays and selected projects* (pp. 233–245). Birkhäuser: Basel/Boston.
Margolin, V. (1998). Design for a sustainable world. *Design Issues, 14*(2), 83–92.
Margolin, V. (2002). *The politics of the artificial*. Chicago: The University of Chicago.
Massey, J. (2009). The sumptuary ecology of Buckminster Fuller's designs. In A. Braddock & C. Irmscher (Eds.), *A keener perception, Ecocritical studies in American art history* (pp. 218–236). Tuscaloosa: University of Alabama Press.
McLuhan, M. (2005 [1966]). The Emperor's Old Clothes. In G. Kepes (Ed.), *The Man-Made Object* (pp. 90–95). New York: George Brazillier Inc. Reprinted in E. McLuhan and W. T. Gordon (Eds.), *Marshall McLuhan Unbound* (20). Corte Madera (California): Gingko Press.
McLuhan, M. (2005 [1971]). Roles, masks and performances. *New Literary History, 2*(3), 517-531.. Reprinted in E. McLuhan and W. T. Gordon (Eds.), *Marshall McLuhan Unbound* (12). Corte Madera (California): Gingko Press.
Meadows, D. H., Meadows, D. L., Randers, J., & Behrens_III, W. W. (1972). *The limits to growth: A report for the Club of Rome's project on the predicament of mankind*. New York: Universe Books.
Morton, T. (2013). *Hyperobjects: Philosophy and ecology after the end of the world*. Minneapolis: University of Minnesota Press.
Ostrom, E. (1990). *Governing the commons: The evolution of institutions for collective action*. Cambridge University Press.
Ostrom, E. (2009). *A polycentric approach for coping with climate change*. Background paper to the 2010 World Development Report.
Otto, F. (1954). *Das hängende Dach*. Berlin: Bauwelt Verlag.
Pang, A. S.-K. (1997). Dome days: Buckminster fuller in the cold war. In F. Spufford & J. Uglow (Eds.), *Cultural Babbage: Technology, time and invention* (pp. 167–192). Boston and London: Faber and Faber.
Papanek, V. (1995). *The green imperative: Natural Design for the Real World*. New York: Thames and Hudson.
Papanek, V. (2011). *Design for the Real World: Human ecology and social change*. New York: Thames and Hudson, Second Edition Completly Revised.
Parent, C. (1971). Face à face: Architecture et Design. *L'Architecture d'Aujourd'hui, 155*, 19–23.
Patrix, G. (1962). Entretien avec Georges Patrix. *Les cahiers de la publicité, 4-4*, 87–102.
Patrix, G. (1973). *Design et environnement*. Paris: Casterman.
Petit, V. (2015). L'éco-design: design de l'environnement ou design du milieu? In *Sciences du Design, 2* (pp. 31–39). Paris: PUF.
Picon, A. (2008). Fuller's avatars: A view from the present. In M. Hays, D. Miller (Dir.), *Buckminster Fuller: Starting with the Universe* (pp. 46-59). New York: Whitney Museum of Modern Art.
Radjou, N., Prabhu, J., & Ahuja, S. (2012). *Jugaad innovation: Think frugal, be flexible, generate breakthrough growth*. San Francisco: Wiley.
Rifkin, J. (2014). The zero marginal cost society. In *The internet of things, the collaborative commons, and the eclipse of capitalism*. Palgrave Macmillan Trade.
Rittel, H., & Webber, M. (1969). Dilemmas in a general theory of planning. In *Policy Sciences 4* (pp. 155–173). Elsevier Science.

Schor, J. B. (2013). From fast fashion to connected consumption: Slowing down the spending treadmill. In N. Osbaldiston (Ed.), *Culture of the slow: Social deceleration in an accelerated world*. Basingstoke: Palgrave Macmillan.
Schwarz, A., Nordmann, A. (2011). The political economy of Technoscience. In M. Carrier, A. Nordmann (dir.), *Science in the Context of Application*, Dordrecht, Springer, pp. 317–336.
Scott, J. (1998). *Seeing like a state*. New Haven: Yale University Press.
Scott, F. (2007). *Architecture or techno-utopia. Politics after modernism*. Cambridge, MA: MIT Press.
Simondon, G. (2014). *Sur la technique (1953–1983)*. Paris: Presses Universitaires de France.
Sloterdijk, P. (2008). The crystal palace. *Publica, 37*.
Sloterdijk, P. (2011). *Neither Sun nor Death*. New-York: Semiotext(e).
Spangenberg, J. H., Fuad-Luke, A., & Blincoe, K. (2010). Design for Sustainability (DfS): The interface of sustainable production and consumption. *Journal of Cleaner Production, 18*(15), 1485–1493
Stahel, W. (2010). *The performance economy*. London: Palgrave McMillan.
Thorpe, A. (2012). Architecture & Design Versus Consumerism: How design activism confronts growth. London/New York Routledge.
Todd, N. J., & Todd, J. (1994). *From eco-cities to living machines*. Berkeley: North Atlantic Books.
Turner, F. (2006). *From counterculture to Cyberculture: Stewart brand, the whole earth network, and the rise of digital utopianism*. Chicago: University of Chicago.
Turner, F. (2009). Buckminster fuller: A technocrat for the Conterculture. In H.-Y. Chu, R. Trujillo (dir.), *New view on R. Buckminster fuller* (pp. 146–159). Standford: Stanford University Press.
Viénot, J. (1948). Du barrage au stylo. *Art Présent. Beautés de la technique*, (pp. 7–8, 9–11 and109–111).
Webster, K. (2015). The circular economy: A wealth of flows.
Winner, L. (1986). *The whale and the reactor: A search for limits in an age of high technology*. Chicago: University of Chicago Press.
Youngblood, G. (1970). *World Game*. Carbondale: Southern Illinois University Press.
Zung, T. T. K. (2001). *Buckminster fuller: Anthology for the new Millenium*. New York: St. Marin's Press.

Part VI
Design, Politics, and Society

Governmentality, Technologies, & Truth Effects in Communication Design

Katherine Hepworth

Abstract This chapter argues that communication design knowledge and artifacts are inherently governmental. As a means of communication that combines aesthetics and function, communication design knowledge is a product and producer of a uniquely pervasive form of governance that has seldom been studied. While several researchers and philosophers have expressed interest in the relationship between power, communication design knowledge and communication design artifacts, the governance inherent in communication design has yet to be seriously investigated. Building on the author's PhD research, this chapter extends Foucault's theories of discursive technologies, truth effects, and governmentality to account for how communication design artifacts and practitioners participate in the discourses surrounding them. Embodied discourse is proposed as the mechanism for this participation. From this perspective, all artifacts are seen as enmeshed in discursive entanglements, continually being imbued with regulatory meaning, and in turn, regulating their viewers and users. Finally, a framework for investigating the technologies implicit in communication design is presented, along with a discussion of the regulatory qualities of communication design artifacts, and of specific processes within communication design practice.

Keywords Communication design · Discourse · Foucault · Truth effects

1 Introduction

Scholars from a range of disciplines have written about the role of power in the field of communication design. Some are interested in how power serves a regulatory role within the communication design profession (Young 2009, p. 124), while others are interested in how power mediates the relationship between communication design artifacts and users (Lupton 1993, p. 7; Lavin 2001, p. 146). The majority of the literature discussing power and communication design artifacts investigates a

K. Hepworth (✉)
University of Nevada, Reno, Nevada, USA
e-mail: Khepworth@unr.edu

© Springer International Publishing AG 2018

P. E. Vermaas, S. Vial (eds.), *Advancements in the Philosophy of Design*, Design Research Foundations, https://doi.org/10.1007/978-3-319-73302-9_23

particular usage of power, known as governmentality. This term refers to the exercise of a specific kind of power called government. First described by Foucault, it is power exercised at a distance, or 'conduct of conduct' (Dean 2015, p. 400). Foucault's concepts of government and governmentality have become influential across the humanities, particularly in archaeology and sociology.

The literature regarding the relationship between governmentality and communication design is fascinating, but small. There has yet to be any serious investigation into the inherently governmental properties of communication design practice and artifacts. Such investigation has been hindered by the academic habit of dividing phenomena into aesthetic or functional categories. Spanning these categories, communication design has not sat comfortably in either. Communication designers' creative agency is routinely exaggerated by scholars invested in an aesthetic perspective on design, and routinely discounted by those with a structuralist, functionally-oriented perspective (See Eskilon 2007, as an example of the former, see Warde 1930 as an example of the latter). These distortions are compounded by the pervasive and enduring influence of modernist ideology within communication design practice and scholarship. This influence dissuades consideration of function in favor of aesthetics, and encourages a view of select designed objects as having unconditional or transhistorical significance. The regulatory, culturally determined nature of communication design knowledge is therefore commonly misunderstood as a rarefied awareness, apart from worldly concerns such as exchanges of power.

This chapter argues that exploring the governmental perspective on communication design will benefit both design and governmentality scholarship. It first defines the field of communication design, then goes on to survey the existing literature on governmental power in communication design practice and artifacts. Lastly, it provides a detailed investigation into how governmental power operates within communication design practice and artifacts. This includes reviewing Foucault's key concepts relating to governmental power, and extending the concepts of technologies and truth effects, using them to conceptualize the different mechanisms by which governmental power operates through communication design practice and artifacts respectively.

2 Communication Design: Practice, Knowledge, Artifacts

Communication design is a small but growing field of design practice, professional knowledge, and scholarship. Historically, it has been referred to by many names, including graphic design, graphic arts, industrial art, and decorative arts (Young 2009, p. 125). These terms reflect the value judgements of early work in the field, in which decorative or artistic renderings were highly prized. In the last few years, rapid growth in the nature and scope of communication design practice has resulted in a plethora of terminology for various aspects of communication design practice, including user experience design, information design, and interaction design. Rather

Hepworth	Ambrose & Harris	Lupton & Phillips
Defining problems	Research	Defining problems
Generating & refining ideas	Idea generation Refinement	Getting ideas
Form creation	Prototyping	Creating form
Preparation for production	Implementation	

Fig. 1 Comparison of stages in communication design practice

than using any of these earlier or later terms, communication design is used here because it most effectively conveys the broad scope of present-day practice.

2.1 Practice

The communication design profession is primarily concerned with 'affecting the knowledge, the attitudes and the behavior of people' (Frascara et al. 1997, p. 3). Communication designers influence knowledge, attitudes, and behaviors by developing effective means of communication for a given audience and context. The audience and context are typically determined by the needs of a client, for a commercial purpose (McCoy 2003, p. 107). In the frequent situation where communication designers are working for a large organization, 'the client' is actually a range of stakeholders with diverse interests. These stakeholders vary depending on the nature of the organization, but often include senior executives, marketers, accountants, production staff, and content providers. Accommodating the diverse expectations of each of these stakeholders within the design process, while also accommodating the needs of users, is a key communication design skill (Colberg 2006, p. 231). Communication designers also address other contextual constraints: the cultural and linguistic norms of the communication channel being used, and the communication norms in the geographical region where the communication will be received.

Communication design practice can be roughly divided into four stages: defining design problems, generating and refining ideas to solve the problem, form creation, and preparation for production (see Fig. 1). While this overall practice process is broadly agreed upon, there is no common consensus on the exact division of these areas; some practitioners and scholars divide communication design practice into fewer stages, while others divide it into more (Ambrose and Harris 2015, p. 6; Lupton and Phillips 2011, p. 5). Historically, when defining design problems and coming up with solutions, designers relied on informal conversations, while for form creation and preparation for production they depended on manual, 'hand and eye' processes of drawing, typesetting, and paste-up composition. Communication design artifacts produced this way were intended to communicate one way, from a client to a large and somewhat anonymous audience (McCoy 2005, p. 279).

Over the last 35 years, the communication channels, methods, and tools available to communication designers have expanded dramatically. This range of options has rendered the relationship between designers, clients, and audiences increasingly interdependent and fluid (Forlizzi and Lebbon 2002, p. 4; Sanders 2006, p. 65). The previously accepted term 'audience', has been replaced in communication design practice by 'users' and 'co-creators', terms that more accurately reflect these recent changes. Over the last two decades, three methods have become particularly common in communication design practice: generative design, where designers use methods that help clients to harness their own creativity; co-design practices, where designers and users build a design solution collaboratively; and user research, which seeks to understand user perspectives and needs (Ambrose and Harris 2015, p. 6; Sanders and Stappers 2012, p. 8; Taffe 2015, p. 39).

Today, defining design problems typically involves user research using formalized design research methods, with the extent and complexity of the user research being determined by the scale and budget of the design project. The second stage of the communication design process, generating and refining ideas, currently includes a wide range of approaches, including the traditional practice of drawing, along with generative design and co-design methods. In current practice, designers use more than one approach, switching from one to the next as they go through various stages of design refinement. For form creation and preparation for production, communication designers now rely on a vast array of methods, most of which employ a combination of computer hardware and software to produce files for digital and print production.

2.2 Knowledge

As a body of professional knowledge, communication design consists of multiple areas of expertise: collaborative communication, cultural and aesthetic sensitivity, aesthetic and design immersion, and familiarity with communicative options (communication strategies, available communication channels, and production processes). Collaborative communication includes a range of strategies aimed at understanding the specific design context, empathizing with user needs and desires in that context, identifying the design problem that needs to be solved, and communicating that problem in a way that all stakeholders can easily relate to (Lupton and Phillips 2011, p. 15). Sometimes this collaborative communication includes a team of designers, sometimes it also includes the client or users.

Cultural and aesthetic sensitivity refers to the designer's ability to identify the cultural and subcultural characteristics of specific users. This includes collecting and studying demographic and psychographic details, as well as using design research methods to gain an understanding of users' levels of familiarity and comfort with the aesthetic sensibilities of various communicative options (McCoy 2005,

p. 280). This aspect of communication design knowledge is typically referred to within professional communication design circles as user research (Ambrose and Harris 2015, p. 33). The cultural and aesthetic sensitivity of user research depends for its success upon the designers' capacity to empathize with users, as well as their level of familiarity with aesthetic and design conventions.

Aesthetic and design immersion is the area most traditionally thought of as communication design knowledge. It involves expert understanding of how aesthetic qualities such as color, line, pattern, and texture can be combined with design elements such as grids, hierarchy, typefaces, and white space to create compositions that are compelling to specific users. It also involves familiarity with the standard graphic conventions in the cultural contexts in which the designer operates (Bowers 1999, p. 10; Meggs 1989, p. 70). For example, a communication designer who frequently works on street signage needs to know the local standards for colors, shapes, pictograms, and compositions of street signs. Similarly, a communication designer who mainly works on books needs to know the conventions of the book format: title page, half title page, end papers, page numbers, running headers, and so on. Understanding historical uses of various aesthetic and design elements is an important aspect of this immersion too, as it allows designers to employ elements that have favorable historical connotations for the specific users and contexts, and to avoid unfavorable connotations (Meggs 1989, p. 134).

The communication strategies that communication designers use consist of identifying, developing, and implementing the most appropriate combination of message (content), aesthetic and design elements (form), communication channels (medium), and production processes (technology) for any given group of users (Frascara et al. 1997, p. 5). Familiarity with communication channels includes awareness of the aesthetic qualities, cost, nature, functioning, and availability of the wide array of options for distributing a given message (Colberg 2006, p. 230). Billboards, flyers, mobile apps, posters, promotional objects, television, text messages, and websites are all communication channels. Each communication channel, or medium, has unique qualities and constraints that inevitably influence the communication they mediate.

Familiarity with production processes includes awareness of the aesthetic qualities, cost, nature, functioning, and availability of the production processes involved in producing work for a given communication channel. For example, the production process options involved in making a website include prototyping software, coding languages, extension languages, compiling programs, content management systems, project workflow systems, and hosting services. In contrast, the production process options involved in producing a printed book include layout software, digital printing, offset printing, letterpress printing, embossing, de-embossing, standard inks, spot color inks, uncoated paper, coated matte paper, coated glossy paper, varnishes, perfect binding, case binding, Canadian binding and so on (Ambrose and Harris 2011, p. 145). The breadth of communication designers' knowledge of all of these areas determines their capacity to produce effective work.

2.3 *Artifacts*

As well as being used to refer to a field and area of practice, the term 'communication design' is also used to refer to artifacts produced in the pursuit of communication design practice. Communication design practitioners and researchers generally understand communication design artifacts as 'visual objects aimed at communicating specific messages' (Frascara 2004, p. 2). These include signage of all kinds; advertising and branding in their many forms; graphical user interfaces of software and operating systems; the front end design of websites; and books and magazines. The field produces an unusually broad array of work, which we all interact with daily (George 2002, p. 1). For the purposes of this chapter, the term 'communication design artifacts' is used to refer to all artifacts that communication designers can reasonably be expected to have designed. This phrasing is used in recognition of the difficulty of ascertaining provenance of communication design artifacts. In contrast with art artifacts, it is extremely rare for communication designers to put signs of authorship on their work, as communication design artifacts depend on the appearance of coming directly from the client for some of their effectiveness.

3 The Form/Function Divide

Investigation of the political nature of communication design artifacts has been hindered by the traditional academic categorizations of phenomena as either aesthetic or functional, sacred or profane, and culturally or politically valuable (Witkin 1990, p. 326). From the seventeenth century onwards, universities have relied on these divisions to render subjects discrete and comprehensible, and popular understandings have tended to follow the same divisions. Artifacts especially, are routinely separated into soft or hard scientific collections (more commonly referred to as social science, and science respectively), with the former containing artifacts of aesthetic or cultural value, and the latter containing artifacts with functional or scientific value (Hodder 1992, p. 11). Power exchanges are primarily studied in political science, the discipline devoted to them; while artifacts used in power exchanges that are judged to constitute art or design of quality are studied in art history or design history. On the rare occasions that political scientists do engage with artifacts, their lack of understanding of basic elements of content and form often leaves their analyses lacking (Rose 2008, p. 37; Sartwell 2010, p. 49). These artificial separations dissuade investigation into the functional aspects of artifacts and phenomena that are usually appreciated for their aesthetic qualities, and vice versa.

Studying communication design within this traditional fragmentation of knowledge is problematic, as the field spans multiple disciplinary divisions. In communication design practice, aesthetic considerations are used *in the service of* functional requirements. However, the functional uses of aesthetic qualities within communication design artifacts are routinely overlooked in the design literature, particularly in histories of communication design artifacts. This is perhaps because these artifacts are primarily information-related and communicative, while design scholars are used to considering artifacts that are primarily physical tools and therefore performative (Hepworth 2012, p. 44; Hodder 1992, p. 12). For example, industrial design artifacts, which are the main focus of design histories, are primarily performative in the sense that they seek to alter behavior by way of physical interaction. In contrast, communication design artifacts seek to alter behavior by way of altering thought, through the means of communication (Hayward 1998, p. 219). Due to the small amount of communication design scholarship, this section covers the entirety of what has been written about governmental power and communication design in the broader design literature. Next, the field of governmentality studies is covered. Finally, perspectives on governmentality and communication design from other fields are explored.

3.1 Design Scholarship on Communication Design and Power

This section surveys work from design studies, design culture, design history, as well as the small sub-field of communication design history. The form/function divide is evident in the design literature's treatment of communication design, which is frequently considered in terms of aesthetic merit, or not considered at all (Blauvelt 1994, p. 206; Carnegie 2013, p. 33; Fallan 2010, p. 8; Julier 2008, p. 49). While there is value in studying artifacts' aesthetic qualities, over-emphasis on aesthetics has led to a systematic devaluing of the field of communication design within the design canon. In relation to communication design artifacts, the divide is compounded by the overwhelming influence of modernism on professional practice and scholarship. Modernism holds that the material substance of the social world can be rationally investigated and explained (Hall 1996, p. 4). The modernist perspective is so inherent in communication design practice that graphic designer Jeffery Keedy has observed that '[i]n graphic design, there is no alternative to modernism' (Keedy 1995, p. 171). He refers to the repeated, seemingly inevitable reversion to modernist methods in communication design practice after short periods of professional experimentation with other influences.

The combined obstacles of the form/function divide and the modernist lens have had an unfortunate effect on the design literature's treatment of communication design artifacts and power. Power is frequently mentioned in a variety of contexts — authorial power, gendered power, political power and corporate power — but these terms are usually used in passing, without clear definition or elaboration (Davidson 2008, p. 86; Julier 2008, p. 1; Simmons 2000, p. 325). The literature's shortcomings in this area are particularly evident in writing on communication design artifacts that have been created and used in the service of establishing and maintaining political power. Certain groups of these communication design artifacts, notably propaganda posters, have been extensively studied and critiqued within the academy. The vast majority of the literature on such artifacts focuses on aesthetic qualities of visual form, frequently to the exclusion of political considerations (Cooter and Stein 2007, p. 184). The literature demonstrates a widespread reluctance to consider communication designers as political agents, or communication design artifacts as political tools, with many scholars opting for a common narrative of communication designer as socially immune, aesthetic savant (for examples, see Eskilon 2007, p. 197; Heller 2008, p. 133; Meggs and Purvis 2006, p. 290; Stanley 1989, p. 16).

Although the majority of the design literature's discussion of power is wanting, there are exceptions. Several scholars who are also communication design practitioners have written nuanced accounts of power in relation to communication design practice. Communication design historian-practitioner Rick Poynor argues that the practice of communication design necessitates reflection on, and awareness of, the role of power in society. He writes: 'It has never been a surprise to [communication] designers that… to portray and frame one's activities and those of others is to assert power and is, ultimately, an attempt to shape the world' (Poynor et al. 2004, p. 184). The rich perspectives on power in some of the scholarly work from practitioner-historians and practitioner-theorists support Poynor's argument.

3.2 *Design Scholarship on Governmentality*

Most practitioner-researchers writing about power and communication design focus on the role design plays in fostering inequitable, gendered divisions of labor. This gender-focused literature describes the governmental effects of power operating through communication design in relation to gendered societal roles, albeit without using the term 'government'. In *Mechanical Brides*, practitioner-historian Ellen Lupton explores the role communication design artifacts have played in constructing and maintaining gendered power divisions. Extending the work of feminist writer Betty Friedan, she argues that printed advertisements for household appliances work in concert with the appliances themselves to reinforce gendered and fundamentally unequal societal norms. Lupton deconstructs how advertisements for household appliances throughout the twentieth century reaffirmed kitchens and laundries as spaces of female work, at the same time as suggesting the burden of such work was being ever reduced by technological advancement. In actuality, the volume of household work for women increased consistently between 1920 and 1960, due to unprecedented standards of household hygiene (Lupton 1993, p. 11, 15).

In a similar vein, cultural historian Maud Lavin has studied how communication design artifacts contribute to public attitudes on abortion. In her essay 'A Baby and a Coat Hanger,' Lavin discusses how the design of public information posters used in the abortion debate in the United States in the 1970s shifted the emphasis of the debate from social and moral issues to a polar divide between foetuses' rights and women's rights. She attributes this shift in large part to 'right to life' campaigns using polemic prose, the use of sonogram imagery of foetuses, sometimes misshapen or dead, along with other provocative design strategies (Lavin 2001, p. 146).

Practitioner-researcher Philippa Goodall critiques the gendered nature of design production and use, describing how communication design practice, knowledge, and artifacts define fields of acceptable gendered behavior and ways of being (Goodall 1983, p. 187, 1990, p. 270). She uses the Foucauldian concept of discourse to frame her argument, describing how the discourse of design is bound up in, and promotes, discourses of womanhood through communication design artifacts. These, in turn, produce unattainable and isolating gendered ideals.

Practitioner-theorist Alan Young also uses a discursive framework to explore the governmental effects of communication design (Young 2009, p. 124). He uses discourse as a way to explore issues of agency and power in communication design practice, focusing on how professional practices serve as a kind of professional self-regulation (Young 2009, p. 128). Other design scholars writing on communication design who describe governmental effects without using the exact term, instead frame their functioning in linguistic terms: argument, enthymeme, and persuasion (Blair 2004, p. 41; Bush 1994, p. 228; Gallagher et al. 2011, p. p.27).

3.3 Governmentality Scholarship on Communication Design

Governmental power is a key preoccupation of scholars working in the field of governmentality studies. Political sociologist Mitchell Dean defines governmentality as 'a perspective on how to investigate diverse practices and regimes of government, the latter understood in the broadest sense as the "conduct of conduct"' (Dean 2015, p. 400). This is an interdisciplinary field made up of human geographers, political philosophers, political scientists, and sociologists who value the insights gained by applying the governmentality perspective to their various objects of study. In governmentality studies, communication design artifacts are infrequently mentioned, and where they are, it is by sociologists working within the field. These scholars demonstrate small, if persistent, interest in the roles aesthetics and material culture play in exchanges of governmental power, and communication design artifacts are invariably mentioned in the context of one of these two themes. When communication design artifacts are mentioned in governmentality studies, the term 'communication design' is not used. Instead, individual artifacts that are routinely produced by communication designers are mentioned by name.

Dean refers to communication design artifacts among a listing of governmental material forms, and describes how the visualizing of peoples, territories and even individual identities is a pre-requisite to their being governed. He writes: 'a map, a pie chart, a set of graphs and tables, and so on. .. all make it possible to "capture" who and what is to be governed' (Dean 2010, p. 41). Dean highlights what several others have observed: communication design artifacts are essential to the exercise of governmental power (Anderson 2006; Kostelnick 2004, p. 215; Rose 2008, p. 36). Sociologist Nikolas Rose also occasionally mentions communication design artifacts. Rose argues that visual communication artifacts are used to give knowledge a sense of stability and definiteness. He refers to visual communication artifacts as 'little machine[s] for producing conviction in others.. . [they are] material techniques of thought that make possible the extension of authority over that which they seem to depict' (Rose 2008, pp. 36–37). He argues that these communication design artifacts are just as capable of and effective at regulating behavior as institutions and practices that are more typically thought of as disciplinary, but that they are overlooked due to their ordinariness (Rose et al. 2006, p. 89).

While the governing qualities of communication design artifacts are of interest to governance scholars, references to these artifacts tend to be in the form of mentions, rather than detailed analyses. In their study of communication design artifacts, governance scholars appear to come up against a barrier of vocabulary and professional understanding. Their research interests provide them with insight into the potent, governing role communication design artifacts play in human interaction generally, but they lack the necessary tools to analyze governance within these artifacts in any great depth.

3.4 Other Perspectives

Apart from the contributions of the design and governmentality literature, scholars from various other fields, including anthropology, archaeology, rhetoric studies, and philosophy, have observed the governmental nature of communication design artifacts. Design anthropologist Dori Tunstall connects Foucault's concept of governmentality with communication design practice and artifacts, identifying communication design's capacity to define acceptable fields of behavior and ways of being. She goes further than other scholars in her investigation of this quality, arguing that designed artifacts and experiences utilize governmental power to construct both individual understandings of acceptable participation in civic life, and personal identities of citizens (Tunstall 2007, p. 4). Her work suggests that communication design artifacts are necessarily more concerted efforts to govern than other artifacts because of their communicative capacity. Postcolonial theorist Benedict Anderson has observed the governing role of maps in the context of European colonization in Southeast Asia during the nineteenth and twentieth centuries (Anderson 2006, p. 171). He attributes detailed Mercatorial maps, in combination with census data, with introducing populations to the previously unheard of term 'country' and instilling a sense of nationalism.

Philosophers Jacques Rancière and Crispin Sartwell both explore the relationship between aesthetics, artifacts, and political power without explicitly referring to governmental power. In *The Politics of Aesthetics*, Rancière argues that aesthetics and political power share the essential quality of determining what is visible and invisible in any given context, and that visibility gained through aesthetic means can give rise to political visibility. To demonstrate, he cites how communication design artifacts (books and posters), and processes (typography, iconography, and page layout) in the twentieth century blurred the existing division between pure art and ornament, overturning previously established representative orders, and thereby fostering new conceptions of the individual and of community (Rancière 2004, p. 15, 19). Sartwell states that artifacts are 'concrete effects of political systems', arguing that political power is inherently aesthetic, as it elicits sensory experiences (Sartwell 2010, p. 2). These aesthetic qualities are felt through the artifacts that are associated with the production and use of political power, from the most sacred to the profane (Sartwell 2005, p. 768, 2010, p. 6).

Rhetorician Charles Kostelnick's work engages design and governance through the lens of visual rhetoric (Kostelnick 2004, p. 215; Kostelnick and Hassett 2003, p. 10). Although he does not mention governmental power by name, Kostelnick is one of the few scholars who study the governmental effects inherent in the minutiae of communication design. By examining the visual qualities of historical documents, and comparing them to contemporary communication design artifacts, he demonstrates the temporally and culturally bound nature of design conventions. He uses the concept of discourse communities in concert with the concept of visual language to explain how certain communication design artifacts and conventions come to be perceived as 'natural' or obvious (Kostelnick and Hassett 2003, p. 30). Kostelnick's work demonstrates that despite their many shapes and sizes, communication design artifacts share the common purpose of changing the activities, attitudes or emotional states of their viewers indirectly, through communication intended to somehow alter their knowledge (Kostelnick 2004, p. 218).

4 Conceptualizing Governmentality in Communication Design

While various scholars have observed the governing properties of communication design artifacts, so far it seems that no-one has yet conceptualized a mechanics of these governing properties. Such a theoretical framework is important for recognizing communication design's influential functioning in society. This section provides a much needed bridge between design and governmentality scholarship. It begins with a brief overview of relevant Foucauldian concepts: power-knowledge; discourse and government; and technologies and truth effects. The original concept of embodied technologies is offered as the means through which communication design artifacts temporarily hold power. These theoretical constructs are then applied to the field of communication design. Communication designers' professional activities are reframed as governmental regulation that combines aesthetic and functional techniques to produce work that resonates within the discursive contexts of its intended users. Communication design artifacts are presented as mediators of power exchanges that depend on their relevance to prevailing societal discourses for their efficacy and commercial success.

4.1 Power-Knowledge

Foucault's work on power emphasizes its inherently dispersed nature and complex manifestations. He writes that power "traverses and produces things, it induces pleasure, forms knowledge, produces discourse. It needs to be considered as a productive network which runs through the whole social body" (Foucault 1977a, p. 119). This perspective is the foundation of Foucault's understanding of governance. In his historical investigations, Foucault studied systems that are extremely hierarchical — the Christian church, medical institutions, prisons — but throughout these investigations, he maintained that power was not centralized at the apex of these hierarchies. Instead, he went to great lengths to show how power is dispersed, web-like, in any given institution or society.

While power is not available to all participants in a given society equally, it is available in some small measure to each individual. He writes: "The summit and the lower elements of the hierarchy stand in a relationship of reciprocal support and conditioning: they 'hold together' (power as a mutual and indefinite 'extortion')" (Foucault 1977b, p. 14). Instead of viewing people at the bottom of the hierarchy as powerless, Foucault examined their relationships to the institutions and knowledges in which they were involved, and found them to have a certain amount of "little powers" (Foucault 1997a, p. 87).

The inherently dispersed nature of power is explained in its relationship to knowledge. Knowledge, Foucault argued, is evidence that power exists and is being utilized. The act of knowing is inevitably a process of being transformed, and in a sense, disciplined, by our own thoughts (Foucault 1997a, p. 8). These thoughts, in turn, are heavily influenced by the societies and institutions we participate in. Foucault considered the link between power and knowledge so strong that he sometimes referred to them as a united entity, "power-knowledge" (Foucault 1980a, p. ix). Anthropologist Juris Milestone elaborates on power-knowledge, stating that human endeavor necessarily organizes and systematizes knowledge, resulting in the formation of knowledge disciplines (for example, science, art, and history). These disciplines are then used to understand and manipulate people as groups (for example, as a citizenry, or as delinquents). It is in this defining of subjects to be dominated or managed that power capacities manifest (Milestone 2007, p. 179).

4.2 Discourse and Government

The knowledge disciplines in Milestone's example are described by Foucault as "principle[s] of control over the production of discourse" (Foucault 1981a, p. 61). Knowledge disciplines then, are productive of discourses, or knowledge patterns that determine appearances of truth. Although he wrote extensively on discourse, there are few succinct, decisive definitions of the term in Foucault's work. However, he does say that discourse is "the difference between what one could say correctly at one period... and what is actually said" (Foucault 1968, p. 63). This somewhat ambiguous definition emphasizes how discourses determine what is knowable in any given context.

With this unique conception of discourses, Foucault sought to emphasize the importance of transformations and process in human thought, society, and knowledge, highlighting the inherently context-bound nature of even the most empirical human knowledge (Foucault 1968, p. 54). He argues that discourses are the means through which some knowledge becomes viewed as common sense, obvious or inherently truthful, while other knowledge remains unknowable or taboo in a given context. Over the course of his career, Foucault vacillated on the relationship between power and discourse. In some works he stated that discourses are not inherently controlling (Foucault 1981a, p. 72). In others, he implied that discourses always have purposes that are advantageous to, and the inevitable result of, exercises of power (Foucault 1971, p. 2).

Discourses are the result of a specific kind of power known as government. Foucault identified three kinds of power: discipline, sovereignty, and government. The first two have much in common with popular understandings of power, relying on human obedience and subjection, and employing force to achieve these ends. Government is a more diffuse kind of power that is associated with satisfactory arrangements of people in relation to their surroundings, including objects, communications, and physical space (Foucault 1978a, p. 211). He writes that government "designate[s] the way in which the conduct of individuals or of groups might be directed... To govern, in this sense, is to structure the possible field of action of others" (Foucault 1982a, p. 790). Discourses are inherently governmental because they engender limits on knowledge, determining the bounds of what is knowable for any particular individual or group, thereby structuring their 'fields of action' within their surroundings.

4.3 Technologies and Truth Effects

Exercise of governmental power relies on tactics that manipulate knowledge on an individual level, effectively shaping each person's attitudes and skills. Foucault sometimes referred to these tactics as 'technologies', and defined them as 'a disparate set of tools and methods' (Foucault 1977c, p. 26). He identified specific technologies, such as 'technologies of sign systems', 'technologies of the self', and 'technologies of sexuality'. This was his shorthand way of referring to the tools and methods that people use to construct their own and each other's understandings of knowledge, themselves, and their sexuality respectively (Stoler 1995, p. 17).

Foucault suggested that the interplay of multiple technologies over time determines the nature of particular discourses, separating phenomena into true and untrue, or knowable and unknowable (Foucault 1982b, p. 225). Although the identifications of truth are invaluable for making sense of the world, the bulk of Foucault's historical investigation shows us that what is perceived as an inalienable truth changes drastically over time, location, and culture (Foucault 1997b, p. 99). Foucault refers to the appearance of truth, the apparent solidity of facts stemming from particular technologies and discourses, as 'truth effects' (Foucault 1997a, p. 119). To convince us of any particular apparent truth, technologies and discourses employ linguistic force combined with other persuasive, strategic forces (Foucault 1967, p. 290, 1976, p. 11).

These truth effects can manifest on various scales: individually, across communities, and across societies. On an individual level, truth effects contribute to self-government, the internalizing of community and societal discourses. On broader, societal levels, truth effects contribute to the social production of attitudes, behaviors, and intentions (Foucault 1981b, p. 93). Societal level truth effects also determine what is visible and accepted in any given society. Much of Foucault's historical investigation emphasizes this shifting visibility and acceptability of various human behavior (See Foucault 1972, p. 40, 1978b, p. 4, 1980b, p. 112).

Technologies and truth effects have proved to be versatile and enduring concepts, frequently used in governmentality studies to analyze a wide range of governmental processes and behaviors. They are used in the following section to examine how discourses operate within communication design. The concept of technologies of sign systems has particular relevance for communication design, since the field relies on textual and visual signs for its communicative capacity. Languages, the alphabets they are written in, visual recognition, and graphic mark making can all be considered technologies of sign systems. These technologies are the cognitive and symbolic foundations that make communication design possible.

4.4 Technologies in Communication Design Practice & Knowledge

Just like other knowledge disciplines, communication design has governmental effects that manifest as technologies (Goodall 1990, p. 269). The technologies involved in communication design operate within academic, educational, and professional communication design literature, in professional tools and spaces, as well as through communication design artifacts. They operate through artifacts over their entire life cycles, from earliest conception to eventual discarding. This section focuses on two technologies that have the most intimate impact on communication design practice: the technology of profession and the technology of production.

The profession of communication design can be seen a 'discourse community' that is influenced by a broad range of factors, of which communication design knowledge is one part (Kostelnick and Hassett 2003, p. 24; Swales 1990, p. 22). Communication design practice is shaped by a combination of broader societal discourses, as well as those specific to communication design knowledge and practice. Communication design courses, professional organizations, the studio environment, computer hardware and software, and informal professional news and communication

Technology of profession		
Forces	**Techniques**	**Devices**
Education Experience Collegial relationships Creative equipment	Knowledge processes Creation practices Visual conventions	Communication design artifacts

Technology of production		
Forces	**Techniques**	**Devices**
Education Experience Production equipment Production materials	Production processes	Production effects within communication design artifacts

Fig. 2 Anatomy of the discourse of communication design

networks all contribute governing qualities to communication design practice. In turn, communication design practice inevitably shapes the governmental nature of communication design artifacts.

The technology of production includes the machinery, materials, software, and coding languages used to produce communication design artifacts. By prescribing the limits of possibility in design production, all of these material and immaterial elements inevitably have governmental effects on the nature of communication design knowledge, on communication designers themselves, and on communication design artifacts. The technology of production changes as rapidly as do the production methods available to communication designers. What was once fundamental communication design knowledge sometimes becomes a historical design production curiosity in a matter of months, as new means of production become available, some traditional approaches are used in new ways, and as others become obsolete.

4.4.1 Anatomy of Professional Technologies

Rose has proposed an anatomy of technologies that is helpful for exploring the governmental qualities of communication design practice. He argues that technologies consist of three kinds of attributes: forces, techniques, and devices (Rose 1996, p. 42). These attributes can be visualized as a hierarchical diagram, moving from most abstract attributes with the most expansive influence on the left, to the most practical and limited in reach on the right (see Fig. 2).

Forces are the most influential attributes of technologies, made up of "an assembly of forms of knowledge" (Rose 1996, p. p.52). They are broad influences that affect society generally, and influence all communication design practice. For example, a force within the technology of profession is creative equipment, including the physical tools used in physical and digital design work: pencil and paper, and computer hardware and software. This equipment delineates the field of possibility for communication designers by defining the visual and physical range in which they operate.

Techniques are the second tier attributes of technologies, operating within the constraints of forces. They consist of the strategies and systems used within a technology to 'produce certain practical outcomes' (Rose 1996, p. 52). The strategies within the technology of profession include knowledge processes communication designers use in their work, such as audience profiling and developing personas, as well as acts of creation involved in the design process, such as drawing and prototyping. The systems include graphic conventions — corporate identity systems, wayshowing systems, promotional campaigns, and content strategies — communication designers frequently utilize to produce particular artifacts. These systems use visual conventions to ensure that related communication design artifacts are perceived as such, thereby strengthening their collective message. These strategies and systems further limit the prospective field of thought and action available to communication designers.

Devices are the forms generated by forces and tailored by techniques within technologies (Rose 1996, p. 53). These forms are frequently physical, but not always, as certain commonly produced communication design artifacts, such as logos and websites, are essentially immaterial. Devices of the technology of profession within communication design include all the various artifacts communication designers produce such as books, magazines, logos, signage, websites, advertisements, and promotional objects.

In terms of the technology of production, production equipment is the force. In the case of printed, physical communication design artifacts, production material includes paper, inks, and printing presses, while in the case of immaterial artifacts such as logos and websites, production materials include the machinery, materials, software, and coding languages used to produce them. By prescribing the limits of possibility in design production, all of these material and immaterial elements inevitably have governmental effects on the nature of communication design knowledge, on communication designers themselves, and on communication design artifacts.

4.4.2 Truth Effects of Professional Technologies

Both technologies of profession and production generate truth effects that are felt directly by communication designers, and indirectly by users of the artifacts they design. The earliest stage of communication design practice, identifying design problems, promotes a view of the world as flawed but diagnosable, and ultimately fixable. Similarly, professional education about, and experience with, communication design creation and production produce truth effects that render the strategies, communication channels, and production processes that are most familiar to a designer as the most correct or necessary solution to a given design problem. No matter how extensive and rigorous the user research is on any given project, the formal education and professional experience of a communication designer invariably skews his or her interpretation of that research.

For example, designers who are trained in a web development-heavy communication design program are far more likely to come up with web-based design solutions for any design problem than designers trained in a program specializing in traditional processes, such as book design and letterpress printing. Work experience has similar truth effects on which design solutions, and which ways to formulate them, are most appropriate for any given situation. For example, communication designers working in a human-centered design firm such as IDEO are more likely to be inclined to trust generative design methods to arrive at the most effective design solution. In contrast, communication designers working in-house for engineering and architecture firms are more likely to place emphasis on working with systems of visual conventions, due to their work environments having a heavy focus on systems design. Each communication designer experiences a unique combination of physical surroundings, colleagues, education, and experience that invariably molds variations in the professional truth effects they experience (Seago and Dunne 1999, p. 16).

These professional attitudes and preferences are true of all professions. However, they are worth elaborating upon in relation to communication design specifically because the truth effects of communication design suggest to communication designers that they are somehow immune to such biases (McCoy in Soar 2002, p. 72). This perception of immunity is fostered by the repetition of thinking patterns and physical practices in a linear process that comes to feel instinctive over time (Hepworth 2014, p. 13; Kostelnick and Hassett 2003, p. 23). Physical actions involved in communication design practice are constantly shaping the attitudes and beliefs of communication designers, and vice versa, in a reciprocal process which goes largely unnoticed or commented upon.

5 Embodied Technologies in Communication Design Artifacts

As discussed earlier, scholars from design, governmentality studies, and other areas have acknowledged that communication design artifacts have governmental effects in society generally. The discursive influence and consequent governmental power of communication design artifacts is conveyed through both their form and content (Cooter and Stein 2007, p. 183). In this section, I present the original concept of 'embodied technologies' as the means by which communication design artifacts can both produce and temporarily hold discourses. I argue that technologies can be retained by artifacts temporarily through the process of embodiment, allowing them to participate in discursive exchanges with their creators (communication designers) and production technologies initially, and later, with their users, environments, and cultural associations across their life cycles. As the context of communication design artifacts change — their role in human interactions, the physical space they inhabit, and the associations they are given — so too do the discourses that become embodied within them.

5.1 Anatomy of Embodied Technologies

Discourses are conceptualized here as becoming embodied within communication design artifacts through two groups of technologies: technologies of form, and technologies of content. Technologies of form are those working through the external and/or physical appearance of artifacts, as well as the form characteristics of individual elements within the artifacts, such as typography. Technologies of content work through the symbolic and textual content of artifacts. (see Fig. 3 for a breakdown of these groupings of technologies). The distinction between technologies of form and content allows for in-depth analysis of how individual design elements contribute to the technologies, which in turn contribute to broader professional and societal discourses and their consequent truth effects.

Rose's anatomy of technologies can also be used to identify and elaborate upon the various aspects of embodied technologies. In the context of embodiment within artifacts, the definitions of forces, techniques, and devices have been adjusted to allow for the characteristics of communication design artifacts. Forces within embodied technologies are the elements of an artifact that determine all its other properties. These include the type of communication design artifact, and the production effects evident in that artifact. The type of communication design artifact determines both the artifact's function and field of action, as well as limiting the nature and quantity of visual elements it can contain. For example, the function, field of action, and visual elements contained within a logo and a poster will be dramatically different. To be used and recognized as a logo or a poster in any given time and place, an artifact must have the characteristics recognized as 'logo-like' or 'poster-like' in that context.

In this conception of power operating through communication design practice and artifacts, the forces of embodied technologies are the same as the devices of technologies of profession and production (see Figs. 1 and 2). This relationship between communication design technologies and technologies embodied within artifacts is hierarchical because each individual artifact is absolutely dependent upon communication design practice, among other societal discourses, for its existence. Every attribute of content and form is influenced by these factors.

In embodied technologies, techniques are the thematic groupings of devices used to coordinate individual elements and consolidate meaning. Techniques include design conventions such as systems of spatial organization, aesthetic groupings, such as stylistic treatments and typefaces, and conceptual groupings, such as themes of reference found within devices. In turn, devices are the smallest elements in a communication design artifact with a recognizable communicative element. They include individual words, symbols, shapes, and colors, as well as typeface accents, cases, and weights.

Technologies of form		
Technology of external form		
Forces Type of communication design artifact Production effects	**Techniques** Composition system Stylistic treatment	**Devices** External shape Colors
Technology of internal word form		
	Techniques Typographic systems Typefaces	**Devices** Type weights Type cases Type accents
Technology of internal image form		
	Technique Stylistic treatments	**Devices** Internal shapes

Technologies of content		
Technology of word content		
Force Language	**Techniques** Phrases Reference themes	**Devices** Words
Technology of image content		
Force Type of meaning	**Techniques** Reference themes	**Devices** Symbols

Fig. 3 Anatomy of discourse embodied within communication design artifacts

While the concept of embodied technologies can be applied to all artifacts, the framework in Fig. 2 was developed specifically with communication design artifacts in mind. Of all the professional and artistic fields that produce visual artifacts, communication design is unusually concerned with the integration of form and content of both words and images. Therefore, technologies embodied within artifacts are grouped according to whether they operate through form or content. Within these form and content groupings, technologies are identified based on whether they operate through the entire artifact, or through specific images or text.

For example, the technology of external form encompasses all qualities that are inseparable from the artifact as a whole. The term 'external form' is used to distinguish these attributes from the form attributes of elements within an artifact. All typographic elements are encompassed in the 'technology of internal word form' because typography relates to the form of words, and the words studied in this context are inside, or internal to, a communication design artifact. The 'technology of internal image form' relates to form qualities of visual symbols within the communication design artifact. The grouping 'technologies of content' contains two further technologies. 'Technology of word content' encompasses use of language within a visual communication artifact, while 'technology of image content' refers to the non-text, symbolic content, along with associated references and meanings. Delineating between these three kinds of form and two kinds of content allows deep investigation into the mechanics of governmental power within communication design artifacts.

I argue that this anatomy of embodied technologies can be used to investigate the discursive qualities of communication design artifacts, providing a cohesive perspective on their aesthetics and functioning in relation to societal power relationships. Studying how each individual device (each unique quality or element) of a communication design artifact relates to a technique, which in turn relates to various forces, reframes communication design artifacts. It shows how each individual artifact is a complex network of multiple meanings and associations that are inextricably connected with governmental power and broader societal discourses, including the professional discourse of communication design.

5.2 *Truth Effects of Embodied Technologies*

The five embodied technologies that can be found within communication design artifacts are felt as truth effects, either reinforcing or challenging our perceptions and beliefs when we view or interact with these artifacts (Riles 2006, p. 136). Sometimes the truth effects are so slight that we do not consciously perceive them, but this does not mean artifacts are neutral. On the contrary, the most powerful truth effects are the ones of which we remain unaware. For example, communication design artifacts used in corporate office environments — logos and branded communications — along with the furniture and decor 'play an important part in calling out "appropriate" attitudes and responses' in workers (Witkin 1990, p. 328).

On a smaller scale, complex data visualizations put viewers in a position of surveillance over the subjects being visualized, providing a truth effect of privilege that is usually unconscious. The visualized subjects appear ordered, comprehensible, and objective (Barton and Barton 1993, p. 146). Reducing the messy complexity of most phenomena into an easily comprehensible graphic is a distinctly persuasive act, and data visualizations are inherently persuasive artifacts. Another truth effect of data visualizations is that, despite presenting partisan information, the format is usually perceived as impartial.

Considering propaganda posters from a Foucauldian perspective, the truth effects they produce are more overt than those of data visualizations. While posters have an informational aspect, we also expect them to persuade; there is no pretense of objectivity inherent in the medium (Sontag 1970, p. 198). Instead, propaganda posters recycle visual styles and motifs from art and design history, combining them with present day calls to action, to evoke an emotional response. This emotional response constitutes a tacit knowledge that contributes to us into altering our beliefs (Polanyi in Reiner 2009, p. 345). The truth effects of propaganda posters are as overt as those of data visualizations are surreptitious.

6 Conclusion

This chapter has demonstrated that despite the pervasive modernist myth in design literature, communication design is far from value neutral. Communication design practice and artifacts are inextricably bound up in both professional and broader societal discourses. These manifest as technologies of communication design practice, and as embodied technologies within communication design artifacts. Technologies of communication design practice channel societal discourses into the production of artifacts, while also generating discursive influences specific to the field.

By investigating the inherently governmental nature of communication design more deeply, scholars studying both communication design and governmentality can enrich their respective fields. Focusing on the power exchanges inherent in communication design practice, and in interaction with communication design artifacts, is one way design scholars can become aware of these biases, and throw off the persistent influence of modernism. The same focus can help scholars in governmentality studies deepen their study of governmental power in material culture, and provide important contributions to the growing literature on the relationships between thought, power, and artifacts.

The original concept of embodied technologies is proposed as one means through which governmental power operates, affecting our thought each time we see communication design artifacts, and our behavior each time we interact with them. The net effect of this repeated exposure through countless daily interactions is an accumulation of truth effects, experienced in every virtual environment, and most physical ones. From the most austere, sacred spaces to the most cluttered and profane, we interact with countless communication design artifacts, and experience their truth effects, usually without even noticing them. These truth effects further the perception that the prevailing attitudes in our societies are natural, or common sense. Like countless mirrors reflecting and collectively exaggerating the dominant attitudes of our time, communication design artifacts extend governmental power into even the most mundane and seemingly innocuous situations and interactions.

References

Ambrose, G., & Harris, P. (2011). *The fundamentals of creative design* (2nd ed.). Lausanne: AVA Publishing.

Ambrose, G., & Harris, P. (2015). *Design thinking for visual communication*. London: Bloomsbury.

Anderson, B. (2006). *Imagined communities: Reflections on the origin and spread of nationalism, new edition*. New York: Verso.

Barton, B. F., & Barton, M. S. (1993). Modes of power in technical and professional visuals. *Journal of Business and Technical Communication, 7*(1), 138–162. https://doi.org/10.1177/1050651993007001007.

Blair, J. A. (2004). The rhetoric of visual arguments. In C. A. Hill & M. H. Helmers (Eds.), *Defining visual Rhetorics* (pp. 41–62). London: Routledge.

Blauvelt, A. (1994). An opening: Graphic Design's discursive spaces. *Visible. Language, 28*(3), 205–216.

Bowers, J. (1999). *Introduction to two-dimensional design: Understanding form and function*. Hoboken: John Wiley & Sons.

Bush, A. (1994). Through the looking glass: Territories of Historiographic gaze. *Visible Language, 28*(3), 219–231.

Carnegie, T. (2013). Design as problem-solving. In E. R. Brumberger & K. M. Northcut (Eds.), *Designing texts: Teaching visual communication*. Amityville: Baywood Publishing Company.

Colberg, S. (2006). Reading minds: The book as a communicational space. In J. Frascara (Ed.), *Designing effective communications: Creating contexts for clarity and meaning* (pp. 229–234). New York: Allworth Press.

Cooter, R., & Stein, C. (2007). Coming into focus: Posters, power, and visual culture in the history of medicine / die Macht des Plakats: Visuelle Kultur in der Medizingeschichte. *Medizinhistorisches Journal, 42*(2), 180–209.

Davidson, J. (2008). Towards an Anarcho-design practise. *Imminent Rebellion: An irregular anarchist journal from deep in the South Pacific, 9*, 85–90.

Dean, M. (2010). *Governmentality: Power and rule in modern society* (2nd ed.). Los Angeles: Sage.

Dean, M. (2015). Foucault Must Not Be Defended. *History and Theory, 54*(3), 389–403. https://doi.org/10.1111/hith.10767.

Eskilon, L. (2007). *Graphic design: A new history*. London: Laurence King Publishing Ltd..

Fallan, K. (2010). *Design history: Understanding theory and method*. Oxford: Berg.

Forlizzi, J., & Lebbon, C. (2002). From formalism to social significance in communication design. *Design Issues, 18*(4), 3–13.

Foucault, M. (1967/1998). On the ways of writing history. In J. D. Faubion (Ed.), *Essential works of Foucault 1954–1984 Volume 2: Aesthetics* (pp. 279–295). London: Penguin.

Foucault, M. (1968/1991). Politics and the study of discourse. In G. Burchell & C. Gordon (Eds.), *The Foucault effect: Studies in Governmentality, with two lectures by and an interview with Michel Foucault* (pp. 53–72). Chicago: University of Chicago Press.

Foucault, M. (1971/1997). On popular justice: A discussion with Maoists. In P. Rabinow (Ed.), *Essential works of Foucault 1954–1984 Volume 3: Power* (pp. 1–36). New York: The New Press.

Foucault, M. (1972). *The archaeology of knowledge*. New York: Pantheon Books.

Foucault, M. (1976). *The history of sexuality*. London: Penguin.

Foucault, M. (1977a). Truth and power. In C. Gordon (Ed.), *Power/knowledge: Selected interviews and other writings, 1972–1977* (pp. 109–133). New York: Pantheon Books.

Foucault, M. (1977b). The eye of power. In *The impossible prison: A Foucault reader* (pp. 9–15). Nottingham: Nottingham Contemporary.

Foucault, M. (1977c). *Discipline and punish: The birth of the prison*. London: Allen Lane.

Foucault, M. (1978a). Governmentality. In *Essential works of Foucault 1954–1984 Volume 3: Power* (pp. 201–222). New York: The New Press.
Foucault, M. (1978b). *The history of sexuality*. London: Penguin.
Foucault, M. (1980a). In C. Gordon (Ed.), *Power/Knowledge: Selected Interviews and Other Writings, 1972–1977*. New York: Pantheon Books.
Foucault, M. (1980b). *The history of sexuality, Volume I: An introduction*. New York: Vintage Books.
Foucault, M. (1981a). The order of discourse. In *Untying the text: A post-Structuralist reader* (pp. 51–78). London: Routledge.
Foucault, M. (1981b). Two lectures. In C. Gordon (Ed.), *Power/knowledge: Selected interviews and other writings, 1972–1977* (pp. 78–108). New York: Pantheon Books.
Foucault, M. (1982a). The subject and power. *Critical Inquiry, 8*(4), 777–795.
Foucault, M. (1982b). Technologies of the Self. In P. Rabinow (Ed.), *Essential works of Foucault 1954–1984 volume 1: Ethics, subjectivity and truth* (pp. 223–251). New York: The New Press.
Foucault, M. (1997a). In P. Rabinow (Ed.), *Essential Works of Foucault 1954–1984 Volume 3: Power*. New York: The New Press.
Foucault, M. (1997b). In P. Rabinow (Ed.), *Essential Works of Foucault 1954–1984 Volume 1: Ethics, Subjectivity and Truth*. New York: The New Press.
Frascara, J. (2004). *Communication design: Principles, methods, and practice*. Allworth Press.
Frascara, J., Meurer, B., van Toorn, J., & Winkler, D. (1997). *User-centred graphic design: Mass communication and social change*. Boca Raton: CRC Press.
Gallagher, V. J., Martin, K. N., & Ma, M. (2011). Visual wellbeing: Intersections of rhetorical theory and design. *Design Issues, 27*(2), 27–40.
George, J. L. (2002). The Functions of Graphic Design: Sociologies, History, and the International Design Conference in Aspen (Ph.D.). State University of New York at Binghamton, New York.
Goodall, P. (1983/1996). Design and Gender. In J. Bird, B. Curtis, M. Mash, T. Putnam, G. Robertson, L. Tickner, & S. Stafford (Eds.), *The block reader in visual culture* (pp. 187–208). London: Routledge.
Goodall, P. (1990). Design and gender: Where is the heart of the home? *Built. Environment, 16*(4), 269–278.
Hall, S. (1996). Introduction. In S. Hall, D. Held, D. Hubert, & K. Thompson (Eds.), *Modernity: An introduction to modern societies* (pp. 1–19). London: Blackwell.
Hayward, S. (1998). "Good design is largely a matter of common sense": Questioning the meaning and ownership of a twentieth-century orthodoxy. *Journal of Design History, 11*(3), 217–233.
Heller, S. (2008). *Iron fists: Branding the twentieth century totalitarian state*. London: Phaidon Press.
Hepworth, K. (2012). *Government emblems, embodied discourse and ideology: An artefact-led history of governance in Victoria, Australia*. Melbourne: Swinburne University of Technology.
Hepworth, K. (2014). History, power and visual communication artefacts. *Rethinking History, 0*(0), 1–22. https://doi.org/10.1080/13642529.2014.932079.
Hodder, I. (1992). *Theory and practice in archaeology*. London: Routledge.
Julier, G. (2008). *The culture of design*. London: Sage.
Keedy, J. (1995). Zombie Modernism. In R. Vanderlans (Ed.), *Emigre: The look back issue* (pp. 171–175). Berkeley: Ginko Press.
Kostelnick, C. (2004). Melting-pot ideology, modernist aesthetics, and the emergence of graphical conventions: The statistical atlases of the United States, 1874–1925. In *Defining Visual Rhetorics*. Mahwah: Lawrence Erlbaum Associates..
Kostelnick, C., & Hassett, M. (2003). *Shaping information: The rhetoric of visual conventions*. Carbondale: SIU Press.
Lavin, M. (2001). *Clean new world: Culture, politics, and graphic design*. Cambridge: MIT Press.
Lupton, E. (1993). *Mechanical brides: Women and machines from home to office*. New York: Cooper-Hewitt National Museum of Design.

Lupton, E., & Phillips, J. C. (2011). *Graphic design thinking* (1st ed.). New York: Princeton Architectural Press.
McCoy, K. (2003). Good citizenship: Design as a social and political force. In S. Heller & V. Vienne (Eds.), *Citizen designer: Perspectives on design responsibility* (pp. 2–8). New York: Allworth Press.
McCoy, K. (2005). Maximize the message: Tailoring designs for your audience in a multicultural era. In S. Heller (Ed.), *The education of a graphic designer* (pp. 279–283). New York: Allworth Press.
Meggs, P. B. (1989). *Type and image: The language of graphic design*. New York: Van Nostrand Reinhold.
Meggs, P. B., & Purvis, A. W. (2006). *Meggs' History of Graphic Design (Fourth.)*. Hoboken: John Wiley & Sons.
Milestone, J. (2007). Design as power: Paul Virilio and the Governmentality of design expertise. *Culture, Theory and Critique, 48*(2), 175–198.
Poynor, R., Crowley, D., & Gallery, B. A. (2004). *Communicate: Independent British graphic design since the sixties*. New Haven: Yale University Press.
Rancière, J. (2004). *The politics of aesthetics: The distribution of the sensible*. London: Continuum.
Reiner, M. (2009). Sensory cues, visualization and physics learning. *International Journal of Science Education, 31*(3), 343–363.
Riles, A. (2006). *Documents: Artifacts of modern knowledge*. Ann Arbor: University of Michigan Press.
Rose, N. (1996). Governing "advanced" liberal democracies. In A. Barry, T. Osborne, & N. Rose (Eds.), *Foucault and political reason: Liberalism, neo-liberalism and rationalities of government* (pp. 37–64). London: Routledge.
Rose, N. (2008). *Powers of freedom: Reframing political thought*. Cambridge: Cambridge University Press.
Rose, N., O'Malley, P., & Valverde, M. (2006). Governmentality. *Annual Review of Law and Social Science, 2*(1), 83–104. https://doi.org/10.1146/annurev.lawsocsci.2.081805.105900.
Sanders, E. (2006). Scaffolds for building everyday creativity. In J. Frascara (Ed.), *Designing effective communications: Creating contexts for clarity and meaning* (pp. 65–77). New York: Allworth Press.
Sanders, L., & Stappers, P. J. (2012). *Convivial toolbox: Generative research for the front end of design*. Amsterdam: BIS Publishers.
Sartwell, C. (2005). Aesthetics of the everyday. In J. Levinson (Ed.), *The Oxford handbook of aesthetics* (pp. 761–770). Oxford: Oxford University Press.
Sartwell, C. (2010). *Political Aesthetics. Ithaca*. Cornell University Press.
Seago, A., & Dunne, A. (1999). New methodologies in art and design research: The object as discourse. *Design Issues, 15*(2), 11–17. https://doi.org/10.2307/1511838.
Simmons, S. (2000). "Hand to the friend, fist to the foe": The struggle of signs in the Weimar Republic. *Journal of Design History, 13*(4), 319–339. https://doi.org/10.1093/jdh/13.4.319.
Soar, M. A. (2002). Graphic Design/Graphic Dissent: Towards a Cultural Economy of an Insular Profession (Ph.D.). University of Massachusetts Amherst, United States – Massachusetts.
Sontag, S. (1970/1999). Posters: Advertisement, art, political Artifact, commodity. In M. Beirut (Ed.), *Looking closer three* (pp. 197–218). New York: Allworth Press.
Stanley, E. H. (1989). The lively poster arts of Rockwell Kent. *The Journal of Decorative and Propaganda Arts, 12*, 6–31. https://doi.org/10.2307/1504054.
Stoler, A. L. (1995). *Race and the Education of Desire: Foucault's History of Sexuality and the colonial order of things*. Durham: Duke University Press.
Swales, J. (1990). *Genre analysis: English in academic and research settings*. Cambridge: Cambridge University Press.
Taffe, S. (2015). The hybrid designer/end-user: Revealing paradoxes in co-design. *Design Studies, 40*, 39–59. https://doi.org/10.1016/j.destud.2015.06.003.

Tunstall, D. (2007). In Design we trust: Design, Governmentality and the tangibility of governance. In Proceedings of the International Association Of Societies Of Design Research, Emerging Trends in Design Research, Hong Kong, 12-15 November 2007.

Warde, B. (1930/2009). The crystal goblet, or why type should be invisible. In H. Armstrong (Ed.), *Graphic design theory: Readings from the field* (pp. 39–44). New York: Princeton Architectural Press.

Witkin, R. (1990). The aesthetic imperative of a rational-technical machinery: A study in organizational control through the design of artifacts. In *Symbols and artifacts: Views of the corporate landscape* (pp. 169–184). New York: Walter de Gruyter.

Young, A. S. (2009). Design as discourse: A Foucauldian approach. In F. Hackney, J. Glynne, & V. I. V. Minton (Eds.), Networks of design: Proceedings of the 2008 Annual International Conference of the design history society (UK) University College Falmouth, 3–6 September (pp. 124–129). Boca Raton: Universal-Publishers.

The Black Book: Emilio Ambasz's University of Design

Matthew Holt

Abstract In 1972 a two-day symposium entitled "Institutions for a Post-Technological Society—The Universitas Project" was held at MoMA and organised by its then curator of design, Emilio Ambasz. Its vision was nothing less than to establish "a new type of University concerned with the evaluation and design of our man-made milieu." Three years in the making Ambasz managed to draw in an incredible list of attendees and participants. A number of these were picked out to respond to a provocation written by Ambasz and forwarded to them as a booklet—it was to become known as the "black book." This chapter will closely examine both the provocation and the responses to it to identify the debate over a nascent philosophy of the artificial and what this may mean for an idea of a new form or approach to knowledge, and therefore of the university. Perhaps it is only now that what was seeded by Ambasz has come to fruition—the possibility of a university of design.

Keywords Emilio Ambasz · The Universitas project · Environmental design · Philosophy of the artificial

1 Introduction

Ever the ugly sister of art history or the wayward brother of engineering, the study of design has struggled to find its own disciplinary certitude. What is the exact nature of the objects it examines? What are the limits of its discursive territory, the extent and breadth of its terminology, the range of its language, the rhythms of its idiom? Does it have a discrete methodology? And perhaps the most anxious question of all: What new knowledge and understanding can it genuinely contribute given the overabundance of relatively new and related disciplines, most particularly science and technology studies (STS) and material culture studies? In other words, can it justify itself?

M. Holt (✉)
Design and Architecture, University of Technology Sydney Insearch, Sydney, Australia
e-mail: Matthew.Holt@insearch.edu.au

© Springer International Publishing AG 2018
P. E. Vermaas, S. Vial (eds.), *Advancements in the Philosophy of Design*, Design Research Foundations, https://doi.org/10.1007/978-3-319-73302-9_24

What follows is not an attempt to provide such a justification. Rather, it endeavours to invert the order and logic of the questions and the perceptions that might demand such a justification. Instead of envisaging design studies as a new field requiring its own methodological, discursive and technical autonomy, this chapter contends that it is other disciplines that need to understand the fact that the world they analyse, the world they attempt to give figure to, and the world in which they attempt to provide meaning for, is a designed world. Whatever one makes of the term "Anthropocene," it does accurately signify a new stage in our relation to the environment, and moreover forcefully implies the designed nature of that environment. The challenges we now face require a new or renewed design philosophy, but also a new or renewed concept of design education.

To that end, I will examine a critical but much understudied event in the redefinition of not only what design is but how it can be considered a distinct form of knowledge, and therefore demanding a distinct institution to explore and teach it: a symposium held in New York in 1972 at the Museum of Modern Art entitled "Institutions for a Post-Technological Society—The Universitas Project." It was organised by the Argentinian architect and product designer, Emilio Ambasz, who was at that time the Curator of Design at MoMA. Ambasz was educated at Princeton where he was also appointed Assistant Professor in 1966. Along with Arthur Drexler he assisted in the foundation of the Institute for Architecture and Urban Studies (1967–1984), which was formed with support from MoMA (Frank 2011). The aim of the Project according to Ambasz was to establish an "experimental type of institution centred around the task of evaluating and designing the man-made milieu" (Ambasz 2006, p. 299). Architectural historian Felicity Scott (2004, 2010) has provided an admirable overview of the context, background and main intellectual topics of the symposium—she also had access to the original documents before they were collated for publication in 2006—but here I would like to focus more specifically on the relation between an emergent philosophy of environmental design the symposium gravitated towards and the proposed university to communicate it. As we shall see "environment" for the organizer and the participants meant the complete environment, incorporating architecture and building, but also systems, information design and planning, as opposed to a focus on singular products or artefacts, or indeed living or biological systems. Nonetheless there was also awareness and discussion among the symposium delegates of the ecological concerns of the day. Above all the symposium heralded an awareness that design knowledge—and therefore design education—is now the most essential form of knowledge, and not a subsidiary or adjunct to other disciplines. The event heralded a "post-technological" and "post-scientific" future for the university.

The symposium was a remarkable gathering of leading experts and intellectuals of diverse fields: cybernetics, systems, planning and organizational theory (Erich Jantsch, Richard L. Meier, Anatol Rapoport), semiotics (Umberto Eco, Thomas Sebeok), sociology (Suzanne Keller), Marxist urban sociology and postindustrial political economy (Jean Baudrillard, Manuel Castells, Henri Lefebvre and Alain Touraine), political theory and philosophy (Hannah Arendt, Ronald Dworkin, Sheldon Wolin), artists, designers, architects and curators (György Kepes, Octavio

Paz, Hans Magnus Enzensberger [who submitted a presentation but did not go to the conference], Gillo Dorfles, Denise Scott Brown, Martin Pawley, Arthur Drexler and Meyer Shapiro). Those who could not attend apparently included Louis Althusser, Roland Barthes, Michel Foucault and Roman Jakobsen (Scott 2004, 2010). Selected individuals were invited to respond to a project working paper in which Ambasz set out the ideas and expectations of the event. The paper was originally delivered bound to these invitees between black covers and thus known as the "black book."

In the book the symposium organizer marks out what he considered to be the new dimensions and characteristics of the designed, artificial world and the possible forms of the academy appropriate to conceptualizing it. The "artificial era" is in the course of replacing the Renaissance world-view. In terms of knowledge, the Renaissance was devoted to a scientific exploration of nature and had in its turn supplanted the medieval, divine milieu. Ambasz believed the pedagogy appropriate to the artificial world could not be established on the scientific model of knowledge because that model is predicated on the discovery of preexistent truths embedded in nature and waiting revelation. Scientific thinking examines what is measurable and quantifiable, while design concerns itself with what "ought to be." In studying the emergent and the embryonic, design and design education require conceptual resources suited to the problematic of dynamic "adaptive systems" not static forms. The goal of design is therefore ultimately one of synthesis not analysis, and the "truths" of design, rather than being theoretically independent of the observer as in scientific ideal are, instead, "dependent on the values of its participants." Thus both conceptually and physically the new university is to be fundamentally different to its predecessors. It is no longer a fixed college but an active, participatory "network." Ambasz understands network to designate a goal-orientated feedback system linking individuals and groups with society at large, as appropriate, he argues, to the "second" technological revolution that creates and controls information rather than "just energy."

The actual details of Ambasz's vision of the Universitas are imprecise, but of course that is intentionally so—the symposium was to debate its possible form and scope. To preface our reading of the project working paper what is most striking about Ambasz's approach to design philosophy and design education is his quasi-cybernetic understanding of both social organization (including the university) and the environment. Physical systems (including "nature"), human artefacts and human-to-human interaction (society) are seen as a single self-regulating unit; a circular system to be sure, but one which learns and adapts as it runs through the cycles of communication (feedback). The capacity to learn from environmental information (rather than just maintain a goal), however, places Ambasz's idea of the Universitas at the level of second-order cybernetics, which Klaus Krippendorff (2007, p. 1383) has defined as the "*cybernetics of participating in systems under continuous reconstruction by their constituents*" or, as Paul Pangaro (n.d.) says, second-order cybernetics "applies the cybernetic frame to the process of cybernetics itself." Of course, such an understanding of systems was by no means unique to Ambasz, and the relation to cybernetics—and/or information theory—to design was already well appreciated (Pask 1963). Also common to the period was a deep and abiding interest from many intellectual fields in similarly conceptualizing self-organizing networks, for example in systems theory, organizational and planning theory, and particularly on the Continent in structuralism itself, for example Claude Lévi-Strauss and anthropology, Christian Metz in regard to film studies, Jacques Lacan and psychoanalysis and most relevant for our context, Roland Barthes on the fashion system (1985) and Jean Baudrillard on "objects" and consumption (1996, 1998). But what Ambasz is attempting to construct here is an intellectual and pedagogical language specific to design, and to conceptualize an institution where additional forms of knowledge and expertise can accrete around the term design in order to arrive at a superior understanding of the artificial environment and how to intervene in it.

The themes, scope, and interdisciplinary nature of the symposium were not only remarkable in ambition but also prescient. The questions raised and debated, especially regarding what may be called a "post-scientific" education relevant to planning and living in an artificial world, have all gained in significance. It is here, at the confluence of thinking between philosophy, planning, sociology, organizational modelling and communications that the theory and pedagogy of design can be sought. In itself this confluence is not necessarily new, and the reader may rightly recognise a common discursive thread that connects the systemic and communicational approach to design belonging to aspects of the Bauhaus through the Hochschule für Gestaltung, Ulm, to similar methodologies on both the Continent and the United States including the design methods and design research movements, contemporary developments in informatics, computing, decision and game theory, and in ecological or environmental thinking (Anker 2010; Martin 2004). But in many respects, especially pedagogically, we have lost the aspiration represented by Ambasz's symposium, or at least lost the sense that these domains can and should be articulated on terms established by design. In the vacuum has stepped "design thinking" and to some degree service design, but both, especially the former, lack rigorous theoretical let alone pedagogical principles. Design thinking has certainly attempted to generalize a version of the design process for predominately business interests and its benefits are its centralization of the user-centred and problem-solving aspects of design. Nonetheless its methodology is a basic step-by-step and quickly exhausted sequence (although see Dorst 2015). Service design, on the other hand, is slowly evolving a specific language and domain of action (see Stickdorn and Schneider 2010). Nonetheless, the environmental concept of design they both represent, however partially, is more significant for contemporary design studies (or a design philosophy) than basic artefact design which is still theoretically and spiritually located in industrial society and the economy of production.

The other reason why developing a design philosophy is so crucial is political. Critical utopian philosophy, for example "Western Marxism," has, despite decades of dissemination through the academy, limited effect on the society it sought to alter. In the so-called first world at least, anything like socialism or a mere alternative to capitalism has had little or no impact. I would argue that, in part, this is due to a lack of understanding of design. Instead of seeing design as the production of commodities, it needs to be understood as inventing, creating and sustaining artificial environments. In the socialist tradition, whether the Arts and Crafts movement or the Bauhaus and beyond, design has always been taken to be an intervention in the shaping of artefacts for industry or refining the manner of their consumption. William Morris, for example, believed properly designed artefacts ameliorate and improve existing conditions, but would not have considered design as counterfactual communication or as planning ephemeral and intangible information and communication services and systems. In a very real sense he could not have. That informational characteristic of political economy (post-handicraft, post-product, even post-object)—the "political economy of the sign" as Jean Baudrillard, a contributor and participant of the 1972 conference, calls it—only effectively transpires in postindustrial societies. In fact it could be argued that design in this sense—"metadesign" (Busbea 2009)—has only come into existence in the mid-Twentieth Century. Design understood in its diverse meanings and above all not restricted to artefact design is a contemporary phenomenon. If alternatives to capitalism—or alternative forms of capitalism—are to be considered, let alone embraced, an acknowledgement and understanding of this environmental sense of design is essential.

2 The Project Working Paper: Inside the Black Book

The project working paper proper is divided into three sections and was originally published in 1971 in the Yale journal *Perspecta* as "I: The University of Design and Development. II: Manhattan: Capital of the Twentieth Century. III: The Designs of Freedom" (Ambasz 1971; an earlier version of the texts made an appearance in the Italian architecture and interior design magazine *Casabella*, Scott 2010, p. 299 n.15). The second part clearly recalls Walter Benjamin's "Paris, Capital of the 19th Century," which was published in an earlier volume of *Perspecta* (1969), alongside another of Ambasz's articles, "Formulation of a Design Discourse." The third section was only to provide context to the working paper but is by far the longest and is, he admits, using the structural anthropologist's term, a "bricolage" of related thoughts. Indeed the black book in its entirety is a montage of ideas and speculation mixed with cogently and logically articulated arguments. As perhaps befits the notion of a design discourse and the pedagogies required to voice it, it is both a poetic and a systematic text.

Ambasz argues that the era of the artificial does not align with a technological world-view, in particular the idea that technology itself can solve the dilemmas and problems it creates. Ambasz maintains that recognizing the artificial essence of contemporary society is neither technological determinism nor technological utopianism. Therefore what Ambasz identifies with a term borrowed and adapted from Jacques Ellul (1954) as the "post-technological" is, in fact, the name of a strategy emphasizing actively informed choice for shaping and "giving meaning" to the future of the city, that is, the design of the future itself. The MoMA curator claims that science is deterministic because it searches for the fundamental atoms, as it were, from which all else is composed. It therefore deals with what is given. For this reason science can operate independently of ethics: understanding what ought to be is not necessary to understanding what is already in place. Design, on the other hand, does not "yield before the facts" because it is primarily a discipline of creation and a discipline inherently concerned with values—what should be created, and why? Ethics cannot be bracketed in the design process or set aside for discussion at a future date. It is in the frame at all times. This ethical dimension inherent to design also means that design is inseparable from a community that authorises its own creations. Designing is fundamentally participatory: the "designer, unlike the scientist, should allow his [sic] view of things as they are to be affected by what he would like them to be" (2006, pp. 21–22). Science is analysis (the breaking down of a phenomenon into fundamental units to be understood), while design is synthesis (the building or assembling of a phenomenon). While not doubting the value of scientific culture Ambasz argues the synthetic nature of design means turning design into a science should be resisted.

Furthermore a designed, artificial milieu for Ambasz signifies the "dialectical interrelations between man-made physical and sociocultural systems," and so any design discourse or mode of thought must, he contends, contain both an "empirical" and "existential" dimension. The empirical is defined as the changing patterns of interaction in the artificial environment and the capacity to act on that environment according to "acceptable" values. The existential deals with the intuition and generation of symbols. The former is the physiological and logical and the latter the ontological and symbolic (p. 23).

Once this field has been staked out Ambasz goes on to ask what would be the foremost issue or problem in the man-made milieu that research should be directed to and consequently whether current institutions are adequately set-up for solutions to that essential problem.

The foremost challenge design discourse should devote itself to is, naturally enough, "urban phenomenon." But Ambasz quickly qualifies this statement: the attempt to create a language appropriate to the entirely artificial milieu must carry with it an explicit critique of past models of urban existence under the conditions of industrial society (which still recognized a fundamental difference between the natural and the created environment). The basic postulate of any design discourse must now be that the natural and the artificial are no longer, at any level, separate.

To the question whether contemporary knowledge is up to the task of formulating this new language of the urban environment—"Are universities presently structured to assume urban commitments?"—Ambasz answers with a firm no. The primary reason is that the current distribution of knowledge in the University remains stubbornly compartmentalized, negating the interdisciplinarity needed to articulate the profile and problems of the complex urban systems we inhabit. Nonetheless Ambasz sees the university of design not as superseding former models of the university—either the "humanistic" (the non-specialized Greek and Middle Ages model) or the "scientific" (the specialized, more empirical von Humboldt model)—but as incorporating and integrating them into, explicitly using a term from Walter Benjamin, a "constellation": "there would thus be, for example, social philosophers in the humanistic university, social scientists in the scientific university, and social designers in the university of design and development" (p. 25). It should be noted here that there is no attempt on the part of Ambasz to explain any of these forms of the university by the socio-economic forces that subtend them, something that Lefebvre, Baudrillard, Enzensberger, and Sheldon S. Wolin in particular will take note of in their responses. It is a more idealistic project than a realist one (as Ambasz himself admits, in fact he calls it a "surrealist" concept, p. 396). Nonetheless, in summary, the university of design is to be interdisciplinary, synthetic in its approach to knowledge formation, not analytic; an adaptive, participatory network (informational, communicational) and above all devoted to the primary problem of the artificial environment—urban space, urban life.

The second section on Manhattan displays in a poetic fashion what Ambasz means by an environmental or ecological understanding of an artificial milieu and instinctively articulates the contemporary interest in informational environs. Manhattan, he writes, is "in essence, a network." An informational dimension maps the physical dimension of the city. For Ambasz, Manhattan is the capital of the Twentieth Century not because it is the apotheosis of modernity, but because it is the city of the "second technological revolution," the revolution of information and communication. Implicitly invoking the work of radical Italian design groups such as Superstudio and Archizoom he was curating for an exhibition at MoMA (Ambasz 1972), Ambasz declares that running through, over and above the existent material structures of steel, concrete, wires and sewerage is an entire "datum plane," an "aerial lattice" of "walking paths, automobile routes, flight patterns, wireless impulses, institutional liaisons, and ideological webs." (p. 27) There is an idea emerging here of the poetry of data. The designed environment is not just a technical, material phenomenon. It is also over-coded and mapped by an informational system that allows for its logical and creative navigation. Any artificial environment is at once physical and spatial, visual and informational. This mapping, however, needs to be qualified. Sometimes the information ecology is not entirely synchronized with the space it is intended to chart. This can lead to a) failed design and/or b) new possibilities, that is, creative slippages. Suffice to say for the moment that design education encourages the student to find in the asymmetries and inconsistencies of informational mapping precisely the creative solutions to environmental design. Ambasz argues that in fact the relative distinction between the bricks and mortar Manhattan and the informational Manhattan means it might be possible to remove this virtual Manhattan and transfer it to other cities—to transfer what he calls its "infrastructure." While he readily admits such infrastructure does not make a city, it can nonetheless be "liberated." In principle informational infrastructure belongs to all. There is an interesting coincidence here between the virtuality of information and politics that, while only loosely hinted at by the text, nonetheless ghosts the entire set of propositions as its own infrastructure, as it were: that the liberal, democratic idea of polity is equivalent in essence to a networked, cybernetic and informational environment. Designed environments contain feedback and therefore are inherently—infrastructurally—communicational and so will integrate and adapt to changes in circumstances, intelligence and actors. Rather than just maintaining a goal, as in a simple first-order cybernetic system, Ambasz will later claim that the goal itself can be modified (p. 37f). This is the value-laden, political and participatory dimension of design and of the Universitas. It is the element of "freedom" in a bound (or networked) system.

Infrastructure is a thought-provoking term to which we will return, a word that really only came into use in the 1950s. For Ambasz, it is not only the material element of a structure but also the informational, data-orientated, and statistical elements that accompanies it. Put differently, it prefigures and supplements the notion of a completed design explicit in "structure" and implicit in "superstructure." In a loose Benjaminian fashion Ambasz describes the portable infrastructure of Manhattan as the framework upon which memories can be hung, imagination explored, and on which intangible experiences can momentarily gel by the "grace of their affinities." This is, as already indicated, poetic discourse to some extent. But it also points to or provokes a path of research to this day undervalued and under-pursued by design education: the metaphoric, atmospheric, quasi-informational, and aesthetic element of artificial urban-scapes. (In other words the element of turbulence and complexity in human and artificial environs: see Carter 2015.) Individual experience of the city is always fragmented, partial and often ambiguous. It is a mosaic of various forms of data-access and data visualisation. While on the contrary, the authorized image of collective experience tends to be rational and abstract: official maps and signs, systems of occupation, prohibition and multiple legal sanctions distributed in spatial form. Experience of the city, as we know, is a continuous montage of both sets of "charting."

Ambasz himself proposes two methods to such research (which echoes the plans for the symposium itself): the "retrospective" approach and the "prospective." The first is a collection or indeed archive of memories of cities as disclosed by authors and artists, again empathizing with the evocative rather than the denotative: from Baudelaire's "fleeting instants" through Debussy's "submerged cathedral" (Ambasz is referring to the piece *La cathédrale engloutie,* prelude for solo piano, 1910) to Frank Zappa's Los Angeles (p. 28). This is again a thoroughly Benjaminian approach to understanding the spatial dimension of an urban environment as it integrates sensation and recollection: the poet discovers within that environment archetypal images that concentrate experiences of place (and to a degree time) to which the pragmatic use of space owes reference. What is discovered is not strictly an alternative space (aesthetic space is not completely dissociated with the utilitarian) but a ghosting or doubling of space—any space is at once its use, its memory, and its potential dis-placement. Poetry does not simply represent a space but configures it (often in the form of a cypher or a code to be decrypted). It locks space and unlocks space; or it adds another dimension or layer to it. It imparts atmosphere.

The prospective phase is the process of building any structure upon the more ambiguous and quasi-material "infrastructure." To this end Ambasz proposes that architectural—and "ceremonial" he says—forms can be understood as "icons," a fusion of the individual designer's partial experience and the inherited, assembled forms of society. For Ambasz the "iconic," a term he borrows creatively from Charles Sanders Peirce (1931–36), is a sign that qualitatively communicates some property of what it represents, and can be recognized as such by its interpretant (an illustration or a model for example). For Ambasz this recognition is a shared undertaking between designer and community. The designer produces and offers icons as a prototype of new values. In itself this says little more than that a designer designs something in context and if it is accepted by the community it potentially "expands" the ethical framework of that community. While a basic understanding of iconic operation, we can accept the point that configuring value in society is intimately tied to configuring structures (or infrastructures). Form and value cannot be separated: every prototype is an argument, every construction potentially a new direction for a community. This essential contiguity between form, rhetoric, information and exemplar provides the foundation of both design education and what it leads to: the Universitas is to become a "Univercity."

The next part "Design as a Mode of Thought" is appended to provide a context for the black book proper, but in fact forms the largest and most detailed section. Ambasz begins by again emphasizing that as a mode of thought science is analytical while design must factor in the values, expectations and desires that accompany the justification and purpose of any constructed environment. Put differently it may be said that the discourse of design is concerned with the language of change—"what if" and not just "what is."

Nonetheless, Ambasz points out that many believe a scientific approach to design to be the most appropriate methodology for understanding design and for teaching it. There are two forms this has taken: the "empirical" and the "normative" (p. 30). For the empiricists, design is basically engineering. Engineering, Ambasz admits, is not strictly a science because it is also concerned with the construction of something with use and purpose and therefore engages with what "ought" to be. Nonetheless this ought is defined exclusively in terms of the finite goal of the exercise (a structure's feasibility, its fabrication, and its ongoing capacity to function as intended), and not what the actual goal should be in the first place. Purportedly design has an ethical level that meditates upon goals, not just pursues them. Without this level, design can indeed be "systematized into scientific discipline" and absorbed easily into engineering. (It can also be added here that design unlike engineering can produce and orchestrate ephemeral structures: services, events and experiences.) Whether design in fact has anything like this freedom—particularly this freedom to engage independently in the question of value prior or simultaneous to an actual brief or contract is debatable, but as we will argue later this reflection on value is contingent upon design offering a counter-vision or counter-discourse to what is, that is, to the status quo. The aesthetic impulse of design is also its ethical impulse. The "freedom" of research is also the freedom to explore the question of value. What, perhaps, defines the goals and curriculum of the university of design is not so much the economic realities of design as the pedagogy of the counterfactual. This critical role was once the domain of the humanities. But what the humanities had in excess, the intricacy and precision of discourse, it lacked in the techniques of speculation and indeed the theoretical, physical and digital tools to produce speculation: scenarios and environments, counter-scenarios and counter-environments.

Design for the young symposium organizer is to be considered more profoundly as a shared meaning-making activity (the "creation of structures") arising from an understanding of the technological world and the capacity to make judgements about its direction and its constitution. Design statements—normative design statements—are therefore not "independent of the observer." They are not neutral, nor are they scientifically "objective." As such it would be wrong to simply confer on the "processes of technology the status of a force of nature." Because design is a synthetic form of knowledge and reflects on the goals to be set by an understanding of artificial, technological society, it must deal—as a matter of course and as definitional—with "alternative futures" (p. 31). Thus while emphasizing the discussion of values is essential to design and sets it apart from other empirical activities, Ambasz points out that this does not merely mean making or ideating under the auspices of commonly entrenched values—values already in place and already accepted. Far from it. He sees the task of design as embodying and thereby proposing new values through form and through "structure." This is the "*symbolic* charge" of design. Design is, literally, a concrete argument.

Ambasz then goes on to align his notion of the iconic substance with the claims he is making regarding the "mode of thought" appropriate for design, and therefore for design education. This alignment comes together through an interpretation of the terms infrastructure, superstructure and structure. The first is the empirical level that the designer organizes, and consists of both the material and the immaterial (information). The second, superstructure, is the level of the normative; it is the arena in which the designer deals with customs, values, and ideas. The third synthesizes in a dialectical fashion the other levels into an "icon" or interface. Though he is not particularly clear about this, it is imagined he means by structure an actual artefact or an environment.

The final part of this section ("The Designs of Freedom") sketches out in more detail the mission of the Universitas and the manner in which it will operate. The goals of the Universitas are to: (1) Actively participate in the creation of the man-made milieu; (2) Actively examine the nature of the social contract, insofar as that contract is now established on the necessity to continuously design and manage the artificial environment, and (3) Mediate between the processes of the infrastructural level (materials and information) and the aims of the superstructure (values and norms).

The model of how these goals are pursued, adapted, and redirected is a thoroughly cybernetic one, intersecting with systems theory, organizational theory, and planning theory; it is what Ambasz calls a "network model." If the Universitas is conceived as a network and comprised of the various levels and functions detailed above, then it is essentially a goal-orientated, communicative entity. It receives feedback from its constituents and its existent programs and projects according to a set goal. That goal in turn can be altered by the feedback, hence the Universitas, even though Ambasz does not say this directly (he says it is an "open" system, p. 38), is a second order cybernetic regime, that is, the goal of the system can be altered by the system itself. For Ambasz, it is of vital importance that the input the individual supplies to the group network at any level of the Universitas is heard by the "code," that is, the programmed goal of the network. In turn that input is "decoded" and fed back to the larger project. This process of constant recognition apparently leads to a "self-modifying ethical system." Again, there is (until the very last page of the black book) no direct politicization of this cybernetic vision of intra- and infra- structural communication, but it is clearly a model of a relatively democratic institution, at least in the liberal sense: the individual's needs and contributions are actualized in concert with the group without being absorbed by it. In the network (the group), the node (individual) still has "independence."

To further describe this feedback mechanism and communicational environment, Ambasz adds to the interaction between the two structural elements—infra- and super-structure—another four "functions" necessary for any group network: (1) The informative, which is essentially a hermeneutic function, decoding previous structures and analyzing them; (2) the postulative, which is the design and/or production of new structures and new codes (this, in general, is the domain of individuals—of "artists, social thinkers, poets, and designers"); (3) decision-making (which is group-based); and lastly (4) the regulatory, which is the implementation process.

But according to Ambasz the university of design is to remain devoted to the first two levels (or networks) of the informational and the postulative. The others, decision-making and regulation, will be left to ancillary group networks, especially the regulatory as the constraints of feasibility would impinge upon the role of analysis and the creation of "icons." This more or less describes the current state of the university, and when shorn of its cybernetic wool, it is also a basic description of the design process: analysis, proposal, decision, and project management. Designers are commissioned, they propose, the client makes the decision on the proposal, the build is then conducted by engineers, the construction company, or the artwork is sent to be printed, etc. Nonetheless, however much of Ambasz's ideas are standard practices given a new language and however apolitical they appear, they do charge design education with a role often ignored or denied to it: the judgment of the goal of urban society and ways in which to organize that goal. The (severe) limits to this ambition are clear though: as a profession and as *universitas*, design has very little political and economic power, very little sway over neoliberal economics or government decision-making; its professional organisations are not significant lobbyists. As such the institutions—networks—comprising "design" do not, in general, participate at the required level of assertiveness needed to influence policy, let alone effectively intervene in the economic structures and imperatives that drive and shape the urban environment. But it is precisely in this domain where postindustrial labour is located. It is precisely in this domain where postindustrial intelligence is located. It is also the domain, increasingly so, where postindustrial pedagogy is situated. One reading of Ambasz's broader concept of design and the Universitas is that it could be merely describing what Gilles Deleuze (1992) would identify as a "society of control." But for Ambasz this utterly designed environment, the city itself, is a source of the possible invention of apparatuses and styles of inhabitation that result in the liberation of capacity, labour and value, not the incarceration of potential: "it is a paradox that any system capable of allowing the greatest possible individual freedom, the fullest possible personal enjoyment of the pleasures of the senses and of the spirit, will not be some sort of unstructured Arcadia but, rather, a highly complex physical and sociocultural artefact" (p. 40).

3 Responses to the Black Book

As could be imagined, the responses—the essays submitted by selected participants and read by all before the symposium started—vary greatly as to what theme or idea they glean from the black book. But in this section I will concentrate on the relation between environmental design and the university (or pedagogical principles/ issues) that may be found in them. As such not all the papers will be considered and not all will be given equal weight; so too the considerable record of the actual sessions will be given minimal attention—unfortunately, because they are fascinating in their own right.

The radical French sociologist and philosopher Jean Baudrillard's contribution, "Design and Environment: Or, the Inflationary Curve of Political Economy" (pp. 50–65), fundamentally reinterprets the entire field of postindustrial political economy itself in terms of environmental design. In complete accordance with Ambasz's provocation he claims the postindustrial milieu is an entirely artificial one that has left behind any remaining distinction between nature and industry. He also claims therefore that political economy too has fundamentally altered. The categories once used to understand industrial society, whether production, class, labour, use- and exchange-value, etc., must all be reinterpreted in the light of an informational, communicational and networked (cybernetic) environment. We have moved, Baudrillard argues, from a "*metallurgic* society" to a "*semiurgic*" one. This coincidence of production and signification—anything and everything is produced as much as a sign (messages, signals, codes, etc.) as an artefact—Baudrillard calls the "political economy of the sign" (see also Baudrillard 1972).

For Baudrillard this utterly artificial, cybernetic environment is more akin to an apocalyptic scenario in which contemporary capitalism extends itself without any gaps or obstacles than one in which a potential utopia can be found. It is not a millenarian state produced by moral laxity, however, but by techno-semiotic profusion: it is an environment in which the distinctions between modalities of artefacts and the experiences associated with them have collapsed (whether religious, ceremonial, technical, symbolic, etc.). Not only is the "great referent," nature, "dead," but so too any alternative mode of experience that is not informational and commodified.

To reply to the pedagogical aspect of Ambasz's text, Baudrillard explains the emergence of this scenario—the political economy of the sign—by reference to the Bauhaus school. By collapsing the distinction between form and function, beauty and utility, art and technology, Baudrillard argues that the Bauhaus rationalizes the entirety of the artificial world—anything can become an object of design, from the "spoon to the city." Using Ambasz's terms infra- and superstructure, he claims that "objects, forms, materials, which heretofore spoke their own group dialect derived only from an idiomatic practice or from an original 'style,' begin to be thought out and written in the same tongue, the rational Esperanto of design" (p. 54). The Universitas for Baudrillard is a direct successor to the Bauhaus and its integrative, environmental vision, insofar as the Universitas is to bring all areas of design together, including and perhaps most significantly, urban planning and architecture, and to treat them in the same manner: as a form of communication. Such, he says, is the…

> … political *ideology* of design, which today is taking on its planetary dimension in the discourse of the environment. From Gropius to Universitas, the thread is unbroken, leading to what we might call a metadesign, a political meta-economy, which is to neo-capitalism what classic liberal economy was to capitalism (p. 63).

This is a lot to pin on the Bauhaus, and it certainly did not go unchallenged by the other participants. Meyer Schapiro, in particular, in the second working session of the symposium, reminded interlocutors that the Bauhaus was thoroughly Romantic and Expressionist in its early years (only later becoming more rational and objective); but more importantly indicated that the legacy of the Bauhaus was its commitment to continuous experimentation and questioning of, specifically, architectural and spatial practice: "an ongoing, consistent, advancing practice" (p. 356). Nonetheless, the problem with the Universitas for Baudrillard, as he elaborates in the same session, is that the university of design is much too rational and there is no room for those practices which are anti-value—the social function of art he argues is be to anti- or counter-productive—that is, practices which destroy value (by which he means are outside the system of utility, function, rationality, and meaningfulness). He points to a paradox: can one create a university devoted to the destruction of value (of functionality and the morality that accompanies it) (p. 343)? But this question is only significant, as is the whole notion of art in contradistinction to design as non-utilitarian sovereign play, if Baudrillard does not believe in his own diagnosis (that there is no longer any distinction between art and design, beauty and utility, aesthetics and functionality).

After granting it is redundant to speak of any firm distinction between nature and technology, urban sociologist Manuel Castells focuses on the meaning of environment, arguing that it is first and foremost to be understood as a social relation. This relation can be expressed in many forms ("morphologies" is Castell's term), "dress and table manners" (p. 66), for example, but it is most noticeably apparent in scale and importance in urban forms. As a social relation, the theoretical interpretation of the environment is conducted through the manner in which urban symbolism can be said to relate to and embody social movements. The environment is encoded action, encoded practice, in other words. While asserting "space has meaning" Castells, however, rejects interpreting space in terms of semiology. He argues that semiology cannot explain the active nature of urban symbolism. It will only see urban phenomenon as a representation to be deciphered. Semiology is thus always historical. Instead he proposes the analogy of information theory in which any urban, architectural form can be considered "senders, relays and receivers" of general ideological praxis. The social space—the environment—is cinematic, not textual. Interpretation of urban environments is a form of de-composition, not hermeneutics. Allowing for the activity of symbolism in turn allows for the introduction of social movements and how they may transform space and "reconstruct the city" (p. 73). Castells gives examples of May 68 and the squatter's settlements (*campamentos*) in Santiago 1970–71. Whatever the actual outcomes, effective transformation of social relations—the environment—involves the effective transformation of spatial conditions.

Which is why universities are still important. They define a different social relation. They practice a different social relation to other "corporate" institutions. For Castells even the new university, Universitas or "Univercity," will retain the three fundamentals of "*research, teaching* and *practical application*" (p. 77). But according to his principle of symbolic action the disinterested expert will not be the source of either pedagogical or research-based innovation. The goals of an innovative institution will, instead, come from the "praxis that is contradictory to the social order" (p. 78). Again we find ourselves left with a similar question as Baudrillard's thoughts on the relative roles of art and design: how to incorporate the alterity of value in an institution?

In Gillo Dorfles' response we find him arguing precedence should be given to what he calls the "*anthropopsychological aspect between man* [sic] *and environment*" (p. 80). Environmental design needs to take into account the psychological, affective dimension of the environment once supplied by regular contact with nature. The Universitas for Dorfles will explore the manner in which memory and environment interact so that the affective dimension can be identified and held onto by individuals; to create, *ex novo*, an "iconic symbology" to recuperate the natural. (Baudrillard would say this recuperation has taken place at the level of the sign: nature and the natural can now only be connoted, never directly experienced.) Such education and awareness is to "stimulate perceptiveness" in the public so it can make informed aesthetic decisions as to the composition and content of the urban environment. This ability is what he calls the "proairetic" aspect of design and is to be at the forefront of design pedagogy: the capacity to determine and communicate preferences through narrative.

In his paper Umberto Eco argues that the classical university originating in Europe in the Twelfth Century does not exist anymore. It may persist, as the Roman Empire lingered for centuries, but it is essentially finished. He claims it is in the process of being replaced by the "open" university, or the "mass" university, which in principle dissolves class distinctions and will function as a universal lyceum or large high school (pp. 347–348) teaching the basic corpus of knowledge (the treasury of science, Eco calls it), while research will move to other institutional forms and locations, precisely the opportunity for the Universitas to exploit.

Eco is also, understandably, concerned with how the postulative dimension of the experimental university connects with research and then to decision-makers. He articulates a common issue: how can artists, social thinkers, poets, translate their research into viable action? (Eco forgets here that they would have already been "trained" as environmental designers by the Universitas.) Eco's proposal is to feed the decision-making and regulative processes back into the Universitas, that is, the network—it is a "circular, rather than linear, model" he asserts (p. 99).

As for teaching and learning practice, he readily accepts that there is a difference in experience between teacher and student but on the whole society sees advanced technicians (artists, architects, city planners) as shaman because the skills of those technicians are not readily accessible (everyone can sing in the shower Eco says, and so can connect with Frank Sinatra, but very few know how to build a home). In order to disperse the shamanic qualities of these professions without actually getting rid of architects the Universitas will include the participation of the residents of the city: "The aim of the Universitas should be to transform the inhabitants of the environment in question, from guinea pigs to advisors and from advisors to collaborators" (p. 104).

Proving that concerns have not radically changed in the last 40 or so years Erich Jantsch (pp. 111–141) praises the Universitas project for addressing design and management together. He offers a "human cybernetics" approach to design education and to theorise the structure of a "transdisciplinary university." He wonders whether design can be taught as an "objective" form of knowledge, but nonetheless links the expanded sense of design (as management, planning and organization) to "noetic" evolution. In other words, knowledge has its own development equal to that of biological or social evolution. Jantsch believes recognizing this equivalence inevitably leads to human systems design that can be studied and implemented through "inter-" and "transexperiential" inquiry. A long and involved explanation of the theoretical underpinnings of this inquiry and how it relates to design and design education begins at this point—Jantsch complains later in the sessions that no one has read his contribution—but it boils down to recommending cybernetics over heuristics and processes over structure in order to form a "cultural designer."

First recognizing that the founding of an institution of higher learning is a "political act," Hans Magnus Enzensberger, poet and activist, member of Gruppe 47, is deeply suspicious of both government and private sector involvement in the university called for by Ambasz. He states the practice and pedagogy of design should be directed to the overall interest of society and not controlled by powerful benefactors. He also is suspicious of Ambasz's idealistic notions of the social contract and the presumed power of the individual to express their design intentions in group networks, believing those ideas disregard categories of "domination, social power, and material interest" (p. 106). He in fact finds the concept of design "deeply ambiguous" (p. 107). On the one hand, it signifies "commodity aesthetics," and on the other points to a profoundly aspirational desire on the part of designers to intervene on a concrete, material level in everyday life. The interests of capital, however, ever defeat those aspirations: "Until today, the history of design has remained a series of defeats, suffered by the high-flying aspirations of the designers in their battle against utilization by *Das Kapital*" (p. 107).

His vision of the university, though of its time, is still radical: reduction of privileges for staff, reduction of the necessity to publish, no tenure for teachers (whose qualifications should come from social and professional experience, rather than "conventional academic standards"); students are to have participatory rights in governing the university (including budgetary responsibility, the hiring of teachers, the content and format of curricula); in fact, teachers are to be elected by students and their learning requirements articulated in discussion with the instructors.

Enzensberger also recognizes that the Universitas should be spatially and architecturally unique. In order to avoid any descent into bureaucracy there is to be no central administration building (and all correspondence is to be undertaken by telephone). There is to be no campus as such; indeed no real distinction between working and living environments, or between work and leisure time. Apartments will be built to embody this new, integrated environment and will be fitted out with "videophone, real-time computer consoles, video recording and cameras, closed circuit TV, and telephone xerography (that is, access to the library via telephone)" (p. 109). Enzensberger insists these combined spaces are to be apartments rather than dormitories (and scaled to need rather than privilege, for example, a student with a family has precedence over a single professor) so that teaching and learning can be hosted in them if required. This vision is indeed one of a "univercity."

Other respondents too focus on the political dimension and ramifications of the project. After noting the Universitas proposal seems to be attempting to reconcile the two traditions of Continental theory and American pragmatism, Henri Lefebvre claims that the word design is asked to bear too much significance and that designers are unrealistically proclaimed modern day "demiurges" (Jivan Tabibian is also concerned about the designer turning into a new, platonic philosopher-king, p. 259). Lefebvre proposes to narrow the focus—but it is not by much in truth—to interventions into urban space, which for him is currently being appropriated and colonized by the severe market conditions produced by capitalist technocracy. Space needs to be contested. It needs to be decolonized. Design and designers are therefore to offer "*another space*" (p. 172). Intervention in the visual, however, is virtually rebuffed here. The visual seems for Lefebvre to be utterly implicated in the "society of the spectacle" (Debord 1995) with little or no hope of redemption. In a similar vein Martin Pawley argues that the university has become increasingly militarized, structurally in terms of administration, and also in terms of its purpose: the total mobilization of "scientific power" for national ends (p. 208). He warns that the troops, the students, are on the verge of mutiny. (Pawley returns critically to the Universitas theme of the "network" in his 1973 publication, *The Private Future*). Sociologist Suzanne Keller also focuses on space but through the lens of communication, outlining what might be the future of communities radically altered by telecommunications technology. Services will no longer be sought out but delivered. These, and "other innovations," she writes, "such as computer medical diagnoses [and] long-distance socializing will transform… the basic design" of communities. "Telecommunities" will de- and re- zone elementary urban activities (p. 145).

In his lengthy contribution Anatol Rapoport defines design as the "concretization of imagined objects or events into plans of action" (p. 219). After a discussion on the nature of goals and the different levels of design, from smaller scale projects wherein goal setting and achievement are technically circumscribed, to the larger scale in which designing concerns the environment itself and therefore political futures, he takes as his theme for the university of design the "creation of a semantic environment" (p. 229). The semantic environment is the only truly man-made environment and is "hospitable to radical institutional changes." In an interesting parallel with the physical environment, he also argues that the man-made milieu, the semantic environment, is subject to pollution (redundant, excessive information; the noise intrinsic to messages). One role of the Universitas is to immunize against this form of pollution. He also points to the problem others at the symposium share—design can imagine the future but it needs the appropriate institutions and assistance to bring the imagined into being—and so the program of the university of design should be "aimed at investigating and publicizing the obstacles to the creation of such institutions" (p. 230). On the other hand Alain Touraine in his contribution does not tackle the question of the definition of design but certainly offers an analysis of the current state of the university in the light of its "crisis" (at least, for him, in France post-68). He argues schools are fundamentally ideological because they (merely) reproduce knowledge. Universities, however, produce knowledge and therefore have the potential to distance themselves from ideology. To do so, Touraine believes the new university should separate scientists from teaching and place them in their own truly independent institution. Students too will have considerable autonomy ("modern techniques should allow self-teaching," p. 274). He also argues that the actual concept of "student" requires rethinking: being a student need not be for a specific period of time or for a specific age group. He envisages a "multiversity" rather than a university. The multiversity will be an oppositional institution allowing the "hidden to come to light, the silent to speak, the forbidden to occur" (p. 275). The political philosopher Sheldon S. Wolin is even sharper in his doubts over design and any utopian aspect it may contain, whether related to the environment or to education. He believes it is a "professional euphemism for control over people and things, a euphemism, that is, for power," and that systems theory is too much akin to a bureaucratic ideal encouraging uniformity and actively discouraging "diversity and deviance" (p. 293).

Whatever the case, that is, whether design and its systems and its pedagogy are mere facsimiles of bureaucratic societies of control or an essential method of conceptualizing and planning the artificial, because it is future-orientated, unreservedly occupied with the artificial environment and devoted to forms of counterfactual communication (and so unites aesthetics and ethics), design cannot be separated from any discussion of the future of university pedagogy, nor of any future reimagining of the university, whether architectural or semantic. Design in this sense is not simply a course one takes within the university but a course the university takes. Indeed Ambasz does not see the university of design as superseding former models of the university but as incorporating and integrating them (p. 25).

4 Universities and Design Pedagogy: The Role and Place of the Counterfactual

In the same decade of the conference Jean François Lyotard also ventured to define the new parameters of knowledge in a postindustrial society. He noted in the influential text *La condition postmoderne* that knowledge is now vast and impersonal, exceeding the capacity of the individual to acquire it, let alone deploy it. Furthermore, the acquisition of knowledge in what he calls "computerized societies" displaces the Humboldtian notion of *Bildung* (training, fashioning) of the individual, particularly schooling the individual to participate in national culture and a shared set of values to which the graduate contributes, promotes and reinforces, in favour of what Lyotard (1984, p. 41f) calls "performativity." Performance is a procedure of legitimating knowledge. Research is legitimated if it not only executes some sort of function, but when its output is maximized and input minimized ("improved performance"). Postindustrialism, furthermore, is a fundamentally global phenomenon. Political economy is no longer tied to the nation-state, and the research produced by universities is to assist and expedite multinational expansion and transactions. In turn the individual a university produces is a knowledge-worker not a citizen. The university today, therefore, is primarily skills-based, not ideals-based: "The transmission of knowledge is no longer designed to train an elite capable of guiding the nation towards its emancipation, but to supply the system with players capable of acceptably fulfilling their roles at the pragmatic posts required by its institutions" (p. 48). The legitimation of knowledge is no longer sought in a story of emancipation (from superstition, from authority), but in the success of the ongoing operation of the system. The question driving education is no longer "Is it true?" but "What use is it?" "Is it saleable?" and "Is it efficient?" (p. 51).

The account of the university as a symbol for the production or reproduction of knowledge mutating inexorably from a cultural institution (*Bildung*) to a technocratic organization is a familiar one. Above all it is a story of loss of autonomy: "higher education" has become a "subsystem of the social system" says Lyotard (p. 48). This narrative is of course by no means incorrect. Nonetheless it also tends to be self-actualizing, and ironically self-legitimizing. There are different modes of legitimation offered in the form of the expanded sense of design that are not contiguous with technocratic forms of knowledge and the maximisation of output. And while Lyotard too points to alternative forms of legitimation (especially paradoxical and dissonant research, all forms of smaller narratives, *petit récit*) he does not refer to design or design education.

As made abundantly clear from the various contributions of the participants of the MoMA symposium, design and design education is uniquely positioned within this new informational, technical and communicational environment of the postindustrial: at once within it and yet able to conceptualize it; at once its most contemporary avatar and the radical proposer of alternatives to it. Unlike the Bauhaus and its successors, which, however brilliantly and however intelligently, primarily furnished products and training for the industrial economy, the university of design is to intervene at another level. Not only is the Universitas to provide design and designers for the postindustrial economy but to illuminate—from the inside, as it were—the fault lines of this economy, expose its shortcomings, decolonize it, and provide it with ethical and aesthetic infrastructure, that is, a critical, pedagogical language sophisticated enough in both content and intent (a language severely lacking in the contemporary institution, including design schools). Thus we can take from these proceedings at least one aspect to focus on: the emphasis placed by many symposium participants, including the organizer, on providing alternative visions to the status quo via the conceptualizing and designing of the artificial milieu, and suggest this is an essential aspect of any new university—or similar institution—devoted to design. The scholarship, research and practice required to envisage and employ the "counterfactual" element of design is, I would argue, the foundation stone of any university education. It is an essential part of the understanding and creation of artificial environments—and the modes of information and communication needed to articulate them.

The counter-factual element is the "utopian" element of all design because it indicates that the current state-of-affairs and therefore the current status quo can be redesigned. Design is concerned with altering what is by providing a vision for what will be. The exploration of the communication of the counter-factual is, partly, the aesthetic dimension of design (the essential element of experimentation)—the form of possibility, as it were. Furthermore, the aesthetic dimension of design is the design of the informational overlay of the created—artificial—environment. It is the cartography of the artificial.

5 Toward a Philosophy of the Artificial (Environmental Design)

The idea of a philosophy of the artificial as a philosophy of the creation of artificial environments and the information mapping that accompanies them is not just to expand the notion of design beyond applied art—and therefore beyond commodity culture and mass production. It is also, ironically, to constrain the notion of design. It is to avoid current claims of a universal design philosophy based on the quasi-metaphysical notion that "everything is designed," which can border on the nonsensical. For instance Nelson and Stolterman (2012) have recently asserted design as a human practice is contiguous with the first use of fire and Tony Fry's notion of "ontological design" too commences with a mythological Ur-scene, in his case of hominids lifting rocks to use as tools (2012). It should be clear by now that design in its expanded sense only effectively begins with the postindustrial.

Following from our readings of the MoMA symposium it can be declared that design is not synonymous with "technics"—including the French sense of *la technique,* which incorporates the industrial arts and artisanal knowledge—and therefore with the philosophy of technology, which interprets the artificial under the category of making (*technē, poiēsis, mimēsis*) and therefore of production. In this view design history becomes the history of devices and their impact: the book, the telephone, plastic, etc. The philosophy of technology, however holistic, does not take into account the mapping (sense-making) of the device to its environment—only the causal effect of the device upon a situation, or, differently put, how technology affects and transforms perception (Ihde 1979). Design, however, does not make things as such. It does not produce artefacts. It assembles, arranges, configures, plans, instructs, informs, and communicates. It proposes (and therefore is potentially counterfactual). It is closer to action, *praxis*, but *praxis* can refer to various forms of reflective goal-orientated activity, not only that which involves foresight and the proposition of the counterfactual. There is no direct Aristotelian category matching the contemporary meaning of design. In fact it could be said that design reverses the classical order and begins with *mimēsis* (simulation), but it simulates without an original model. It simulates (prototypes) counter-factually. Design knowledge and methodology: simulation as praxis.

Design, then, in a very concrete sense, is "post-technological." To recognize this is to move us on from debates on the deceptive nature of technology and the influence that has had on the philosophy of design, for example, Vilém Flusser's (1999) summary of design as a form of trickery against nature (through technology we outfox nature, and therefore the implication is that we are to be punished for our transgressions). As he acknowledges this is the ancient idea of technology (or craft: *technē*) being in excess of the limits and permissions of nature (*physis*). The relevance of this view is questionable. Rather, today, the artificial order is fabricated as it is understood; it is known and altered at the same time. Design, artifice, can only be deception or trickery against a background of a stable ontology of nature (and therefore of myth, metaphysics, religion, etc.). "The words," Flusser argues, "*design*, *machine*, *technology*, *ars* and *art* are closely related to one another, one term being unthinkable without the others, and they all derive from the same existential view of the world" (p. 18). This is not the case anymore. Design philosophy instead points to a post-metaphysical understanding of our milieu, and is therefore not comprehensible at the level of the view that divides the world into *technē* and *physis*. Furthermore, as Baudrillard suggests in his response to the black book, in a postindustrial political economy it may no longer be possible to distinguish between art and design at the level of either functionality or aesthetics: no "object" can be said to be an end-in-itself, all objects must signify. Be that as it may, this is still an artefactual interpretation of art and design. A post-technological understanding of art would not be as grounded in the representational (the mimetic, *mimēsis*) as in what curator Nicholas Bourriaud calls the "relational" (2002); indeed there is an increasing intersection of design and art at the level of "participation" (Holt 2015).

This does not mean that design and design education have nothing to do with artefacts, or the material qualities of the world of things. In fact, Paul Carter (2015) has recently proposed that design research has a sensitivity to materials not normally given precedence in the theoretical sciences, and can call upon habit, technique and experience to augment the speculative. While clearly evoking a vision of things based in craft, he sees this sensitivity instead as a form of cartography: mapping and making are simultaneous. This accords with recent calls from engineering design to better know the artefact under theoretical discussion (Kroes 2012, p. 158: "getting your hands dirty" he says), and to do so is to recognize the equality of the informational and the material element of any design—that the artefact is always placed and understood in an environment.

Intentionally echoing Herbert Simon's *The Sciences of the Artificial* (1981) Victor Margolin (2002) also argues that the field to which design pertains coincides with the artificial:

> If we consider design to be the "conception and planning of the artificial"... then its scope and boundaries are intimately entwined with our understanding of the artificial's limits. That is to say, in extending the domain within which we conceive and plan, we are widening the boundaries of design practice. To the degree that design makes incursions into realms that were once considered as belonging to nature rather than culture, so does the conceptual scope of design practice widen.

But by limiting this definition to the conception and planning of an artificial *environment*, we can literally enclose the scope of the meaning of design allowing for the particular field or intentionality peculiar to that field to come into better visibility. (Making visible is itself a form of information mapping. It is a process of singling out or identifying a phenomenon, experience or agent from among others, but also illuminating the limits or edges of an environment; graphics is also a form of spatial demarcation.) Thus in designing a service, the client or need is circumscribed—a medical service, for example—by the history of forms available to be activated or re-configured. For instance, design for health involves a specific, operational concept of wellbeing, an institutionalized model of the human body, an idea of a spatial and informational system that represents and supports it, including architecture and services, and an array of real and imaginary users (clients, beneficiaries, a public...). Unlike other disciplines that may study the same area, design philosophy and/or design education not only analyses the interaction between values, services, information and spatial forms particular to that area but also "postulates" counterfactual possibilities to the existent understanding of it, whether in the form of modifications or even radical alternatives. And again different from other disciplines design (literally) has the skills and equipment to develop these alternatives and to communicate them and to put them into practice. This practical, interventionist feature distinguishes design from similar approaches to circumscribed fields or networks of knowledge and activity, for example, Michel Foucault's archaeological or genealogical approach to the study of health and its institutions (1963). Design philosophy ensues from an understanding that any demarcated field of activity or knowledge (what we are calling environmental) is artificial.

Originally meaning "circumstance" (Pearce 2010), the word environment may suggest there is nothing linear about design philosophy and therefore design studies or design education. There is no natural evolution of forms to be studied, no chronology of influential artefacts to be learnt; and, if there is, then the very mapping of that chronology, the design of the timeline as it were, already displaces the linearity of narrative or sequence into environmental (or atmospheric, situational) form. That timeline will have its own limit, place, and context. It will have its own history. It is an ecological not a linear phenomenon; one cannot explain the flower simply by the fertilizer as Gaston Bachelard (1969) remarked.

The environment may be defined as a designed and therefore circumscribed matrix of space and information. By space and information I do not mean real and virtual, or real and ideal (space can be virtual, information spatial). The spatial could be defined by all manner of forms of space, not just physical: for example, "hodological," ambient, personal (space occupied by our bodies), "peripersonal" (space surrounding our bodies and literally within reach), perceptual, conceptual, semioticized, objective and subjective, poetic, existential, cognitive, etc. Or, put differently, in David Harvey's matrix (2006, p. 135): absolute, relative, relational on the y-axis, and on the x-axis material (experienced), representations of space (conceptualized space), and spaces of representation (lived space). But also information or communication can be spatial in these senses—information can, for instance, be hodological, conceived as a passage or a path.

The difference between design philosophy and any other form of philosophy is that it extrapolates from the assemblage (the assembled, circumscribed "environment," whether, for example, health, leisure, food and foodways, fashion, transport and communication, etc.). It does not derive its theses from a pre-artificial, pre-environmental point of departure, be it anthropology, a thesis on being, or a theory of subjectivity or consciousness.

Furthermore if design is primarily environmental design (in our expanded sense of the term) then it is fundamentally interior design—architecture as much as graphics, planning as much as textiles. The "interior" is the primary figure of an enclosed, artificial environment. To use Ambasz's terms, design philosophy is therefore the philosophy of infrastructure, not superstructure (ideology: form without spatial and graphic dimension) or structure (simple architectonics: architecture without informational cartography). Design is the creation of artificial environments and the information ecologies that accompany them, that is, it is not the simple production or creation of discrete artefacts: the design element is the element of information, the element of instruction (code). Design embeds the instructions of how to use the environment it creates. As Krippendorff (2006) has argued design is always design semantics. But in fact the semantic element is not added to an environment; the semantic element forms the environment (as Dorfles maintained). This element of information can pre-exist the operation of the design, as in planning, or is contiguous with it (instructions), or it can delimit the environment (on/off). There are in fact a whole series of modalities of such circumscription. It can also be unconscious and unplanned (without an author, as it were).

The Bauhaus and others understood this environment as a primarily biological, natural event or circumstance that could be reproduced at an artificial level (Anker 2010). But the artificiality of the environment does not come from the fact that it is unnatural (or merely technological instead of natural), but by the fact that it signifies: that any organised space is encoded, overlaid with a map in which to understand it; or, equally, to interact with it. This defines designed space. Designed space means "environment." It is the task of design/the designer to create the informational ecology that encloses and integrates the environment, whether spatially, figuratively, semiotically, visually, graphically, atmospherically, etc., and the role of the design history to decode this ecology.

An understanding of design is fundamental to contemporary knowledge practice, just as science was for the Renaissance worldview and hermeneutics for the medieval. Urban, spatial and environmental systems are to be understood using design methodologies that are not only analytic but also synthetic. The aesthetic and the ethical are intertwined: with the freedom of aesthetic research comes the revaluation of the values of urban, artificial life.

6 Conclusion: Univercity

It is difficult for a number of reasons to judge the influence of the symposium on future debates about the meaning and scope of design; perhaps even more difficult to determine its influence on design practice, although, as an early and active advocate of "green" sustainable design, it clearly carried into Ambasz' own industrial design and architectural work (which he would leave MoMA to pursue in 1976).

First, the immediate practical outcome was never realized. With the theoretical framework established by the symposium, Ambasz planned to found an actual city—a "Univercity" in New York State. His wish was that it be hosted by the State University of New York and populated initially by some 30,000 inhabitants. It was to be established through the Land Grant Act—which helped found American universities in the Nineteenth Century—with the patronage of then Mayor of New York and long-term trustee of MoMA, Nelson Rockefeller. In an interview in 1993 on his time as curator of design at MoMA, Ambasz said that this city was to be a place "… where we could experiment with transportation systems, preventive health care, technologies, everything which I felt really didn't have only to do with the design of the physical environment, it also had to do with the design of a quality-of-life situation" (Ambasz 1993, p. 15). Ambasz describes the idea of the Univercity in its entirety as a fable written at the close of the conference in 1972 (appended to the symposium documents published in 2006, pp. 505–507) and from the viewpoint that it had already been established; it had become the "much talked-about showcase its founders had hoped for" (p. 506). There are suggestions that the idea had got as far as some "retreats" in the Hudson Valley (Bach 1999, p. 11), but Ambasz himself makes no mention of them in the 1993 interview, and there certainly was no university of design created as such.

Second, the documents, transcripts and other collateral were not published until thirty or so years after the conference. In the same interview Ambasz expressed regret about not publishing the material straightaway, and gave the reason as sensitivity to the criticism from a number of the participants that the project was attempting to impose a philosopher-king model on the academy (because, in the main, of scale: the university of design was to be *universitas*, the "whole," and literally a "univercity"). Nonetheless, as he points out, he was not interested in forming a philosophic, nor scientific university in the Greek or Humboldt traditions, arguing "I was interested in defining the notion of a *universitas* that would deal with the problems of designing and managing the manmade milieu that we have invented and it has now become to occupy a place next to the natural milieu, the natural environment" (1993, p. 6). In hindsight this seems a perfectly reasonable ambition.

Third, the contributors and participants absorbed the conference provocations in different ways into their future work. The articulation of the questions of design and environment in a postindustrial context deeply influenced the direction of Baudrillard's thought, assisting a turn from analysing consumer society from a Marxist perspective to a post-Marxist one in which design categories enter and govern his thinking—codes, simulation, modelling, cybernetics and communication. But other participants did not follow through the consequences of the discourse around the university of design in any direct manner. For example, Umberto Eco who had a life-long professional interest in interdisciplinary education and architecture and urban planning, did not return to the theme of a university of design in any significant way, and in outlining the essentials of postindustrial knowledge economies Manual Castells' volumes on the "network society" (2010) do not centralize the design discourse formulated by his contribution to the symposium as might be expected. This seems to be the case too for Hannah Arendt, who attended the first, third and fourth "working sessions" (see Ambasz 2006, pp. 339, 372, 376–378). During one discussion, she revisits her example of the table from *The Human Condition* (1998 [1958], p. 58)—the table is used as a simile for the material dimension of public life, as that which both joins and separates people—and makes the argument that the solidity of three-dimensional form created by the architect is primarily a mode of security against the temporal nature of human life (Ambasz 2006, p. 377). She also draws attention to a scale of permanency of objects, from the immediately consumed (food) through the persistent (shoes) to the near-immortal (works of art) in order to disabuse some of the contributors of the notion that all objects have become commodified. There is no particular reconsideration, however, of the material element of the political in the light of what we are calling "environmental" thinking in her subsequent work to the conference (for example, the lectures that appear in the later editions of *The Life of the Mind* 1978). Having said this, it would be fascinating to know how much, if any, of the conference ideas and her experience there fed into her considerable input into the Structured Liberal Education (SLE) program at Stanford University. The actual syllabus does not suggest so (Stanford University n.d.), but the communal, live-in structure of its organization might.

In terms of education universities are still to catch up with this vision, not necessarily to match it or implement it, but at least to begin the debate anew about what a "post-technological" society might mean and the manner in which to offer pedagogical tools and a pedagogical framework to both comprehend and intervene in it. In retrospect and in the sense of potential still to be unlocked, Ambasz's symposium may yet be seen to be a foundational event of the informational, communicational and environmental approach to design which, today, whether through participatory design, design thinking, systems and service design, and advanced urban planning, is undergoing a renaissance, as if the ethical and pedagogical dimension of the design for and of the artificial is only finally beginning to dawn on us.

References

Ambasz, E. (1969). The formulation of a design discourse. *Perspecta, 12*, 57–70.

Ambasz, E. (1971). I: The university of design and development. II: Manhattan: Capital of the twentieth century. III: The designs of freedom. *Perspecta, 13*(14), 359–365.

Ambasz, E. (Ed.). (1972). *Italy: The new domestic landscape. Achievements and problems of Italian design*. New York and Florence: The Museum of Modern Art and Centro Di. Catalogue.

Ambasz, E. (1993). Interview with Sharon Zane. The Museum of Modern Art oral history program. https://www.moma.org/momaorg/shared/pdfs/docs/learn/archives/transcript_ambasz.pdf. Accessed 10 Apr 2016.

Ambasz, E. (Conceived and directed). (2006). *The universitas project, solutions for a post-technological society*. New York: The Museum of Modern Art.

Anker, P. (2010). *From Bauhaus to ecohouse: A history of ecological design*. Baton Rouge: Louisiana State University Press.

Arendt, H. (1978). *The life of the mind*. San Diego: Harvest.

Arendt, H. (1998 [1958]). *The human condition*. 2nd ed. Chicago: Chicago University Press.

Bach, C. (1999). Making space. *Américas, 51*(1), 6–15.

Bachelard, G. (1969). *The poetics of space*. Trans. Maria Jolas. Boston: Beacon Press.

Barthes, R. (1985). *The fashion system*. Trans. Matthew Ward and Richard Howard. London: Cape.

Baudrillard, J. (1972). *Pour une critique de l'économie politique*. Paris: Gallimard. English edition: Baudrillard, J. (1981). *For a critique of the political economy of the sign*. Trans. Charles Levin. Telos Press.

Baudrillard, J. (1996). *The system of objects* (J. Benedict, Trans.). London: Verso.

Baudrillard, J. (1998). *The consumer society*. Trans. Chris Turner. London: Sage.

Bourriaud, N. (2002). *Relational aesthetics*. Paris: Les presses du réel.

Busbea, L. (2009). Metadesign: Object and environment in France, c. 1970. *Design Issues, 24*(4), 103–119.

Carter, P. (2015). *Turbulence. Climate change and the design of complexity*. Sydney: Puncher & Wattmann.

Castells, M. (2010). *The rise of the network society* (2nd ed.). Malden: Wiley-Blackwell.

Debord, G. (1995). *The society of the spectacle*. Trans. Donald Nicholson-Smith. New York: Zone Books.

Deleuze, G. (1992). Postscript on the societies of control. *October, 59*(Winter), 3–7.

Dorst, K. (2015). *Frame innovation. Create new thinking by design*. Cambridge, MA: MIT Press.

Ellul, J. (1954). *La technique ou l'enjeu du siècle*. Paris: Librairie Armand Colin. English edition: Ellul, J. (1964). *The technological society*. Trans. John Wilkinson. New York: Vintage.

Flusser, V. (1999). *The shape of things. A philosophy of design*. London: Reaktion Books.

Foucault, M. (1963). *Naissance de la clinique*. Paris: Presses Universitaires de France. English edition: Foucault, M. (1986). *The birth of the clinic. An archaeology of medical perception*. Trans. A. M. Sheridan. London/New York: Routledge.

Frank, S. (2011). *IAUS: The institute for architecture and urban studies. An insider's memoir*. Bloomington: AuthorHouse.

Fry, T. (2012). *Becoming human by design*. London/New York: Berg.

Harvey, D. (2006). *Spaces of global capitalism: Towards a theory of uneven geographical development*. London/New York: Verso.

Holt, M. (2015). Transformation of the aesthetic: Art as participatory design. *Design and Culture, 7*(2), 143–165.

Ihde, D. (1979). *Technics and praxis: a philosophy of technology*. Dordrecht: D. Reidel.

Krippendorff, K. (2006). *The semantic turn. A new foundation for design*. Boca Raton: Taylor & Francis.

Krippendorff, K. (2007). The cybernetics of design and the design of cybernetics. *Kybernetes, 36*(9/10), 1381–1392.

Kroes, P. (2012). *Technical artefacts: Creations of mind and matter: A philosophy of engineering design*. Dordrecht: Springer. https://doi.org/10.1007/978-94-007-3940-6_1.

Lyotard, J-F. (1984). *The postmodern condition: A report on knowledge* (G. Bennington and B. Massumi, Trans.). Manchester: Manchester University Press.

Margolin, V. (2002). *Politics of the artificial: Essays on design and design studies*. Chicago: University of Chicago Press.

Martin, R. (2004). Environment, c. 1973. *Grey Room, 14*(Winter), 78–101.

Nelson, H. G., & Stolterman, E. (2012). *The design way. Intentional change in an unpredictable world* (2nd ed.). Cambridge, MA: MIT Press.

Pangaro, P. (n.d.). *Cybernetics—A definition*. http://pangaro.com/definition-cybernetics.html. Accessed 2 Jan 2015.

Pask, G. (1963). The conception of a shape and the evolution of a design. In J. C. Jones & D. G. Thornley (Eds.), *Conference on design methods* (pp. 153–167). Oxford: Pergamon Press.

Pawley, M. (1973). *The private future: Causes and consequences of community collapse in the west*. London: Thames and Hudson.

Pearce, T. (2010). From "circumstances" to "environment": Herbert Spencer and the origins of the idea of organism-environment interaction. *Studies in the History and Philosophy of Biological and Biomedical Sciences, 41*, 241–252.

Peirce, C-S. (1931–36). *The collected papers*. Volumes 1–6. Charles Hartshorne and Paul Weiss (Eds.). Cambridge, MA: Harvard University Press.

Scott, F. D. (2004). On the "counter-design" of institutions: Emilio Ambasz's universitas symposium at MoMA. *Grey Room, 14*(Winter), 46–77.

Scott, F. D. (2010). *Architecture or techno-utopia. Politics after modernism*. Cambridge, MA: MIT Press.

Simon, H. (1981). *The sciences of the artificial*. Cambridge, MA: MIT Press.

Stanford University. (n.d.). *Structured liberal education syllabus archive*. https://undergrad.stanford.edu/programs/residentialprograms/sle/about-sle/syllabus-archive. Accessed 23 Apr 2016.

Stickdorn, M., & Schneider, J. (2010). *This is service design thinking: Basics, tools, cases*. Amsterdam: BIS publishers.

The Design of Nothing: A Working Philosophy

Paul A. Rodgers and Craig Bremner

Abstract In an era of digital production and disruption this chapter probes how design might now best labour under a philosophy of nothing. Nothing is a pronoun for something and nothing is now the derivative project for design. As such design requires a new form of inquiry to produce new insights and a new working philosophy from the design of nothing. Design was inserted into the digital stage called social commerce as an essential component for the production and exchange of nothing and lies at the core of service. That is, the production of nothing requires, and produces, nothing but the logistics of nothing. The design and production of nothing has disrupted the philosophy of design – its histories, its apprenticeship to the project for a 'better world', Simon's (The Sciences of the Artificial, MIT Press, Cambridge, MA, 1969) 'preferred state', its devotion to fashion, and so on, and having dismantled 'industry', nothing has produced its most beguiling product yet, the fixation of capital production without a product. In the era of the production of nothing design, a discipline of now questionable utility, product of the derivative, operating in conditional space, has collapsed its scope into the belief in universal innovation, and must now develop a new operative philosophy.

Keywords Design · Nothing · Crisis · Philosophy

1 Introduction

Philosophy (noun) is defined as the study of the fundamental nature of knowledge, reality, and existence (especially when considered as an academic discipline). So, a philosophy of design would include the study of the theoretical basis of the art or action of conceiving of and producing a plan or drawing produced to show the look and function or workings of a building, product, or other object before it is built or

P. A. Rodgers (✉)
Imagination, Lancaster University, England, UK
e-mail: p.rodgers@lancaster.ac.uk

C. Bremner (✉)
School of Communication and Creative Industries, Charles Sturt University, Bathurst, Australia
e-mail: cbremner@csu.edu.au

© Springer International Publishing AG 2018

P. E. Vermaas, S. Vial (eds.), *Advancements in the Philosophy of Design*, Design Research Foundations, https://doi.org/10.1007/978-3-319-73302-9_25

made. Galle (2002), however, suggests that a definition of the philosophy of design is neither feasible nor welcome. Rather he says that the philosophy of design is whatever philosophers of design do. And what philosophers of design do is think and write about various aspects of design, and gain insights about them via rational reflection rather than empirical observation. Thus, the philosophy of design can be described as "*...the pursuit of insights about design by philosophical means.*" Galle (2002: 216). This begs the question – what is the point of a philosophy of design? If we agree that insight is an end in itself, then given the above description, a philosophy of design is useful because it offers us insights about design, which we could not obtain otherwise. Similarly, Lawson reminds us that designers usually design things not just because they enjoy doing it but also because they are fascinated by what they do. Also, this interest in creating things generates a collection of attitudes that Lawson calls "philosophies" that in turn effects how designers go about things. Lawson's illuminating book "*How Designers Think: The Design Process Demystified*" recounts a number of insightful philosophies (attitudes) of design including "*...[the designer's] job is to give the client... not what he wants but what he never even dreamt he wanted*" (Lasdun 1965) and "*...in designing for building every architect is involved in foretelling what is going to happen*" (Price 1976). The insight (or working philosophy) that this paper will make is that in an era of digital production and disruption, the design of nothing might require a philosophy of nothing. We accept that as a postulate a philosophy of nothing can never be demonstrated but in this paper we present the case that it can be understood.

Taking Galle's and Lawson's leads in this paper, we have thought and now write on an aspect of design. The insight we have gained, through critical reflection, and present here is that where once we bonded around the production of something (tangible goods), we now specialise in the production of nothing and this design of nothing has produced a new working philosophy. Design is neither a product nor a service. Design occurs in relationship to everyone and everything – it describes and shapes relationships. But if we really care about design, what should we do? Is design about the design of nothing or should we be designing nothing? A philosophy for the design of nothing needs to ask – in the search for coherence what do my actions look like today if I look through the lens of design? On the basis of a philosophy of self-understanding what needs to be done?

This paper will argue that the design and production of nothing has disrupted the platform of design and its decaying dream of the 'preferred state' (Simon 1969). Design's project for a 'better world' has lapsed and in its place design now only manages to serve both the fashionable diffusion of products and services, and the suppression of the reduction of consumption. Having dismantled 'industry', production and consumption are no longer projects for the masses. Nowadays, services are 'designed' to appear individual so the individual can be increasingly engaged in the production of their own 'service' (Manolis et al. 2001). Consequently, the production of nothing has produced new forms of consumption and lots of it, which design continues to imagine it has some control over if only to restore some relevance to itself. So, is the alternative to overproduction and debt-fuelled growth, manifest in the current economic crisis we find ourselves in, once again to slow it down and become lazy as autonomist theorists (Gorz 1989; Cox and Bazzichelli 2013) and practitioners have stated for some time now? If this is indeed the case, can design manage to be lazy? This paper presents the case that to operate in an era of digital production and disruption, design might best labour under a working philosophy of nothing.

2 The Exchange of Nothing

As Bifo Berardi, the Italian Marxist theorist and activist in the autonomist tradition, explains with typical clarity: "*the dismantling of industry is unstoppable for the simple reason that social life does not need industrial labour any more*" (Berardi 2011: 152). That is to say social life is now defined by being served, and because service is being designed to resemble a meeting between friends, with extended encounters and exchanges, where customers believe that the service provider is actually interested in them, giving them a sense of temporary social connectedness, social commerce, capitalism's vein in social media is remaking its fortune on the back of this desire to feel connected in a relationship and belong. However, the reality is that there is nothing going on here but paying to be serviced on the capital stage.

This exchange of *nothing* is at the core of service, its staging, and its design (Bhaskar et al. 2012). We know that a service cannot be manufactured and can only be staged. Hence the rise of the service economy produced the rise of the staging of brand, into which design has been inserted as an essential component for the production of *nothing*. That is, the production of *nothing* requires, and produces, *nothing* but the staging of *nothing*. And social life is now mostly just entertainment served digitally before it is swallowed.

By the middle of the twentieth century very few people were involved in the production of anything, industrialisation 'produced' everything. At the same time marketing made sure it was nothing, of no concern, for everyone to imagine 'consuming' everything. Now, courtesy of the digital, everyone is involved in the project of producing *nothing*, but that *nothing* is consuming every imagining. Even worse than restoring production to everyone, the digital has imposed productive capacity on everyone coercing the partial production (deceptively termed co-production) of almost every pay-for-use encounter, but only the promise of encounter is the new material of design. Once upon a time design was a serious project, and that project was to persuade us to produce a better world (Warman 2011). Given this new digital project, instead of persuading a system of production might we now need to persuade ourselves that we can imagine a world in which we want to live?

Bruno Latour proposed that having privileged ideas over aesthetics, design today can be anything (Latour 2008). But if design can be anything then it can also be *nothing* and this perhaps is the biggest challenge that design now faces? Similarly, in his 2002 essay "Junkspace", Rem Koolhaas states: "*If space-junk is the human debris that litters the universe, junk-space is the residue mankind leaves on the planet.*" (Koolhaas 2002: 175). And junk-space is the product of all architecture unwittingly producing a single conditional building – a stage for the production and consumption of *nothing*. This production of nothing is not isolated to the world of design. Boris Groys has commented similarly by proclaiming that: "*Advanced contemporary art is basically art production without a product.*" (Groys 2012: 11).

So it is rather easy to see that the design and production of *nothing* has disrupted the philosophy of design – its pursuit of insights into its histories, its apprenticeship to the project for a 'better world' embodied in Simon's (1969) 'preferred state', its devotion to fashion, and so on. And having dismantled 'industry', *nothing* has produced its most beguiling product yet – the brand – the fixation of capital production without a product. Of the many conundrums triggered by the production of nothing, for design the most comical is the production of nothing still requires the design of nothing. This farcical situation requires new insights, a new postulate is necessary because a new foundation is required for this new labour.

3 Proving Nothing by Philosophy

The idea of a philosophy of design is not impossible to countenance but design philosophy might well be based on a platform that is not what is being designed. As we know, the word "design" is both a noun and a verb. As a noun, design is generally used to denote a plan or drawing produced to show the look and function or workings of something before it is built or made (*e.g. the design team have just revealed their design for the new museum*). Design, therefore, embodies purpose, planning, or intention that exists or is thought to exist behind an action, fact, or material object. But design can also be used as a verb to describe the action of conceiving of and producing such a plan or drawing. That is, deciding the look and function of something by making detailed drawings of it (*e.g. important changes to the new health service were designed to provide greater end user choice)*. So, design is to do or plan something with a specific purpose or intention in mind. However, if we consider Vilem Flusser's (1992) meditations on the term "design" then his forensic etymology defines design as a means of cheating nature and ultimately, cheating culture itself. If the noun/verb generalities are what we have focused on knowing about design – *i.e.* a philosophy built on global cultural enhancement – then it is quite obvious this isn't happening by design, or more accurately, what is happening by design is not cultural enhancement. That is, design philosophy might not be about what is being designed.

Ever since idea was split from manufacture in the first orbits of the industrial revolution ideas had to be communicated to systems of production and gave birth to a language commonly called design. The split between idea and manufacture means that design spins itself into existence by virtue of its ability to do, in Flusser's terms, two duplicitous things – persuade someone to make something and then augment that fabrication by cheating as many people as possible into buying this thing. And all this persuasion/cheating takes place in an increasingly virtual environment resulting in a paradox for design; one action brings design recognition while the other robs it of relevance. But it is difficult for a philosophy of design to determine which action produces which result. In that sense, design is similar to the paradox that Slavoj Žižek describes in capitalism:

> *The paradox of this virtualization of capitalism is ultimately the same as that of the electron in particle physics. The mass of each elementary particle is composed of its mass at rest plus the surplus provided by the acceleration of its movement; however, an electron's mass at rest is zero, its mass consists only of the surplus generated by the acceleration, as if we are dealing with a nothing which acquires some deceptive substance only by magically spinning itself into an excess of itself.* (Žižek 2012: 246)

The excess of recognition is the result of spinning the many disciplines now grouped under the word design into a new state while simultaneously robbing design of its relevance. And the excessively cited "*preferred state*" by Herbert Simon (1969) characterises, rather than defines, a state that design once dreamt of – a better world that is, paradoxically, only better if you already own it. The infinite possibilities promised by the pathway to the "*preferred state*" became the infinite modularisation of design actions in the chain of production with no product, while the infinite responsibilities of maintaining that "*state*" came to nothing long before the production of nothing. Is Simon's "*preferred state*" the spinning from nothing into something indicative of a new trend? Are we witnessing the disruption of the design of something for the designing of nothing – design continuing to serve production in the era of no products?

In this paper we approached our answer to these contentious questions from the perspective that design is largely a process of the management of its own spin, and management is the only product of the mature neo-liberal service economies. Our perspective is clearly framed by Bifo Berardi's incisive clarification that social life no longer needs industrial labour. If Bifo is correct then social life is now just mass-entertainment projected digitally before it is staged, served and swallowed. Because there is no product to profit from services, they need to be performed and patronised repeatedly, so to perform services repeatedly service 'makers' had to develop methods to manage the provision of nothing. The service sector continues to refer to what they provide as 'products' even though there are no products being manufactured here, but it is the only way the service providers can talk about the nothing that is the service. Given this predilection for calling their 'stuff' products, the service industry has employed the logic of Taylorist scientific management techniques to prove they can adhere to new standards to guarantee services – nothing – can be reproduced infinitely (Bremner 1995).

These inherent paradoxes have not hindered the rise and rise of the neo-liberal service economy whose major 'products' are now virtual financial and professional services (the stock exchange, banking, and insurance, all backed by a monstrous legal machine). And while these exchange services contribute most to neo-liberal GDP, design is mostly co-opted into designing services to resemble meetings between friends in what are termed "*boundary open*" encounters (Mars and Nicod 1984). As we have stated these are extended service encounters and exchanges where customers are conditioned to believe that the service provider is interested in them as a person (not just a customer). The aim is to produce the feeling of a relationship or a sense of social connectedness and this relational fiction has been extended into every service avenue via branding – drinking coffee, air travel, hotel booking, keeping fit, and so on. The brand is the service product 'par excellence' now determining service development and design by aesthetic rather than financial or operative decisions (Carmagnola 2009). With the emergence of social media, neo-liberal capitalism opened the boundaries of a new business opportunity branded 'social commerce' remaking its fortune on the back of the desire to feel connected, attached, in a relationship, to belong. To do this it co-opted design into creating the conditions to stage all these fake sensations. By applying "*lean ethnography*" (Tonkinwise 2014: 30) to this new design opportunity designers appear to have been diverted to observing the conditioned performance of consumers/producers resulting in nothing but unfulfilling services requiring mystifying and contracted payment (think of telecommunications contracts as an example).

Is it possible to reconcile the inherent contradiction(s) of a philosophy of design in an era of the production of nothing? For example should design be reconceived given the current fashion for disruption? Disruption in the way design is currently viewed as a profession with an identity crisis (Richardson 1993; Friedman 1994), disruption in the way design works (or more accurately doesn't work) economically (McGuirk 2011), or disruption in the way design has been usurped by the digital (Hight and Perry 2006). It is widely acknowledged that design currently exists in a climate of disruption. The authors, in their recent work (Bremner and Rodgers 2013; Rodgers and Bremner 2013), have highlighted that these forces emanate from professional, cultural, technological, and economic dimensions. For instance, the crisis of the design profession in the contemporary, challenging and dynamic world where disciplinary boundaries are continually blurring and where growing evidence suggests that design itself is in the middle of a great transformation (disruption?) inasmuch as the market-driven years of the 1980s and 1990s have given way to a more people-centred or service-centred era. Paul Atkinson believes this change is not so much a crisis for design, but a crisis for the design profession when he states: "*Post-industrial manufacturing necessitates a new kind of designing that has the potential to create a different role for the designer.*" (Atkinson 2009). The economic disruption facing the design industry in general and solo designers in particular is reflected in the patterns of financial and royalty payments' meltdown. This has been well documented and is especially prominent in Justin McGuirk's recent exposé on the fiftieth anniversary of the Milan Furniture Fair where he revealed that the vast majority of the designers exhibiting there are barely able to afford to pay their own rent. The evidence clearly highlights that more than 2700 furniture leading brand manufacturers exhibited their work at the Salone Internazionale del Mobile, Milano in April 2011, but many of the lamps and chairs shown there are prototypes produced by designers for free, in the hope they will make their money back in royalties or in the form of future commissions. As McGuirk points out, however: "*Only the lucky few ever do. I spoke to one young designer who has five items in production with a respected Italian manufacturer—no small achievement. 'My royalty check last year came to 600 Euros,' he said. [That's] half a month's rent.*" (McGuirk 2011).

Moreover, in Peter Kester's critique of Design in general and Exhibitions in particular, he asks what the proliferation of exhibitions and events for designers really means? Also, what does it mean for the profession? Like McGuirk's analysis of the economics of contemporary design, Kester highlights the substantial changes that have taken place over the past three decades and asks pointedly: "*What (design) shows are really interesting? It is strange, even disturbing, in a time where everything changes, where economic, ecological, social, and cultural issues prevail, the design festival reigns supreme.*" Kester suggests the spectacle predicted by Guy Debord (1977) seems to have become reality. Today, design is entertainment and may well become something akin to a shopping spree (Kester, 2010). This downgrade of the financial stock of designers coincides with the "financialization" of the global economy, which has turned all exchanges into a derivative, a form of insurance against change. For a profession predicated on change, this development is potentially terminal. The final disruption we will discuss here (although there are likely many more that we have overlooked) is how explosive developments in digital technologies have disrupted how design is communicated, connected, and configured. These huge advances in digital technologies have disrupted conventional creative processes and procedures that enable individuals to engage in a form of digital design and production that calls into question their familiar relationship with the design, production, and consumption of products. As a result we have been conditioned into the role of "prosumers" engaged in the incremental co-production of our own services (anything online from banking to games and anything in real time/space from tourism to fitness). All of this production has managed to co-opt design into its production but really produces nothing. And all of this non-matter consumes all manner of energy and effort. For a philosophy of design the design of nothing has added an array of new vectors shaping the imaginary based on the images suspending virtual experiences and contractual realities.

4 Nothing but Crises for Design…

Revisiting some of the authors' previously published work in this area (Bremner and Rodgers 2013; Rodgers and Bremner 2013), we wish to re-state how *nothing* has disrupted the platform of design – as stated its histories, its apprenticeship to the project of 'better', its devotion to fashion, and so on. Moreover, having dismantled 'industry', *nothing* has designed the best product imaginable – design management. When production and consumption were manufactured design worked for manufacture. Now that the digital has coerced everyone into the production and consumption of nothing, design can only manage or persuade – an iniquitous position design education adopted with relish. So, instead of concerning ourselves with projecting 'what-might-become', the digital has hoodwinked us into producing the design of an 'other' world where the project is to archive 'what-was'. And, as is widely acknowledged, we have all contributed (and continue to contribute) to a world that we no longer want or need, or 'what-might-not-become' (Rodgers and Bremner 2013). There are many statistics that provide clear evidence for this blunt statement, but the three most significant figures clearly indicate how we are collectively destroying many of the features of society that we claim to hold most dear. These are:

- ***Our Ecological Crisis*** – We continue to deplete and degrade our natural world on a massive scale, using up more resources every year. We live as if we have more than one planet at our disposal, using the equivalent of 1.5 planets just to meet our current rate of consumption. Over the past 40 years, we have seen one third of our agricultural land disappear and we regularly see rapidly falling water tables that bring us on a path towards food supply crises, food riots, and an expected doubling of food prices by the year 2030.
- ***Our Social Crisis*** – 2.5 billion people on our planet live in poverty. There have been many successes at lifting people out of poverty, but this number has not changed much over the past few decades.

  and
- ***Our Spiritual Crisis*** – According to the World Health Organization, 3 times as many people die from suicide as die from homicide or in wars. Although men and women are fighting many wars around the world at present, 3 times more people kill themselves than kill others. This inner crisis also manifests itself in many other forms including rapidly growing figures for burn out and depression that both indicate an increasing gap between our exterior activities and our interior sources of creativity and presence.

These three dimensions (and there will be others that we have overlooked) of collectively creating results that nobody wants constitute the most significant institutional, leadership and design failure of our time. This failure of relying on 'old', 'conventional', and/or 'historical' ways of designing and managing things is patently not fit to address the significant challenges we now face (Scharmer 2011; Norman 2010; Marshall and Bleecker 2010). Given this appalling failure, we wish to present a philosophy for design that shares Kenya Hara's notion of "Exformation" (Hara, 2007). In short, understanding and acknowledging that we might know nothing, and comprehension and recognition of nothingness is necessary for the beginning of any future design project. Hara proposes that "known" and "understood" are horrible concepts, which usually mean that your work (designs) have nothing new to offer the world. Žižek might describe this as an example of what he posits as an "unknown known" – that which we intentionally refuse to acknowledge that we know (Žižek 2004). As such, we present five ideas for a working philosophy for the design of nothing over the next five sections.

4.1 *The Derivative of Nothing*

We have touched on the derivate as the preferred instrument of modern finance – preferred even though its operative obscurity has eminent economists concerned about its potential impact on the foundations of capital – and the derivative's similarity to what was information that is now called big-data (the scrapping up of everything) reveals something about how design might need to think about a new relationship to the machine. When Paul Mason writes: "The end of capitalism has begun" he cites the fate of the machine as one of the key indicators:

> *Once you understand that information is physical, and that software is a machine, and that storage, bandwidth and processing power are collapsing in price at exponential rates, the value of Marx's thinking (the creation of an "ideal machine") becomes clear. We are surrounded by machines that cost nothing and could, if we wanted them to, last forever.* (Mason 2015)

For the entire period of the practice of design the machine, and technology in general, has been championed and this reached its zenith with Reyner Banham's book first published in 1960 "Theory and Design in the First Machine Age" (Banham 1980). This devotion to the machine failed to account for an obvious inversion of value. From the inception of design the machine instantly turned the idea of design into an image of itself making ideas derivatives of the machine. This could go unnoticed for the period that the machine required massive investment and attendant fidelity. But now as Mason makes clear, due to the relentless operations of market and relentlessly cheaper digital flows of capital the machine now costs nothing. And yet design hails the birth of yet another machine dependent on the same digital flows – 3D printing – whose imminent ubiquity is being sold as the holy grail of design philosophy – democratic production. No longer will design be subject to the tyranny of machines owned by others, the designer like the ancient artisan, will reclaim his or her own production. This is not democracy. It is simply another stage where the idea is being converted into the image of a new machine. The financial derivative works by insuring against change. The machinery of design works by making design a derivative.

4.2 *The Discipline of Nothing*

Design just like Fine Art before it has undergone something of a significant transformation in recent years. Design, too, has refocused its lens to privilege ideas over aesthetics. As such, today, design can be anything. Bruno Latour famously claimed that:

> *...design has been expanding ferociously from the design of objects that we use on a daily basis to cities, landscapes, nations, cultures, bodies, genes, political systems, the way we produce food, to the way we travel, build cars and clone sheep.* (Latour 2008)

Moreover, if you study how design is celebrated nowadays by the likes of the UK's Design Council then its "winners" routinely range from things like drugs that enhance sexual performance (Viagra) to business software. Stuart MacDonald (2012) describes this new creative landscape as a "*post-modern soup*" in which cultural, economic, social and educational issues are swimming and where "*mongrel*" or "*hybrid*" institutions will flourish. Typically, Andrea Branzi (2008) sees this same fluidity from a different perspective, not design as a sum of devices but design as a sum of deceptions:

> *The difference between entrepreneur, designer, and consumer tends to disappear, all are authors of choices of behaviour and creative gestures, no one repeats 'that which already exists' but are forced to interpret the opportunities offered by the market in a personal manner. In this 'creative society' the active organisms are therefore multiple and they generate a diffuse energy that isn't easily governable, that moves without 'end' and without 'an end'.* (Branzi 2008)

And the 'creative society' Branzi refers to is not the failed promise of mass creativity or the curse of the 'creative industries', nor is he referring to the idealise diffusion proposed in Ezio Manzini's recent (2015) book "Design, When Everybody Designs" (Manzini 2015). Branzi's "*creative society*" consists of, as he says, "*authors of choice*" forced now by deceptive digital flows to choose between everything and nothing to fashion their personal project in which case design has no choice but to be anything. But if design can be anything then it can also be nothing and this perhaps is the biggest challenge that design now faces? As the title of Arthur Danto's essay goes (if we substitute the word 'art' for 'design'), "After the End of Design" what does it mean when Design can be anything? Perhaps nothing?

4.3 The Space of Nothing

In his essay "Junkspace" Rem Koolhaas proclaims:

> *If space-junk is the human debris that litters the universe, junk-space is the residue mankind leaves on the planet.* (Koolhaas 2002)

Junkspace, Koolhaas claims, is what remains after modernization has run its course. Modernization had a rational program – to share the blessings of science, universally, but Junkspace is its apotheosis. It is "the product of an encounter between escalator and air-conditioning, conceived in an incubator of Sheetrock (all three missing from the history books)". Long before Koolhaas entered Junkspace into the history of architecture Cedric Price's project "Fun Palace" depicted the contingency and non-solidity of buildings as changeable, and expendable life-value sources, which can be used, reused, misused, or disused. Price was not prefiguring Junkspace but space based on the recognition of the increasing capacity of society to change its mind (Vodanovic 2007). One can already see the results of society changing its mind producing Junkspace in the making with the decline of high street shopping and the rise in virtual and online shopping. Online commerce is the fastest growing retail market in the world. In Europe alone, regular sales and unceasing growth is expected to continue from €185.39 billion in 2015 to €219.44 billion in 2016 – an increase of nearly 20%. Virtual shopping figures will soon surpass high street shopping and these giant shopping malls strung together with Sheetrock and escalators and dependent on air-conditioning will all become conditional space… Junkspace. Like Koolhaas, Franco Berardi, aka "Bifo" informs us that we have created a world that is seriously ill prepared to deal with the mounting environmental, social, economic, and spiritual crises we face because we have an overriding obsession with accumulation, property, and greed, and strive for continual expansion and social well-being (Berardi 2010). And the authors of "Global Insanity Redux" detail delusional activities:

> *…that are strategically designed (typically based on an ideal) to improve the human condition via technology and industry (that) very often cause as much if not more harm than good, in the form of unintended ("entropic") consequences resulting from ignored and/or unperceived realities that are incompatible with the intended ideal.* (Coffman and Mikulecky 2015: 2)

This strategic design fallout illustrates what Koolhaas called:

> "*…conditioned space" which "inevitably becomes conditional space; sooner or later all conditional space turns into Junkspace…*" (Koolhaas 2002)

The moral of this story is the design of nothing is conditional on strategies to fashion increasingly deceptive conditions for unnecessary pay-for-use encounters resulting in more and more junk.

4.4 The Appearance of Nothing

It is precisely because nothing ever really works, that the useful and the useless are interchangeable, but design stubbornly strives to be useful. Designers never achieve perfection nor should they even attempt to do so. It is a pointless pursuit. David Pye eloquently convinced us of this ridiculous notion in his wonderful book "The Nature and Aesthetics of Design" (Pye 1978). We are exposed to the products, systems, services, and spaces of design all day long. There is hardly anything in our daily lives that has not been designed. However, most of these useful things do useless things that no one wants them to do. Alessandro Mendini also warned of this folly in a typically cryptic essay in list form entitled "The Utility of Uselessness" (Mendini 1979). He concludes his short list with this statement:

> *It is useful to think of the uselessness of the useful.* (Mendini 1979)

Following Mendini's advice it might be useful for design to think of the uselessness of when and how it had been useful – i.e. when design signed up to the project of imagining a better world on the one island we share. Design has certainly made a better world, but only for those who already own it, for whom Mendini might have written "*it is useless to think of the usefulness of equality*". And it might be truly useless to imagine a role for design in the construction of equality as we witness the values at the heart of the Universal Declaration of Human Rights – "*liberty, equality, and fraternity*" – be replaced by – "*comfort, security, and sustainability.*" In a Faustian contract with liberal capital design has willingly devalued its currency for a role in all three – designing in wealth while designing out crime as if societal and environmental collapse could be thwarted if everyone behaved sustainably. As David Pye suggests:

> "*The concept of function in design... might be worth a little attention if things ever worked. It is, however, obvious that they do not.*" He goes on: "*Nothing we design or make ever really works. We can always say what it ought to do, but that it never does.*" (Pye 1978)

Planes occasionally fail, our computers crash regularly, our trains break down, our dinner table should be impervious to scratches and be self-cleaning – but they are not of course, and our motor cars, refrigerators, air conditioning units, and homes all consume valuable resources like a hungry animal. "*Never...*", Pye declares, "*...do we achieve a satisfactory performance...* [but] *If we cannot have our way in performance we will have it in appearance.*" (Pye 1978). How prescient was Pye? The design of nothing is now only performed for appearance.

4.5 The Regulation of Nothing

Enzo Mari, one of the most thoughtful and intellectually provocative Italian designers of the late twentieth century, in his famous "*Vaffanculo*" talk defined creativity as the door of hell. In this famous talk, Mari draws a straight line on a blackboard with a piece of chalk from left to right. At the right end of the line Mari states: "*... lies the maximum known quality – individuals such as Bach, Mozart, Piero della Francesca, etc...*" At the start of the chalk line on the left, exclaims Mari: "*...lies the student of today.*" Mari pronounces that the student "*...wants to get there* [the right hand end of the line]..." *He [the student] will work hard all his life, but he doesn't make it [he stops half way]. It's very hard to reach those* [Bach, Mozart, Piero della Francesca, etc....]. *Someone will say 'poor guy, he is really incapable'... I don't know. I say that even the student that arrives only half way does something good. But what do we teach in our design schools today? What do our pseudo-artists teach? Freedom, creativity... There is no word more obscene and unhealthy than the word creativity. We don't say anymore 'go and work hard and gain that', that's the reality. It's the only reality we have. We say that we are creative. Like this... like this...* [Mari scribbles on the blackboard]... *We produce the nothingness... The shit with the word creativity.*" Mari reserves some of his more stinging criticism for the annual lavish Milan Furniture Fair when he says:

> *The Salone del Mobile is standing on a word that I think is the gate of hell – 'creativity'. All of these idiots decide to make the creative world. What is the problem today? Everyone is looking to patent something – a spider, an ant run, a fart, only to have his five minutes of*

advertising in total ignorance. But the problem today is to eliminate 99% of this stuff... look at them one by one and say Vaffanculo, Vaffanculo, Vaffanculo!!!.

What Mari seems to be vehemently criticising is the question of what constitutes an idea, which is still assumed to be the basis of all design thought and action, and upon which any and all philosophies of design are dependent. What he is saying is it is obvious to him that gone are the constitutive ideas, having now been replaced by the regulative; that in our design schools talent (which has always resided in the few) be replaced by creativity (which has been ruled universal). And with the willing acceptance of this new regulation – that we are all creative and its corollary that everyone can practice design – the idea has been consumed by its derivative – innovation, that is in itself a process constitutive of nothing but the search for its own regulations.

5 A Philosophy of Nothing

In his keynote address to the XIX International Society of Human Ecology Conference in Canberra, Australia in February 2013, Hiroshi Komiyama, Chairman of the Mitsubishi Research Institute Inc., and President Emeritus of the University of Tokyo, made the case that the industrialised world is saturated with products. He illustrated his case for focusing on new demands (ecology, aging, urbanism) using some simple examples (Komiyama 2013). In the industrialised world the production of cement is now stable and China will reach saturation in 5 to 10 years (*i.e.* continued economic growth will not come from construction); the number of houses already outstrips the number of households leading to millions of vacant houses (*i.e.* continued economic growth will not come from constructing houses or selling mortgages – as the 2008 subprime mortgage crisis already demonstrated); and per capita car ownership stabilised in Europe, the USA and Japan almost 10 years ago (a statistic supported by growing public subsidies to the world's car manufacturers) and China will reach saturation in 5–10 years. As André Gorz confirms in his "Critique of Economic Reason":

It is no longer true that producing more means working more, or that producing more will lead to a better way of life. The connection between more and better has been broken; our needs for many products and services are already more than adequately met, and many of our as-yet unsatisfied needs will be met not by producing more, but by producing differently, producing other things, or even producing less. (Gorz 2001)

When industrialisation split idea from manufacture, design became a process of instructions to persuade us what to make and then consume. However, with the advent of the digital both idea and manufacture were reconciled and celebrated for turning everyone from consumers into producers, and designers into precariously employed facilitators cum entrepreneurs managing micro-elements of macro-service-products. Also, because this evolutionary crisis is not yet considered in the history of design it will remain difficult to understand, making it easy for capital to continue to manipulate design. Soon design must realise that before it is creative or innovative it is playing bit parts in derivative services – at best conditioning 'users' to repeat gratuitous performances for social commerce. As we have presented here, in the absence of industrial labour social life is now social commerce. Whereas once we bonded around the production of something, now that we produce nothing the desire to feel part of something can be sold to us via social commerce. The design of nothing is therefore producing a new working philosophy.

Design is neither a product nor a service. Design occurs in relationship to everyone and everything (like gravity it is not a thing in itself – it describes and shapes relationships). And because design is not a service it is also not a solution to anything – if anything it is a global problem. It is a global problem because we know we like to design (or claim to), but we know we don't quite know why we design. We don't know that we don't know what we should be designing, and we don't want to admit we know what we are really designing. Therefore, we have to admit we have never devoted time and energy to the project for a better world (and not just better for those who already own the world), but in its and our damaged state we have no choice but to design for the world and ourselves, and each other. And we need to understand that the project for a better world has never been about the design of an ideal space and time, but about the search for a better philosophy for the design for what is here and now.

So if we really cared about design, what should we do? Is design about the design of nothing or should we be designing nothing? A philosophy for the design of nothing needs to ask – in the search for coherence what do my actions look like today if I look through the lens of design? In the search for familiarity how do I understand my actions today if I look through the lens of design? What are we willing to admit of our self-understanding of our feelings about design? On the basis of a philosophy of self-understanding what needs to be done?

As we have demonstrated in this paper it is easy to see that the design and production of *nothing* has disrupted the platform of design and its decaying dream of the 'preferred state'. The historic subscription taken out by design for the project for a 'better world' has lapsed and in its place design now only manages to serve both the fashionable diffusion of products, and services called products, and the suppression of the reduction of consumption. Having dismantled 'industry', production and consumption are no longer projects for the masses. Services are 'designed' to appear individual so the individual can be increasingly engaged in the production of their own 'service' (Manolis et al. 2001). The production of *nothing* has certainly produced new forms of consumption and lots of it, which design continues to imagine it has some control over if only to restore some relevance to design.

If it was overproduction and debt-fuelled growth that created the present economic crisis, then is the alternative once again to slow it down and become lazy, as autonomist theorists and practitioners have stated for some time now?[1] If this is the

[1] See, for example, the Italian movie *Lavorare con lentezza* (2004); released in English with the title *Working Slowly (Radio Alice)*, directed by Guido Chiesa and written together with the Wu Ming collective.

case can design manage to be lazy? This paper presents the case that to operate in an era of digital production and disruption, design might best labour under the philosophy of nothing.

References

Atkinson, P. (2009). "Boundaries? What Boundaries? The Crisis of Design in a Post-Professional Era," in *Proceedings of the 8th European Academy of Design Conference* (Aberdeen: The Robert Gordon University), pp. 34.

Banham, R. (1980). *Theory and Design in the First Machine age* (2nd ed.). Cambridge, MA: MIT Press.

Berardi, F. (December, 2010). 'Bifo', "exhaustion and senile utopia of the coming European insurrection" *e-flux* Journal, 21, pp. 1–8.

Berardi, F. (2011). Bifo. In G. Genosko & N. Thoburn (Eds.), *After The Future*. AK Press.

Bhaskar, R., Naess, P., & Høyer, K. (Eds.). (2012). *Ecophilosophy in a world of crisis: Critical realism and the Nordic Contributions*. London: Routledge.

Branzi, A. (2008). *Introduzione al Design Italiano: Una Modernità Incompleta*, Baldini Castoldi Dalai Editore, Milano, pp. 187, (translation author's own).

Bremner, C. (1995). Picturing contemporary management. In S. Rees & G. Rodley (Eds.), *The Human Costs of Managerialism: Advocating the Recovery of Humanity* (pp. 243–248). Sydney: Pluto Press.

Bremner, C. and Rodgers, P.A. (Summer 2013). "Design without discipline", Design Issues, Vol. 29, Issue 3, pp. 4–13.

Carmagnola, F. (2009). *Design: La Fabbrica del Desiderio*. Lupetti: Milano.

Coffman, J. A., & Mikulecky, D. C. (2015). Global insanity Redux. *Cosmos and History: The Journal of Natural and Social Philosophy, 11*(1).

Cox, G., & Bazzichelli, T. (2013). How to do things with business: An introduction. In T. Bazzichelli & G. Cox (Eds.), *Disrupting business: Art and activism in times of financial crisis* (pp. 7–21). Brooklyn: Autonomedia. http://www.autonomedia.org.

Debord, G. (1977). *Society of the spectacle*. Detroit: Black and Red.

Flusser, V. (March 1992). "On the term 'design'", Artforum.

Friedman, D. (1994). *Radical Modernism*. New Haven: Yale University Press.

Galle, P.. (May 2002). "Philosophy of design: An editorial introduction", Design Studies, Vol. 23, No. 3, pp. 211–218.

Gorz, A. (1989). Critique of economic. *Reason*. Available at http://www.antenna.nl/~waterman/gorz.html.

Gorz, A. (2001). *Critique of Economic Reason: Summary for Trade Unionists and Other Left Activists, Chapter 3: 1.2: The Crisis of the Work Ethic*, Global Solidarity Dialogue, Last Updated: 25 Dec 2001. http://globalsolidarity.antenna.nl/gorz.html

Groys, B. (2012). Under the gaze of theory, *e-flux* Journal, 35 May 2012, pp. 11.

Hight, C., & Perry, C. (2006). Collective intelligence in design. *Architectural Design, 76*(5), 5–9.

Komiyama, H. (2013). *Sustainability of the Earth's environment*. Canberra: XIX International Society of Human Ecology Conference. http://ocw.u-tokyo.ac.jp/lecture_files/gf_20/13/notes/en/13komiyama_eng.pdf.

Koolhaas, R.(Spring 2002) "Junkspace," *October*, Vol. 100, pp. 175–190.

Lasdun, D. (April 1965). An Architect's approach to architecture, *RIBA* Journal, Vol. 72, No. 4.

Latour, B. (2008). "A cautious Prometheus? A few steps towards a philosophy of design (with special attention to Peter Sloterdijk)" *Keynote Lecture for the Networks of Design meeting of the Design History Society, Falmouth, Cornwall, 3rd* September 2008.

Mari, E. (May 1, 2012). http://www.youtube.com/watch?v=X49crKOX9Js; http://www.youtube.com/watch?v=E2T_1SQHlQk#t=04m29s

Manolis, C., Meamber, L. A., Winsor, R. D., & Brooks, C. M. (2001). Partial employees and consumers: A postmodern, meta-theoretical perspective for services marketing. *Marketing Theory, 1*(2), 225–243.

Manzini, E. (2015). *Design, when everybody designs: An introduction to Design for Social Innovation*. Cambridge, MA: MIT Press.

Mars, G., & Nicod, M. (1984). *The World of Waiters, London*. Sydney: Allen and Unwin.

Marshall, J., & Bleecker, J. (2010). Undisciplinarity. In P. A. Rodgers & M. Smyth (Eds.), *Digital Blur: Creative Practice at the Boundaries of Architecture, Design and Art* (pp. 216–223). Oxon: Libri Publishers.

Mason, P. (2015). 'The end of capitalism has begun', *The Guardian*, 17 July 2015, <http://www.theguardian.com/books/2015/jul/17/postcapitalism-end-of-capitalism-begun>

McGuirk, J. (2011). "Designs for life Won't make you a living," *The Guardian*, 18 April 2011.

Mendini, A. (1979). Utilità dell'inutile. *Modo, 21*.

Norman, D. (2010). "Why Design Education Must Change", *Core77*, 26th Nov 2010, [retrieved from: http://www.core77.com/blog/columns/why_design_education_must_change_17993.asp on Tuesday 11th Sep 2012.

Price, C. (1976). Anticipatory design *RIBA* Journal, Vol. 84, No. 7, July 1976.

Pye, D. (1978). *The nature and aesthetics of design*. London: The Herbert Press. https://www.youtube.com/watch?v=X49crKOX9Js

Richardson, A. (1993). The Death of the Designer. *Design Issues, 9*(2), 34–43.

Rodgers, P.A. and Bremner, C. (June 2013). Exhausting discipline: Undisciplined and irresponsible design *Architecture and* Culture, Vol. 1, Issue 1, pp. 138–158.

Scharmer, C.O. (2011). "Leading from the emerging future", *Minds for Change – Future of Global Development Ceremony to mark the 50th Anniversary of the BMZ Federal Ministry for Economic Cooperation and Development*, Berlin, 13 Nov 2011.

Simon, H. (1969). *The sciences of the artificial*. Cambridge, MA: MIT Press.

Tonkinwise, C. (2014). Design studies – What is it good for? *Design and Culture, 6*(1), 5–44.

Vodanovic L. (2007). "Obsolescence and exchange in Cedric Price's dispensable museum", *Invisible Culture*, Issue 11, Dec 2007. [accessed at: http://www.rochester.edu/in_visible_culture/Issue_11/vodanovic/vodanovic.html

Warman, M. (2011). "Dieter Rams: Apple has Achieved Something I Never Did," *The Telegraph*, June 7, 2011,. www.telegraph.co.uk/technology/apple/8555503/Dieter-Rams-Apple-has-achieved-something-I-never-did.html (Accessed 13 Mar 2012).

Žižek, S. (2004). "What Rumsfeld Doesn't know that he knows about Abu Ghraib", *In These* Times, 21 May 2004.. http://inthesetimes.com/article/747/what_rumsfeld_doesn_know_that_he_knows_about_abu_ghraib

Žižek, S. (2012). *Less than nothing: Hegel and the shadow of dialectical materialism*. London: Verso.

The manufacturer's authorised representative in the EU is Springer Nature Customer Service Centre GmbH, Europaplatz 3, 69115 Heidelberg, Germany. If you have any concerns regarding our products, please contact ProductSafety@springernature.com

Printed and bound by CPI Group (UK) Ltd, Croydon, CR0 4YY
01/10/2026
02225901-0009